C0-ALS-395

A BIOLOGICAL APPROACH TO HEALTH

LIFE SCIENCE AND MAN

Kenneth L. Fitch
Illinois State University

H. Chandler Elliott (deceased)
University of Nebraska

Perry B. Johnson
The University of Toledo

HOLT, RINEHART AND WINSTON, INC.
New York Chicago San Francisco Atlanta
Dallas Montreal Toronto London Sydney

Copyright © 1973 by Holt, Rinehart and Winston, Inc.
All rights reserved
ISBN: 0-03-091520-1
Library of Congress Catalog Card Number: 72-91075
Printed in the United States of America
3 4 5 6 0 7 1 1 2 3 4 5 6 7 8 9

PREFACE

A generation ago a textbook for health courses would likely use the term *hygiene.* The term has fallen into disrepute, partly because it seems to evoke suggestions of disinfectants, pasteurization, sterile packaging, and other antiseptic practices which are now so commonplace as to require no indoctrination. Also, it suggests a wearying reiteration of public school admonitions about brushing teeth and regular bathing.

Etymologically, however, the concept of hygiene embraced the ancient belief that continuing health could be expected if one lived wisely. This was the cult of Hygeia, the goddess who watched over the health of Athens. In the classical world Hygeia symbolized the maxim *mens sana in corpore sano*—that a rational life and physical vitality are indispensable adjuncts of one another.

However, man has always found it more expedient to depend on healers than to live wisely. The legendary Greek physician Asclepius did not concern himself with wise living but achieved immortality by the skillful use of the scalpel and a knowledge of the curative powers of certain plants. With the rise of Asclepius, Hygeia was relegated to a subservient position along with the goddess Panakeia whose followers were preoccupied with the omnipotent power of drugs, and whose cult is alive today in the search for a panacea or universal remedy for all ills.

The cults of Hygeia and Asclepius persist to the present as the antipodal views of medicine. The Hygeists are represented by those who believe that health is part of the natural order of things, and that the primary function of medicine is to discover and teach the natural laws that will ensure a healthy mind in a healthy body. The Asclepians are represented by those with the less idealistic view that the function of medicine is principally to treat disease and to repair the imperfections of accident or birth.

It is not surprising that the cult of Asclepius has the greater appeal to the general public—the view that health and fitness are to be considered the normal state and a sort of birthright, that it is the vicissitudes of daily life that account for misfortunes of ill health. This attitude is only slightly less pervasive of the medical profession—departments of public health and preventive medicine generally play a small role in the medical curriculum.

This attitude is reflected also in the communications media. News of organ transplants, which are great scientific achievements but of little practical importance to mankind, receive far more attention than news relating heart disease to cigarette smoking or to poor dietary practice; the possibility of a drug that might cure schizophrenia is more newsworthy than a suggestion that the abnormal stresses of our society are the major cause of mental illness.

All of us who teach college health courses will continue to espouse the precepts of Hygeia and do our best to inculcate the principles of wise living that are most conducive to health and the good life. But we are usually confronted with the problem of teaching not only the principles of Hygeia but also a basic introduction to the methods of the modern Asclepius—and all usually in a one-semester course of two or three credit hours. A problem usually exists, then, as to what we should emphasize and what we should not.

Some authors have expanded the concept of hygiene to encompass virtually all aspects of human welfare. But can an author, within the limits of a standard text, expect to treat in a scholarly manner such subjects as the enjoyment of life, self-confidence, and the development of meaningful life goals? These are philosophical concepts that must be developed by the individual himself through his whole educational experience.

Some authors seem to feel that a discussion of disease processes is too technical—that good nutrition, regular medical and dental consultation, and proper exercise are of sufficient concern for the layman, and that further knowledge, classified as "clinical matters," is best left to the doctor who will decide what is best for us. No reasonable person can dispute the doctor's role, but tolerating vague suppositions through unwarranted generalizations can result in either ill-advised complacency or needless apprehension. We are most competent to deal with what we understand. And a reasonably informed person can best cooperate with his doctor.

We believe, therefore, that such "clinical matters" as the use and misuse of antibiotics and hormones, the mechanism of action of various contraceptive techniques, the complications of pregnancy, or the physiological basis of circulatory shock are proper considerations for health courses even though the explanations may require some basic clinical physiology. Many such topics can be covered only imperfectly without becoming excessively technical, but an honest attempt should be made.

The answers are generally of little consequence when there is only vague understanding of how the body is put together, and how the basic physiological functions are carried out. This does not require a college course in physiology and anatomy, but it does require a basic tachnical vocabulary. After all, much of biology and medicine is simply an understanding of terminology, and terminology cannot be escaped if one is to be the "educated layman" to whom so much current writing is directed. For example, the word "skin" is not sufficiently descriptive in discussing the effects of burns or the deposition of fat in obesity. Inflammatory processes cannot be described adequately by simply using the term "white blood cells." Discussions on immunity and tissue transplantation are futile if we are restrained from the use of such terms as "lymphocyte" and "cortisone-type drug." Such terms are used regularly by news magazines. Surely we can expect as much of persons who are to be college graduates.

This does not mean that any useful purpose is served by presenting a long list of digestive enzymes and their exact function, or by naming individual bones or muscles. Specific enzymes, bones, and muscles may be described, but only when such information is useful or indispensable to the discussion. Otherwise, knowledge of how muscles and enzymes work in general should suffice.

This is not intended to be a reference book. For that reason tables and graphs have been kept to an essential minimum; and we have used restraint in the use of statistics. Also, the text is not meant to be a source of home remedies; in fact few are given. Our purpose is rather to provide an understanding of the physiological basis of health and disease and, in some cases, the rationale for treatment. This we hope will promote a clearer appreciation of the physician's obligations and responsibilities, and promote intelligent cooperation.

In summary: we wish to present neither an ineffective eulogy of healthful living nor a text for professional students of health sciences. Our aim is rather to summarize those practical facts concerning public and personal health that should be known to every educated person. We pay homage to both Hygeia and Asclepius, but with the hope that you will not merely give lip service to the former and depend too largely on the latter.

January 1973

K. F.
P. J.

CONTENTS

PART ONE
INTRODUCTION

1
THE NATURAL HISTORY OF HEALTH AND THE STATISTICS OF MORTALITY

In 1866 Florence Nightingale was asked to participate in the dedication of a new children's hospital in Manchester, England. Miss Nightingale was not, by nature, particularly tactful nor given to flowery oratory, and she replied tersely that, "building more children's hospitals is not the proper remedy for infant mortality and sickness—the true remedy lies in improving the children's homes [Brierley]."

In the century since Miss Nightingale made this observation, we have continued our emphasis on the clinical aspects of health. Our successes would seem to have proven Miss Nightingale wrong. We have learned to change the natural history of disease, to alleviate suffering, to postpone untimely death, and to prevent crippling. Wonder drugs like penicillin now provide immediate cure for diseases such as lobar pneumonia, which, 40 years ago, claimed a quarter of its victims.

But there has been, for the last two decades, a steady leveling off in our health progress and, curiously, this change in trend has coincided with an enormous expansion in our national medical research program. Federal expenditures for medical research have grown from about $1 million in 1947 to over $1 billion in 1965.

This enormous expansion in medical research during a time when the indices of health show a lag is a paradox of modern medicine—we know more and more about health and disease and yet little progress is being made in improving the health of the population. How can these things be measured, and what are the explanations? Was Miss Nightingale right after all?

Progress in the nation's health is measured in part by life expectancy. Life expectancy at birth is a theoretical estimate in a particular year of the average length of life of a newborn baby. Life expectancy of a newborn baby is calculated on the assumption that he would be exposed throughout his lifetime to the age-specific death rates for that particular year. Life expectancy is, then, an expression of the hazards imposed upon any particular age group by all causes of death in all the years of life remaining for that group, assuming that the causes of death in the future will be the same as those of the year of the estimate. It is obviously an artificial sort of estimation, but a useful one.

Table 1-1 gives life expectancies as calculated in 1870, at the beginning of the century, and at the end of each decade since then. Expectancies for men and women differ somewhat, and therefore the sexes are listed separately. The difference is partly due to the more hazardous life led by most men; but of greater importance seems to be the fact that women are simply hardier than men—their death rate being lower than that of men in all age groups.

TABLE 1–1 Changes in Life Expectancy 1870–1960

	Males		Females	
End of Decade	*Expectancy in Years*	*Increase over Decade*	*Expectancy in Years*	*Increase over Decade*
1870	39.25[a]			
1900	48.23[a]		51.08	
1910	50.23	2.00	53.62	2.54
1920	56.34	6.11	58.53	4.91
1930	59.12	2.78	62.67	4.14
1940	62.81	3.64	67.29	4.62
1950	66.31	3.50	72.03	4.74
1960	67.60	1.29	74.20	2.17

[a] Males and females combined.
SOURCE: *Vital Statistics of the United States*

The most remarkable aspect of this table is the total gain in life expectancy over the last 90 years. For males there was an approximate ten-year gain in life expectancy for each 30-year period starting in 1870; females did even better—the gains made by females exceed those by males in all decades except 1911–1920. Significant also is the slowing of the rate of increase in life expectancy in the decade 1951–1960.

Data for 1969 and 1970 were not yet available at the time of this writing, but those for 1968 show that life expectancy at birth has declined one year (to 66.6) in males and 0.2 year (to 74.0) in females. It is very likely that life expectancy at birth has, at best, reached a plateau, and may be undergoing a decline.

What are the causes of these changes—what improvements in medical care or changes in our way of life so dramatically increased life expectancies during the span of years between 1870 and 1950? Why has a leveling off occurred?

The first question can be answered in part by examining the causes of death in selected years over the last century. Table 1-2 lists the 15 leading causes of death in 1870, 1901, 1937, and 1967.

TABLE 1–2 Changes in the Leading Causes of Death since 1870

Rank	*1870*	*1901*	*1937*	*1967*
1	*Tuberculosis*[a]	*Tuberculosis*	Heart disease	Heart disease
2	*Pneumonia*	*Pneumonia*	*Pneumonia* and *Influenza*	Cancer
3	Accidents	*Diarrhea* and *Enteritis*	Cancer	Stroke
4	*Diarrhea* and *Enteritis* (inflammation of the intestinal tract)	Heart disease	Stroke	Accidents
5	*Scarlet fever*	*Nephritis* (inflammation of the kidneys)	Accidents	*Influenza* and *Pneumonia*
6	*Infant cholera*	*Diseases of infancy*	*Nephritis*	*Diseases of infancy*
7	Circulatory disease	Stroke	Diabetes mellitus	Arteriosclerosis
8	*Encephalitis* (inflammation of the brain)	Accidents	Suicide	Diabetes mellitus
9	Convulsions	Cancer	*Diseases of infancy*	Cirrhosis of the liver
10	*Measles*	*Bronchitis*	*Appendicitis*	Suicide
11	*Whooping cough*	*Meningitis*	*Syphilis*	Emphysema
12	*Malaria*	*Diphtheria*	Hernia and Intestinal obstruction	Congenital malformations
13	*Diphtheria*	*Typhoid fever*	Congenital malformations	Homicide
14	Cancer	*Influenza*	Cirrhosis of the liver	Hypertension
15	Stroke	Paralysis	Homicide	*Nephritis*

[a] Causes of death in italics are diseases that are wholly or in large part infectious in nature.

SOURCE: Data from 1870 taken from United States Bureau of the Census. Data for 1901, 1937, and 1967 from *Vital Statistics of the United States*.

A completely satisfactory comparison of these years is not possible since there have been many changes in the naming and classification of diseases and causes of death since 1870. For example, the 1870 report lumps together under circulatory disease causes of death now listed separately under heart disease, arteriosclerosis, and hypertension. The terms "convulsions," "paralysis," and "enteritis" describe signs of disease rather than diseases. As late as 1929 cancer mortality was divided into only six specific types, with the balance of deaths lumped together under "other unspecified organs." Today cancer mortality is separated into 56 different categories.

Nevertheless the major items of the earlier data enable us to list the principal causes of death over the last 100 years and to make some realistic generalizations.

Remarkable in this compilation is the decreasing importance of infectious disease as a cause of death. The period between 1870 and 1901 was in many respects a golden age of medical discovery. The causative organisms for tuberculosis, lobar pneumonia, cholera, malaria, diphtheria, typhoid fever, dysentery, gonorrhea, and leprosy as well as those for many other diseases of less importance were discovered, making possible the advent of preventive medicine during the early years of this century. Also, the development of sterile and antiseptic techniques during the late 1860s coupled with the general adoption of anesthesia meant great advances in surgery. (In fact, as a practical matter, antiseptic surgery depended to a great extent upon painless surgery—as one surgeon of the era put it, "With anesthetics ended slapdash surgery; anesthesia gave time for the theories of Pasteur and Lister to be adopted in practice.")

It is not surprising then that starting in 1901 deaths from infectious disease began a steady decline. Two-thirds of the causes of death in 1870 and 1901 were from infectious disease; this was reduced to one-third in 1937 and to one-fifth in 1967. But this does not tell the whole story: Tuberculosis and pneumonia accounted for over 20 percent of *all* deaths in 1901; in 1967 they accounted for less than 6 percent. Furthermore, pneumonia no longer kills the young and healthy that it once did; deaths from pneumonia now are more often the final stage in other noninfectious lung diseases. By 1937 the long list of infectious diseases that were the major killers of children had come under control.

The changes in the causes of death since 1937 reflect, more than anything else, the effectiveness of antibiotics in the treatment of lobar pneumonia, nephritis, syphilis, and appendicitis.

As the infectious diseases have come under control, and a greater proportion of the population lives on to late middle age and old age, it should be expected that degenerative diseases would come into greater prominence. Little progress has been made in retarding the seemingly inevitable changes that lead to poorer circulation, failing regenerative powers, and abnormal tissue formation (cancer) that seem to be the fate of all of us who live long enough to suffer them. This is another way of saying that lifespan, the time from birth to life's upper biological limits, has not changed much since ancient times. Heart disease, cancer, stroke, arteriosclerosis, cirrhosis of the liver, emphysema, and hypertension are all concomitants of old age, and their rise in importance as causes of death cannot be ascribed to an "epidemic" but to the fact that the overwhelming majority of people dying each year in this country are old people. It is easy for us to forget that the leading causes of death in the not too distant past were killing people during their infancy (cholera, enteritis), childhood (scarlet fever, measles, whooping cough, diphtheria, lobar pneumonia), and young adulthood (tuberculosis). It is true that certain indulgences, especially cigarette smoking and overeating, hasten the onset of degenerative diseases, making many of us old before our time, but, in general, the rise in incidence of degenerative disease is really a reminder that we are mortal (Table 1–2).

How does this change in causes of death affect life expectancy determinations? In general, infectious disease is a hazard of the first two decades of life, and when these are largely eliminated as causes of death, it is not surprising that the life expectancy at birth rises dramatically. When a four-year-old child survives lobar pneumonia and lives on into adulthood, he makes a significant contribution to the average life expectancy. Curing a 65-year-old person of cancer so that he may live on to age 70 when he dies of heart disease will have relatively little effect on life expectancies at birth.

A measure of our success in treating degenerative diseases lies in the changes in life expectancies of those in the older age groups. As an example, a person of age 40 in 1870 could expect to live, on the average, 26.4 more years. In 1967 this had increased to 31.4 for males and 37.37 for females—an increase of 5 and 11 years, respectively. Put in this way, our medical achievements over the past 100 years do not seem so spectacular.

We noted earlier that the data for 1968 showed a slight decrease in life expectancy at birth. This was reflected also in the general death rate, which rose from 9.4 per thousand population in 1967 to 9.7 per thousand in 1968—the highest death rate since 1951.

Is it possible that over the last two decades, when the lengthening of life expectancy slowed and finally came to a halt, that we could have done better? One method of determining this is to compare our performance with that of other countries. Table 1-3 lists the 24 countries that lead the world in life expectancy at birth.

In the interval 1959–1968, the life expectancy of males in the United States dropped from 13th to 21st place; for females it remained the same, in 7th place. Other data from the same source (*United Nations Demographic Yearbook*, 1968) show that in the "middle years of life" (ages 15 to 44), the death rates in the United States are higher than those in Sweden for every major grouping of causes of death except one. The only exception is suicide, which is one and a quarter times higher in Sweden. But we more than make up for this slight advantage in suicide rates with our excessive homicide rates: nine times as high for women and seven times as high for men. (Among males age 15 to 34 in the United States the three leading causes of death in 1967 were in order: accidents, homicide, and suicide; for females it was accidents, cancer, and suicide—homicide ranked 5th.)

It is probable that one of the most sensitive indices of the quality of our health care is infant mortality rate. The improvement in infant survival in the first half of this century was spectacular. In 1915 about 100 of every 1000 babies

(a) **0 TO 1 YEAR**

First month of life is the most hazardous, particularly to premature infants. Recent decline in death rates in large part can be attributed to improved treatment of respiratory and gastrointestinal diseases. Death rates in this age group are much higher among the poor.

(b) **1 TO 4 YEARS**

Leading causes of death are congenital malformations, accidents, and cancer (particularly leukemia); pneumonia persists as a major hazard to life in this and in the oldest age group.

(c) **5 TO 14 YEARS**

Because of the virtual eradication of infectious diseases as causes of death this age group now has the lowest death rate of any group. Deaths from cancer and congenital heart disease are consequently showing a relative increase. Accidents of all kinds are the leading causes of death.

(d) **15 TO 24 YEARS**

This is the age group that used to die in large part from tuberculosis. Now the leading causes of death are traffic accidents, homicide, suicide, and cancer. Traffic accidents kill over four times as many males as females: suicide is more than twice as common among males. Cancer is a far greater hazard in females.

FIG. 1–1 Causes of death from birth to old age. (Vital Statistics of the United States. U. S. Public Health Service)

(e) 25 TO 34 YEARS

Automobile and other traffic accidents, suicide, and homicide continue to be leading causes of death in both sexes, but females continue to show higher rates for cancer.

(f) 35 TO 44 YEARS

Causes of death for males and females show a marked divergence. Males die predominantly from heart disease, lung cancer, and accidents: females from breast cancer, uterine cancer, and heart disease. A propensity for suicide seems to be shared by the two sexes.

(g) 45 TO 64 YEARS

Leading causes of death in both sexes are heart disease, cancer, stroke, and accidents. Lung cancer is most prominent in males, breast cancer in females.

(h) 65 YEARS AND OLDER

Leading causes of death are heart disease, stroke, and cancer, but a wide variety of degenerative diseases come into prominence.

TABLE 1–3 Expectation of Life at Birth: Males and Females

Males	Years of Life	Females	Years of Life
1. Sweden	71.60	Netherlands	76.1
2. Netherlands	71.1	Norway	75.97
3. Norway	71.03	Sweden	75.70
4. Israel (Jews)	70.41	France	75.4
5. Denmark	70.1	United Kingdom	74.9
6. Switzerland	68.72	Denmark	74.7
7. United Kingdom	68.7	United States	74.2
8. East Germany	68.47	Australia	74.18
9. New Zealand	68.44	Canada	74.17
10. Canada	68.35	Switzerland	74.13
11. Japan	68.35	Japan	73.61
12. France	68.2	Israel (Jews)	73.59
13. Ireland	68.13	East Germany	73.53
14. Australia	67.92	New Zealand	73.75
15. Bulgaria	67.82	Czechoslovakia	73.57
16. Belgium	67.73	West Germany	73.57
17. West Germany	67.62	Belgium	73.51
18. Czechoslovakia	67.33	Austria	73.41
19. Italy	67.24	Finland	72.6
20. Greece	67.46	Italy	72.27
21. United States	67.00	Ireland	71.86
22. Hungary	67.00	Hungary	71.83
23. Austria	66.57	Bulgaria	71.35
24. Finland	65.4	Greece	70.70

SOURCE: *United Nations Demographic Yearbook,* 1968.

born alive died before their first birthday. Now fewer than 23 in 1000 die in their first year of life in the United States.

However, the improvement in the infant mortality rate has slowed down in the past two decades. In 1958 the United States ranked 7th in the world in its infant mortality rate of 27.1. In 1968 it ranked 14th with a rate of 22.6, and the rate of 22.6 is still considerably above the rate of Sweden two decades ago. Sweden's 1952 infant mortality rate was 20.0; in 1958 it was 15.9; in 1968 this had decreased to 13.1!

However, statistics by themselves are not very informative—they raise more questions than they answer. What are the reasons that our life expectancies are lower and our infant mortality rate higher than a number of other countries? For example, why does Sweden have such a low infant mortality rate? Sweden and the United States use the same World Health Organization criteria for the definition of a live birth. One medical scientist who examined the use of these criteria in Sweden concluded that "Swedes can count dead babies as accurately as we can."

Swedish doctors themselves give several reasons for their low infant mortality rate [Rutstein]. Ninety percent of pregnant women in Sweden pay their first visit to their physician during the first three months of their pregnancy. In New York City just under half (47 percent) of the women delivered in municipal hospitals and slightly more than one third (34.5 percent) of those delivered in the wards of voluntary hospitals in 1961 had not had the benefit of a single visit to a physician before their delivery.

There seems little basis to dispute the general concensus that the United States is the center of medical research. But likewise there seems little to dispute the conclusion that we are behind several other countries in the ability to deliver medical care.

The lesson here is that the health of a nation's population is dependent upon factors other than the skill of its physicians and the quality and abundance of its hospitals. Florence Nightingale recognized that advantageous social conditions were requisites for good health. Certainly a decline in life expectancy and a rise in death rates at a time when medical knowledge is advancing is a sign of social ills. The rising incidence of violence, crime, alcoholism, and drug abuse as well as the effects of degraded environments and unsanitary living conditions are taking their toll. Perhaps Florence Nightingale was right. To build a hospital amid slums is putting the cart before the horse.

The rise in life expectancy during the last half of the nineteenth century was largely made possible by social progress. By this we mean such matters as providing adequate quantities of safe drinking water, the proper disposal of human excreta, better nutrition, and the breaking up of the overcrowded slums created by the industrial revolution. The great advances in life expectancies that occurred between 1900 and 1950 were the result of the application of scientific knowledge to medical care. Medical progress has continued since 1950, but not at a rate sufficient to offset the debilitating effects of a troubled society. If the health of our population is to improve it will be in large part dependent upon factors other than medical science.

A partial solution is to see to it that adequate medical care is available to all—and this may necessitate a change in our medical care delivery system. But at least as great a challenge will be whether or not we can rehabilitate an environment that is increasingly a threat to both our physical and mental health.

REVIEW QUESTIONS

1. What is meant by life expectancy?
2. In what way have the causes of death changed over the past 70 years?
3. What is the relative importance of infectious diseases as contrasted to degenerative diseases as causes of death today?
4. Can we reasonably expect that life expectancy will significantly increase in the next two or three decades? Justify your answer.

5. How does the United States rank in its health statistics in comparison to other countries?
6. How do you think Florence Nightingale would view our present efforts in health care?

REFERENCES

BRIERLEY, J. K., *A Natural History of Man.* Englewood, New Jersey: Fairleigh Dickinson University Press, 1970.

RUTSTEIN, D. D., *The Coming Revolution in Medicine.* Cambridge, Mass.: M. I. T. Press, 1967.

Additional Readings

DUBOS, R., *Man Adapting.* New Haven, Conn.: Yale University Press, 1965.

GOLDSTEIN, S., "The biology of aging," *New England Journal of Medicine,* **285:** 1120–1129 (1971).

LEWIS, I. J., "Science and health care—the political problem," *New England Journal of Medicine,* **281:** 888–896 (1969).

MARSTON, R. Q., "To meet the nation's health needs," *New England Journal of Medicine,* **279:** 520–524 (1968).

MCNERNEY, W. J., "Why does medical care cost so much?" *New England Journal of Medicine,* **282:** 1458–1465 (1970).

ROBSON, A. M., "The British National Health Service and some lessons to be learned from it," *New England Journal of Medicine,* **280:** 754–761 (1969).

SHANNON, J. A., "Medicine, public policy and the private sector: organizational deficiencies," *New England Journal of Medicine,* **281:** 135–141 (1969).

2
THE NATURE OF LIFE

Protoplasm is the stuff of life whether leaf, grasshopper, bacterium, or man, and is remarkably similar from whatever source. It is a gelatinous substance composed largely of water (71 percent) and protein (25 percent) with small amounts of sugar, fatty materials, and numerous minerals, such as calcium, sodium, potassium, and iron.

Proteins are made up of carbon, hydrogen, oxygen, and nitrogen with sulfur and phosphorus usually present. It is the carbon, however, which makes possible the marvelous versatility of proteins, which in turn makes possible the versatility of protoplasm. Carbon atoms have the unique ability to link themselves into long chains or into rings, and then to add to these the atoms of other elements. Since the character of a molecule can be completely changed by the addition of a single atom, the infinite possibilities for variation in form and function of protoplasm are made possible.

The field of organic chemistry makes use of this characteristic of the carbon atom. From coal tar—the fossil remains of prehistoric life—the chemist is able to create new chemicals with unlimited variation in their properties—perfumes, artificial sweeteners, headache remedies, dyes, plastics, and hundreds of other substances we consider vital to modern society. Likewise the long chains of carbon atoms which form cellulose (wood pulp and cotton) can be modified with the addition of other elements, such as nitrogen and sulfur, to make explosives, plastics, and paints.

Illustrated in Figure 2–1 are some of the ways in which carbon atoms link together and form familiar substances.

FIG. 2–1 Common organic molecules.

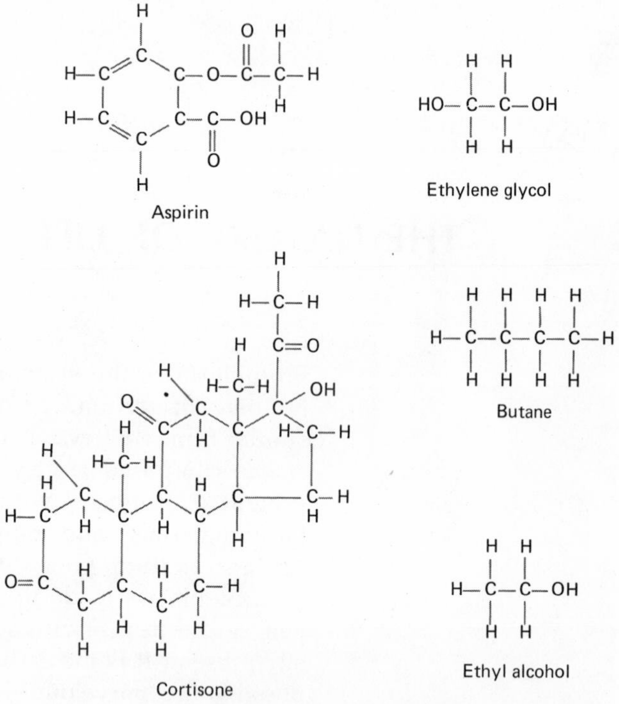

Aspirin

Ethylene glycol

Butane

Cortisone

Ethyl alcohol

Organization of Protoplasm

Up until about 300 years ago the tissue of living organisms was thought to be composed of structureless pulp—that is, that skin was skin by virtue of its total structure and could not be subdivided into smaller units. The invention of the microscope made possible the study of the basic organization of living things.

The first person to look at the microscopic side of life was a Dutch dry goods dealer named Anton Leeuwenhoek who lived during the latter part of the seventeenth century (Fig. 2–2). He was truly an explorer, the first to enter a weird and almost unbelievable new world. He became so absorbed in what he saw that his explorations became a lifelong passion. The lenses which he designed and manufactured himself were so advanced for his time that it was 200 years before his discovery of bacteria was supported by others. His persistence and scientific curiosity equaled his technical skill. He wrote, "In narrowly scrutinizing three or four drops of water I may do such a great deal of work that I put myself into a sweat." He was the first to discover that ordinary pond water contained thousands of tiny forms of life—little animals, or "animalcula," as he called them. He found that blood was not the red liquid it appeared but rather was composed of a clear fluid in which were suspended pale pink discs, the corpuscles. In his description of sperms he first explains that the material for study was obtained as a surplus from marital intercourse, without committing any sin (a consideration of some magnitude in 1680). He described the semen as "crowded with an infinity of animals like tadpoles."

FIG. 2–2 Anton Leeuwenhoek, a Dutch merchant of the seventeenth century, described protozoa and bacteria he observed through lenses he manufactured himself. It was 200 years before comparable microscopes were developed by others. (Parke-Davis)

Cells and Tissues

Much later it was realized that bacteria, sperms, and such simple animals as the amoeba and malarial parasite are composed of a single unit of protoplasm called the cell (Fig. 2–3), and that the tissues which make up our muscles, skin, liver, and other organs are collections of millions of cells arranged together as sheets, tubes, or more complex assemblages. Whereas the protoplasm of the amoeba must function as a Jack-of-all-trades, the protoplasmic units or cells of higher animals become specialized for particular activities. Collections of such specialized cells are called tissues. The various tissues in turn are combined into organs.

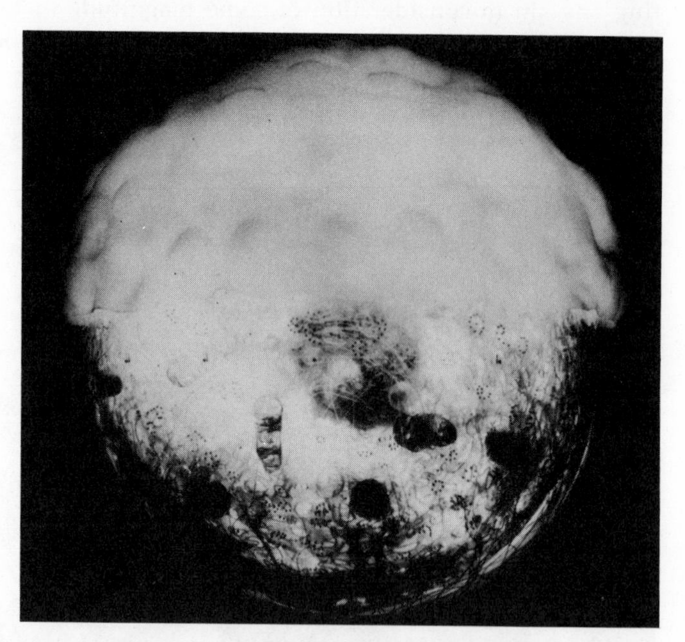

FIG. 2–3 Model of a cell. (The Upjohn Company)

However, we should not forget that the most complex animal starts out as a single cell, the fertilized ovum. This cell divides almost endlessly, and the resulting cells follow hundreds of paths to specialization. All the cells, in fact, become specialized for some particular function with the exception of the sex cells, which are set aside at an early stage of development for the purpose of reproduction.

Obviously, the cells of the very early embryo are not much different from one another. However, in a matter of a few weeks the human embryo is composed of millions of cells specializing in some function helpful to the whole. Some cells produce a cement, complete with reinforcing rods, which binds them together in a supporting function as cartilage or bone; some become transparent as in the cornea and lens of the eye; some become impregnated with waterproofing material to form skin; some become factories for producing digestive enzymes; some become specialized in the transmission of electrochemical signals and collectively become the nervous system. All these and many more specializations arise from a single cell. This process of differentiation of cell types is still little understood, and it is one of the great mysteries of biology.

Rudolf Virchow (1821–1902), the great pathologist, likened the human body to a "cell-state" with a specialization of labor and social organization. Although the arrangement brings about a sophistication of achievement otherwise impossible, it carries with it the hazard that some specialized groups of cells become indispensable, and a strike on the part of this relatively small group paralyses or destroys the entire community. Thus, the purposeful contraction of the heart is dependent upon a specialized bundle of cells within the heart muscle. If these few cells cease to function, the heart no longer pumps efficiently, and every other cell of the body is put in jeopardy through lack of oxygen. The amoeba, with its single all-purpose cell, faces no such hazards.

Just as the molecule is the fundamental unit in chemistry, the cell is the fundamental unit in all life. So it is the cells which use the food, drink the water, and depend upon the oxygen of the air. The complex anatomy of the human body is, in large part, an elegant mechanism which serves to bring to the cells the food, air, and chemicals they require. But in spite of their specializations the cells of all types of tissues have characteristics in common, the most basic of which are shared with *all* cells whether bacterium, amoeba, or plant cell.

The Nucleus

Near the center of the cell is a more or less spherical body called the *nucleus.* The substance of the cell outside the nucleus is called the *cytoplasm.* Enveloping the cytoplasm is a *cell membrane* which maintains the integrity of the cell.

The special constituent of the nucleus is a material called *chromatin,* which appears as a complicated interlacing of fine threads (Fig. 2–4). Chromatin is the working phase of the *chromosomes,* which appear as discrete forms only when the cell divides. Chromosomes are apparently tightly coiled versions of the chromatin, in much the same sense that a greatly stretched spring becomes much shorter and thicker when relaxed. The chromatin, or uncoiled chromosomes, is composed of linearly arranged *genes,* probably 10,000 to 50,000 in number. The genes are now known to be segments of a molecule called *deoxyribose nucleic acid,* the *DNA* written about so much in current literature. Within the DNA molecule is a simple code which uses four "letters." The letters are chemical subunits called nucleic acid bases, and are designated as A, G, C, and T which stand for adenine, guanine, cytosine, and thymine, respectively. Each sequence of three code letters is a "word" or codon, and codes for one of the amino acids, the structural units of proteins. A whole protein, then, would

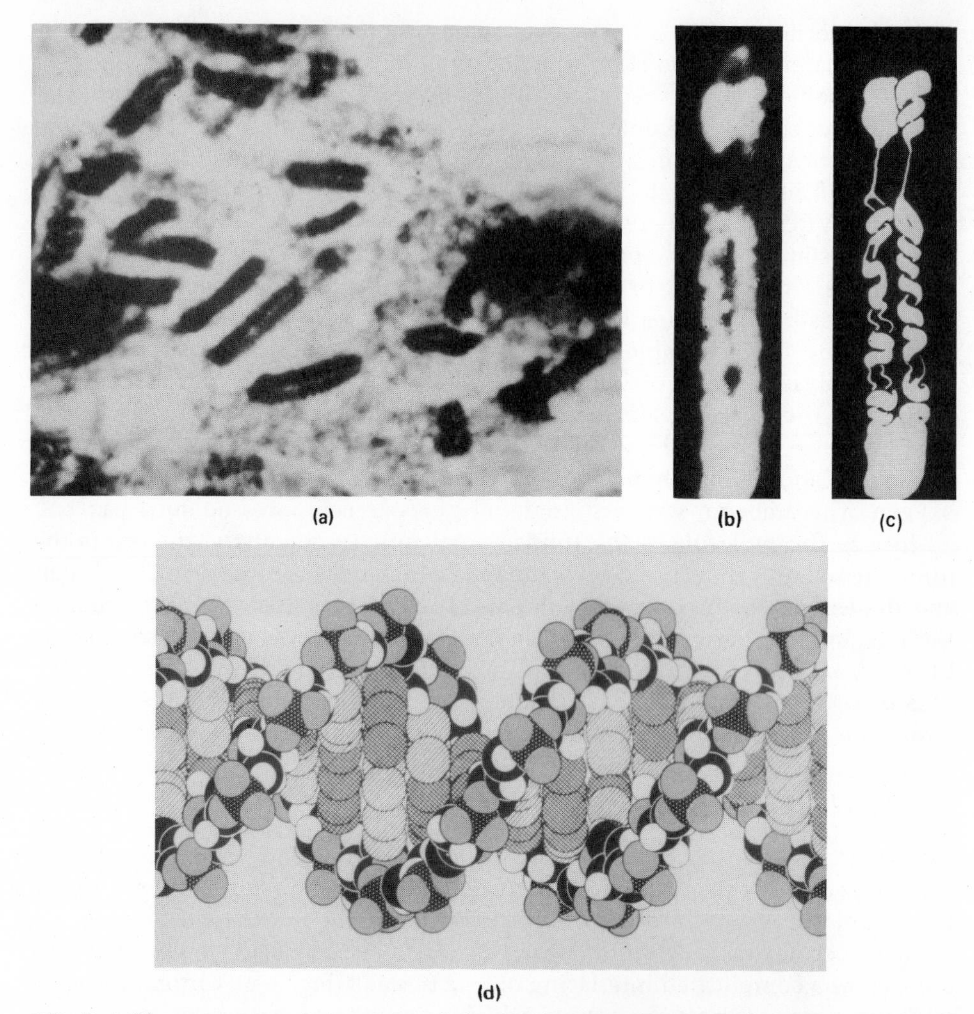

FIG. 2–4 Chromosomes, chromatin, and the DNA molecule: (a) A chromosome as seen under the light microscope. (b) The chromosome in (a) greatly enlarged. (c) A drawing of (b) to show coiled structure. (d) Model of the DNA molecule. The coiling of the chromosome in (c) would seem to be related to the helical structure of the DNA molecule, but this has not yet been proven. (a, b, and c from G. B. Wilson, Michigan State University; d from L. D. Hamilton, Sloan-Kettering Institute)

be coded or determined by a long series of codons. All the possible proteins which an animal is capable of synthesizing are represented within the coding of this chromatin.

The nucleus exercises its control of the cell by sending messengers—which are another form of nucleic acid, *ribonucleic acid or RNA*—which carry information from the genes to the protein-manufacturing machinery of the cytoplasm. Obviously, only a few genes are used by a relatively inactive bone cell, for example, whereas a greater number of genes are used by a stomach cell. The nature of a cell—by this we mean, ultimately, the nature of its proteins—is determined by which genes are "turned on"; but how this turning on is accomplished is not clear. French scientists Francois Jacob, André Lwoff, and Jacques Monod won the Nobel Prize in 1965 for suggesting a mechanism for this turning-on process. (An American, James D. Watson, and two British scientists, Francis H. C. Crick and Maurice H. F. Wilkins, shared a Nobel Prize in 1962 for describing the structure of DNA within the chromosomes.)

The Cytoplasm

If we think of the nucleus as the control unit of the cell, the cytoplasm can be thought of as the part which is concerned with the everyday work. In the specializations of cells for different kinds of work, such as contraction in muscles, the secretion of digestive juices by cells of the stomach, pancreas, and intestines, and the transmission of nervous impulses by the nerve cells, it is the cytoplasm which takes on the characteristic appearance related to its activity. In fact the pathologist makes use of the appearance of the cytoplasm to determine whether a cell is healthy or sick and unable to function.

A number of structural parts, called *organelles,* are present in the cytoplasm. The *mitochondria* (Fig. 2–5) are tiny cucumber-shaped organelles scattered through the cytoplasm, and are the power plants of the cell. The enzymes related to oxidation of food and the release of energy are now known to be arranged along shelflike folds within the mitochrondrion. As would be expected the cells which are metabolically very active, such as gland cells and muscle cells, have great numbers of mitochondria. For example, mitochondria number about 2500 per cell in the liver.

A complex system of narrow channels is distributed throughout the cell and eventually connect to the outside. The membranes that form the walls of these channels are called the *endoplasmic reticulum.* Fine granules, called *ribosomes,* line the cytoplasmic side of the membranes and are the site of protein synthesis (the target of the RNA messengers from the nucleus). Other membrane-lined spaces within the cytoplasm, called *Golgi bodies,* are believed to be packages of protein secretion in the process of delivery out of the cell.

Basic Tissue Types

Cells are classified, sometimes rather arbitrarily, into four types: epithelial, connective, muscular, and nervous. These terms also apply to tissues since, by definition, tissues are collections of these four types of cells. Sometimes terms

FIG. 2–5 Model of a mitochondrion. Mitochondria are just barely visible under the light microscope, but the electron microscope has revealed their highly complex anatomy. Mitochondria are responsible for the combustion of glucose and the mobilization of the released energy in all the cells of the body. (The Upjohn Company)

such as liver (hepatic) tissue and kidney (renal) tissue are used, but are, strictly speaking, inaccurate, since several cell types are present in each.

Epithelial Tissue

Epithelial tissues cover surfaces. Thus skin and the mucous membranes lining the nose, mouth, windpipe, intestines, and excretory passages are epithelial tissues. Because gland cells almost invariably develop from epithelial membranes, they are also logically classified as epithelium (Fig. 2–6)

Epithelial cells may be flattened like pavement stones (squamous), cube-shaped (cuboid), or column-shaped (columnar). Epithelial cells always rest upon a *basement membrane,* a structureless layer of material secreted by the epithelial cells. In the respiratory passages and in certain parts of the genital systems (Fallopian tubes in the female and in the beginnings of the sperm ducts in the male) the columnar cells are capped by hairlike *cilia.* The cilia whip back and forth so that their surfaces appear like a wheat field undulating in a breeze. The waves, however, travel in only one direction so that particles (dust and other small particles in the air passages—sperms and eggs in the genital passages) are moved toward the outside.

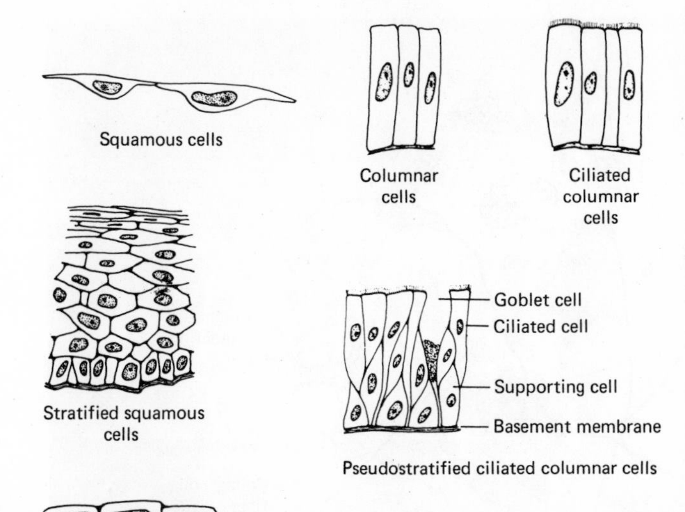

FIG. 2–6 Types of epithelium. Squamous epithelium forms delicate membranes which line blood vessels and body cavities; columnar epithelium forms the lining of most of the gut; ciliated columnar epithelium lines the genital passages; stratified squamous epithelium forms the epidermis of the skin and lines the mouth, esophagus, and vagina; pseudostratified ciliated columnar epithelium lines the respiratory passages; cuboidal epithelium forms the tissues and ducts of most glands.

Connective Tissue

Connective tissue (Fig. 2–7), as its name implies, fills in spaces not otherwise occupied and, in addition, serves a supportive function. Also vital is its function in tissue repair. Cells of the nervous and muscular tissues lose the power to divide soon after birth. Therefore when these cells are destroyed by injury or disease, the defect must be repaired, and this repair is a function of connective tissue.

Connective tissue is composed of a variety of cell types which produce a great amount of extracellular material—the glassy substance of cartilage and the salts of bone, as well as several types of fibers. The commonest of the cells, the *fibroblast,* produces the tough *collagen fibers* which form the reinforcing rods in bones and cartilage, the binding strands of ligaments, the tendons of muscles, and the fibrous tissue of scars.

The fibroblasts readily respond to a variety of irritative stimuli, either chemical or physical, to produce fibrous (scar) tissue. Even though the reaction is undoubtedly protective in nature the result is often a disease in itself. For example, a high intake of alcohol stimulates the overgrowth of fibrous tissue in the liver (cirrhosis) to the extent that the epithelial (gland) cells are replaced. Poisons such as carbon tetrachloride and chloroform have a similar effect. Fibrosis in the lungs is one of the disabling results of cigarette smoking (Chapter 28).

In addition to bone, cartilage, and fibrous tissue, connective tissues include fat (adipose tissue), elastic tissues, blood, and blood-forming tissues.

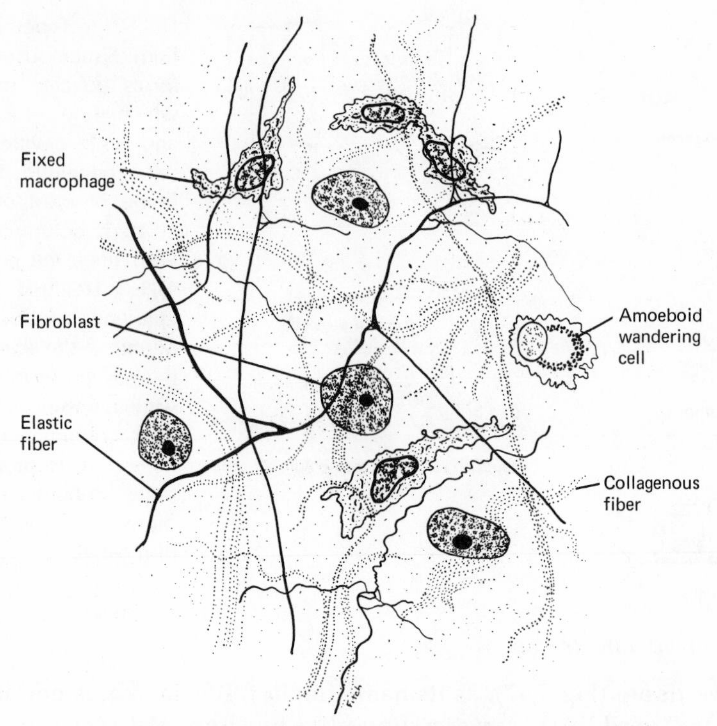

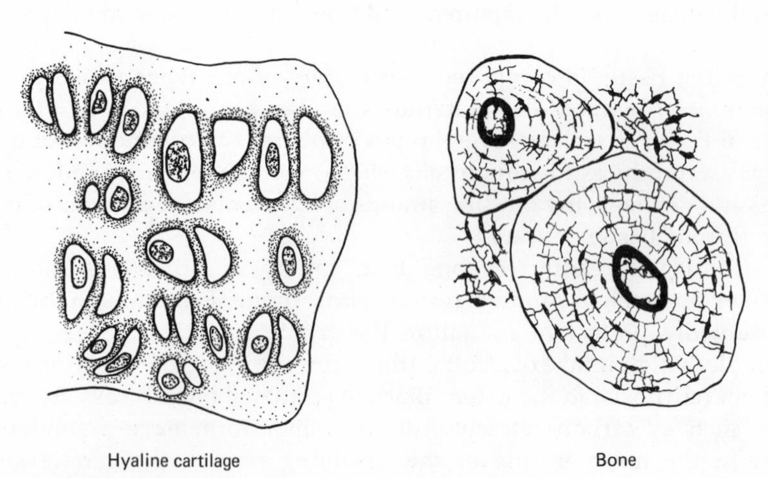

FIG. 2–7 Connective tissues, bone, and cartilage. Loose connective tissue acts in many ways as a "filler" tissue, occupying the spaces between other tissues and organs.

Muscular Tissue

Muscular tissue (Fig. 2–8) is composed of cells specialized for contraction, and exists in three forms: (1) smooth or involuntary; (2) striated, skeletal or voluntary; and (3) cardiac.

Smooth muscle cells are slender and tapering with ends which interlace with one another. Smooth muscle contracts sluggishly, and for long periods without fatigue, and for this reason is particularly adapted for visceral func-

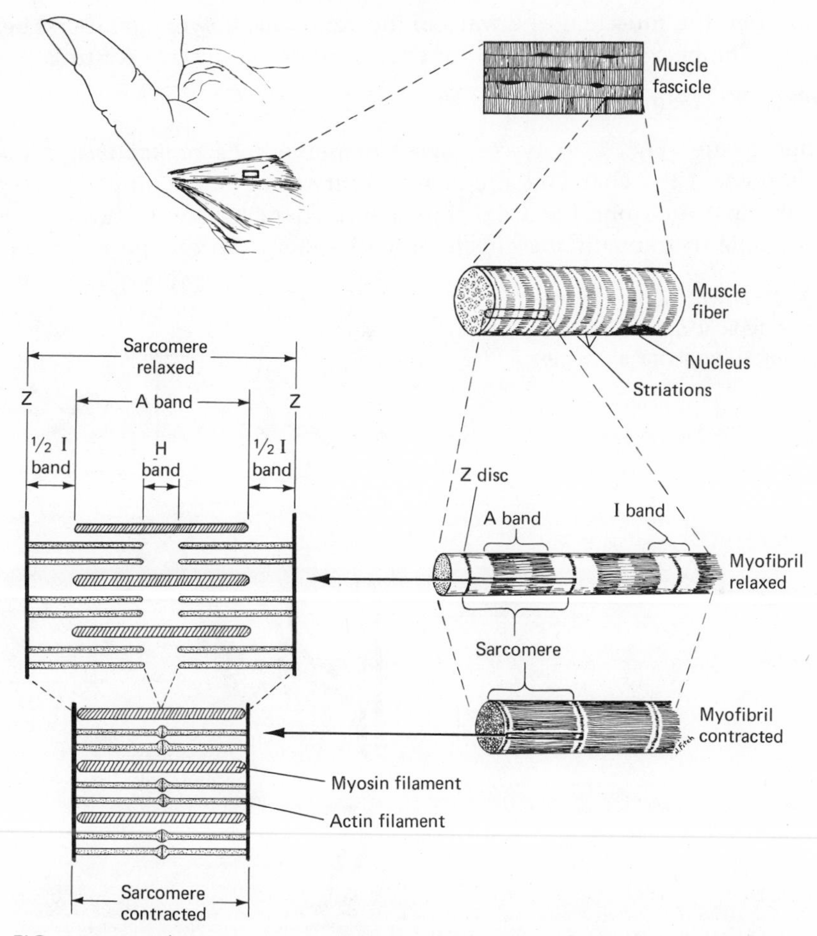

FIG. 2–8 Muscle tissue as seen from the gross to the microscopic level. (From Bloom and Fawsett, *A Textbook of Histology*, 9th ed., W. B. Saunders, Philadelphia, 1968.)

tions, for example, the walls of the digestive organs, blood vessels, and respiratory and excretory passages.

Striated or voluntary muscle is composed of *muscle fibers* formed by great numbers of cells which have become fused together. Consequently, a muscle fiber is quite large and has great numbers of nuclei. These are the muscles employed in our everyday activities—since they are the ones under voluntary control performing body movements. Voluntary muscle is discussed at more length in the chapter on the locomotor system.

Cardiac muscle is the muscle of the heart. Like voluntary muscle it is composed of muscle fibers, but the fibers in this case are interconnected so that the walls of the two atria (upper chambers) function as if they were a single large muscle fiber; the muscle of the walls of the ventricles (lower chambers) behave similarly. The heart muscle is not, of course, under voluntary control.

Nervous Tissue

Nervous tissue (Fig. 2–9) is composed of nerve cells or *neurons*. Although all cells possess the characteristic of irritability (that is, the ability to respond to a stimulus), neurons have developed this characteristic to the extent that they are able to transmit the effects of a stimulus from one part of the cell to

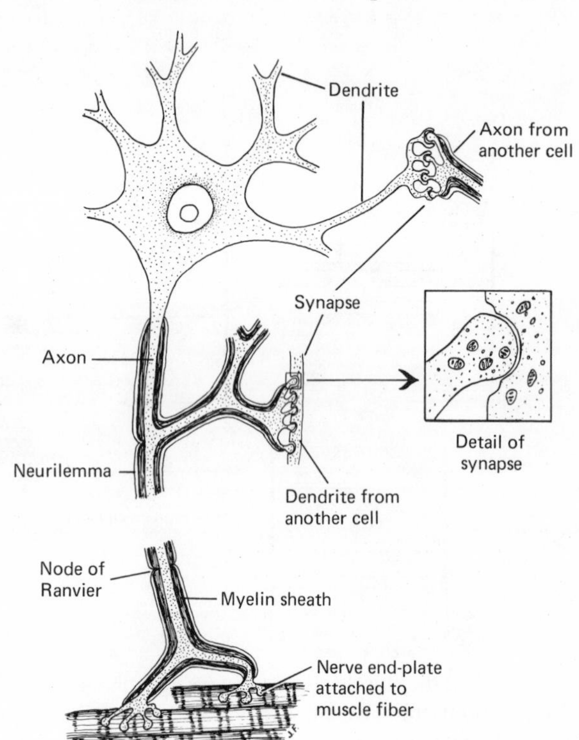

FIG. 2–9 Shown here are the parts of a typical neuron, in this case a neuron supplying a muscle.

all other parts, even over a length of several feet—the length of some nerve cells which connect the spinal cord with distant parts of the body.

These remarkable cells are responsible for the unique characteristics of the human organism and, fittingly, are considered in some detail in connection with the nervous system (Chapter 33).

It is not difficult to see that we are what our cells are. While we may feel that we can have no control over such an infinitely small part of our body, it should be obvious that there are in reality many ways in which we do have an influence upon our cells. The food we eat or do not eat, the water we ingest or lose and do not replace, the vitamins and minerals we consume in foods or pills, the smoke we inhale into our lungs, the drugs we may take or reject, the exercise we take or ignore—it is precisely these things and many more that exert a very real effect on the health and fitness of our cells and, thus, our mind and body and life.

REVIEW QUESTIONS

1. In what way are bacteria, amoebae, parasites, insects, and other living things all similar?
2. What is the purpose of the nucleus of a cell? the cytoplasm?
3. What is the relationship between the genetic code and protein synthesis?
4. What are the principal organelles of the cytoplasm and what is the function of each?
5. What kinds of body structures are made up of epithelial tissue? connective tissue? What are the two other kinds of basic tissues?

Additional Readings

Breneman, W. R., *Animal Form and Function,* 3rd ed. Waltham, Mass.: Blaisdell, 1966.

Grollman, S., *The Human Body,* 2nd ed. New York: Macmillan, 1969.

Johnson, W. H., L. E. Delanney, E. C. Williams, and T. A. Cole, *Principles of Zoology.* New York: Holt, Rinehart and Winston, 1969.

Tuttle, W. W., and B. A. Schottelius, *Textbook of Physiology,* 16th ed. St. Louis, Mo.: Mosby, 1969.

3
HEREDITY AND HEALTH

Human heredity is the study of the "inborn" qualities of human beings. These qualities include both physical and mental characteristics. It must include also a study of characteristics which distinguish humans from nonhumans as well as those which distinguish a small group of people, or families, or races. Thus human heredity ultimately comes down to the study of differences among humans, the causes of these differences, and the way in which they are passed on from generation to generation.

The broad study of heredity in general is termed *genetics.* The word is derived from the Greek root *gen* which means to become or grow into something. So genetics concerns not only the study of the inheritance of differences among all living things, but also the manner in which the hereditary material causes the development of these differences and how these differences become expressed.

Although man long ago developed empirical methods of plant and animal breeding—practically all present-day breeds of dogs were developed long before anything was known about the science of genetics—the real mechanisms of inheritance have only become known over the past 60 years. The basic patterns of inheritance, discovered through a series of extraordinary experiments performed by an Austrian monk, Gregor Mendel, in the 1860s, were rediscovered in the early 1900s. These basic principles, called *Mendelian genetics,* are now known to apply to all living things and are relatively simple to understand. The work of the last 60 years, and particularly that of the last decade, has elucidated the manner in which these principles operate, and we are now on the threshold of understanding the exact nature of how specific traits are inherited and expressed.

Man is not a favorable subject for genetic study. The diversity from individual to individual is great, and yet the geneticist requires numerous crossings of exactly alike individuals and often needs to know the result of a cross between diverse genetic types. In man reproductive unions are entered into without any intent to serve an experimental plan. Also for the human geneticist knowledge of the inheritance of one or more characteristics over many generations is often necessary, and knowledge of more than four generations is seldom available to him.

However, many of the obstacles have been overcome. Although the mechanisms of genetics have been discovered through the study of rapidly breeding laboratory animals, the applicability of these mechanisms to human beings has been established by the collection of data from marriages which happen to fit the necessary requirements. And the immense size of the human population presents millions of unions for study.

In the next few decades much effort will be devoted to the limitation of population growth, and in these efforts attention will be drawn to the possibility of directing the inheritance of desirable characteristics. Knowledge of human genetics will not only satisfy our desire to better know ourselves, it will also form a basis for the practical socio-genetic decisions which are to come.

Human Mendelian Genetics and the Chromosomal Theory of Inheritance

In Chapter 2 we discussed the nature of the cell nucleus. The nucleus is composed of a material called chromatin which appears under the microscope as a complicated interlacing of fine threads (Fig. 2–4). Chromatin is the functional state of the chromosomes which appear as 46 discrete forms only when the cell divides. Chromosomes are apparently tightly coiled versions of the chromatin. The chromatin, or uncoiled chromosomes, is composed of linearly arranged genes, probably 10,000 to 50,000 in number.

Twenty-three pairs constitute the 46 chromosomes. In sperms and eggs only one of each pair is represented, so the chromosome number is reduced to 23. The union of the egg and sperm restores the number to 46, and the 46 chromosomes present in all the cells of the new individual represent two of each kind of chromosome: One of each pair is from the mother and one from the father.

Inherited characteristics are transmitted by the genes which are arranged linearly along the chromosomes. At a given point on a given chromosome, called a *locus,* there always occurs a gene for a particular trait. Several different possible genes may be able to occupy this particular locus, and these several genes are called *alleles.* The expression of a gene often depends upon the nature of the allelic gene (that is, the gene at the same location on the other chromosome of a pair). In some pairs of allelic genes both may be expressed, but more often one is expressed and the other is not. The gene which is expressed is called the *dominant gene* and the expressed characteristic is the *dominant trait.* The other gene is suppressed or "hidden" and so is called the *recessive gene* and the characteristic which it would express, if allowed, is called the *recessive trait.*

An example of a simple dominant gene is that for normally pigmented skin. The allele for this gene is albinism or absence of normal pigment development (really, the inability to manufacture the pigment melanin). The normal skin trait is represented by the letter *A,* the capitalization indicating that the gene is dominant; albinism is represented by the lower case *a,* indicating a recessive gene. If these were the only two possible alleles for this locus, an individual could have only three possible combinations of the two genes: *AA, Aa,* or *aa.* The *AA* individual received identical genes from his parents, and is said to be *homozygous* for the trait of normal skin pigmentation; the *Aa* individual received an *A* gene from one parent and an *a* from the other. (Actually there are two possible combinations: *A* from the father and *a* from the mother; and *a* from the father and *A* from the mother. However, in either case the result is the same.) These *Aa* individuals are said to be *heterozygous* for this trait, and they exhibit normal pigmentation since the dominant *A* is expressed and the recessive *a* is not. The *aa* individual received identical genes from his parents and is said to be homozygous for albinism. In this example, then, there are three different types of gene combinations or *genotypes,* but only two possible trait expressions or *phenotypes,* that is, normal pigmentation or albinism. Thus an individual with normal skin pigmentation may be either homozygous *AA* or heterozygous *Aa*—these two genotypes give identical expression or phenotype because of the dominant nature of gene *A.* The albino individual we know to be *aa,* since for a recessive gene to be expressed it must occupy the appropriate locus in both of the chromosomes.

We can illustrate the inheritance of albinism by considering the mating of two individuals of genotypes *AA* and *aa.* We shall say that the *AA* is the father and the *aa* the mother. All the sperms produced by the father will carry *A* since this gene is present in both chromosomes carrying this trait; all the eggs produced by the mother will carry *a* for the same reason. The *A* sperms will fertilize *a* eggs and all the offspring of such a union will be heterozygous (*Aa*) for normal skin pigmentation, that is, they will be phenotypically the same as the father. In genetics the offspring of this first cross would be called the F_1 generation.

A crossing of members of the F_1 generation, that is, two individuals heterozygous for this trait, will produce members of the F_2 generation. In this case, the germ cells will be of two kinds: one-half will carry *A* and one-half will carry *a* (one-half of the eggs produced by the mother will carry *A* and one-half *a*; one-half of the sperms produced by the father will carry *A* and one-half *a*). The F_2 offspring will consequently be of four types as shown in Figure 3–1. These four combinations will be expected to occur in equal numbers. But since *Aa* is represented twice, the genotypic ratio is actually 1:2:1, and if four children are produced we would expect 1 *AA*, 2 *Aa*, and 1 *aa*. However, only two phenotypes will be present: 3 of the 4 children would be of normal skin pigmentation (one homozygous for the trait and two heterozygous for the trait) and one would show albinism:—a phenotypic ratio of 3:1.

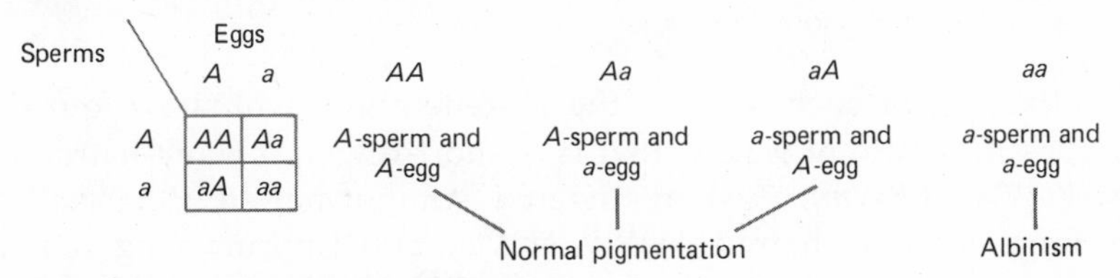

FIG. 3–1 The inheritance of albinism.

We can now consider the behavior of two separate characters determined by different loci on different chromosomes. For this purpose we can take as an example the inheritance of two dominant traits; a type of short-fingerness called brachydactyly is determined by the dominant gene *B* and normal finger length by the recessive allele *b*; wavy hair is generally determined by a dominant gene *W* and straight hair by the recessive allele *w*. Thus a person whose genotype is *BBWW* would have short fingers and wavy hair, the person *bbww* would have normal fingers and straight hair. We may illustrate the pertinent chromosomes and the F_1 generation as shown in Figure 3–2.

No matter how the assortment of chromosomes occurs, each sperm and each egg will be alike in regard to these two characters. All the F_1 individuals will consequently be heterozygous for both traits and will resemble their

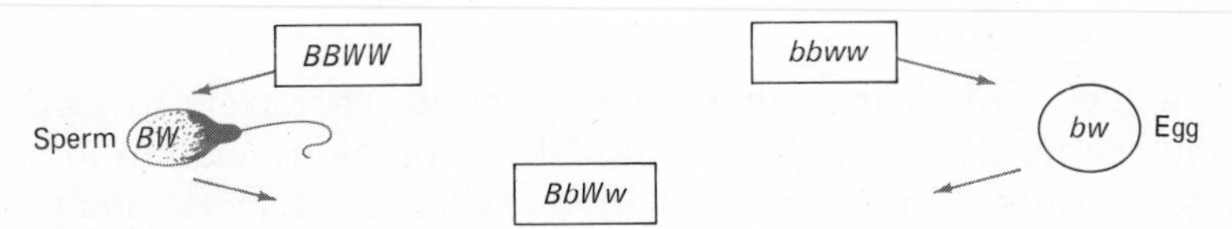

FIG. 3–2 The inheritance in the first (F_1) generation of two traits: wavy hair and short-fingerness.

father in these traits, showing short-fingerness and wavy hair. The F_1 individuals are each able to produce four different types of germ cells as shown in the table in Figure 3–3.

BbWw F_1 sperms \ BbWw F_1 eggs	*BW*	*Bw*	*bW*	*bw*
BW	*BBWW*	*BBWw*	*BbWW*	*BbWw*
Bw	*BBWw*	*BBww*	*BbWw*	*Bbww*
bW	*BbWW*	*BbWw*	*bbWW*	*bbWw*
bw	*BbWw*	*Bbww*	*bbWw*	*bbww*

FIG. 3–3 The inheritance in the second (F_2) generation of two traits: wavy hair and short-fingerness.

The offspring of such a cross, the F_2 generation, will show four different phenotypes in a ratio of 9:3:3:1: that is, 9 short-fingered wavy-haired (*BBWW, BbWw, BbWW, BBWw*); 3 short-fingered straight-haired (*BBww, Bbww*); 3 normal-fingered wavy-haired (*bbWW, bbWw*); and 1 normal-fingered straight-haired (*bbww*). Since few families will number 16 or more children, the ideal ratio would seldom be obtained. But the chances of a given phenotype occurring through a single pregnancy could be calculated. For example, the heterozygous parents (F_1) would have a slightly greater than 50-50 chance (actually 9:7) of producing a child with short-fingerness and wavy hair, and only about a 6 percent chance of producing a child with both normal fingers and straight hair.

When three pairs of alleles, each located on a different chromosome pair, are considered, the triply heterozygous individual might be indicated by *AaBbCc*. Such an individual would produce eight different kinds of germ cells: *ABC, ABc, AbC, Abc, aBC, aBc, abC,* and *abc*. If two such heterozygous individuals were crossed, eight phenotypes would be expected in the offspring in a ratio of 27:9:9:9:3:3:3:1. A square such as used in Figures 3–1 and 3–3 (called a Punnett square) would have to contain 64 boxes. It is obvious that to follow a number of inherited traits simultaneously presents considerable statistical difficulties.

Mendelian Heredity

Gregor Mendel studied the inheritance of certain characters in garden peas, and found the results as we have outlined them here in relation to dominant and recessive traits in humans. He referred to genes as "factors" and postulated that factors exist in pairs and that the factors are responsible for inherited characteristics. He also postulated that in the formation of germ cells, the factors are separated or segregated so that each sperm or egg gets only one factor of each pair. This is *Mendel's Law of Segregation of Characters.*

He further postulated that when two factors occur together (as in a heterozygous individual) one factor will be expressed (dominant) and the other suppressed (recessive). Mendel also observed that when he followed several characteristics through several generations that the traits involved were inherited independently of each other. For example, he found that flower color (white or red) and stem length (short or long) in garden peas were inherited in all possible combinations. This has been observed to be true in humans as we showed in the inheritance of two dominant traits (short-fingerness and wavy hair). This is *Mendel's Law of Independent Assortment of Characters.* Mendel knew nothing of chromosomes, and we now know that the law of independent assortment does not apply when the genes occur on the same chromosome.

Between the time of Mendel's work, published in 1866, and the rediscovery of his laws of inheritance in 1900, much was learned about cell structure, cell division, and the nature of germ cell production. It was discovered that all cells of animals and of higher plants contain a definitive number of chromosomes characteristic of the species. This characteristic number was referred to as the diploid or $2n$ number. Table 3–1 shows the chromosome number of a selected group of animals.

TABLE 3–1 Diploid Chromosome Number of a Few Selected Animals

Name	*Diploid Number*
Man	46
Cat	38
Dog	52
Mouse	40
Chicken	18
Leopard frog	26
Fruit fly	8

Each diploid set of chromosomes is composed of twice the number of the different kinds of chromosomes. In man the diploid number is 46, and there are 23 paired kinds of chromosomes. The pairs of like chromosomes are referred to as *homologous chromosomes;* one of each pair is derived from the father through the sperm and one from the mother through the egg.

The constancy of the chromosome number depends upon the reduction of the chromosome number sometime during the process of reproduction. If this reduction did not occur, the chromosome number would double with each generation. The reduction of chromosomal number occurs in the development of sperms and eggs; this process is known as *meiosis.*

Meiosis is illustrated in Figure 3–4. Only two pairs of chromosomes are

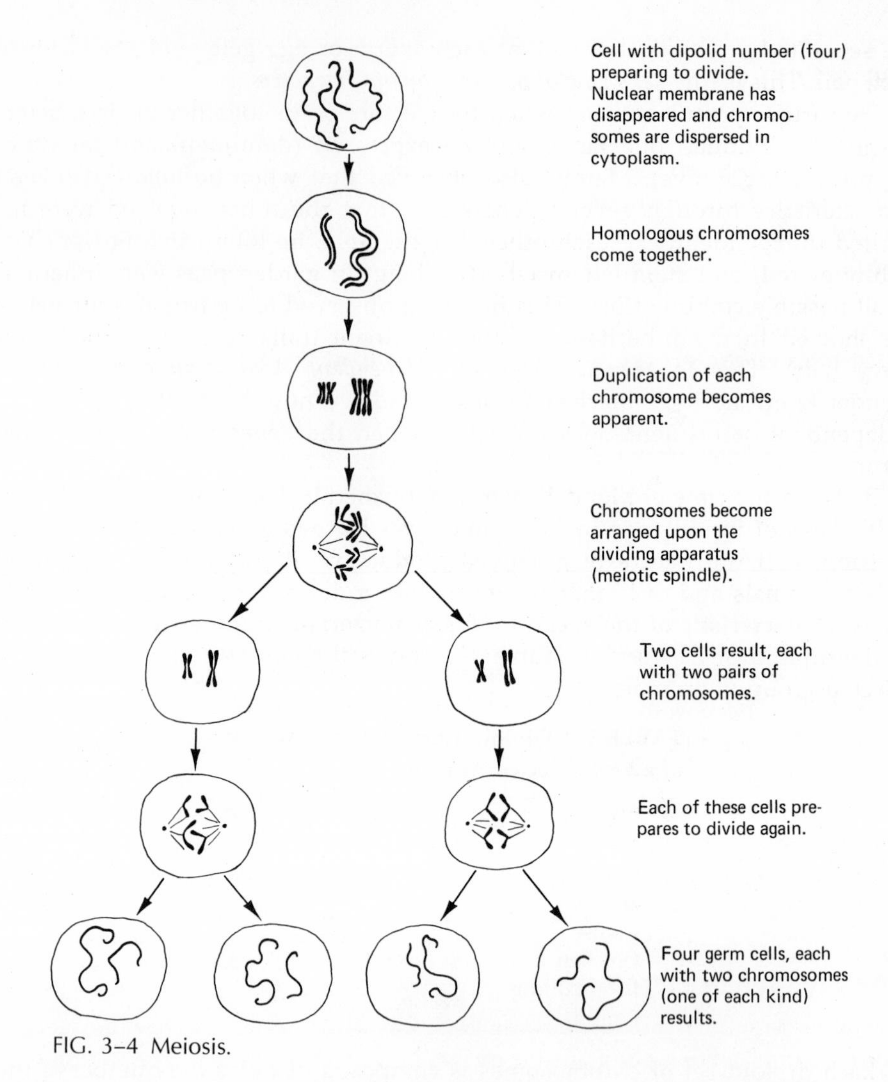

FIG. 3–4 Meiosis.

illustrated here, whereas 23 pairs of chromosomes undergo meiosis in the human.

The cells of the testis and ovary which divide to produce the sperms and eggs are diploid. The meiotic process involves a coming together of homologous chromosomes and then their division. So, for a time, the actual number of chromosomes is twice the diploid number, but there follows two successive divisions, called simply the first and second meiotic divisions, which result in four daughter cells, each with but one of each chromosome—in Figure 3–4, two chromosomes. Two is the *haploid* number in an animal with four chromosomes. The haploid number in humans is 23, and all normal sperms and eggs contain this number of chromosomes.

The process of meiosis is essentially the same whether sperms or eggs are produced, but whereas four sperms are produced by the two divisions, only one egg is produced. The same divisions occur, but in the egg one daughter cell of each division retains practically all the cell material; the other cell is obviously quite small and serves no further function, and is called a polar cell. The egg cell, then, is quite large in comparison to the sperm cell, and provides cytoplasm for several cell divisions after fertilization before nourishment is received from the mother.

In the meiotic process we see the mechanism of Mendel's postulates. The chromosomes exist in pairs and Mendel's factors were in pairs; the paired chromosomes separate at meiosis, and Mendel postulated the segregation of factors in gamete formation. When more than one pair of chromosomes are present, they will assort themselves at random during the meiotic division, maternal chromosomes with other maternal chromosomes or with paternal chromosomes, in all possible combinations, and Mendel postulated independent assortment of factors. When these observations were put together, the *chromosomal theory of heredity* was established.

The phenomenon of independent assortment of chromosomes during meiosis, even without other factors operating, would account for a very high variation in the possible kinds of sperms or eggs produced by a single individual. The number of possible combinations of 23 different chromosomes with completely random assortment is 8,388,608 or 2^{23}. Since both the man and the woman each have this variation in possible types of germ cells, the diversity of offspring becomes immense, and the possibility of two identical sperms fertilizing two identical eggs comes to about one in 64 trillion times. This shows that we are all quite unique.

Non-Mendelian Heredity

Although Mendel's postulates are generally true as far as they go, several modifications were added in the early years of this century. For example, in many different experiments in both plants and animals it was discovered that when two individuals homozygous for different alleles were crossed, the expression of the trait in the F_1 was unlike either parent but rather was intermediate in nature. This apparently means that *both* genes are expressed. Examples of this lack of dominance are difficult to establish in humans, although they probably exist. (As we shall see later the A and B factors of the ABO blood groups are both expressed, neither being dominant over the other.) An example among plants is found in four-o'clocks. When red four-o'clocks are crossed with white ones, the F_1 individuals are all pink-flowered. The F_2

generation shows 1 red, 2 pink, and 1 white. The genotypes of such a cross are shown in Figure 3–5.

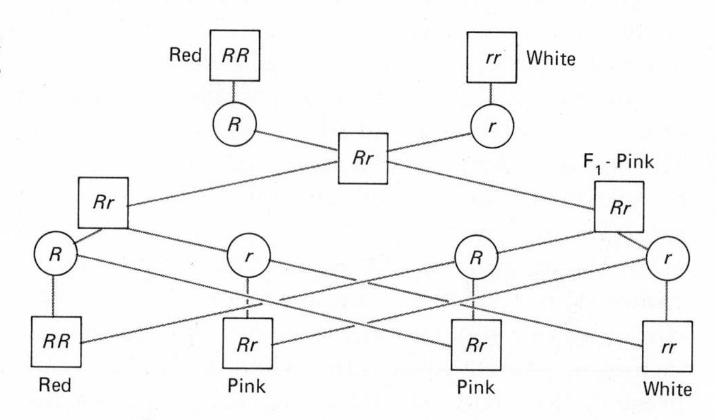

FIG. 3–5 Inheritance of a trait showing incomplete dominance.

Also, early geneticists were puzzled by certain traits which were not inherited as a few alternate forms. For example, a number of characters such as egg production in chickens, milk production in cows, and traits in humans such as size, skin color, and mental ability show a continuous gradation in their expression. The explanation for this type of inheritance involves the concept of multiple factors whereby more than one locus of one chromosome pair is involved. For example, if two loci were involved, four genes could operate; in the case of three loci, six genes, and so on. Recent work on the inheritance of skin color indicates that perhaps five or six gene pairs are involved. Since skin color serves as an excellent example of cumulative gene action, we can discuss it at some length.

Several different skin pigments are present in all races, but the intensity of pigmentation is largely dependent upon the concentration of two dark brown pigments, melanin and melanoid. The amount of the synthesis of these pigments appears to be determined by the number of the kind of pigment-producing genes present. The genes which appear in the Negroid race are more potent pigment producers than the alleles which appear in the Caucasian race. For simplification we can indicate the genotype of "full Negro color" as *AABBCCDDEE* and that of the "full white color" as *aabbccddee*. It should be emphasized that "white" genes produce both the melanin and melanoid pigments, but the amount of the pigments produced is less and the resulting color is much less intense. The nature of the melanin and melanoid pigments in *all* races appears to be identical.

The mating of a Negro with a Caucasian gives rise to a mulatto whose genotype would be *AaBbCcDdEe*, and whose pigmentation would be the result of a blend of pigment production by both "black" genes and "white" genes. If an independent assortment of genes occurs, it is clear that 32 (2^5) different kinds of eggs or sperms would be produced by these heterozygous individuals.

A sperm or egg might possibly carry all "black" genes or all "white" genes, but more probably would contain a mixture of both. Figures 3–6 and 3–7 show how the offspring of two mulattos could be either darker or lighter than either parent.

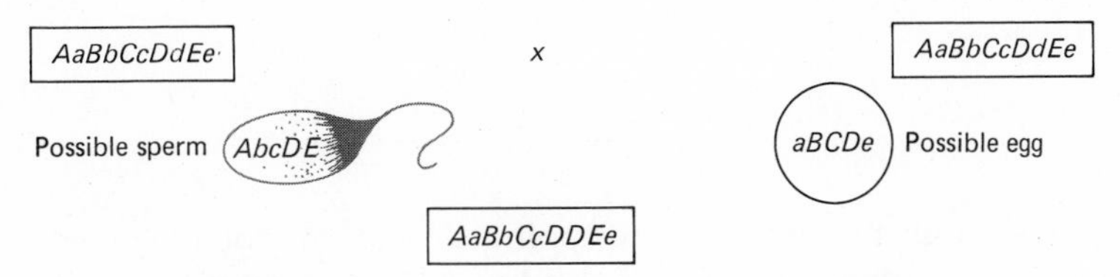

FIG. 3–6 Inheritance of skin color in the first (F_1) generation.

Since the child has more "black" genes than either parent (six "black" genes and four "white" genes), he would be darker in color than either parent.

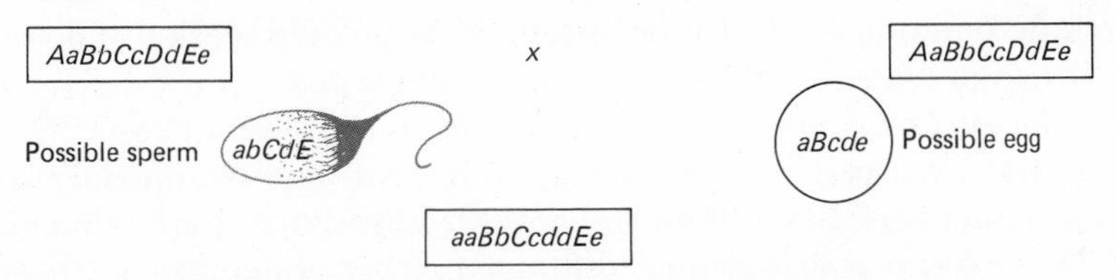

FIG. 3–7 Inheritance of skin color in the second (F_2) generation.

This child has three "black" genes and seven "white" genes and the pigment production would be less than either parent and his skin color would be lighter.

A slight possibility exists that sperms and eggs would each carry all "white" genes. Since one sperm in 32 will carry all "white" genes and one egg in 32 will carry all "white" genes, the chance that one such sperm would fertilize one such egg is $1/32 \times 1/32$, or once in 1024 fertilizations. The same likelihood exists for the inheritance of all "black" genes for skin color.

Inheritance of Blood Types

We mentioned earlier that the inheritance of the ABO blood factors is an example of alleles which exhibit a lack of dominance. The genes *A* and *B* stimulate the development of protein factors *A* and *B*, respectively, in the red blood cells. If both genes are present, neither is dominant over the other, and both protein factors are present. A third allele, *O* is recessive to either the *A* or the *B* gene. Combinations of these three genes make possible four phenotypes and six genotypes as shown in Table 3–2.

The possible kinds of offspring from various matings can be predicted, and this knowledge is now used in many states in law courts where cases of disputed parentage are involved. For example, cases of the accidental exchange of babies in hospital nurseries or disputed parentage involving illegitimacy can often

TABLE 3–2 Phenotypes and Genotypes of the ABO Blood Groups

Phenotypes	*Genotypes*
A	*AA, AO*
B	*BB, BO*
AB	*AB*
O	*OO*

be resolved through the use of blood typing. A man of type *O* and a woman of type *B* could not be the parents of a child of type *A* or type *AB*, but they could be the parents of a child of type *B* or type *O*. A woman of type *A* who accuses a man of type *AB* of being the father of her type *O* child obviously has made a mistake. If the man were type *A* or *B* or *O*, the blood typing would determine little more than that the man *could* be the father. So, blood typing can sometimes exclude parentage, but cannot prove it. You could work out possibilities in the matings between each pair of the six genotypes.

The Rh factor has a more complicated pattern of inheritance. When the Rh factor was first discovered, it was thought that Rh-positive individuals had a genotype of either Rh-Rh, or Rh-rh and that Rh-negative individuals a genotype of rh-rh. However, it was later found that a number of degrees of Rh-positiveness exist, and that these degrees are attributable to eight different alleles, each of which controls the presence or absence of three different Rh factors. The eight alleles are RH^{CDE}, Rh^{CDe}, Rh^{Cde}, Rh^{cDE}, Rh^{cDe}, Rh^{cdE}, and the Rh-negative allele Rh^{cde}. The *C*s, *D*s and *E*s representing the production of the specific protein factors, or *c*s, *d*s, and *e*s the absence of production. A person of genotype Rh^{cdE}-Rh^{cde} would be only weakly positive in comparison to a person of genotype Rh^{CDE}-Rh^{CDE} (the Rh-negative individual has a genotype of Rh^{cde}-Rh^{cde}).

Only about one-sixteenth of the cases where the mother is Rh-negative and the child Rh-positive result in erythroblastosis fetalis. This suggests that perhaps one of these factors (*C, D*, or *E*) incites a more violent reaction than the others.

About 85 percent of white persons are Rh-positive and 15 percent Rh-negative. Among black people the ratio is about 92 percent positive and 8 percent negative; Orientals (Chinese, Japanese, American Indians) are practically all Rh-positive.

The Sex Chromosomes

Of the 23 pairs of human chromosomes 22 pairs are alike in the sense that both of each pair could be present in either sex. The 23rd pair are not alike. Every normal woman has in her cells two of what are called X-chromosomes. A man has one X-chromosome, and its mate is a smaller, distinctly different,

Y-chromosome. The X- and Y-chromosomes are called the *sex chromosomes;* the other 22 pairs of chromosomes are called *autosomes.*

We have already noted that in the production of eggs or sperms each receives one of each kind of chromosome. In the production of eggs, then, each will receive an X-chromosome. But when the male forms sperms, half will receive the X-chromosome and half the Y-chromosome. At fertilization an X-bearing sperm will result in a girl (XX) and a Y-bearing sperm will result in a boy (XY). The sex of the child is determined, then, through the father in his production in equal numbers of two kinds of sex determining sperms.

From a purely statistical consideration one would expect an equal number of boys and girls to be conceived. However, this does not appear to be the case. Not only are more boy babies born—at a ratio of about 105.5 boys to 100 girls—but far more male fetuses die during gestation. McKusick estimates that about 130 boys are conceived for every 100 girls! Boys continue to succumb at a higher rate during infancy and childhood so that by the age of sexual maturity the ratio becomes about 100 boys to 100 girls. By age 30, however, females outnumber males, and continue to extend their advantage throughout life.

Nature has apparently offset the biological weakness of the male during growth and development by favoring the conception of boys. How this is accomplished is unknown. It has been speculated that the Y-bearing sperm is lighter in weight (since it carries less genetic material) and can swim faster in the race through the female genital tract to the awaiting egg; or that the egg exercises some sort of chemical preference for the Y-bearing sperm. However, these remain mere speculations. The operating mechanism will probably yield to further investigation, and it is quite possible that Y-bearing and X-bearing sperms will be separable so that parents will be able to select the sex of their offspring (a debatable blessing).

Sex-Linked Inheritance

Those traits which are transmitted through genes located on the sex chromosomes are referred to as sex-linked characters. The small Y-chromosome carries few genes so that most of the loci on the X-chromosome are represented only once in the male. This means that the inheritance of traits carried on the X-chromosome follow a different pattern from those carried on the autosomes.

Color-Blindness

Probably the most prevalent of the hereditary sex-linked conditions is color-blindness. In its common form the individual is unable to distinguish between red and green when these colors are of the same intensity (they usually have a different intensity as in traffic lights). The normal retina contains three kinds of color receptors, called *cones,* which are sensitive to red, green, and blue. All the

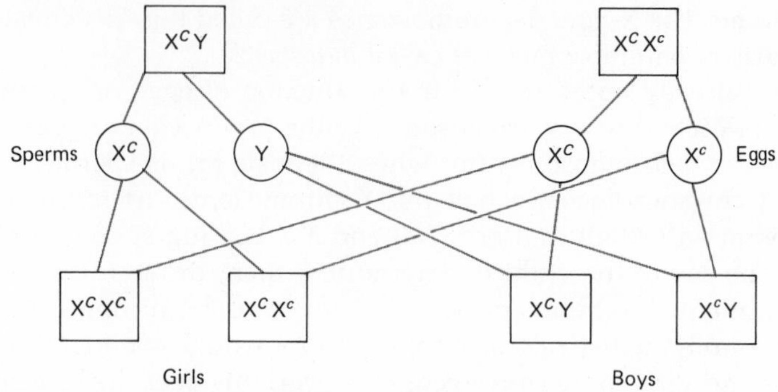

FIG. 3–8 Inheritance of color-blindness.

gradations of color normally seen are due to a differential stimulation of these three different kinds of cones. The color-blind individual is lacking in one kind of cone—usually of the red or green and rarely of the blue.

The gene for normal color vision may be represented by X^C and its recessive allele by X^c. A woman heterozygous for color-blindness would have the genotype X^CX^c, and would have normal vision. A color-blind woman would have the genotype X^cX^c. A man, having only one X-chromosome, is either X^CY (normal vision) or X^cY (color-blind).

Far more males are color-blind than females—from 5 to 9 percent of American men and less than 1 percent of the women. Why this is so can be understood by following this trait in the offspring of a heterozygous (*carrier*) female and a normal male. (Fig. 3–8).

Of four children we can expect, on the average, two girls and two boys. Both girls will have normal vision, but one will be a carrier for color-blindness like her mother. One boy will be normal and the other color-blind. Put in another way we can say that in such a marriage half the girls will probably be carriers and half the boys color-blind.

The offspring of a color-blind man and a normal (noncarrier) woman would be as shown in Figure 3–9.

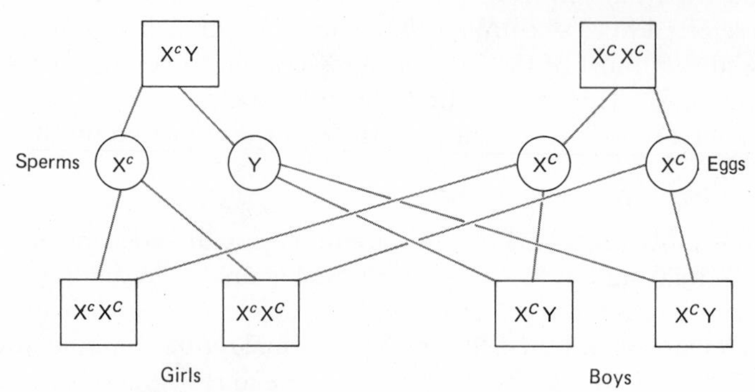

FIG. 3–9 Inheritance of color-blindness (continued).

Here we see that *all* the daughters will be carriers and *all* the sons normal. This demonstrates that sex-linked characters *can not* be transmitted from father to son, but only from father to daughter to grandson.

Color-blind women can be expected among the offspring of a color-blind man and a carrier woman as shown in Figure 3–10. We can expect that half the daughters will be carriers and half color-blind; half the sons color-blind and half normal.

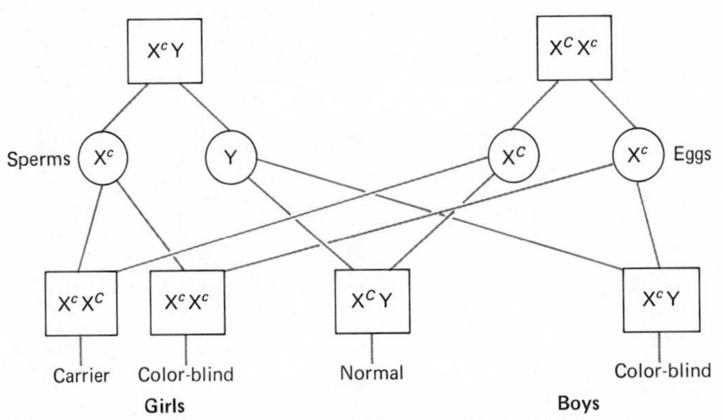

FIG. 3–10 Inheritance of color-blindness (continued).

Hemophilia

A far more serious and more dramatic sex-linked condition is hemophilia. A defective, recessive gene on the X-chromosome results in the lack of a clotting factor essential for normal coagulation of the blood.

Hemophilia has great historical interest because of its presence in the royal families of Europe (Fig. 3–11). The defective gene apparently arose spontaneously in Queen Victoria or in one of her parents. (One gene may be changed to another by small alterations in the genetic code. Such changes can be induced by irradiation, but also seem to occur without apparent cause. These spontaneous changes are called *mutations*.) Three of Victoria's daughters are known to have been carriers of the gene and one of her sons was afflicted. Iltis has shown that the present royal family of England is free of the hemophilia gene, but the carrier daughters of Victoria introduced the gene into several of the royal families of the continent—notably those of Spain, Germany, and Russia. The gene may well have been a contributing factor in bringing on the Russian revolution inasmuch as the suffering of the Czarevitch Alexi led the Czar and Czarina to seek the aid of the religious mystic Rasputin. Rasputin's corrupt influence on the royal family brought on a demoralization and indignation which helped to precipitate the collapse of the government.

About 80 percent of hemophiliacs are of the classical form, now called hemophilia A. A second form, hemophilia B (also called Christmas disease after

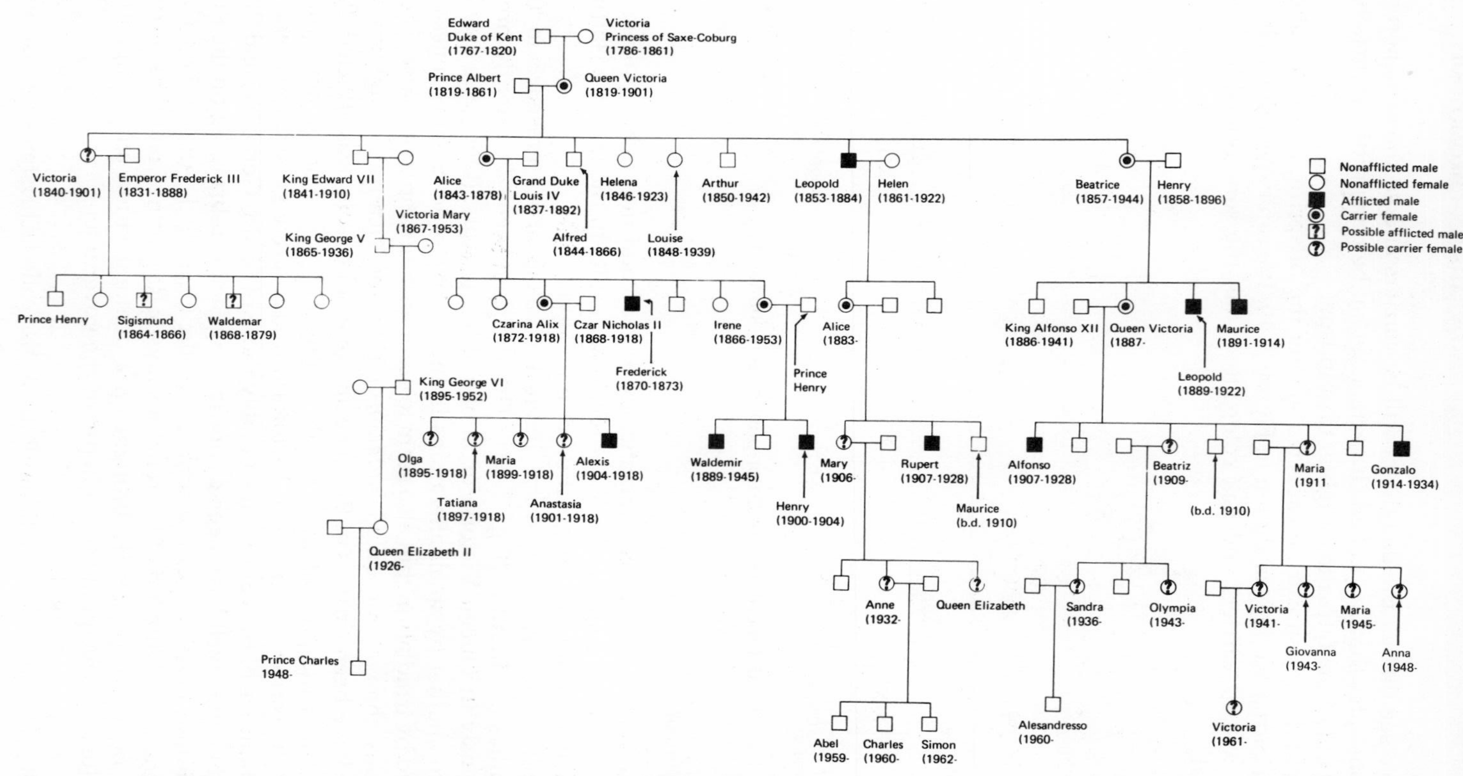

FIG. 3–11 Hemophilia among the royalty. Queen Victoria was a carrier of the gene for hemophilia, and passed the disease on to at least one son (Leopold). Daughters Alice and Beatrice also inherited the gene and were carriers, and introduced hemophilia into the Russian and Spanish royal families, respectively. Although Queen Elizabeth II and her husband, Prince Philip, are both direct descendants of Victoria, the hemophilic gene is apparently absent from each, according to Iltis.

the family in which it was first discovered), accounts for most of the remaining cases. It is quite interesting that hemophilia B is also sex-linked. The transfusion of blood from a person suffering from hemophilia A will temporarily correct the clotting deficiency of a patient with hemophilia B and vice versa. This demonstrates that different clotting factors are deficient in the two major forms of the disease.

Female hemophiliacs are theoretically possible if a hemophiliac male has daughters by a carrier female. They do occur but are rare. In 1951 a 24-year-old English woman almost bled to death following childbirth, and she was found to have a hemophiliac father and a hemophiliac brother—which demonstrated that her mother was a carrier. A few other cases are also known.

Chromosome Abnormalities

Chromosomes are small structures even at a cellular level and when 46 are present they can present a rather uncertain picture if techniques of their preparation are not perfect. In the 1920s the geneticist T. S. Painter, using a difficult technique, suggested that the number of human chromosomes was 48. In 1956 two young scientists, J. H. Tjio and A. Levan [see Levitan and Montagn], made use of a new technique which revealed the different shapes, sizes, and distinguishing characteristics of the human chromosomes so that they could not only be accurately counted but also identified and numbered (Figs. 3–12 and 3–13). In the next few years came a series of discoveries which explained a number of human abnormalities.

Down's Syndrome

Down's syndrome or Mongolian idiocy had long been a baffling phenomenon. These unfortunate individuals (see Fig. 3–14) suffer a developmental defect which results in very low intelligence along with a number of characteristic physical defects familiar to everyone: They usually are small with weak muscular development; the tongue is thick and tends to protrude from the mouth; the forehead is large and the bridge of the nose is sunken, and the ears are usually small and malformed. Skin folds at the inner angles of the eyes give them a somewhat Mongolian appearance, although they are in no other way like Orientals unless they are born to them.

The new chromosomal techniques soon revealed that Mongolian idiots show 47 chromosomes in their cells rather than 46. The extra chromosome, later identified as No. 21, is apparently produced by the "sticking together" (nondisjunction) of the two No. 21 chromosomes during the process of egg formation so that both chromosomes go into one egg. Such eggs would have 24 chromosomes, and when fertilized by a normal sperm will result in a fertilized egg containing 47 chromosomes. This extra chromosome (in this case called

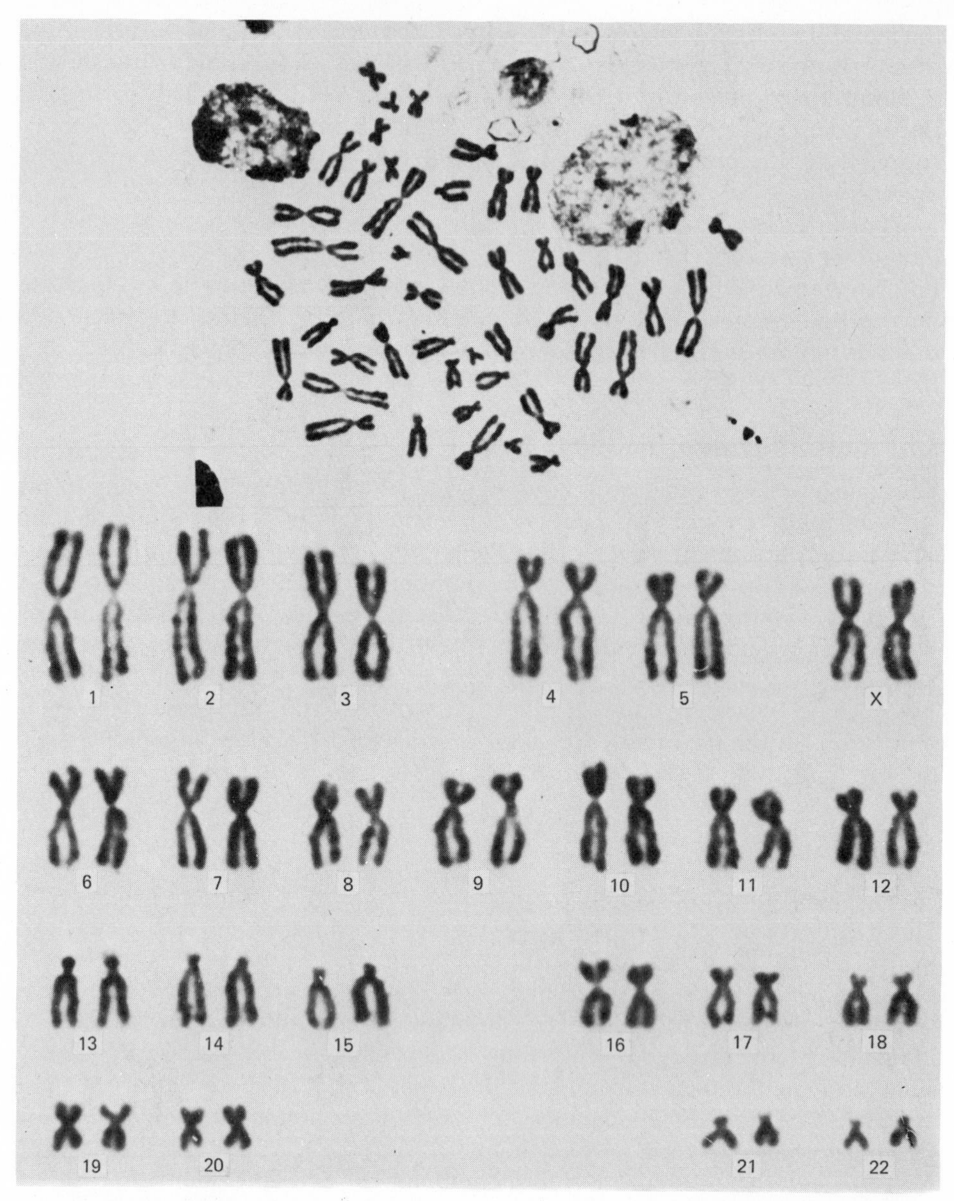

FIG. 3–12 The human female chromosomes as they appear *above,* in a photograph of a cell preparation, and *below,* after they have been matched and rephotographed. Note the two X chromosomes. (Dr. Maimon M. Cohen, State University of New York at Buffalo School of Medicine)

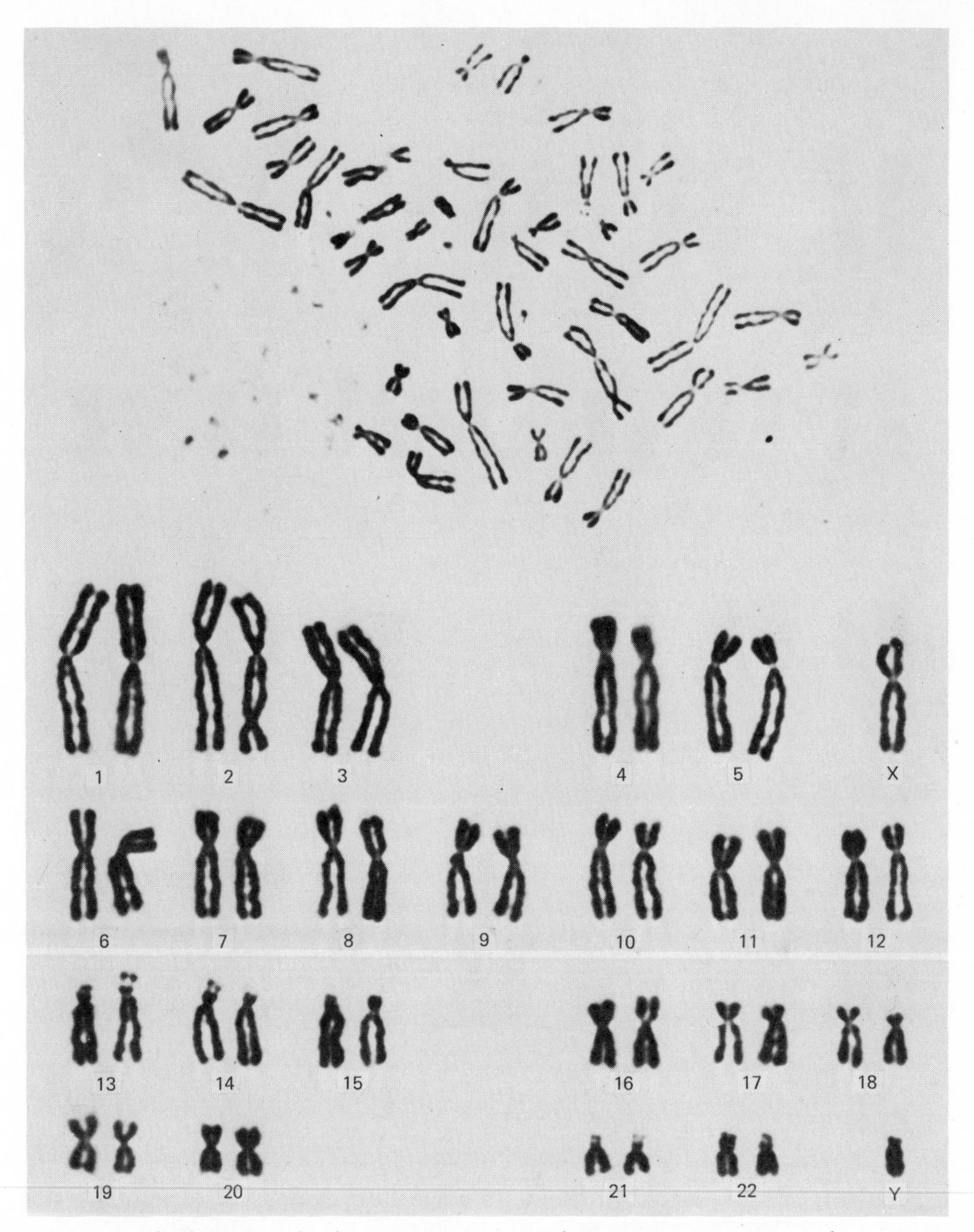

FIG. 3–13 The human male chromosomes, prepared as in Figure 3–12. Note the presence of one X chromosome and one Y-chromosome. (Dr. Maimon M. Cohen)

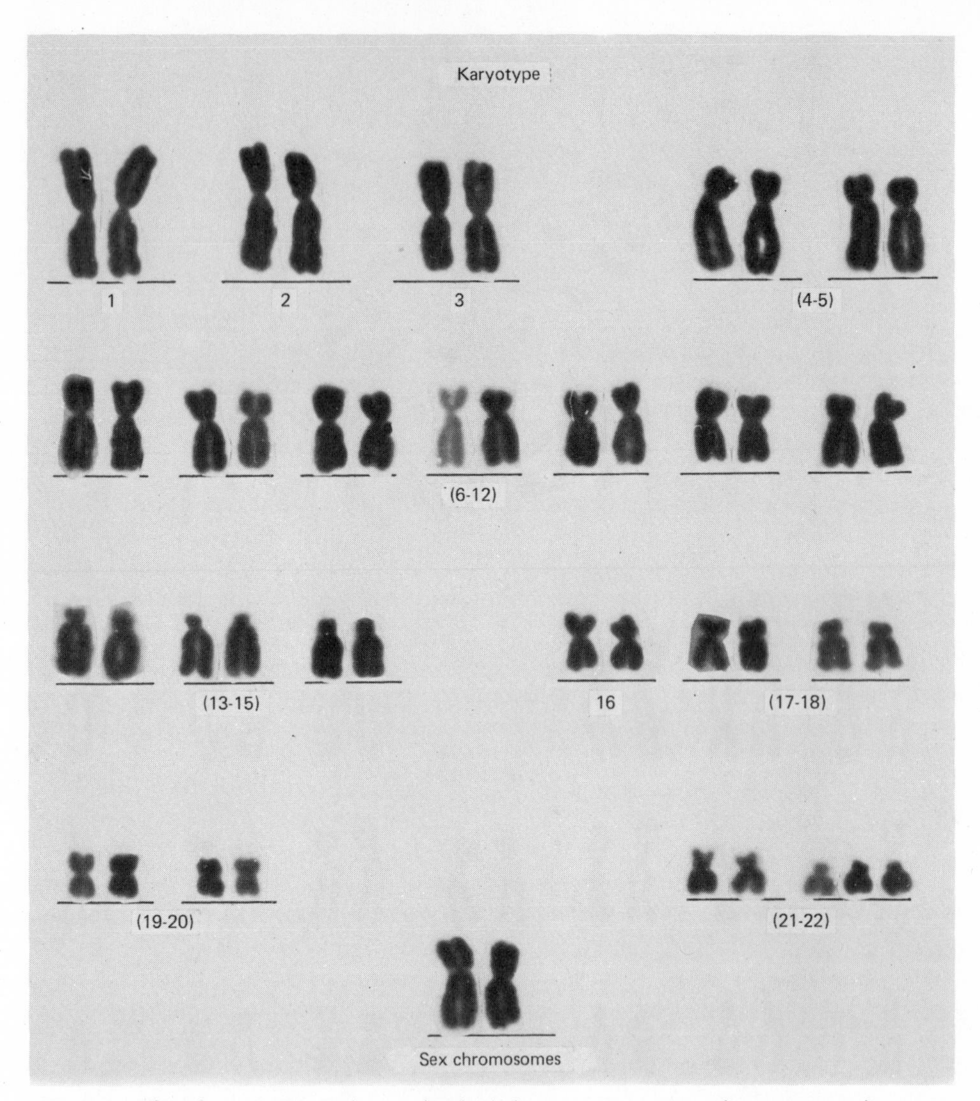

FIG. 3–14 The chromosomes of an individual showing Down's syndrome (mongolism). An extra chromosome, probably of No. 22, is present. (Dr. Maimon M. Cohen)

trisomy 21) in some way interferes with normal development and results in the abnormality. The trouble seems to lie in egg production rather than in sperm production because the age of the mother has a strong relationship to the incidence of the defect. The incidence is less than 1 per 100 births among mothers under age 34; 3 per 100 among mothers 35 to 39; 7 per 100 among mothers 40 to

44; and 30 per 100 among mothers 45 or older. How age affects the mechanism of meiosis is completely unknown.

It is estimated that there are over 50,000 cases of Mongolian idiocy in the United States, and they constitute about 5 percent of all institutionalized mental defectives. Many mongoloids show particular susceptibility to respiratory and heart disease, and to leukemia. Less than half survive to the age of 10.

Turner and Klinefelter Syndromes

The Turner and Klinefelter syndromes result from an abnormal number of sex chromosomes, and show not only sexual abnormalities but also defects seemingly unrelated to sex.

The Turner syndrome (see Fig. 3–15) results from an *XO* genotype which, in this instance, indicates but one X-chromosome and no Y. These individuals show the sexual characteristics of an immature female. The pubic and axillary hair is scanty or absent, the breasts fail to develop, and the uterus and vagina remain infantile. They are also usually of small stature, and frequently of subnormal intelligence. Occasionally a Turner female is fertile and these individuals have been found to be chromosomal mosaics. By this is meant that some of the cells are XO and some are XX. Apparently during the embryonic two-cell stage an X-chromosome is lost by one of the cells, and all of its cell descendants lack a second X-chromosome. The degree of sexual development in these cases would be determined by the proportion of ovarian tissue made up of the normal cells.

The Klinefelter syndrome (see Fig. 3–16) occurs most often when the individual has an XXY chromosome combination although XXXY, XXXXY, and XXYY combinations are also known. The meiotic process which leads to these genotypes is not known with certainty, although it is apparent that the egg or sperm or both carry one or more excess sex chromosomes. The presence of one Y-chromosome, even in company with several X-chromosomes, swerves the individual toward male development, although he is usually, if not always, sterile. He sometimes shows slight eunuchoid bodily characteristics (disproportionately long arms, sparse body hair), and is frequently mentally defective.

The Inheritance of Intelligence

No one questions assertions that some breeds of dogs are more intelligent than others or exhibit markedly different personality traits. Yet when these same characteristics are said to be inherited in humans there is immediate controversy. Very high intelligence is almost certainly the result of a fortuitous combination of a great number of "intelligence" genes. The evidence suggests that

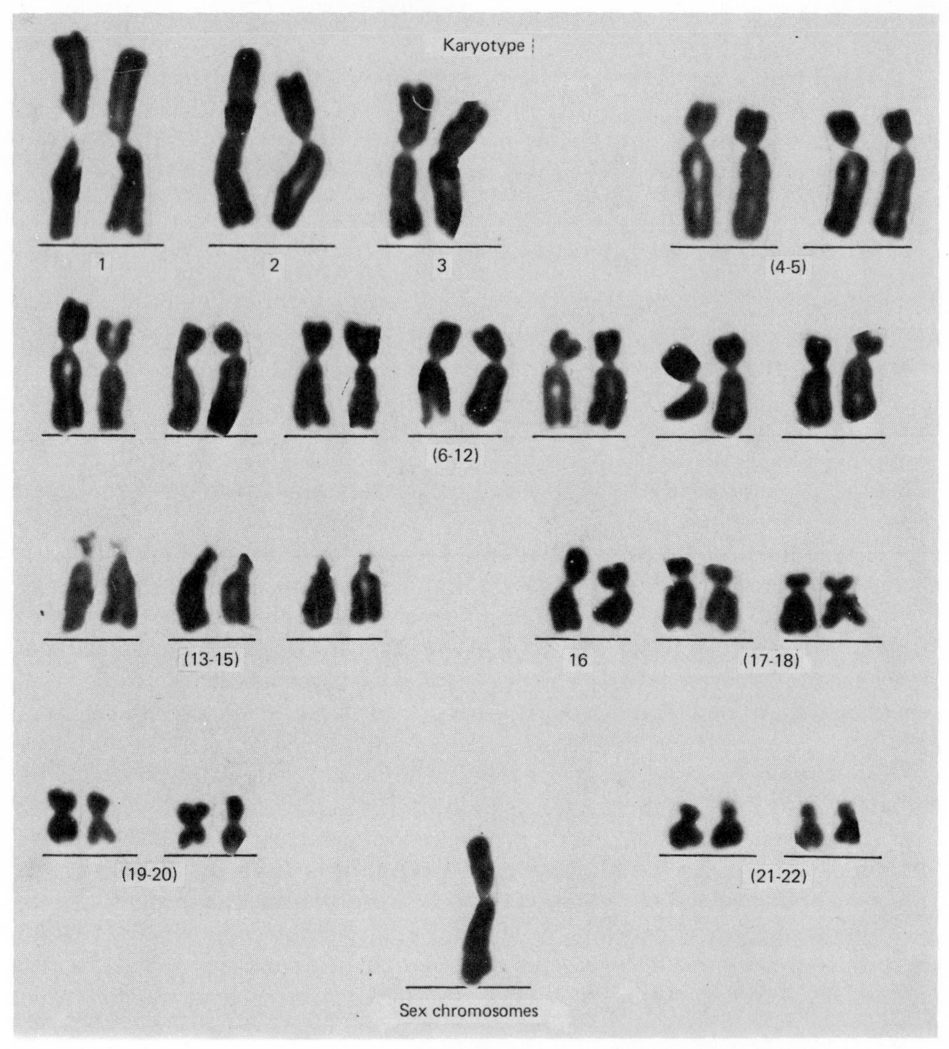

FIG. 3–15 The chromosomes of a female with Turner's syndrome. Only one sex chromosome (an X) is present. (Dr. Maimon M. Cohen)

a man and woman, both of quite ordinary intelligence, can very likely have between them the genes which, if they happen to be present in the sperm and egg which unite, produce great genius. History abounds with such individuals: da Vinci, Shakespeare, Spinoza, Franklin, Lincoln. That these individuals developed their great genius as a result of environment or special training is conceded to be unlikely indeed.

The inheritance of personality traits and artistic talents are just beginning to be understood. Yet evidence from studies on identical twins reared apart

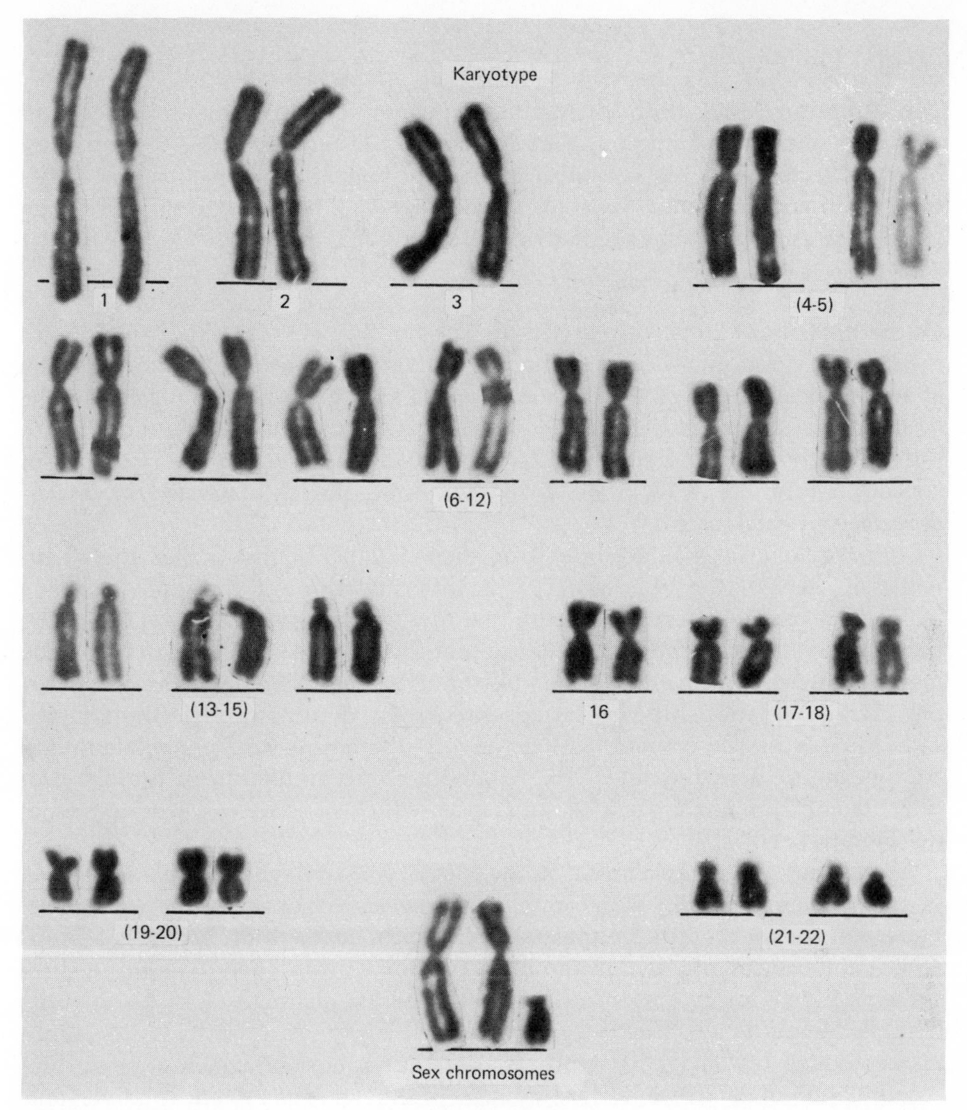

FIG. 3–16 The chromosomes of a male with Klinefelter's syndrome. Two X-chromosomes and one Y-chromosome are present. (Dr. Maimon M. Cohen)

suggests that these too are largely inborn, but involve such a large number of genes that their inheritance is complex and largely unpredictable.

We can little concern ourselves here with the inheritance of these traits other than to say that high intelligence, artistic talent, and favorable personality characteristics are most likely to be passed on by parents who exhibit these traits.

Mental Deficiency

The American Association for Mental Deficiency defines a mentally deficient person as one "with a mind so retarded as to make him incapable of competing on equal terms with his normal fellows or of managing himself or his affairs with ordinary prudence." On this basis some 3 percent of the American population is judged to be mentally deficient.

Severe Mental Retardation

Of the total about 25 percent are in the class of *severely mentally deficient* with intelligence quotients (I.Q.'s) in the range of 0 to 50. These individuals usually show physical defects as well as mental ones and the impairment can often be traced to glandular defects, birth trauma, viral disease during pregnancy, or to rather clearly defined hereditary factors. Only those due to the latter are of concern here, and they include Down's syndrome, already discussed, phenylketonuria, and galactosemia.

Phenylketonuria is rare (about 4 in every 100,000 births) but of special interest in view of the fact that (1) it is inherited through a simple recessive gene, (2) it can be detected after the first few weeks of infancy by a test of the baby's urine, and (3) there is evidence that the disease can be treated and the mental deficiency prevented. Phenylketonuria is an example of the one gene-one enzyme relationship. In the absence of the dominant allele the afflicted individual lacks the enzyme which converts the amino acid phenylalanine to another amino acid tyrosine. This results in an accumulation in the blood of phenylalanine and other related substances which in some way lead to mental deterioration.

Restriction of phenylalanine in the diet in these children reduces the level of phenylalanine in the blood and may prevent the onset of the symptoms. However, the diet is quite unpalatable, and proper nutrition becomes a problem. Another difficulty lies in the identification of true phenylketonuria since some relatively minor conditions will give positive urine tests.

Galactosemia is also due to a simple recessive gene and results in the failure to metabolize the sugar galactose, one of the monosaccharides of milk. The accumulation of galactose leads to damage of the nervous system and liver, and eventually to growth failure, mental retardation, and death. If the condition is identified at an early age, milk can be replaced in the diet by foods free of galactose and normal development occurs.

Moderate Mental Retardation

The remaining 75 percent of the mentally deficient are judged *moderately mentally deficient* and show an I.Q. of between 50 and 70. Generally, these persons are aclinical, that is, they show no physical abnormalities and their condition cannot be traced to a specific disease or metabolic defect. A few of these mentally retarded individuals have afflictions little related to hereditary factors, and are the consequence of nutritional deficiency, hearing impairment, or emotional disturbance. These are becoming a smaller proportion of the total with improvements in diet, medical care, and education.

The vast majority of the moderately retarded individuals are not, then, the result of some over-all abnormality as is most often the case with those severely retarded, but rather seem to be the products of defective or deficient "intelligence" genes. They most often are the products of parents who also show some mental "slowness." (Two moderately retarded parents are no more likely to have a severely retarded child than are two brilliant parents.) Supporting these generalizations are the following observations:

More than 75 percent of the moderately retarded are offspring from that part of the population which is also moderately retarded; when both parents are moderately retarded the chances for any given child also to be moderately retarded is about 65 percent; when one identical twin is moderately retarded, the other almost always shows the same intelligence patterns; among fraternal twins the second twin is apt to be moderately retarded in only about 30 percent of the cases.

The evidence is clear that moderate mental retardation is often hereditary, but no simple Mendelian inheritance is involved. It is probable that a fairly large number of genes act in a plus or minus fashion to produce all grades of intelligence from moderate mental retardation to genius, and a preponderance of "minus" genes will more likely occur in the products of mentally slow individuals than among more clever ones. It is quite reasonable to suppose that two dull parents have between them enough "plus" genes to produce a child of high intelligence, which would explain the variation within a given family.

Inherited Predisposition

The control of infectious diseases, better nutrition, and generally improved living conditions have been the major factors in the increase in life expectancy in this country over the last 60 years. The result has been a marked increase in deaths from diseases of old age, namely, cardiovascular disease, cancer, and diabetes mellitus. What we are doing is eliminating or modifying many of the environmental hazards, and the more we do this the more we bring out the *inherent differences* in individuals. As the environmental factors become more constant for everyone the difference in the incidences of these degenerative diseases will be more and more dependent upon heredity.

Heredity probably plays at least a minor role in all diseases; for example, some people appear to be more susceptible to certain infectious organisms than others; and nutritional needs, particularly the requirements for vitamins and minerals, show a great variation which is probably at least partly hereditary. Hence deficiency diseases afflict some and not others in given situations.

However, of greatest importance is that we need to know about the inheritance of the major causes of death. Our attention here will be directed to the role of heredity in the incidence of the three of most importance: cardiovascular diseases, cancer, and diabetes melitus.

Cardiovascular Disease

We are including a number of conditions, especially those related to coronary atherosclerosis and hypertension. Changes in the arteries generally reflect changes in the body's metabolism which are to some degree under genetic control. The development of atherosclerosis, for example, involves the deposition of fatty plaques in the walls of arteries. A defect in fat metabolism is suggested, and we know that such defects can be inherited. The condition hypercholesterolemia (high blood cholesterol) is an inherited disease in which blood cholesterol levels become so high that cholesterol crystals are deposited in the blood vessel walls. (Diets disproportionately high in fats will also increase blood cholesterol, but not to such a marked extent. Dietary factors related to high blood cholesterol and atherosclerosis are discussed in Chapter 23.) Other clearly genetic conditions involve the inability to store fat. Thus even slight defects in fat metabolism could very well increase the fat levels of the blood and promote atherosclerosis.

Hypertension, when not the effect of other conditions, is more likely to occur in an individual where the family has a history of the disease. An hereditary predisposition for hypertension may mean its development from emotional or physical stresses which would not ordinarily be of consequence.

Cancer

Heredity does not play a significant role in most cancers. For example, identical twin studies show that the commonest cancers are the result of environmental factors. However, a few cancers are definitely hereditary. A well-known example is the childhood tumor of the eye, *retinoblastoma,* which accounts for about 2 percent of all childhood cancers. A person surviving this cancer generally passes the condition on to half his children. This identifies the condition as a Mendelian dominant. Those who survive this cancer should seek genetic counseling.

Further, as noted earlier, children with Down's syndrome have an unusually high incidence of leukemia—an incidence some 20 times that of normal children. However, these two examples are exceptional, since more and more cancers are now being traced to rather specific environmental factors (Chapter 12).

Diabetes Mellitus

Diabetes mellitus is almost certainly an inherited disorder. However, it exists in at least two distinct forms, and the relationship, if any, between the forms is unknown. The more serious juvenile form of diabetes seems to be, in many pedigrees, a simple recessive trait (Fig. 3–17).

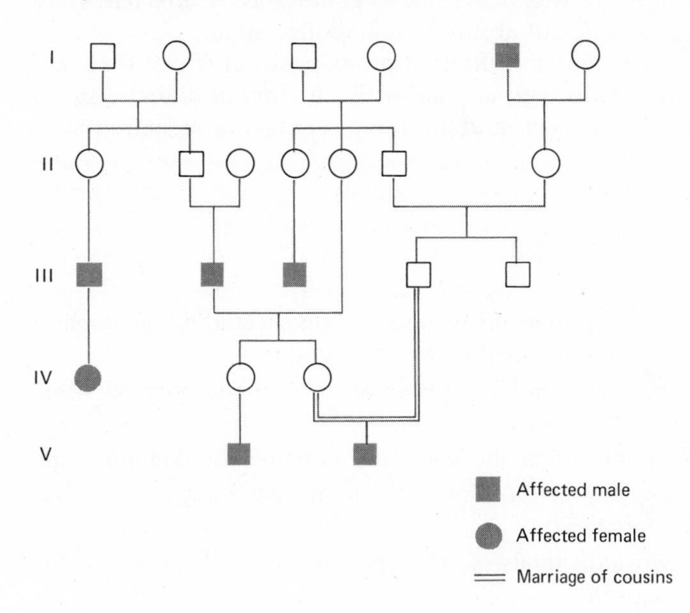

FIG. 3–17 Diabetes mellitus with recessive inheritance. Note that the condition has a tendency to skip generations.

The adult form of diabetes seems on the other hand to be an inherited predisposition for the disease which is transmitted as a simple dominant trait (Fig. 3–18).

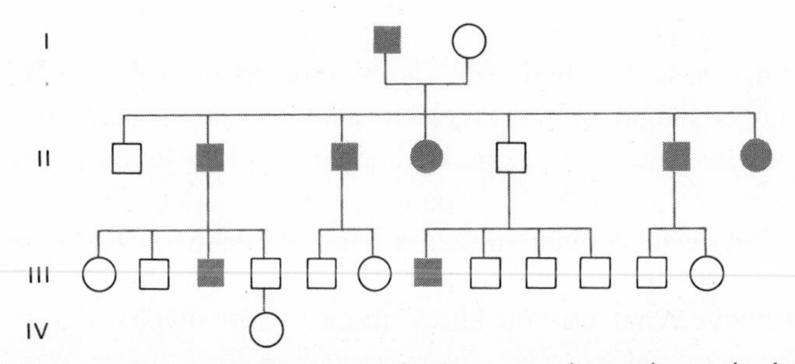

FIG. 3–18 Diabetes mellitus with dominant inheritance. In this pedigree the fourth son in the second generation would appear to have inherited a predisposition for the disease which he passed on to one son; but never developed the disease himself. Otherwise, the pedigree shows the usual pattern of dominant inheritance, that is, a tendency to occur in each generation.

In the adult type of diabetes we have the foremost example of inherited predisposition. Numerous environmental factors govern the time of onset of the symptoms and its severity, including diet, pregnancy, injuries, and possibly, emotional traits. One report records the history of a pair of 59-year-old identical twins. One of them had been diabetic since the age of 38, but had always been an alcoholic. The other was clinically free of the disease, but had lived a more temperate life. He was nevertheless genotypically affected, since he showed, when tested, a deficient ability to metabolize sugar.

We should assume, then, that an inherited predisposition is just that, and not a predestined certainty. However, anyone with a history of diabetes in his family would do well to view dietary and alcoholic indulgence as more than a trivial gratification.

REVIEW QUESTIONS

1. What is genetics?
2. Is plant and animal breeding dependent upon an understanding of genetics?
3. Why is the genetics of man particularly difficult to study?
4. What is a locus on a chromosome? What is an allele? What is the relationship between the two?
5. Explain an example in man that illustrates the principle of dominant and recessive characters. Explain the meaning of the terms heterozygous, homozygous, genotype, and phenotype.
6. In a cross between two individuals of genotype *AA* and *aa* what will be the genotype of the offspring? What is this generation called?
7. In a cross between two individuals of genotype *Aa* and *Aa*, what will be the genotypes of the offspring? What is this generation called?
8. A normal man marries an albino woman and their first child is an albino. What are the genotypes of each of these three persons? Is it possible that the second child could be normal?
9. If short-fingerness (brachydactyly) is determined by a dominant gene *B*, and normal finger length by the recessive allele *b;* and if wavy hair is generally determined by a dominant gene *W*, and straight hair by the recessive allele *w:* (a) What are the phenotypes of people with the following genotypes: *BbWw; BBWW; Bbww?* (b) A man with wavy hair and brachydactyly marries a woman with wavy hair and normal fingers. They have a child with straight hair and brachydactyly. What do you know about the genotypes of each of the three persons?
10. The problem in Question 9 is an application of Mendelian heredity. Explain how Mendel's laws apply.
11. What is the purpose of meiosis? How does it aid in explaining Mendel's laws?

12. How is skin color inherited?
13. What are the possible genotypes of children born to parents with the following phenotypes (assume *A* can be either *AA* or *AO*, and *B* either *BB* or *BO*): *A* and *B*; *O* and *A*; *AB* and *O*; *A* and *A*?
14. A woman of type *A* has a child of type *O*. What are the possible phenotypes of the father? Males of what phenotypes could be excluded as fathers?
15. Can you think of a reason based on genetics why males seem to be biologically inferior to females?
16. Males more often suffer from inherited diseases that are sex-linked and recessive. Why?
17. Can a man with hemophilia who marries a woman who is not a carrier pass the disease on to his sons? to his daughters? What proportion of his daughters will be carriers? What proportion of his daughters' sons (his grandsons) will have hemophilia?
18. What are chromosomal abnormalities? Briefly discuss some conditions that are caused by abnormal chromosomes.
19. What is meant by inherited predisposition? Give some examples.

REFERENCES

ILTIS, H., "Hemophilia, 'the royal disease' and the British royal family," *Journal of Heredity*, **39:** 113–116 (1948).

LEVITAN, M., and A. MONTAGU, *Textbook of Human Genetics*. New York: Oxford University Press, 1971.

MCKUSICK, V. A., *Human Genetics*, 2nd ed. Englewood Cliffs, New Jersey: Prentice-Hall, 1969.

Additional Reading

SCHEINFELD, A., *Your Heredity and Environment*. Philadelphia: J. B. Lippincott, 1965.

PART TWO
SEX AND REPRODUCTION

4
THE STUDY OF HUMAN SEXUALITY

It appears well within bounds to affirm that many of our present beliefs concerning average sex experience and normal sex life have the status of surmises standing on foundations no more secure than general impressions and scattered personal histories. It is time we began building on detailed case records running through lifetimes in series counted in tens of thousands. In view of the everlasting gonadal urge in human beings, it is not a little curious that science develops its sole timidity round the pivotal point of the physiology of sex. Perhaps this avoidance—not of the extreme, of the abnormal and the diseased, but of inquiry into the general usage and physical sex conduct of mankind—perhaps this shyness is begotten by the certainty that such study cannot be freed from the warp of personal experience, the bias of individual prejudice, and, above all, from the implication of prurience. And yet a certain measure of opprobrium would not be too great a price to pay if we rid ourselves thereby of any basic fallacy.

Our protests against the sensual detail and the exaggerations and credulities of pornographic pseudo-science lose force unless we ourselves issue succinct statistics and physiological summaries of what we find to be average and believe to be normal, and unless we offer in place of the prolix mush of much sex teaching the simple statements called for in any sane instruction. Considering the inveterate marriage habit of the race, it is not unreasonable to demand of preventive medicine a place for a proper section on conjugal hygiene that might do its part to invest with dignity certain processes of love and begetting.[1]

[1] R. L. Dickinson, *Human Sex Anatomy*. Philadelphia: Williams & Wilkins, 1933.

No single physiological process causes as many anxieties, tensions, and other emotional disorders as our sexual functions. Yet physicians and others who have sought to counsel and advise on sexual matters have found it difficult to obtain strictly factual information which was not biased by moral, philosophic, or social interpretations. Until recent years the scientific understanding of human sexual physiology and sexual behavior was more poorly established than the understanding of almost any other function of the human body.

Dickinson was one of the pioneers in the investigation of human sexual anatomy and physiology, and yet his major work was published in the thirties. It is perhaps not coincidental that enlightenment in sexual matters has accompanied the emancipation of the female in Western society. As women were finally allowed free expression in social matters, there followed the development of contraceptive methods for women, and at the same time a freer discussion of sexual matters generally. This more liberal atmosphere made possible the significant scientific studies which were to follow.

An educated approach to such problems as sexual inadequacy in marriage, sex education of children, and premarital sexual outlets depends upon a knowledge of sex which represents conclusions independent of preconceived notions. These independent conclusions must be based upon an exhaustive analysis of the actual sexual behavior of thousands of people and the interrelationships of that behavior with their social histories. This was Dickinson's plea after nearly half a century as a practicing gynecologist who was confronted repeatedly with sexual problems based upon ignorance or folklore, and with treatments dependent upon conjecture.

Dr. Alfred C. Kinsey and his associates at Indiana University were the first to undertake a study as envisioned by Dickinson, but not without considerable skepticism on the part of his colleagues. They argued that it would be impossible to induce such a large number of people to talk candidly about their sex life, and that if such a large sample were obtained that they would represent an abnormal group of exhibitionists. Kinsey found, instead, a widespread sympathy with the possibilities of his study, and had little difficulty in enlisting volunteers. In fact, after the study had been in progress a few years and knowledge of its existence spread, he and his co-workers had far more volunteers than they could process. The result was the two volumes, *Sexual Behavior in the Human Male* and *Sexual Behavior in the Human Female,* which represent the analysis of detailed interviews of more than 16,000 men and women.

It was not Dickinson's challenge, however, which stimulated Kinsey to begin his investigations into sexual behavior. As a teacher of biology he had been asked by his students a number of questions about sex. Many of the questions required answers which were not readily available, and in some cases he found that generalizations were based on research techniques which would be considered grossly inadequate in a scientific investigation of any other nature. He found also that a good many of the published studies were "confusions of moral values, philosophic theory, and the scientific fact."

However, the study was not without its difficulties. Some medical groups objected on the grounds that Kinsey and his associates were practicing medicine without a license; attempts were made to persuade Indiana University to stop the study and dismiss Kinsey from the faculty. One high school teacher who cooperated in the study lost his job because of disapproval by the school board. Some psychologists contended that sexual behavior was primarily a psychological problem; some sociologists felt that the problems were social in nature and that neither biologists nor psychologists were qualified to study them. A group of physicians claimed that the taking of histories constituted clinical practice and that any such study should be conducted only by physicians inside clinics. Some leaders in sex education contended that human sexual behavior was primarily a question of emotions, and that since no scientific study could measure emotions, any results would be invalid; and even if they were valid that they should not be published until at least 100,000 case histories had been tabulated. Some scientists urged that "moral evaluation" be included or that only "normal" behavior be studied. But the number of professional persons who objected to the study was far exceeded by those who offered cooperation; these included not only physicians, psychologists, and sociologists, but historians, lawyers, and others who saw the immense value of such a study.

As it turned out Kinsey was an ideal scientist to undertake leadership in the investigation. By training he was an entomologist and was particularly interested in the characterization and classification of certain wasps. This doesn't seem much related to sexual behavior of humans, but the techniques of investigation were appropriate. His previous research demanded the ability to recognize the uniqueness of individuals and of the wide range of variation which may occur in any population. This involves the measurement of *variation* of great numbers of specimens within the species being studied. The major problem in any study which seeks to determine the incidence of any type of behavior in a population is that of sampling and the meaning of the variation encountered. Individuals must be chosen in a fashion that eliminates all bias, and must include persons from the whole population. The analysis of the results must include comparison between populations of individuals with different backgrounds that may account for the differences observed.

Such research methods mean that the results can be expected to be the same (generally within 1 to 5 percent) in any subsequent study using the same population (in this case the whole United States). This is quite different from the results obtained by some "social scientists" who hobnob as tourists in some particular social environment, and acquire from the experience some "impressions" and "hunches" of a whole group. The latter method saves time, but the generalizations have little scientific validity.

The physiological aspects of human sexual activity have just begun to be investigated. Dr. William H. Masters and his research associate (and wife), Dr. Virginia E. Johnson, have recently published the results of the first extensive study of directly observed physiological reactions of humans during sexual arousal. Their books, *Human Sexual Response* and *Human Sexual Inadequacy,* provide the answers to many long-asked questions, and also disprove a number of fallacies.

The works of pioneering scientists, such as Dickinson, Kinsey, and Masters and Johnson, will perhaps make possible a sound approach to sexual maladjustment. As Masters so properly states, "Without adequate support from basic sexual physiology, much of psychologic theory will remain theory and much of sociologic concept will remain concept."

It is possible that many aspects of sexual maladjustment will not be amenable to medical or psychological treatment. But little progress can be expected as long as sex is dealt with in a confusion of secrecy, ignorance, folklore, and exploitation. These can only lead to fear, mental and physical disorders, shame, and a loss of human dignity.

Knowledge and its appropriate and skillful dissemination are indispensable ingredients of any effort to deal with problems of human sexuality, whether viewed within a physiological, psychological, or sociological context.

REFERENCES

DICKINSON, R. L., *Human Sex Anatomy.* Philadelphia: Williams & Wilkins, 1933.

FEDERMAN, DANIEL D., "Disorders of sexual development," *New England Journal of Medicine,* **277,** 351–359 (1967).

KINSEY, ALFRED C. et al., *Sexual Behavior in the Human Female.* Philadelphia: W. B. Saunders, 1953.

KINSEY, ALFRED C. et al., *Sexual Behavior in the Human Male.* Philadelphia: W. B. Saunders, 1948.

MASTERS, WILLIAM H., and VIRGINIA E. JOHNSON, *Human Sexual Inadequacy.* Boston: Little, Brown, 1966.

MASTERS, WILLIAM H., and VIRGINIA E. JOHNSON, *Human Sexual Response.* Boston: Little, Brown, 1970.

5
THE FEMALE

A dictionary defines sex as a state of being male or female. What are maleness and femaleness? What are the anatomical and physiological factors which determine male and female behavior? These are the questions we attempt to answer in this and the following four chapters.

The Reproductive Process

Man and all higher animals reproduce sexually, the male producing sperms and the female producing eggs or ova. The union of the sperm and egg is fertilization. The new cell or fertilized egg must undergo a great number of cell divisions and growth before it can become an independent organism. The egg carries the food supply or yolk which is necessary for the early part of this development, which means that it must be particularly large, and consequently immobile.

Fertilization requires, then, that the sperm come to the egg, and must be motile. The structure of the sperm reflects its basic requirements. It carries in its head end the hereditary material and just sufficient food for a few days of activity; the remainder of the sperm is the locomotor apparatus ending in a long flagellum giving it the appearance of a long-tailed tadpole. It requires no more than this—for it need live only long enough to travel from its point of release to the awaiting ovum. If it fails in this one task for which it is meant, there is no longer any reason to survive.

Human Reproductive Organs

The paired organs in which ova or sperms are produced are called *gonads.* In the female they are called *ovaries;* in the male, *testes.* The other sexual organs, such as those in the male concerned with the production of semen and the delivery of sperms and those in the female concerned with reception of the sperm and the nourishment and growth of the embryo, are called *accessory reproductive organs.*

Certain other adult features of both sexes not directly related to the sexual function, such as a deeper voice and facial hair in the male and enlarged breasts and hips of the female, as well as the development of axillary and pubic hair in both sexes, are dependent upon male and female hormones, and are therefore called *secondary sexual characteristics.*

The Female Reproductive System

The Ovaries

Each ovary is about the size and shape of an unshelled almond, and lies suspended from the inner wall of the pelvis by a fold of peritoneum (membrane lining body cavity).

The eggs or ova develop from cells which cover the surface of the ovary (the *germinal epithelium* in Fig. 5–1). Clumps of these cells penetrate into the deeper substance of the ovary and form *egg nests.* Only one of the cells of the egg nest is destined to become an ovum; it enlarges, and the other cells proliferate as *follicle cells.* The ovum and follicle cells together form a *primary follicle.* In later stages the follicle cells greatly increase in number and a space appears among them. The appearance of this space marks the beginning stage of a *Graafian follicle.* The Graafian follicle is a highly complex collection of many cell types which, if it grows to maturity, will be the size of a child's marble and project like a boil or blister from the surface of the ovary.

The follicle cells which adhere directly to the surface of the ovum form the *corona radiata;* these cells will remain with the ovum when it leaves the ovary. The outer cell layers of the follicle are composed of *theca cells.* These cells produce the hormone *estrogen* during the maturation stages of the Graafian follicle.

The fully ripened follicle finally ruptures, spilling the ovum and its surrounding corona radiata into the peritoneal cavity. This event is termed *ovulation,* and in humans occurs approximately midway between the menstrual periods.

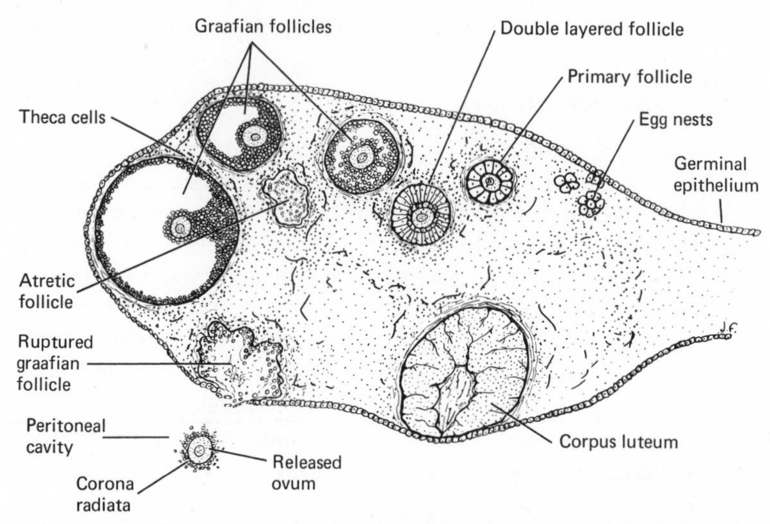

FIG. 5–1 Idealized representation of the human ovary drawn to show the stages in the development of the ovum and its supportive structures. Beginning with the egg nests derived from the germinal epithelium, which actually occurs during embryonic life, shown at the upper right, progressively older stages are drawn counterclockwise. See text for further explanation.

In humans only one Graafian follicle generally grows to full maturity at any one time; the others which begin to develop at the same time degenerate and become *atretic follicles.* Animals, such as cats, which have their young in litters mature many Graafian follicles and ovulate an equivalent number of ova. In humans and other mammals which bear single offspring, some mechanism operates to suppress the growth of all but one (or occasionally two or three) of the Graafian follicles; how this occurs is unknown.

The cells of the old Graafian follicle left behind in the ovary after ovulation multiply and enlarge to form a new structure called the *corpus luteum* (yellow body). The corpus luteum functions as a temporary hormone-producing (endocrine) gland. It continues and increases the production of estrogen begun by the Graafian follicle and, in addition, produces a second ovarian hormone, *progesterone.* Estrogen and progesterone play an integral part in a monthly cycle of events of which menstruation is the most overt sign. This sexual cycle, or *menstrual cycle,* is discussed later in the chapter.

Uterine Tubes

The *uterine tubes* (or oviducts or Fallopian tubes in Fig. 5–2) transport the ova from the peritoneal cavity to the uterus. A fringed, trumpet-shaped mouth more or less envelops the ovary and ensures the movement of the ovum into the tube. (Recent evidence suggests that at ovulation the fringe of fingerlike extensions actually "search" the area, locate the ovum, and participate in a "swallowing" process that thrusts the ovum in the tube.) The ovum is moved along the tube by peristaltic or wavelike contractions of the muscular wall and also by the beat of cilia. The uterine tubes open medially into the cavity of the uterus.

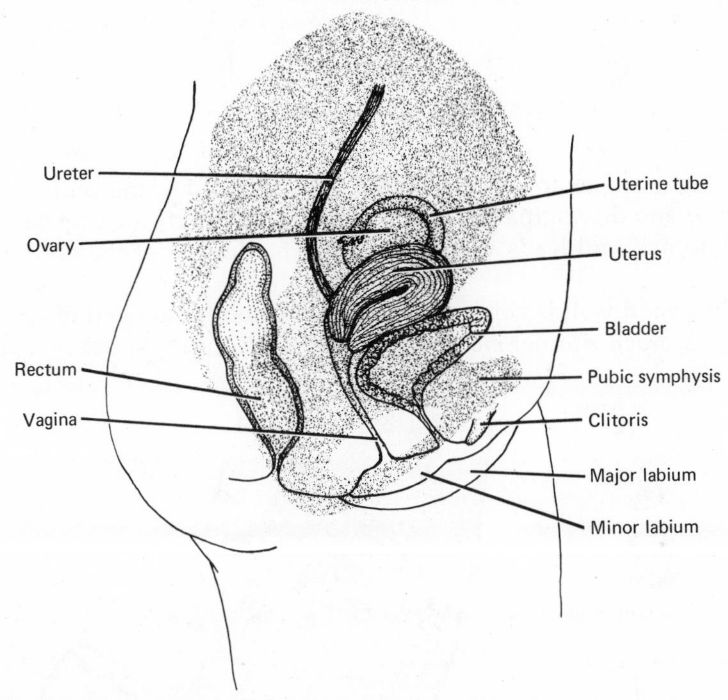

FIG. 5–2 Female sexual anatomy in lateral view.

The Uterus

The nonpregnant uterus is a hollow, pear-shaped organ about 3 inches long and about 1 inch thick and consists of two main parts, the *body* and the *cervix* (Figs. 5–2 and 5–3). The body rests upon the bladder, and the narrow, lower, cervix bends downward and projects through the anterior wall of the vagina

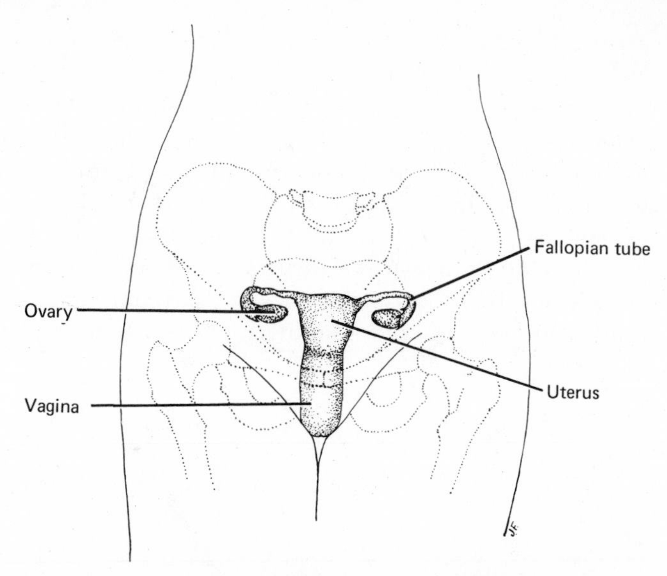

FIG. 5–3 Female sexual anatomy in frontal view.

almost at right angles to it. This should be contrasted to the uninformed idea of the uterus and the vagina being in line with each other, since the latter condition is abnormal and leads to prolapse, that is, descent of the uterus into the vagina.

The thickened wall is composed largely of smooth muscle (the *myometrium*) (Fig. 5–4). A much thinner *endometruim* forms the lining of the uterine cavity.

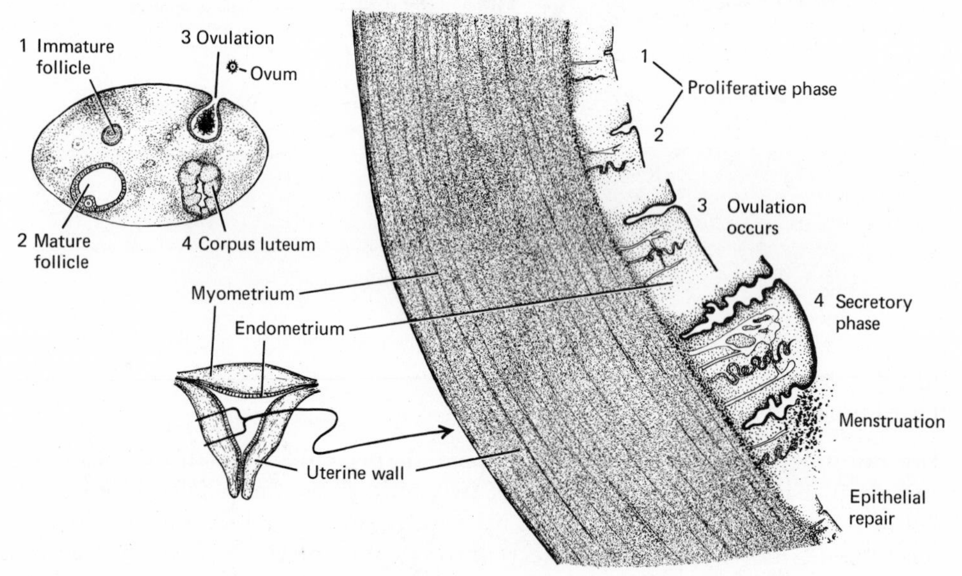

FIG. 5–4 Changes in the ovary and in the uterus during the menstrual cycle.

It seems quite remarkable that this compact structure can accommodate such great change as occurs in pregnancy. Before pregnancy the capacity of the uterine cavity is about 3 to 4 milliliters (about one teaspoon). At the end of pregnancy the capacity has increased to 5 or 6 liters. This increase in size is due to both *hypertrophy* (increase in size of cells) and to *hyperplasia* (increase in number of cells).

The endometrium consists of a single layer of closely packed cells which line the uterine cavity and of an underlying bed of vascular connective tissue. Numerous fingerlike extensions of this cell layer penetrate the connective tissue nearly to the myometrium, forming the tubular *uterine glands.*

Two types of arteries provide the blood supply to the endometrium, and serve to divide the endometrium into two functional divisions. One type of artery is distributed to the deeper part or *basal layer* of the endometrium. The second type is highly coiled as it enters the endometrium, and supplies only the much thicker *superficial layer.*

Menstruation

Menstruation is the discharge of blood which occurs when the superficial layer of the endometrium is shed. We shall see that menstruation is the terminal stage of the cyclic events which are controlled by hormones, but for the present we are principally concerned with the anatomical changes which occur.

Just prior to menstruation the coiled arteries begin to contract, and the blood supply to the superficial layer is greatly reduced. As a consequence these tissues die and small parts are sloughed off with minor hemorrhage from the broken arteries, and menstruation or the *menstrual phase* begins (Fig. 5–4). The dead tissues, blood, and mucus exudate from the uterine glands, are gradually expelled through the cervical opening into the vagina over a period of four to six days. The sloughing passes as a wave over the endometrium, so that some areas are completely denuded of superficial endometrium at the time that other areas are in the beginning of the process. The basal layer remains intact during menstruation, and repair begins at once. Cells from the edges of the broken glands grow over and cover the exposed surface, and when this healing is complete, menstruation stops. During the next 11 or so days the superficial layer grows until it becomes at least three times as thick as the basal layer. This is the *proliferative phase* of the menstrual cycle. When growth is complete, the uterine glands begin to enlarge and secrete, and the *secretory phase* is said to have begun.

The fully grown endometrium remains in the secretory phase for about 12 days. We can summarize the typical menstrual cycle of 28 days as having, then, a proliferative phase of 11 days, a secretory phase of 12 days, and a menstrual phase of 5 days. This typical or "ideal" 28-day cycle, however, is only the most common cycle length, and is not to be thought of as the only normal one.

The Vagina

The vagina is a blind pouch about 3½ inches in length which lies between the bladder and urethra in front and the rectum behind (Fig. 5–2). The upper end of the vagina forms a furrow called the *fornix* around the cervix. The deeper, posterior part of this furrow is the *posterior fornix.* Note in Figure 5–2 that the tip of the posterior fornix lies high in the pelvis which places it in close relation to the lowest part of the peritoneal cavity. This means that injuries to the upper end of the vagina may bring serious infection to the body cavity membrane (peritonitis).

The walls of the vagina are largely muscular with a skinlike lining of stratified and flattened cells. Most of the vagina develops in common with the uterus; the lower end, however, forms from a shallow inpocketing of the skin. The two parts fuse, and an opening breaks through between them. This opening is constricted to varying degrees by the remnants of the former partition. These remnants persist as the *hymen.* The hymen is usually torn or stretched by the entering penis during first coitus. It may, however, be disturbed by tampons, by a physician in a premarital pelvic examination, or in other ways. Thus it is not an absolute indication of virginity.

Female External Genitalia

The area between the thighs extending from the pubic bone in front to the tail bone or coccyx behind is termed the *perineum.* The anus occupies the posterior part of the perineum, and genital and urinary structures, the anterior part.

In the standing position the only parts of the female perineum which are visible are the outer lips, or *major labia* (Figs. 5–2 and 6–6), and their anterior area of fusion, the pubic mound or mons, which together form an inverted U. Both major labia and mons are swellings of skin caused by underlying pads of fat. The labia flatten out and fuse in the area in front of the anus. The pubic mound and major labia become covered with hair at puberty.

The inner or opposed surfaces of the major labia are smooth and sparsely covered by delicate, rudimentary hair. Numerous sebaceous and sweat glands keep the inner surfaces moist. Separation of the major labia reveals a space, the *vulva,* and the pair of inner lips or *minor labia.*

The minor labia appear more truly as lips than do the major labia (which are actually pendulous swellings). They are hairless, but moist through the action of great numbers of sebaceous glands—typical of all surfaces within the vulva. The minor labia are longest at their midpoint, becoming progressively shorter posteriorly until they become reduced to a shallow transverse fold behind the vagina. Anteriorly the two minor labia fuse to form a collar around the *clitoris.* This collar, called the *prepuce of the clitoris,* is much more prominent than the clitoris itself, and, indeed, usually conceals the clitoris. The clitoris, like the penis in the male to which it corresponds, is composed of erectile tissue and is capable of enlargement or erection by engorgement of blood during sexual arousal. Unlike the penis it is not traversed by the urethra.

The space or cleft enclosed by the minor labia is the *vestibule,* and contains the openings of the urethra, vagina, and the ducts of two large *vestibular glands* (Bartholin's glands) which secrete mucus. The urethral opening is located behind the clitoris, just in front of the vaginal orifice.

Two elongated masses of erectile tissue are embedded in the lateral walls of the vestibule, underlying the minor labia. These become engorged with blood during sexual arousal, and result in a perceptible spreading and gaping of the vestibule and vulva.

The clitoris, like other erectile tissue, has a rich nerve supply, and it is not surprising that clitoral contact brings about a rapid sexual response. The erectile tissue underlying the minor labia is also highly sensitive, and probably can play an important role in coitus. (See discussion of orgasm in Chapter 7.)

The Female Breast

Anatomically the breasts (mammary glands) are part of the skin. However, their function is reproductive, and their growth during puberty is dependent upon changes occurring in the ovaries. For these reasons they are functionally a part of the female reproductive system (Fig. 5–5).

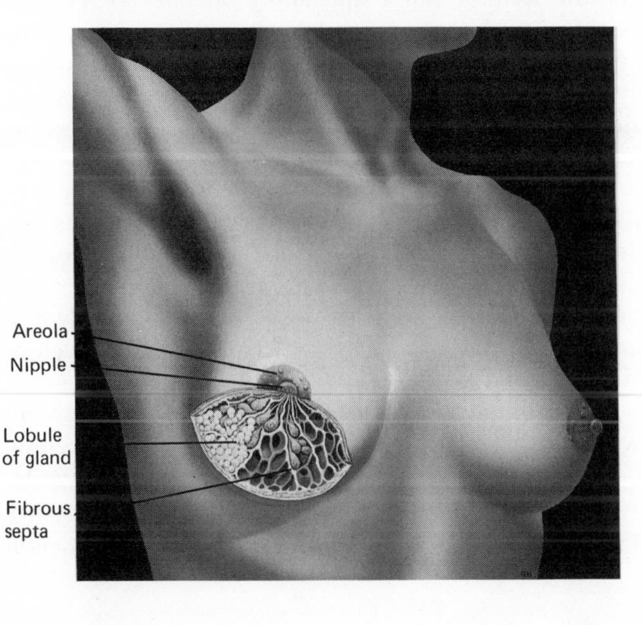

Fig. 5–5 The female breast. (Adapted with permission from Hamilton, *Surface and Radiological Anatomy,* 4th ed., Baltimore: Williams & Wilkins, 1958)

The breasts of both sexes undergo identical early development. In the embryo 15 to 20 groups of epidermal cells burrow into the deeper layers of the skin. Each of these groups of cells divides and subdivides many times, ultimately forming a branching duct system or lobe opening onto the surface at the nipple—the site of the original epidermal ingrowth. The area around the nipple develops a deeper pigmentation and is called the *areola.*

Development never goes beyond this stage in the male. At the time of puberty in the female growth is resumed. This involves development of more extensive duct systems and the deposition of fat and connective tissue in the spaces between the ducts. Most of the characteristic enlargement at this time then is due to fat rather than to any significant increase in functional tissue. In fact, the milk-secreting parts do not fully develop until pregnancy occurs. Other changes occurring at pregnancy as well as processes concerned with lactation (milk-secretion) are discussed in Chapter 9, Pregnancy and Birth.

The Menstrual Cycle and Puberty

The onset of puberty is dependent upon changes in the anterior pituitary gland which cause it to alter the kind and amounts of hormones it secretes. Prior to this time it produces large amounts of *growth hormone,* and as long as secretion of this hormone continues, some growth will occur. At commencement of puberty the anterior pituitary decreases its production of growth hormone and begins the production of gonadotropic (gonad-stimulating) hormones.

Under the influence of the latter—*follicle-stimulating hormone* (FSH) and *luteinizing hormone* (LH)—the ovary enlarges and begins to function. This functional activity of the ovary results in the obvious physical changes which we associate with puberty—growth of pubic and axillary hair; fat deposition in the breast and hips; and modifications of the muscular and skeletal systems which we associate with the feminine form.

We need now to discuss the interactions of the ovarian and pituitary hormones and how they affect the sexual functions. The FSH stimulates the immature Graafian follicle to enlarge. As mentioned earlier, only one follicle generally ripens in humans. Perhaps it suppresses development of other smaller follicles; if so, then two or three follicles developing at once would result only when there are that number at exactly the same stage of development—in a dead heat, so to speak. This is only speculation; all we do know is that the development of two or more follicles at one time is uncommon, corresponding to the low incidence of fraternal twins or triplets. (Recently, gonadotropic hormones from the pituitary, presumably including FSH, have been used in the treatment of infertility of women. In some cases they have worked too well, as newspaper reports of quintuplets, sextuplets, and even septuplets attest.)

The enlarging Graafian follicle begins to produce several hormones collectively called estrogens, which we shall simply refer to as estrogen. The amount of estrogen produced steadily increases as the follicle grows to maturation. At puberty estrogen induces the changes in other reproductive organs from their infantile state to a mature functional state. This includes enlargement of the uterus, the vagina and the external genitalia, and the development of the secondary sexual characteristics mentioned earlier. Estrogen also appears to play a role in the termination of growth, that is, once sexual maturity is reached, the growth rate tapers off. Generally, girls who reach puberty at an early age will fail to grow as tall as their later maturing sisters.

After puberty estrogen maintains these changes which it has initiated, and also takes on additional activities which are part of the menstrual cycle. Principal among these is stimulation of growth of the endometrium following menstruation. It is responsible, then, for the proliferative phase of uterine activity. Estrogen also suppresses the secretion of FSH so that by the eighth to tenth day of a typical cycle FSH reaches its highest level, and for the remainder of the cycle, progressively decreases.

As FSH begins to decrease, LH, the second gonadotropic hormone of the anterior pituitary, begins to be produced. Exactly how LH is related to ovulation is unknown, but when LH reaches its peak level—at about the 14th day of a 28-day cycle—ovulation occurs; following ovulation, its production shows a steady decline to the end of the cycle. The transformation of the ruptured Graafian follicle into the corpus luteum is dependent upon LH—in fact, this is its most obvious function, hence its name, *luteinizing hormone.*

The corpus luteum continues the production of estrogen and also begins production of the second important ovarian hormone, *progesterone.* Progesterone has little to do with the development of female sexual characteristics; instead it is primarily concerned with putting the uterus and breasts into a functional state for pregnancy: It causes the uterine glands to secrete, bringing in the secretory phase of the cycle and also induces formation of the milk-secreting cells of the breasts. These effects are transitory and quite minor since progesterone is secreted for such a short period of time in a typical cycle; during pregnancy progesterone is secreted at high levels for many weeks, and the effects are pronounced.

The drop in production of LH during the secretory phase of the menstrual cycle is probably due to inhibitory effects of progesterone and estrogen on the anterior pituitary. In any case this drop in LH level has one dramatic effect: The corpus luteum is sustained only as long as adequate gonadotropic hormone (in this case LH) is available. So, as the level of LH falls off, the corpus luteum tends to dry up and, consequently, the production of estrogen and progesterone begins to taper off. Without the sustaining influence of estrogen and progesterone, the degenerative changes in the endometrium leading to menstruation are set in motion. Progesterone is apparently required for proper function of the coiled arteries, and when it becomes deficient, these arteries undergo a spasm which cuts down blood flow, leading to virtual starvation of the superficial part of the endometrium. These changes usually begin two or three days before menstruation (on the 26th or 27th day of a 28-day cycle), and irrevocably lead to menstruation.

The drop in level of estrogen and progesterone secretion relieves the inhibition on the anterior pituitary so that by the 24th or 25th day of the cycle FSH secretion is once more on the increase. This means that by the time of menstruation FSH is at a substantial level, and enlargement of a Graafian follicle is underway. Before the end of menstruation the Graafian follicle is producing sufficient estrogen to stimulate the proliferation of the endometrium.

So we have come full circle—a new ovum is maturing and the uterus is once again being prepared for pregnancy. But how is the cycle interrupted when pregnancy occurs, that is, how is menstruation and the consequent loss of a growing embryo prevented? To follow the course of events leading to pregnancy we must back up to the 14th day of the cycle when ovulation occurred.

Fertilization and Implantation

An ovum can be fertilized by a sperm only for 8 to 24 hours after ovulation. So if coitus occurs at or near the time of ovulation the sperms will have to make their way through the cervical opening, traverse the uterus, and enter the uterine tubes. Fertilization occurs, then, somewhere near the ovarian end of the tube.

The first three days of human development are spent in descending the uterine tube to the uterus, and during this time the fertilized egg divides into eight closely packed cells (Fig. 9–1). The next two or three days are spent in the cavity of the uterus where it develops to a stage called the blastocyst. By the end of the sixth day (20th day of the menstrual cycle) the blastocyst attaches itself to the endometrial lining of the uterus, and begins to sink into the soft tissues beneath.

Once embedded in the wall of the uterus it starts to develop the embryonic part of the placenta. More details of this process are discussed in Chapter 9; here we are primarily concerned with the fact that these placenta-forming cells begin the production of *chorionic* (Greek for placenta) *gonadotropic hormone*—in other words a gonad-stimulating hormone produced by the placenta.

The chorionic gonadotropic hormone stimulates the corpus luteum to continue the production of estrogen and progesterone which are essential if pregnancy is to continue. Thus chorionic gonadotropic hormone substitutes for LH which is normally at a suppressed level of production at this time in the cycle.

These are the normal sequence of events which lead to pregnancy. It would seem to be a rather touch-and-go affair inasmuch as fertilization has to occur during a rather short period of time. It is obvious that fertilization is not necessarily followed by pregnancy: If the journey to the uterus is completed too soon, the embryo may be too immature to become embedded; if the journey is delayed, which would happen if fertilization occurred as much as a day after ovulation, the embryo might not be able to begin production of chorionic gonadotropic hormone soon enough to stay the degeneration of the corpus luteum.

Health and Disease of the Female Reproductive System

Menstrual Problems

When one considers the complexity of factors which govern the menstrual cycle, it should not be surprising that some of the 350 or so "periods" during a lifetime might be other than typical. The following are a few of the menstrual disturbances which may be experienced.

Dysmenorrhea

Dysmenorrhea refers to the "cramps" or pain which sometimes accompanies menstruation. Dysmenorrhea may be caused by abnormalities of the uterus, but more commonly it occurs without apparent cause, in which case it is called *primary dysmenorrhea.* In a study [Miller and Behrman] of 800 young college women it was found that 50 percent experienced some degree of dysmenorrhea, but only about one-third of these considered the discomfort worthy of complaint. In three percent of the 800, symptoms were severe enough to cause them to miss classes.

Primary dysmenorrhea tends to disappear with maturity, and particularly after birth of a child. Also, the recent discovery that painless menstrual cycles are frequently experienced by women taking contraceptive pills has revealed an effective therapy for women who suffer severe menstrual pain. Mild physical activity has been very helpful in relieving dysmenorrhea in some girls and women.

Amenorrhea

The absence of the menstrual flow is known as amenorrhea. Amenorrhea is most commonly associated with pregnancy and later in life with menopause. However, insufficient secretion of ovarian hormones can be a cause, particularly in young women. In this case it can frequently be corrected by treatment with synthetic hormones. Also, a woman may experience an amenorrhea as a consequence of factors which alter her living routine, as, for example, in going off to college or during a long vacation. Generally, such "missed" periods are of little concern.

Premenstrual tension

Most women occasionally experience mental depression, irritability, and nervousness five to seven days before the onset of menstruation. These psychological disturbances are commonly accompanied by fluid retention and a consequent weight gain. Collectively, the symptoms are referred to as premenstrual tension. The symptoms disappear and the mood improves with the onset of menstruation. Considerable urine is excreted during menstruation and the loss of fluid brings about a return to normal weight. The fluid retention is believed to be the primary problem and the other symptoms are secondary. Progesterone is probably the cause of the excess fluid since similar symptoms occur during early pregnancy when progesterone levels are high, but this has not been definitely established.

Various tranquilizing drugs and sedatives, as well as diuretics, are often used by the physician to treat premenstrual tension, but understanding and compassion by those around her are probably of greatest benefit.

The Menopausal Syndrome

The gradual decline in ovarian activity is termed the *climacteric;* the actual permanent cessation of menstrual bleeding is *menopause.* Thus menopause is simply the final episode of the climacteric and means the termination of reproductive capability. Most women pass through the period of the climacteric with only inconsequential discomfort. However, a few suffer from nervousness and "hot flashes" (flushing sometimes followed by chilling and trembling). These are phenomena associated with blood circulation which are presumably caused by an oversecretion of FSH by the anterior part of the pituitary gland. The failure of the ovaries to respond to FSH stimulation with the maturation of a Graafian follicle and the production of estrogen means there is no repression of FSH. Exactly how excess FSH affects blood vessels is not known.

Menopausal complaints may be serious enough to justify medical attention. If so, the new synthetic hormones available often make possible effective treatment by substituting for the lost ovarian hormones.

It is important to recognize that menopause causes but one significant and certain change—that of child-bearing capacity. It does not in itself lead to obesity, mental disorder, loss of sex drive, or loss of femininity.

Leukorrhea

The discharge of any nonbloody fluid from the vagina is leukorrhea. It may be a greater than normal secretion by the uterine glands, particularly those of the cervix, during the latter part of the menstrual cycle, and if so, is probably of little consequence. However, the cause may be infection by the organism *Trichomonas vaginalis,* a large unicellular animal (protozoan). Trichomonas infections are frequently accompanied by intense local itching (pruritis vulvae), and this should be reason enough for most women to consult a physician.

Abnormal Uterine Bleeding

By abnormal uterine bleeding is meant excessive menstrual flow or nonmenstrual bleeding. Either is cause for concern. Excessive menstruation can lead to serious anemia and deserves medical attention. Nonmenstrual bleeding may be a sign of a tumor, which may or may not be cancerous. Examination by a physician is essential.

Tumors and Cancers

The uterus has a proneness toward benign tumors called fibroids. About half of all women of age 50 or older have one or more of these growths in the wall of the uterus. Most often they are small, produce no symptoms, and ordinarily stop growing after menopause. However, they may interfere with pregnancy in women of child-bearing age, and in such instances the surgeon may be able to simply remove them.

Malignant growths of the uterus most commonly arise in the cervix, and occur almost entirely in women who have had sexual experience and especially if they have given birth to one or more children. This suggests that irritation to the cervix through childbirth, abortion, and coitus may be predisposing factors.

The "Pap" smear or Papanicolaou test is a procedure for identification of cancer of the cervix (Fig. 5–6). Dr. Papanicolaou was originally seeking a method for determining the stage of the menstrual cycle by the appearance of cells from the lining of the vagina. This is a routine technique in a number of laboratory animals, but he found it to be impractical in the case of the human female. He did discover, however, during his examination of smear preparations from hundreds of women, that abnormal cells were occasionally observed from the cervical area of some women. The most obviously abnormal cells were found in women harboring cancers of the cervix. He found that he could identify very early stages of cancer—tumors so small that they were not otherwise detectable.

His observations, first made in the 1920s, were not adopted by pathologists for general use until the late 1940s. Now the procedure is routine. A cotton-tipped applicator is used to swab a few cells from the cervix which are then placed on a microscope slide and stained. Early cancers seem to shed profuse

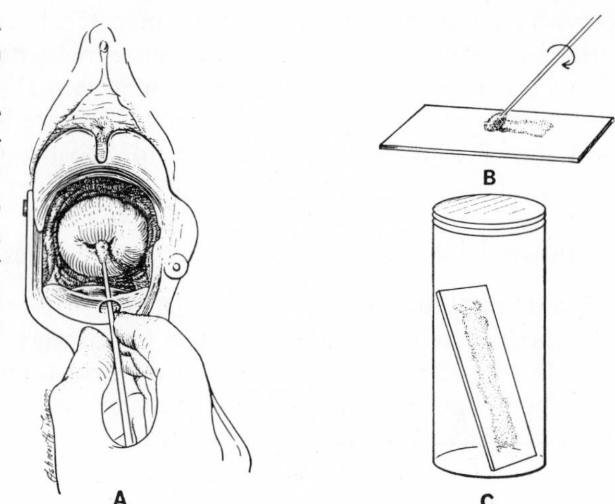

FIG. 5-6 The "Pap" smear. (A) A cotton applicator is used to obtain cells and mucus from the cervix. A vaginal speculum separates the walls of the vagina. (B) A smear of cells and mucus is produced by rolling the applicator on a glass slide. (C) The smear is placed in a mixture of ether and alcohol for preservation. A pathologist will later examine the smear for the possible presence of cancer cells. (Ralph C. Benson, M.D., University of Oregon Medical School)

numbers of cells with large, deeply staining nuclei which make them stand out against a background of normal cells.

Estimates are that cervical cancer (which killed over 9000 American women in 1970) would be virtually eliminated if all adult women would present themselves biannually for "Pap" smears. Early detection makes possible almost certain cure. (It is not, perhaps, out of place to mention here that Dr. Papanicolaou and others have devised techniques for the cellular detection of cancers from many sites by the smear method. Lung cancer has been detected from sputum; stomach cancer from stomach washings; cancer of the colon from cells in feces; cancers of the kidney, ureter, bladder, and prostate from cells in the urine.)

Disturbances of the Breasts

Premenstrual swelling and pain are probably the most common breast disturbances, and are part of the general tendency toward fluid retention during the last part of the menstrual cycle. The cause is probably excessive progesterone secretion, although the matter is in dispute. Physicians frequently treat the problem by prescribing a drug (diuretic) which increases urinary excretion.

However, these and other disturbances of the breasts shrink into insignificance when compared with the overwhelming importance of breast cancer. While uterine cancer death rates have shown a steady decline, mortality from breast cancer has continued quite stable for the last 30 years. These are discouraging statistics, for they indicate that a rather constant proportion of breast cancers is incurable at the time of diagnosis. Obviously, *effective treatment and cure of breast cancer are vitally dependent upon early detection.* Since there are not enough physicians to examine all women as often as necessary, it becomes the responsibility of each woman to become aware of the principal danger signals of breast cancer and to learn the techniques of self-examination (page 175). These techniques are also described in pamphlets and in motion picture films prepared and distributed by the American Cancer Society. These films and other materials are available to any interested group of women on request from local or State divisions of the Society.

REVIEW QUESTIONS

1. What is the definition of sex?
2. What is meant by secondary sexual characteristics? Give some examples for both males and females.
3. Trace the development of an ovum from its first appearance to the time of ovulation.
4. What cells produce estrogen? progesterone?
5. What is the shape of the uterus? What are its two main parts?
6. What are the three stages of the menstrual cycle? What are the principal events occurring in the ovary and uterus in each stage?
7. What is the position of the vagina in relation to the uterine cervix, the rectum, and the peritoneal cavity?
8. What structures comprise the external genitalia of the female?
9. What is the source of the gonadotropic hormones? What do they do?
10. Name the ovarian hormones? What is the function of each?
11. What is meant by the terms dysmenorrhea and amenorrhea?
12. What information is obtained from a "Pap" smear?
13. What can be done to reduce the mortality from breast cancer?

REFERENCES

MILLER, N. F., and S. J. BEHRMAN, "Dysmenorrhea," *American Journal of Obstetrics and Gynecology*, **65:** 505–514 (1953).

PAPANICOLAOU, G. N., and H. F. TRAUT, "The diagnostic value of vaginal smears in carcinoma of the uterus," *American Journal of Obstetrics and Gynecology*, **42:** 193–219 (1941).

Additional Readings

CORNER, G. W., *The Hormones in Human Reproduction*, Princeton, New Jersey: Princeton University Press, 1946.

WILLIAMS, R. H., *Textbook of Endocrinology*, 4th ed. Philadelphia: W. B. Saunders, 1968.

ZACHARIAS, LEONA, and RICHARD J. WORTMAN, "Age at menarche: Genetic and environmental influences," *New England Journal of Medicine*, **280:** 868–875 (1969).

6
THE MALE

The reproductive responsibilities of the male are the production and delivery of sperms. The structures which perform these functions compose the *male genital system.*

The principal parts of the male genital system are the *testes* which produce the sperms and the male sex hormones; the *vasa deferentia* (sing. *vas deferens*) or **sperm ducts** which transport the sperms; the *penis* which serves to introduce the sperms into the vagina of the female; and the semen-producing glands, which are the *seminal vesicles, prostate,* and *Cowper's glands* (Fig. 6–1).

The ovaries and testes make their appearance during the sixth week of embryonic life. They both arise from primitive structures which appear the same at this stage. Also, a bit later two sets of genital ducts arise. If the fetus is going to be a girl, the female set develops and the male set regresses; if the fetus is going to be a boy, the male set develops and the female set regresses. Remnants of the regressed set persist into adult life.

In both sexes the gonads (ovaries and testes) move gradually from their original site of develop-

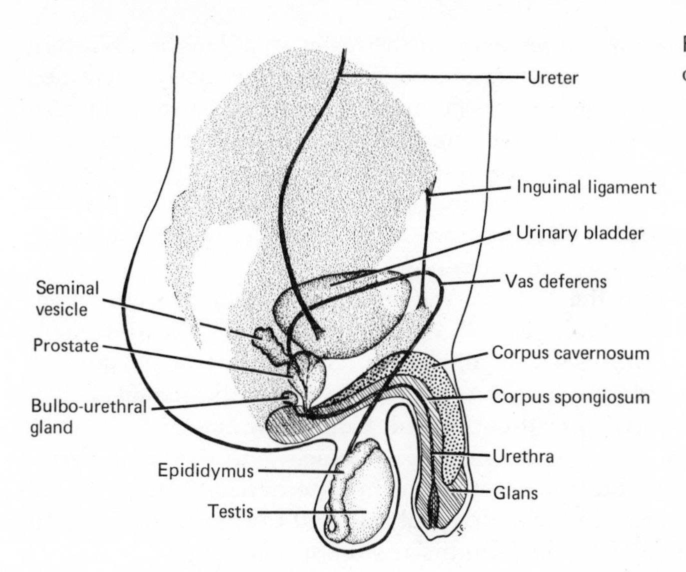

FIG. 6–1 Male sexual anatomy in lateral view.

ment, high in the abdominal cavity, to a position in the pelvis. The ovaries remain here in the female fetus. But in the seventh month of male fetal life the testes descend still farther and, by the time of birth, have taken up their permanent position in the scrotum (Fig. 6–2).

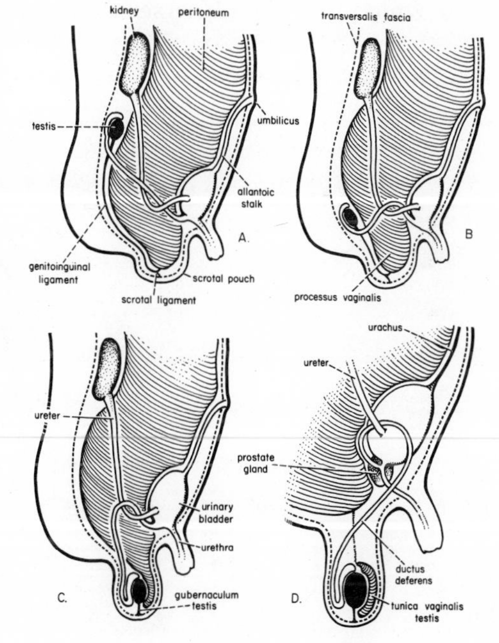

FIG. 6–2 Descent of the testis. (A), (B), and (C) represent stages in the descent of the testis from its site of development near the kidney to the scrotum. (D) represents the normal relationships after birth. The communication between the peritoneal cavity and its extension into the scrotum (the tunica vaginalis testis) is closed, but remains a potential weakness in the abdominal wall. Most inguinal hernias follow this route. (From Berger, *Elementary Human Anatomy,* New York: John Wiley, 1964.)

The usual explanation for the presence of the testes in a dangling scrotum is that the scrotum provides a temperature below that of the body. Increased temperatures are known to slow down sperm production and also to shorten their life. The reflex behavior of the scrotum seems to bear this out: On cold days the muscles of the scrotal wall contract and raise the testes to a position closer to the body; on warm days the muscles completely relax so that the testes hang far from the body; also, the scrotum is abundantly supplied with sweat glands which presumably aid in cooling.

In order to descend into the scrotum the testes pass along a route which carries them through the abdominal wall. Actually, they push ahead of them the obstructing layers of muscles and connective tissue so that extensions of the deeper parts of the abdominal wall form lining layers of the scrotal wall. A passageway, the *inguinal canal,* is thus formed, connecting the scrotum with the abdominal cavity. In it are the nerves, blood vessels, and the *vas deferens,* which were drawn down into the scrotum during the descent. The inguinal canal is a potential weakness in the abdominal wall and can cause problems at birth and later in life (*inguinal hernia;* see discussion on page 80).

Structure of the Testis

The testis (Fig. 6–3) is made up largely of tightly coiled seminiferous (sperm-forming) tubules. These tubules, several thousand in number, are arranged into groups called lobules (little lobes). The tubules empty into a network of channels which terminate in a series of parallel ducts. These ducts leave the testis

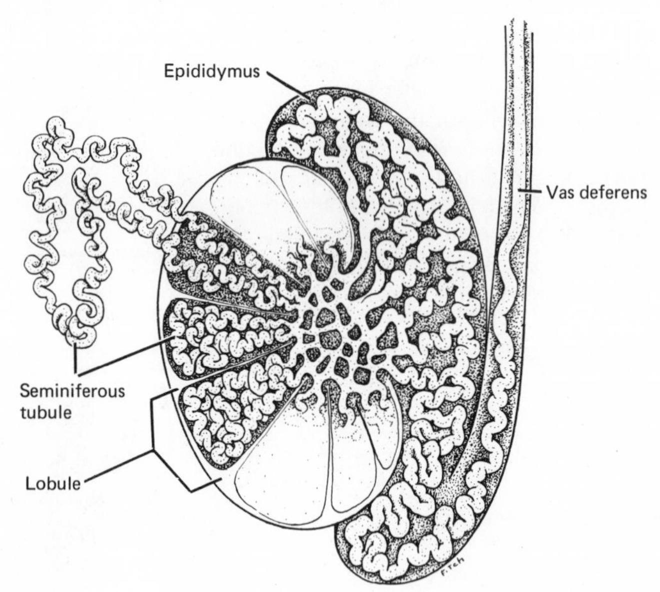

FIG. 6–3 The structure of the testis.

and deliver sperms into the *epididymus* which forms the first part of the sperm duct (vas deferens).

Figure 6–4 is a section through a part of a testis showing the internal structure of several seminiferous tubules. Each seminiferous tubule is composed of concentric layers of cells which represent, centripetally, stages in the maturation of the sperms. The outer layer is composed of unspecialized male germ cells which undergo a series of divisions which reduce the number of chromosomes to one-half, that is, from 46 to 23. The union of the sperm and the 23-chromosome egg at fertilization returns the number to 46, and makes possible the blending of the parental characteristics into the offspring.

Sperms are produced continuously and in prodigious numbers—a typical healthy male produces 100 million or more daily. The sperms reach maturity by a series of transformations which reduce them to the absolute essentials—hereditary material and a locomotor apparatus. Late in this process of maturation they are freed into the fluid-filled lumen of the tubule and moved along toward the *epididymus,* a journey which is believed to take several weeks. During this passage the final maturation takes place, and the sperms are stored in an inactive state in the epididymus.

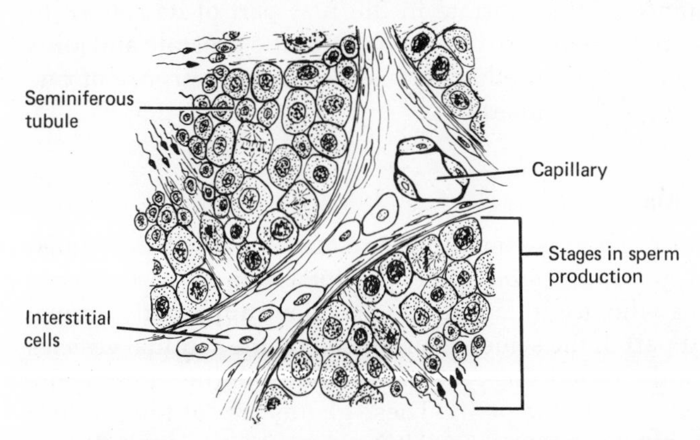

FIG. 6–4 The seminiferous tubules and interstitial cells.

Occupying much of the space among the coils of the seminiferous tubules are clusters of *interstitial cells.* These cells produce the male sex hormone, *testosterone,* which is responsible for the sexual changes occurring at puberty.

We have already discussed the roles of the pituitary gonadotrophic hormones FSH and LH in the female. These hormones also function in the male. The FSH is responsible, along with testosterone, for the growth and activity of the germinal epithelium, that is, the production of sperms. The LH is called, in the male, *interstitial cell-stimulating hormone* (or ICSH) and, as its name implies, stimulates the production of testosterone by the interstitial cells. Since testosterone functions with FSH in sperm production and in turn is under the control of ICSH, it is obvious that the functions of these hormones are interdependent.

The Transport of Sperms

The epididymus is a much coiled tube some 20 feet long which forms a comma-like cap over the posterior surface of the testis. At the lower pole of the testis the tail of the comma turns upward, gradually unwinds, and becomes the much straighter *vas deferens.*

The vas deferens together with the accompanying blood vessels, nerves, and the coverings referred to earlier form the *spermatic cord,* and can be felt through the skin at the base of the scrotum. The spermatic cord continues upward over the front and top of the pubis, where again it can be felt, not under the pubis as one might suppose from the position of the testis. It then passes through the inguinal canal into the abdominal cavity. Once in the abdominal cavity the vas deferens leaves the accompanying structures and passes along the inner wall of the pelvis making its way to a point below and behind the urinary bladder. Here the vas enlarges and joins with the short duct from the seminal vesicle to form the *ejaculatory duct* (a poor name, since it has little to do with ejaculation).

The base of the bladder rests upon the prostate gland, and the male urethra passes through the substance of the prostate in the first part of its course. In order to reach the urethra, the ejaculatory duct penetrates the prostate and joins the urethra, and from this point the urethra serves the double purpose of carrying both urine and semen to the outside.

Semen-Producing Glands

In spite of their importance, sperms constitute so small a part of the semen that their absence would not appreciably affect the total volume of a seminal sample. Hence individuals who are in fact sterile may have apparently normal ejaculate. Most of the fluid part of the semen is produced by the seminal vesicles and prostate. Smaller glands, called *Cowper's glands,* lie on either side of the urethra just above its entrance into the penis. These glands are said to produce a small amount of thick mucus before ejaculation, but their exact function seems to be in dispute.

The seminal vesicles were at one time thought to store sperm, hence the name. However, sperms are rarely found within them, and it is probable that their principal function is to produce certain substances related to sperm nutrition. These substances include ascorbic acid, fructose, and five amino acids. Exactly how they function in relation to sperm activity is quite unknown.

The prostate secretes a thin, milky, alkaline fluid containing enzymes and other substances apparently vital for successful fertilization. The alkaline characteristic of the prostatic fluid serves to neutralize the normally acidic nature of both the male urethra and the vagina. Since sperms require an alkaline environment to become optimally motile, this function of the prostatic secretion would seem clear enough. The function of the numerous other substances is unknown.

External Genitalia of the Male

The external genitalia of the male includes simply those structures which are visible—the *scrotum* and the *penis.* The pubic area is not as prominent in the male as in the female because fat deposition is not as extensive; likewise there is practically no fat in the walls of the penis or scrotum, so the skin of both structures appears quite loose. The pubic hair extends onto the base of the penis, and often far up the abdomen. The scrotum is sparsely covered with hair. Development of pubic hair does not, as is commonly supposed, have anything to do with fertility.

The penis (Fig. 6–1) is composed largely of three columns of erectile tissue. Two of these erectile bodies, the *corpora cavernosa,* arise along the inner margin of the pubic arch and join with the third, the *corpus spongiosum,* which lies in the midline and receives and carries the urethra. The corpus spongiosum is quite narrow in the shaft of the penis, but enlarges terminally into the *glans penis.* The corpora cavernosa end behind the glans penis which caps them.

The loose skin of the penis is attached around the base of the glans, but a foldlike collar extends over the tip of the glans penis as the *prepuce* or foreskin. The prepuce is adherent to the glans penis during development, and it is only at birth or two or three weeks later that the prepuce becomes free enough to be easily retracted.

The Mechanism of Erection

The erectile tissues of the corpora cavernosa and the corpus spongiosum are the same. They are composed of a spongelike system of expanded blood vessels which lie between the arteries and veins. In the relaxed condition these spaces contain little blood. Upon sexual arousal the arteries entering the erectile tissue relax allowing a greatly increased blood flow. The increased blood pressure has the effect of closing the veins so that blood accumulates in the spongelike spaces. The result is a marked enlargement and turgidity of the whole organ. Full erection also causes the penis to rise from its pendant state to a position well above the horizontal plane (Fig. 6–5). The return of the erect penis to the flaccid state is probably due to a contraction of the arteries which decreases the blood pressure that in turn diminishes blood flow and eases the pressure on the draining veins.

No subject is more prone to folklore than penis sizes—both flaccid and erect dimensions—and their relation to sexual prowess. One of the results of the Johnson and Masters study was to expose a number of these "phallic fallacies." For example, it is generally assumed that full erection of a larger penis provides

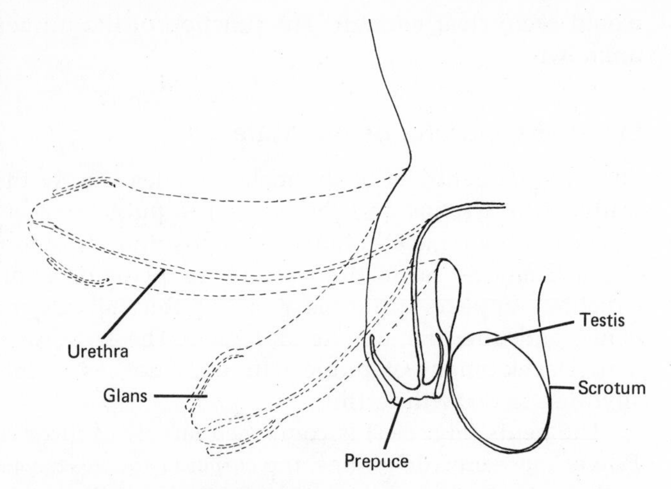

FIG. 6–5 Erection of the penis. The size of the flaccid penis is no indication of the final erect size—small flaccid penises tend to become proportionately larger, so that there is less variation in the size of erect penises than is commonly supposed.

a proportionally greater penile size. Masters and Johnson found, on the contrary, that penises with as much as 2 inches difference in length (3 inches as compared to 5 inches) showed little difference after erection. Both tended to increase to about 6 to 7½ inches, that is, a 100 percent increase for the smaller organs and a 50 percent increase for the larger ones. Furthermore, the almost universally established folklore that sexual performance is related to the size of the penis, either flaccid or erect, has little basis in fact. To be sure there are certain rather rare diseases which may render a male "less masculine" and which prevent final maturation and growth of the penis, but in no other instance is the size of the penis in any way related to sexual prowess. We might also add that the common male supposition—that women prefer a large penis—is equally erroneous. Women seem to have very little concern one way or another about penis size, and regard it as a male "hang-up."

Examples of actual anatomical or physiological "incompatibility"—that is, a penis too large or too small for the vagina—are rare. Physiologically, both the penis and vagina, at the *proper state of arousal,* are infinitely well-suited for satisfactory, compatible coitus regardless of their original "size." Nearly every case of so-called anatomical incompatibility is traced to emotional causes, and even these are relatively rare.

Comparison of Male and Female External Genitalia

Although quite distinct when fully developed, the external genital organs of the two sexes arise, in large part, from common origins. For example, the clitoris and the glans penis develop from the same structure, as do the scrotum and major labia. That this is so becomes evident when we examine the early stages of human development (Fig. 6–6). At about seven weeks of age the fetus can be identified as neither male nor female (except by the sex chromosomes); in-

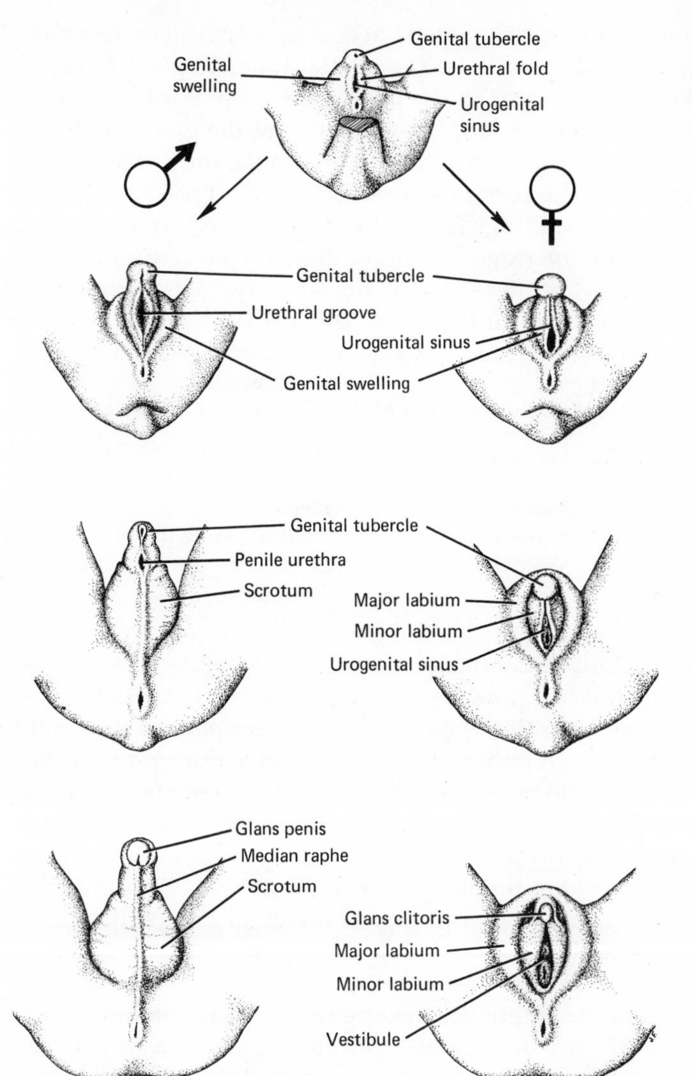

FIG. 6–6 Development of the external genitalia in the male and female.

stead it has certain structures which will undergo different courses of development depending upon the sexual fate of the individual. These structures consist of a pair of *genital swellings,* a pair of smaller *urethral folds,* and a single midline enlargement anterior to the urethral folds called the *genital tubercle.* The urethra and vagina open into a common chamber (the *urogenital sinus*), which in turn exits to the outside in the space between the urethral folds. This space is called the *urethral groove.*

If the fetus develops into a female, the genital swellings enlarge to become the major labia; the urethral folds become the minor labia; the common chamber into which the urethra and vagina opens becomes everted, enlarging the urethral groove into the female vestibule; the genital tubercle persists as the clitoris. More extensive changes occur in the male. The urethral folds undergo a fusion which encloses the urethral groove and elongates the groove onto the genital tubercle. The groove is gradually closed into a canal, the *penile urethra,* whose walls are composed of the fused urethral folds and whose termination is the glans penis. The genital swellings fuse to become the scrotum into which the testes descend. A prominent ridge on the midline of the scrotum marks this line of fusion. On the basis of these observations embryologists have been able to designate certain structures in the sexes as equivalent or homologous (meaning from a common origin). Table 6–1 shows a few of these homologies.

TABLE 6-1 Homologous Structures of the Male and Female Genitalia

Indifferent Stage	*Female*	*Male*
Genital swellings	Major labia	Scrotum
Urethral folds	Minor labia	Walls of penile urethra
Urethral groove	Vestibule	Cavity of penile urethra
Genital tubercle	Clitoris	Glans penis

Whether the fetus develops into a male or female is dependent, of course, upon the genetic makeup; but just how these hereditary factors influence male or female expression are not clear. It appears that hormones play an important role, and hormonal imbalances in either the mother or fetus during pregnancy can cause intermediate expressions. For example, a genetic female may show an enlarged clitoris and a persisting urogenital sinus, so that the vagina terminates internally. Deficient expression in the male may result in incomplete fusion of the urethral folds so that the urethra opens to the outside at the base of the penis. If this is accompanied by incomplete fusion of the genital swellings, such an individual would appear similar to the female above, and the sex of neither could be determined with certainty by mere gross inspection. Fortunately, it is now possible to determine genetic sex by examination of hair follicles, skin cells, or blood cells, and if need be, minor corrective surgery can usually alter the principal defects. Also fortunate is the fact that these deformities have caused physicians to take note of the dangers inherent in the use of certain hormones by pregnant women.

Male Genital Disorders

Inguinal Hernia

The passage of the testis through the abdominal wall leaves a potential weakness through which other visceral structures might pass. Any manifestation of such a weakness is termed an *inguinal hernia.* Particularly troublesome are certain structures suspended within the peritoneal cavity which have a tendency to pass into and through the inguinal canal, which in turn would carry them into the scrotum. Normally, the inguinal canal closes during the first few months after birth. But some remnant of it persists, and the degree of patency of this remnant determines the probability of a later inguinal hernia.

Although inguinal hernias are far more common in men, they may occur in women. As noted above, the major labia of the female correspond embryologically to the scrotum of the male, and contain a rudimentary inguinal canal even though no structures pass through it.

Any increase in intraabdominal pressure will tend to force abdominal structures into a persisting canal; defecation and lifting of heavy objects are two common aggravating activities. The physician, when examining for inguinal hernia, asks the individual to cough; this also increases intraabdominal pressure. If the inguinal canal is open, the doctor will be able to feel the movement of the visceral structures against his fingers at each cough (Fig. 6–7).

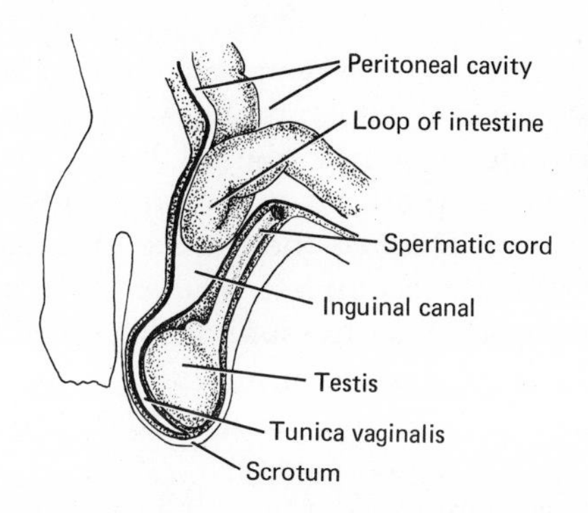

FIG. 6–7 Inguinal hernia. In most instances of inguinal hernia, portions of the viscera tend to follow the route of the descent of the testis.

We have described the commonest type of inguinal hernia; another type exists which bypasses the inguinal canal. But in either case the movement of a loop of intestine or whatever into the canal causes great pain, and if untreated, may become lodged in the canal so tightly that the blood supply is cut off. This is called a strangulated hernia, and is exceedingly dangerous since the starved tissue may become gangrenous. The only cure is surgical repair of the weakness in the abdominal wall, an operation usually quite free from complications and with a high degree of success.

Hernia is the extrusion of a visceral structure into or through any part of the abdominal wall, and can therefore occur anywhere a weakness exists. Quite unrelated to sexual structures, hernias of the umbilicus (umbilical hernia), the esophageal entrance into the abdomen (esophageal hernia), and the entrance and exit points of the large blood vessels of the lower extremity (femoral hernia) can occur, and all are amenable to surgical repair.

Cryptorchidism

Occasionally, the testes fail to descend into the scrotum and instead are retained in the pelvic cavity. If this cryptorchidism (meaning hidden-testes) involves both testes, the individual is sterile. Even though sperms are produced they apparently are unable to survive the few higher degrees of temperature within the pelvis. However, the interstitial cells are unaffected and testosterone is produced in the normal amounts to permit the male secondary sexual characteristics to develop normally and to enable the individual to be typically male in other respects.

The testes of the newborn male infant are sometimes located in the inguinal canal or in the upper part of the scrotum, and complete the descent only after several weeks or months. In fact, the testes at this time may move up and down the inguinal canal in response to temperature changes. In the normal course of events the testes finally settle permanently in the scrotum.

Prostate Enlargement

The prostate gland has a tendency to enlarge in later years, usually after the age of 60. The reasons are not clear although it may be connected with decreasing testosterone production. Since the prostate surrounds the urethra at the base of the bladder, enlargement causes it to exert pressure on the urethral wall and also to bulge upward into the bladder wall. This has the effect of not only decreasing bladder volume, but also of making complete bladder evacuations impossible. The urine retention causes a back pressure which may cause kidney damage as well as permitting stagnation of the urine with a good chance of infection. Surgical removal of the prostate is the usual treatment. If done early, serious damage to the urinary system is prevented.

Cancer of the prostate gland is also frequent in older men. In fact, it is second only to lung cancer among fatal cancers in males. Also, among all men over 50 autopsied from any cause of death, over 15 percent show the presence of a prostatic cancer.

Frequent routine rectal (proctoscopic) examinations is the best means of identifying prostatic abnormalities, and cancer authorities recommend that this should be a regular part of a physical examination of all men of middle age and older.

Phimosis

Phimosis is the condition in which the prepuce or foreskin forms such a tight collar over the glans penis that it cannot be retracted. This irretractibility of the prepuce is a frequent reason given for its surgical removal—circumcision—in newborn male infants. As noted earlier, the prepuce is normally adherent to the glans for some days or weeks after birth so that phimosis may be regarded as normal at this time. Usually it will become freely retractible by six weeks of age, especially if the mother makes a practice of gently attempting to expose the glans at each bathing.

Many physicians believe that all male babies should be routinely circumcised for reasons of hygiene, since mucous gland secretions (smegma), and even semen can collect between the glans and foreskin to cause irritation and infection in men who do not cleanse themselves thoroughly and regularly.

Eunuchism and Castration

Eunuchism occurs when the testes fail to develop or when they are surgically removed (castration) during infancy or childhood. The eunuch generally has disproportionately long extremities, a marked tendency toward obesity, and poor muscular and skeletal development which leads to poor posture. The genitalia remain infantile, and no secondary sexual characteristics develop. Thus he is most often beardless, never develops acne, and the voice is either high-pitched or remains in the "breaking" stage.

In Mediterranean countries, especially Italy, boys were castrated to provide future singers for the choirs of the great churches. In the sixteenth, seventeenth, and eighteenth centuries these "castrati" were known for their art of *bel canto,* and parts in operas and oratories were written especially for them. When highly trained these individuals were able to sing a full octave above the ordinary woman's vocal range with the clear timbre characteristic of boy voices, but with the power and control of the adult male. These vocal characteristics were sustained for decades, and many earned great incomes. Late in the eighteenth century as many as 2000 children a year were believed to have been castrated, and it is said that Joseph Haydn was saved from such a fate only by a last minute parental change of mind. In the latter part of the nineteenth century, several popes condemned the practice, but the castrati persisted as an important part of the Vatican chapel into the twentieth century. The last surviving castrato died in 1922, and recordings of his voice reportedly exist in private collections.

Testicular deficiency among adolescent males today is usually a congenital defect. Proper hormone replacement therapy can frequently stimulate development of secondary sexual characteristics and enlargement of the genitalia. A normal sexual life may be possible. However, the later the age at which hormone therapy is begun, the smaller is the likelihood of success.

Loss of both testes during manhood is usually the result of a gunshot wound or other injuries which make removal of the testes necessary. Many times these men show little immediate loss of virility and may continue a normal sexual life for some years. However, hormone (testosterone) replacement therapy is usually required if regression of secondary characteristics and impotence is to be prevented.

Sterility and Impotence

The sole reproductive function of the male is the successful delivery of living sperms to the cervical area of the vagina. This requires the production of immense numbers of sperms, the performance of coitus and of ejaculation. Incapacitation at any point will result in procreative failure, or sterility.

Male sterility is estimated to be the problem in fully a third of childless marriages. The most frequent cause is some previous infection in the male genital ducts. Gonorrhea, particularly, can have this effect. Occasionally, the seminiferous tubules may be partially or totally destroyed by mumps, typhus, or irradiation. Some males simply have deficient sperm production without known cause. It would seem that relatively few sperms would be sufficient inasmuch as only one is required to fertilize the egg. However, on the basis of empirical observation, it appears that the ejaculated semen must contain around 150 million sperms in order for fertilization to occur. (The normal male releases about 500 million sperms at each ejaculation.) It has been suggested that the great excess of sperms provide an enzyme called *hyaluronidase* that disperses the follicular cells (corona radiata) that cover the surface of the ovum. Without a sufficient number of sperms to provide the enzyme the follicular cells presumably block the approach of all sperms.

Impotence is the inability to achieve an adequate erection and maintain it to the completion of a satisfactory sexual act. Although impotence may have an organic basis, such as malformation of the penis or drug addiction, it is more commonly a psychological disturbance. Since the impotent male is usually potentially fertile, that is, has a sufficient number of normal sperms, he is not considered sterile in the usual sense, even though he is unable to make successful delivery of sperms.

Impotence is often transitory, especially when it occurs at the first coital experience, when overly fatigued or worried, or after excessive indulgence in alcoholic beverages. Although often amusing in retrospect, such occasions have been known to terminate in severe depression and even suicide. Of more serious concern is the impotence due to psychologic preconditioning such as violation of the mother image. Such disturbances require psychotherapy.

REVIEW QUESTIONS

1. What is the approximate route taken by the testis when it descends from the pelvic cavity to the scrotum? What is the relationship of the descent of the testis to inguinal hernia? to cryptorchidism?
2. What are interstitial cells and what do they do?
3. Trace the pathway of sperms from the seminiferous tubules to the outside.
4. What is the source of the seminal fluid?

5. What structures in the male have a common origin with the major labia, minor labia, vestibule, and clitoris, respectively, in the female?
6. What are some arguments you might advance for and against circumcision?
7. Distinguish between sterility and impotence.

REFERENCE

MASTERS, W. H., and V. E. JOHNSON, *Human Sexual Response.* Boston: Little, Brown, 1966.

Additional Readings

BISHOP, P. M. F., "Intersexual states and allied conditions," *British Medical Journal,* **1:** 1255–1262 (1966).

BOLANDE, R. P., "Ritualistic surgery—circumcision and tonsillectomy," *New England Journal of Medicine,* **280:** 591–595 (1969).

FEDERMAN, D. D., "Disorders of sexual development," *New England Journal of Medicine,* **277:** 351–359 (1967).

WEISS, C., "Ritual circumcision. Comments on current practices in American hospitals," *Clinical Pediatrics,* **1:** 65–72 (1962).

7
SEXUAL AROUSAL, COITUS, AND THE BIOLOGICAL NATURE OF SEXUAL BEHAVIOR

Although the biological function of sexual intercourse is procreation, it is obvious that most coital activity is initiated for the gratification and pleasure of one or both partners. In fact, the biological function is often thwarted through use of some type of contraception. Thus, in most instances, it has a gratifying role which is separate and distinct from reproduction; and in marital relations this function is probably of equal importance to the reproductive one in making marriage a success.

In spite of the great significance of compatible sexual adjustment in marriage, little is done to instruct the young adult properly in the nature of sexual arousal and the sexual act itself. When two people seek this most intimate relationship with only the crudest notion of what they are about, the act can become little more than a clumsy and disappointing physiological palliative for the male, and a physical and emotional frustration for the female.

It is not the intent here to instruct in the details of coital technique, but perhaps it is fitting to note here that since successful relations can be achieved through a variety of mutually stimulatory techniques—all elaborately and appropriately described in a number of "marriage manuals"—some difference of opinion may exist between sexual partners as to what constitutes "normal" sexual behavior. Most authorities, psychologists, physicians, and even many ministers now agree that almost any form of sexual expression agreeable to both participants not only is *not* perverse but is normal and healthy. Atypical sexual experiences become abnormal only if an obsession develops and leads to emotional problems. Since this is not usually the case, there is little cause for avoiding enjoyable experimentation and deviation from the typical genital-to-genital, male-prone sex act. However, the "experimenter" may need to proceed with some reserve and genuine concern, since such deviation may come into conflict with the beliefs and background of the partner. Nontypical activity is only desirable when it benefits and pleases *both* partners; and you will be better able to consider intelligently the psychological and physiological aspects of human sexual behavior once you understand more precisely what is meant by coitus and sexual arousal.

The Nature of Sexual Arousal

The response of the male to sexual stimuli is a rapid erection of the penis. The penis may reach full erect size from a flaccid state in three to eight seconds following the onset of sexual stimulation. Accompanying erection are other physiological changes such as an increased breathing rate and heart rate as well as a flushing of the skin through the dilation of cutaneous blood vessels.

The sexual response in the female is generally slower and more complex. The vaginal response is the most rapid; a "sweating" of the vaginal skin develops within seconds of sexual stimulation and serves as vaginal lubrication. The walls of the vagina undergo an expansion so that both the diameter and the length are increased. Both the vagina and the labia become engorged with blood, causing a swelling and flattening of the vulva and a greater exposure of the vaginal entrance. In full arousal the breasts increase in size and the nipples become erect. The increased heart rate and breathing rate and the flushing of the skin noted in the male occur also in the female.

Coitus

The act of sexual intercourse is made possible when the female presents herself in an accommodating position. In most instances she lies supine with her thighs spread and knees bent. The coital act is initiated when the glans of the erect penis gains access to the vaginal entrance. Unless partially blocked by the hymen, the stiffened organ glides easily through the vestibule and into the lubricated corridor of the vagina. The vagina, enlarged and lubricated by sexual arousal, easily accommodates the entire shaft of the penis; and the terminal glans, at full penetration, pushes ahead of itself the cul-de-sac of the posterior fornix, and lies close to the dimpled opening of the cervix. The highly pleasurable tactile stimulation is maintained by repeated withdrawals and penetrations which cause an ever-increasing level of tension and excitation and terminates in the peak of sexual activity, the *orgasm*. The orgasm is an abrupt release of all tensions, the individual plunges into a series of muscular spasms which spread through the entire body and are accompanied by a moment of great ecstasy. In a minute or less the individuals return to a normal or even subnormal physiological state.

All available evidence suggests that the orgasm is essentially similar in the two sexes. Certainly there is no measurable difference in the physiological reactions. Two differences, however, should be noted: The female may regularly experience several orgasms during a single coital experience; this is rare in the male. In the male orgasm is accompanied by ejaculation or release of seminal fluid; no similar response occurs in the female.

The sequence of events from arousal to orgasm are the rule in the male, but this may not be the case in the female. Kinsey reported that only about half of the women interviewed experienced orgasm in the first month of marriage, and considerably less than half experienced orgasm consistently (that is, over 90 percent of the time) even after five years of marriage. This has a significance which we shall note later.

Factors in Sexual Arousal

Tactile stimulation from touch or pressure is the sort of stimulation which most often brings a sexual response. Certain areas of the body, particularly the mouth and genitalia, are richly supplied with end organs for touch (nerve endings) and have long been recognized as "erogenous zones." In spite of the importance of the female breasts in patterns of American sexual behavior, only about half of the women studied by Kinsey indicated that such stimulation was pleasurable, and probably fewer than half are sexually aroused by such stimulation. The male probably derives the greater share of erotic pleasure from the handling of the breast and it should be obvious that the female libido is in no way related to the size of the breasts as some men apparently believe.

Generally, some physical contact is required for sexual arousal in the female. By this we mean that psychological factors, which are so important in the sexual response in the male, are of less significance in the female. Males appear to be more often conditioned by previous sexual experience. Just as Pavlov's dogs were conditioned so that they salivated upon hearing a dinner bell, most human males become sexually aroused when a stimulus brings to mind a sexual experience. Males, therefore, become sexually aroused upon exposure to a variety of stimuli; for example, the anticipation of renewing a sexual experience, observing females, reading material or viewing pictures related to sex, discussing sex. Nearly all males are aroused to the point of erection several times per week, and in adolescent males it is frequently several times per day.

The average female is not so easily aroused, nor as often. Psychological stimuli, such as romantic movies, may cause a sort of sexual arousal, but the feeling may not carry with it an urgent desire for sexual activity as it does in the male. Whatever arousal she has seems to be satisfied simply by affection and attention. (Kinsey found, however, a few females—some two or three percent—who were more intensely aroused by psychological stimuli than any males. These women were able to bring themselves to orgasm simply through fantasy, with no physical stimulation whatever. This only rarely, if ever, occurs in males.)

The Biological Nature of Sexual Behavior

The relative roles of physical and psychological stimuli in the two sexes does much to explain the different attitudes toward sex which are so easily observed. A male may be sexually aroused in observing a given female and may be fully prepared for coitus with her simply for the pleasure he anticipates from the act. Males can divorce other interpersonal considerations from the sex act and might not even particularly care for the female involved. However, this is not a characteristic of the minority of males, as some literature would lead us to believe. The inclination for casual and promiscuous sexual behavior is probably present in all men.

The female, on the other hand, finds it much more difficult to become sexually aroused without some sort of emotional involvement. In fact, she can discuss sex with detachment, without any motivation for sexual activity whatever. This is frequently difficult for males, especially younger ones. Since she becomes aroused principally through physical stimulation, this, of necessity, means that some emotional attraction toward the male is required before such stimulation can occur.

Hence the male can respond simply to the "idea" of sex and may be inclined to a variety of partners. In fact, many males can respond sexually to perhaps a majority of the suitable females he encounters. With her requirement for emotional involvement, the female is more inclined to be satisfied with one male partner for any given period of time.

The reasons for these differences in libido, or sexual drive, are unclear. From a biological point of view the explanation perhaps lies in the differences in the physiology of the two sexes. No ejaculation occurs in the female because she has neither prostate gland nor seminal vesicles. In the male these glands are continually active, become engorged with their secretions, and call for relief. This can develop into a state of considerable urgency, and would account for the ready response of the male to psychological stimuli—a physiologist would say that the triggering mechanism has a lowered threshold. Certainly, a satiated male responds less readily to psychological stimuli and is less interested in females unattractive to him.

Other Sexual Outlets

The absence of coitus would incline the male to a sexual outlet through other means. He may physically stimulate himself to orgasm and ejaculation (masturbation); it may occur as a result of stimulation by a sexual partner during "petting;" or the release may occur in a dream (nocturnal emissions). Regardless of how it occurs, some sexual outlet exists. The urgency may be so great that the male masturbates simply to seek relief from the sexual tensions that build up. Although masturbation does occur in females, it is not nearly as widespread and is not as regularly indulged in as in the male.

The traditional advice to the adolescent male so frustrated by the nagging urgency of his sexual drive is to participate in strenuous and exhausting exercise or to "take a cold shower." The normal male finds the advice useless. In fact, recent evidence suggests that strenuous exercise tends to increase sexual tensions in many males. The cold shower might serve to relieve an erection, but does nothing to prevent the onset of another. Masturbation becomes the logical outlet, and if he is admonished and made to feel shame for the activity, considerable emotional harm can develop. Kinsey and his co-workers found that almost 90 percent of all males masturbate more or less regularly during the period of late adolescence and before marriage, and remarked that "It would be difficult to show that the masturbatory activities have done measurable damage to any of these individuals, with the rare exception of the psychotic who is compulsive in his behavior." They found, to the contrary, that many boys lived in continual conflict because of fear of social disgrace, a sense of sin, or worry over the effect of such behavior on their ultimate sexual capacities, as a result of the traditional teaching concerning masturbation. Kinsey concludes, "For the boys who have not been too disturbed psychologically, masturbation has, however, provided a regular sexual outlet which has alleviated nervous tensions; and the record is clear in many cases that these boys have on the whole lived more balanced lives than the boys who have been more restrained in the sexual activities."

Nocturnal emissions or "wet dreams" are experienced by practically all college-age, single males. In view of the rarity of spontaneous ejaculation in males through psychological stimuli alone, it seems surprising that erotic dreams should be so effective. The fact that orgasm during dreams is much less common in women gives further evidence that a physiological process is in effect in the male that is absent in the female, namely, the accumulation of seminal fluids that demand release.

In comparison with older generations, today's young males less often worry about the significance of nocturnal emissions, and accept them as a normal part of the male sexual experience. This is probably a healthy development.

Premarital physical contacts between males and females that stop short of actual sexual intercourse are referred to as heterosexual petting. Although these may go no further than kissing or simple caressing, they may involve considerable sexual arousal. If no sexual arousal is elicited the behavior should not be referred to as petting.

The precoital activities of married persons have much in common with premarital petting. The difference, of course, is that the marathon petting activities commonly a part of dating among high school and college students is simply a *substitute* for coitus; and the problems arising from the activity come principally from the frustration which develops from lack of the normal consummation. Petting is often continued by the male in the hope that the female will finally succumb; but if he fails in the seduction, his protracted erection can be followed by considerable physical discomfort (the cause of which is unknown). The female, on the other hand, can often derive substantial satisfaction from the affection and attention she receives without feeling the driving need for coital climax (although some females also experience physical discomforts from prolonged arousal without orgasm). She is usually a good deal more passive in her petting activities, the male assuming the principal role in the relationship. With proper manual stimulation of the female genitalia, principally the minor labia and clitoris, she will likely experience a satisfactory orgasm while the male continues in his frustration. Of course, when the female actively reciprocates by handling the male genitalia, orgasm and ejaculation would normally occur quite quickly.

In spite of the fact that our society generally disapproves of such intimacy, it is prevalent in young unmarried adults. And if we acknowledge its existence, we can at least point out the lesser of the evils involved. Certainly, it can be effectively argued that it is better not to reach such a state of affairs in the first place; and most parents hope that their daughters (and to a lesser extent, their sons) will avoid this degree of intimacy. When it does occur, however, and the partners proceed to the point that they become highly aroused, they come ultimately and irrevocably to a decision. The following choices are obvious: (1) to have coitus, (2) for one or both to reach orgasm by means other than coitus, or (3) to abruptly cease the activity and suffer the frustrations involved. The choice is not an especially happy one, when society as well as parental influence prohibit the natural inclination; and the alternatives which can bring sexual release have about them a suggestion of perversion, whether real or not. An unconsummated sexual excitement is the only acceptable alternative for many. It is obvious that very intense and intimate petting often yields less net enjoyment than more moderate indulgence.

Premarital Coitus

We have implied that coitus in such instances should be avoided, and we believe that most counselors of young people must, in conscience, give this advice. This is principally a problem for the female, since she is ultimately the one who will have to make the decision whether or not to have coitus. It is true that once young unmarried women reach a stage of life when they are able to live apart from their parents and to support themselves, they are better able to control the circumstances of the event, and at this point premarital coitus becomes socially more tolerable. Certainly, a woman in her midtwenties who has a normal and complete heterosexual relationship, with a mature understanding of the implications, suffers little from the experience.

The argument is a good deal less convincing when we consider the situation of a typical college coed. She lives in a dormitory, more or less supervised, and may be expected to keep "hours." Her coital activities, when they do occur, are likely to take place in less than auspicious circumstances, both in regard to time and place. Also, if sexual fulfillment is a problem in young women, as Kinsey suggests, then a satisfactory experience is unlikely to occur in an automobile where a comfortable arrangement is difficult and when the chance of discovery is possible. Thus there is small likelihood of her achieving any of the indescribable delights promised by romantic novels (written largely by men). Even in the best of circumstances, when the event occurs in the relative privacy of the girl's home or in a motel, other factors come into consideration (apart from the obvious consequences of pregnancy and venereal disease which we shall discuss separately).

The average woman, perhaps even more so in the case of the young woman, cannot easily separate emotional involvement from sexual matters. She is generally unable to participate in the act simply for the pleasure she will derive from it. Sex is all bound up with love, marriage, home, and children, which is not nearly so often the case with the male. When she has coitus, this emotional involvement becomes even stronger.

Simone de Beauvoir expresses the woman's feelings in this way:[1]

> But there is no doubt that for man coition has a definite biological conclusion: ejaculation. And certainly many other quite complex intentions are involved in aiming at this goal; but once attained, it seems a definite result, and if not the full satisfaction of desire, at least its termination for the time being. In women, on the contrary, the goal is uncertain from the start, and more psychological in nature than physiological; she desires sex excitement and pleasure in general, but her body promises no precise conclusion to the act of love; and that is why coition is never quite terminated for her; it admits of no end.

[1] From *The Second Sex*, New York: Alfred A. Knopf, 1953.

This total-involvement characteristic of the female makes it difficult for her to terminate a sexual relationship with the detachment of the male. The dialogue one might find in a "true confessions" magazine relating to the breakup of an affair has considerable validity:

He: "Well, you seemed to enjoy it just as much as I did."
She: "Yes, I enjoyed it, but I thought it meant more to you than just sex."

In other words, the male is quite able to participate in an affair with only a minor emotional commitment. And when the novelty of the sexual experience wears off, his emotional involvement is liable to cool also. (Obviously, this is not always the case.) But the female must reconcile herself to the fact that the sexual side of the relationship may be the primary concern for the male, whereas this is not often the case for the female.

This means that the usual premarital sexual experience cannot be for the female the casual affair which it most often is for the male. And the teenage girl is frequently quite unable to cope with realization that her involvement is not reciprocated; she will likely have feelings of rejection, betrayal, and of deep remorse.

The young female, then, has little to gain by premarital coitus. She likely will not enjoy it, and will probably suffer considerable regret. With the attainment of a certain amount of maturity and independence, these considerations will diminish in their importance, but for the average college girl they are matters of serious concern.

Love, Sex, and Marriage

Anthropologists tend to agree that the strictures placed upon premarital coitus by Western society have resulted in an artificial preoccupation with sexual matters. This preoccupation, aggravated by the mass media which have consistently placed a high premium upon sexual attractiveness, stimulates desires not normally permitted to be enjoyed. The result has been a generally unhealthy synonymy in the minds of the young of sex and marriage—the principal objective in marrying is the unlimited pleasure of coitus with one's mate. The "love" which has become the foundation of so many contemporary marriages is predominantly a desire for coitus with an intensely sexually attractive partner. It is perhaps ironic that limitations on premarital sexual activity, as imposed by Western religions, have had the effect of accentuating the aspects of romantic love in marriage, especially when this aspect is given no important significance in either the Old or New Testaments. This emphasis is in no small part responsible for the high divorce rate—some 25 percent of all marriages in the United States today. When the chief motivation for marriage is legalized cohabitation, it is not surprising that little foundation for an enduring partnership remains after the sexual relationship loses its glamor.

The love of an enduring marriage includes a satisfactory sexual expression, but equally important are common interests, mutual respect, and mutual devotion which must survive the ebb and tide of sexual arousal. The trouble lies, of course, in the fact that romantic love blinds us to an objective view of the nonsexual relationship. It is perhaps enlightening to realize that the so-called arranged marriages of previous generations were remarkably successful. In fact, today, in those parts of Western society in which parents still actively participate in the mate selection of their children, the divorce rate is at a very low level. If nothing else, this demonstrates the importance of objective criteria of cultural and intellectual compatibility as prerequisites for the good life that marriage can provide.

We do not intend to moralize here. What we do wish to emphasize is that marriage amounts to much more than sexual bliss. And getting married primarily for the convenience of sexual union is committing oneself to not only a probable disenchantment but also to a personal and social obligation from which there is no easy retreat.

Homosexuality

No discussion of sexual behavior can ignore homosexual activity since it constitutes a major sexual outlet for a significant proportion of the population. Kinsey and his co-workers found that 40 percent of adult American men and 13 percent of adult American women have had homosexual contact to the point of orgasm, and that 4 percent of men are exclusively homosexual throughout their lives, whereas among women, the figure is about 2 percent.

To understand these figures we need to define our terms so that there is no misunderstanding of what is meant by homosexual behavior. The term *homosexual* refers to sexual relations with individuals of the *same* sex. The stem word *homo*, then, is derived from the Greek word meaning *the same*, not the latin word *homo* meaning man, as has sometimes been inferred. The word homosexual is the antithesis of the word *heterosexual*, which applies to relations between individuals of different sexes. The words homosexual and heterosexual describe the sexes of the participating partners and not the nature of the act.

This definition restricts the use of the term and avoids ambiguity. Although some persons would refer to mouth-genital contacts between males and females as homosexual acts, they definitely are heterosexual. On the other hand, mutual masturbation between two males is a homosexual act regardless of the psychic nature of the response. Likewise some males who are passive in sexual acts with other males contend that they are heterosexual because they fantasize themselves in contact with a female. Again, the act is homosexual, regardless of the motivation or psychic response.

These considerations account for the tendency in much of the recent literature to refer to homosexual acts rather than to homosexuality and to classify those who participate in homosexual behavior according to a scale of hetero-sexual-homosexual activity. Following precise definitions, Kinsey and his associates rated adult males and females on a scale of 0 to 6 as follows:

0: Exclusively heterosexual with no homosexual experience.
1: Predominantly heterosexual, only incidentally homosexual.
2: Predominantly heterosexual, but more than incidentally homosexual.
3: Equally heterosexual and homosexual.
4: Predominantly homosexual, but more than incidentally heterosexual.
5: Predominantly homosexual, but incidentally heterosexual.
6: Exclusively homosexual.

The important and obvious consideration here is that males (or females) cannot be simply divided into two discrete populations (heterosexual and homosexual) that preclude differences in the psyche of those who have performed homosexual acts. Class 1 includes individuals who may only have had a single experience, or who have infrequent homosexual experience out of curiosity or when they are drunk. Class 2 includes those who recognize that they are aroused by homosexual stimuli, but who respond more strongly to those of the opposite sex. Class 3 individuals accept heterosexual and homosexual experiences with no strong preferencs for either. Class 4 is the opposite of 2, and class 5 is the opposite of 1. These categories have the advantage of taking into account both the individual's overt experiences and his psychosexual reactions.

It is clearly unrealistic to classify an individual as homosexual on the basis of a single act, as is often the case with some Army and Navy officials and administrators of schools and prisons. Such judgments are the result of the normal human tendency to put facts into convenient pigeonholes without the recognition that practically all aspects of human behavior are a continuum.

On the basis of the 0-6 scale of heterosexual-homosexual behavior the incidence of homosexual activity cited in the first paragraph of this section becomes more meaningful. Other data compiled by the Kinsey group are the following:

Twenty-five percent of the male population has more than incidental homosexual experiences (ratings 2–6) for at least three years between the ages of 16 and 55.

Thirteen percent of the male population has more homosexual than heterosexual experiences (ratings 4–6).

Eight percent of the males are exclusively homosexual (rating of 6) for at least three years between the ages of 16 and 55 (as compared to 4 percent who are exclusively homosexual throughout their lives).

The data for females are less than half that for the males in all categories.

The Causes of Homosexuality

What are the causes of homosexual activity? Obviously, it becomes more common among males when females are unavailable or in short supply, as in prisons or in the armed services. (The converse is not nearly as common—in the absence of males, females are more likely to be abstentious.) When a preference for homosexual activity occurs, other explanations must be sought.

Starting with Freud, psychologically oriented investigators have tended to look for the solution in childhood experiences. Man has a long period of sexual immaturity when he is dependent upon parents or other adults for the conditioning that will determine his sexual behavior. Early influences that are said to affect gender identification in a young boy are: the absence of a father, or a tyrannical father who vents hatred against the mother, so that the child falls in love with his mother and later flees from the specter of incest and hence rejects all women; particularly strong attraction to the father so that he seeks the father in other males (again a fleeing from incest); and excessive puritanism in which the woman is pure and untouchable and therefore nonsexual.

However, some studies of twins have suggested a possible genetic factor. Kallman found that among 44 homosexual male twins with identical twin brothers, all co-twins were also homosexual. But among 51 male homosexuals with fraternal twin brothers, only 13 were also homosexual. Additionally, there was a good deal of concordance among the identical twins even in the degree of homosexuality according to the Kinsey scale. This was not the case with fraternal twins. In addition to this research on the genetic factors, evidence has recently appeared that suggests that biochemical differences may exist that distinguish male homosexuals from male heterosexuals [Dewhurst; Kolodney].

It is not necessary to view the conditioning and genetic theories as incompatible. Even if one assumes that all who exhibit homosexual inclinations as adults share certain kinds of experiences as children (an assumption far from proved), it does not follow that all who have had such experiences will become homosexuals. Is it not possible that carriers of certain hereditary endowments are more vulnerable to the conditioning influences that precipitate the homosexual personality? It would seem difficult to believe that we all respond alike to childhood experiences or to the sexual attitudes of our elders.

Treatment of Homosexuality

When we speak of treatment or cure we are assuming that a disease exists. Many homosexuals, probably a majority of those of the Kinsey class 5 and class 6, regard themselves as constitutionally different, and therefore unresponsive to psychotherapy. This view is understandable inasmuch as a concurrent view by the general public would remove the stigma associated with voluntary "deviant" behavior. The validity of this viewpoint is a source of disagreement among psychiatrists, but true or not, homosexuals are very difficult to change. In fact, some psychotherapists consider therapy "successful" if an exclusive homosexual becomes able to enjoy both heterosexual and homosexual relations [Oliven]. Even this degree of adjustment requires a very long period of treatment (200 to 350 hours) and many thousands of dollars. With present therapeutic methods it is unlikely that a large number of homosexuals will ever be effectively treated. As a consequence, some psychiatrists argue for identification and treatment during the formative years of childhood and, for adults, counseling that will assist the homosexual in achieving a reasonably normal life, as free as possible from such self-destructive compulsions as promiscuity.

Social Aspects of Homosexuality

Homosexual behavior apparently occurs in all societies, and is viewed with almost all levels of approval and disapproval. In ancient Greece "Socratic" love was a fitting activity for scholars and gentlemen; some American Indian tribes openly accepted it; in Hitler's Germany it was severely punished. Although laws against homosexual behavior have a long history in Western society, there has been a gradual relaxation of their enforcement and even some liberalization in the statutes. The American Law Institute has recommended that private sexual behavior between consenting adults be removed from the list of crimes, regardless of how the behavior is morally considered. This viewpoint is gaining in acceptance, and the State of Illinois has followed the recommendation.

Homosexual behavior is a definite social problem, but one arising as much from the animosity it usually inspires in most heterosexual people as by the actions of the homosexuals themselves. It is true that many homosexuals develop antisocial traits, forming tightly knit cliques in which their way of life is glorified as that of an elite group. But it is also true that many have achieved outstanding careers, not only in art and literature but also in business, education, and politics, often by sublimating their instinctive drives and leading the life of an ascetic. Between these two extremes lie the vast majority of homosexuals who must lead a life in which their inclinations must be disguised, and who are liable to blackmail, persecution, and gross indignities. As long as society forces upon them this "underground mentality" they will lead unhappy and frustrated lives.

REVIEW QUESTIONS

1. What are some of the characteristics of sexual arousal in the male? in the female?
2. In what ways is orgasm different in the male and female? In what ways is it the same?
3. Studies seem to indicate that men are much more readily aroused through

psychological stimuli than are women. Some people argue that this is simply because men have been conditioned to respond in this way while women have not. Others argue that this is instinctual. What is your viewpoint on this?

4. Almost 90 percent of the males interviewed by Kinsey said they had masturbated more or less regularly during late adolescence. This study was made about 30 years ago. Do you think this behavior is different today? Why?
5. Under what circumstances do you believe that premarital coitus might be appropriate and socially tolerable? What are the arguments you might make for premarital chastity in the female? in the male?
6. What, to you, is the purpose of marriage?
7. What is meant by homosexual behavior?
8. In considering Kinsey's rating scale of homosexual activity (0–6) which class(es) would you regard as homosexual?
9. What arguments can you advance in favor of legalized homosexual activity? against legalization?

REFERENCES

de Beauvoir, S. *The Second Sex,* New York: Alfred A. Knopf, 1953.

Dewhurst, K., "Sexual activity and urinary steroids in man with special reference to male homosexuality," *British Journal of Psychiatry,* **115:** 1413–1415 (1969).

Kallman, F. J., *Heredity in Health and Mental Disorder; Principles of Psychiatric Genetics in the Light of Comparative Twin Studies.* New York: W. W. Norton, 1953.

Kinsey, A. C. et al., *Sexual Behavior in the Human Female.* Philadelphia: W. B. Saunders, 1953.

Kinsey, A. C. et al., *Sexual Behavior in the Human Male.* Philadelphia: W. B. Saunders, 1958.

Kolodny, R. C. et al., "Plasma testosterone and semen analysis in male homosexuals," *New England Journal of Medicine,* **285:** 1170–1173 (1971).

Oliven, J. F., *Sexual Hygiene and Pathology,* 2nd ed. Philadelphia: J. B. Lippincott, 1965.

Additional Readings

Bergler, E., *Homosexuality.* New York: Collier, 1962.

Blood, R. O., *Marriage,* 3rd ed. Glencoe, Ill.: Free Press, 1969.

Bowman, H. A., *Marriage for Moderns,* 5th ed. New York: McGraw-Hill, 1965.

Byer, C. et al., *Dating, Marriage, and Human Reproduction.* Glencoe, Ill.: Free Press, 1969.

Dalrymple, W., *Sex is for Real: Human Sexuality and Sexual Responsibility.* New York: McGraw-Hill, 1969.

Fox, R., "The evolution of human sexual behavior," *The New York Times Magazine,* March 24, 1968.

Gagnon, J. H., and W. Simon, "Prospect for change in American sexual patterns," *Medical Aspects of Human Sexuality,* **4:** 100–117 (1970).

HALLECK, S., "Sex and mental health on the campus," *Journal of the American Medical Association*, **200:** 684–690 (1967).
MEAD, B. T., "The case for chastity," *Medical Aspects of Human Sexuality*, **4:** 8–15 (1970).
MORRIS, D., *The Human Zoo*. New York: McGraw-Hill, 1969.
MORRIS, D., *The Naked Ape*. New York: McGraw-Hill, 1967.
PACKARD, V., *The Sexual Wilderness*. New York: David McKay, 1967.
PETERSON, J. A., *Education for Marriage*, 2nd ed. New York: Charles Scribner, 1964.
REISS, I. L., *Premarital Sexual Standards in America*. Glencoe, Ill.: Free Press, 1964.
WEST, D. J., *Homosexuality*, rev. ed. London: Aldine, 1968.

8
CONTRACEPTION

Men and women have always longed for both fertility and sterility, each at its appointed time and in its chosen circumstances. This has been a universal aim, whether people have always been conscious of it or not.[1]

The problems of overpopulation in the world today are becoming more obvious. That these problems will become intolerable if the present rate of population increase is continued is beyond doubt. Argument exists not so much in regard to the reality of the problem, but rather in regard to how the problem should be approached.

The most obvious solution to overpopulation is control of births. Here, again, the desirability of birth control is generally accepted. Some birth control methods have been sanctioned by various religious groups while other methods have been condemned. It is not the place of this book to discuss the morality of any type of contraception. It is fitting, however, to describe contraceptive methods currently in use. Estimates are that at least half the married women of child-bearing age in this country use some method to avoid conception. An understanding as to the effectiveness of the various techniques as well as their mechanism of action would appear desirable.

[1] Norman E. Himes, *Medical History of Contraception.* New York: Gamut Press, 1963.

The Rhythm Method

This method depends upon the couple abstaining from coitus on those days of the menstrual cycle during which conception appears to be possible. As discussed earlier ovulation is believed to occur 14 days before the onset of the following menstrual period. Unfortunately, ovulation may not occur precisely on the 14th day, but may occur as early as the 12th day or as late as the 16th day. Furthermore success of the technique depends upon the absence of sperms from the female genital system well enough *ahead* of the estimated ovulation date to be sure that no "live" sperms are present on that day, and also for a period of time *after* the date to be sure that the ovum is no longer fertilizable to the extent of causing pregnancy. This "fertilizable" period for the ovum is probably less than a day. Sperms are probably able to fertilize an ovum for no longer than 48 hours (although they may remain motile for as long as five days). The "unsafe period," then, lies between the time that previously deposited sperms might survive and the date that fresh sperm might meet a fertilizable ovum. In a 28-day cycle this would mean the 10th through the 17th day, inclusive (eight full days), leaving the remainder of the cycle to be the "safe period" (Table 8–1).

The problem in the determination of the "unsafe period" is the identifica-

TABLE 8–1 The Rhythm Method for Regular 28-Day Cycle

1	2	3	4	5	6	7
8	9	(10)[a]	(11)[a]	(12)[b]	(13)[b]	(14)[b]
(15)[b]	(16)[c]	(17)[c]	18	19	20	21
22	23	24	25	26	27	28
1						

[a] Coitus on the 10th or 11th days may leave live sperm in female genital tract which could fertilize egg ovulated 1 to 3 days later.

[b] Ovulation may occur on the 12th, 13th, 14th, or 15th day.

[c] Fertilizable egg may still be present in uterine tube.

Uncircled numbers are "safe days" when conception is less likely to occur.

Circled numbers are "unsafe days" when conception is more likely to occur.

This is a fertility calendar for a woman who shows a consistent "ideal" cycle of 28 days. Such a woman would likely be fertile only on the days shown encircled. However, a woman who shows any irregularity in her menstrual cycles should use Table 8–2. A woman with entirely regular cycles of less or more than 28 days determines her first unsafe day simply by subtracting 18 from the total number of days in her cycle, and her last unsafe day by subtracting 11 from the total number of days of her cycle. Thus, in a woman whose cycles are consistently 26 days in length the first unsafe day would be day 8, and the last unsafe day would be day 15.

tion of the day of ovulation. This is easily calculated with hindsight; that is, when menstruation starts, the woman can look back 14 days on the calendar and approximate her day of ovulation. The difficulty lies in determining the date of ovulation in a cycle of unknown duration, since few women show absolutely regular cycles. Most women have 5 to 6 different period durations which occur more or less at random (although the 28-day cycle is usually the commonest of these). For example, a typical woman might have cycles as short as 25 days and as long as 31 days. In the shortest cycle she would ovulate on about day 11, and in the longest cycle, on about day 17. Since a woman cannot predict the length of the cycle she is in, she must abstain from coitus during the "unsafe" days of *all* her possible cycles. In this case it would mean from the 7th to the 20th day, inclusive. If the period of menstruation is excluded from the days available for coitus, as is frequently the case with married couples, this leaves 11 days in the "safe period" (Table 8–2).

TABLE 8-2 The Rhythm Method for Women with Irregular Menstrual Cycles

No. of Days in Shortest Menstrual Cycle	*First Unsafe Day of Any Period*	*No. of Days in Longest Menstrual Cycle*	*Last Unsafe Day of Any Period*
21	3	26	15
22	4	27	16
23	5	28	17
24	6	29	18
25	7	30	19
26	8	31	20
27	9	32	21
28	10	33	22
29	11	34	23
30	12	35	24
31	13	36	25
32	14	37	26
33	15	38	27
34	16	39	28

Since most women have menstrual cycles that vary slightly in length, the unsafe period must be calculated to accord with both the longest and shortest cycles as observed over a period of at least a year. (At the beginning of a given cycle she doesn't know what length the cycle will be except within the limits she has observed.) Since the unsafe period will occur earlier in a short cycle the first unsafe day is determined by subtracting 18 from her shortest cycle; likewise, the last unsafe day will occur later in longer cycles, so 11 is subtracted from the length of her longest cycle. Thus, the unsafe days of any given cycle must be calculated as comprising all the unsafe days of all her different cycles. For example, a woman whose shortest cycle has been 25 days and whose longest cycle has been 31 days will use the table to determine that her unsafe days are day 7 to day 20 inclusive.

Effectiveness of the rhythm method depends, obviously, on determining the variability pattern of a woman's menstrual cycles. Usually, cycles covering a full year are desired if an accurate rhythm calendar is to be established. For the newlywed, this means consulting her physician considerably in advance.

How good is the rhythm method of birth control? From a statistical point of view it does, in fact, fall short. A chemist, A. J. de Bethune, applied the principles of probability to the rhythm method. If p is the probability of failure (that is, conception) in any one month, then $(1-p)$ is the probability of success. Success for a given number of months requires success in each month of the period. The probability of success during m months would be $(1-p)^m$.

As an example let us suppose that the probability of success for one month is 95 percent (from available evidence the probability is more likely closer to 85 percent). Let us see what happens when we extend the period of exposure. The probability of success during two consecutive months is $(0.95)^2$, or 90 percent; during three consecutive months it is $(0.95)^3$, or 85.5 percent. The probability of success for an entire year (13 cycles) is only about 51 percent. If we use a more realistic figure of 85 percent as the probability of success for one month, the percentage diminishes to about 12 percent! When the calculations include the number of coital acts per cycle, the mathematics become rather complicated, but de Bethune has worked these out. His conclusions are: "Even if the fertile period is as brief as 12 hours, these laws show that a couple who desire a 2-year spacing of children are limited, statistically, to two acts of coitus per cycle. Couples who desire a 4-year spacing are limited to a maximum of one act of coitus per cycle."

Condoms

Condoms are sheaths of latex or rubber which envelop the erect penis and prevent the deposition of semen in the vagina. The theoretical long-term effectiveness of condoms is quite high. But, as is the case with all contraceptive "contrivances," the high level of effectiveness is dependent upon use 100 percent of the time.

This is the only contraceptive technique which is presently available to the male, and for this reason will probably continue to have importance, particularly in coital acts outside marriage. Also, it is the most effective prophylaxis against venereal diseases.

Vaginal Diaphragms

The vaginal diaphragm is a soft rubber cap which serves to partition off the anterior part of the vagina so that sperms are prevented access to the opening of the cervix (Fig. 8–1). The diaphragm is usually used with some sort of spermicidal cream or jelly which helps to seal the cap in place, and to provide

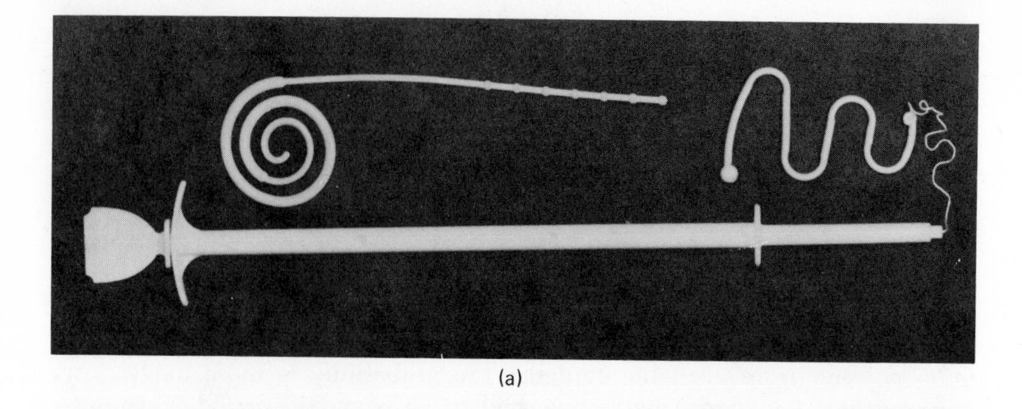

(a)

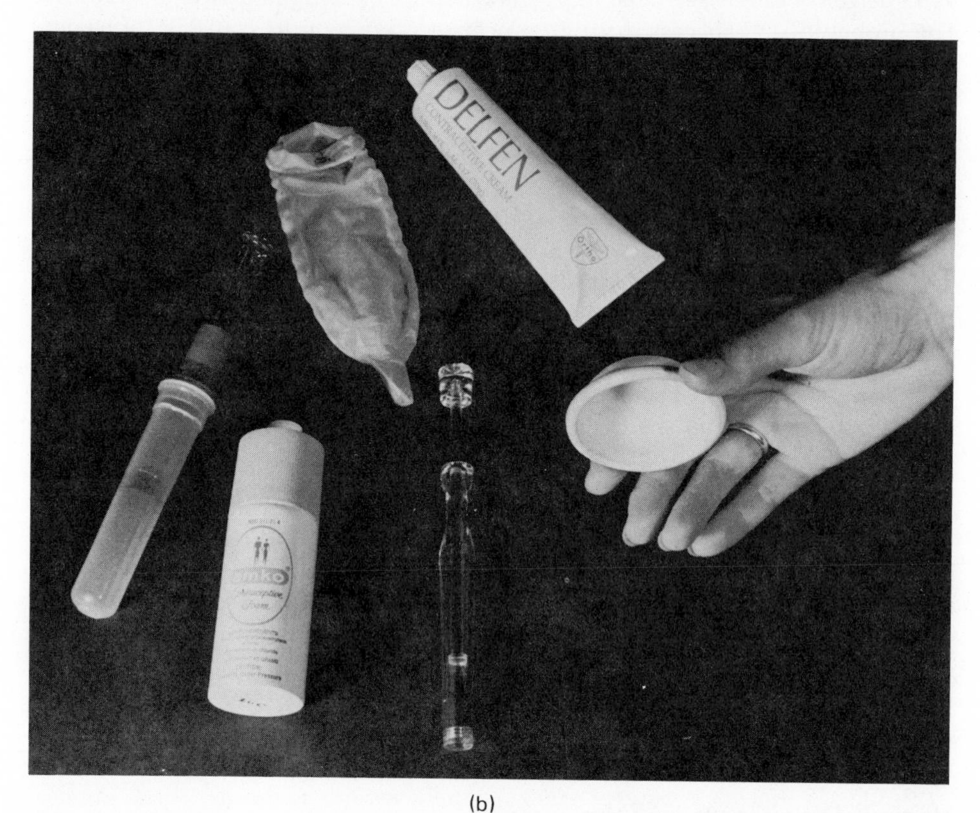

(b)

FIG. 8–1 (a) Shown here are the two commonest forms of intrauterine devices. In the spiral, at left, a beadlike extension protrudes from the cervical opening. The loop on the right is shown with its dispenser. (b) Contraceptive methods that require application with each act of coitus. Shown are cream with applicator, condom, diaphragm, and foam with applicator.

a chemical barrier as well as a physical one. Diaphragms come in a variety of sizes because vaginal dimensions differ. If the wrong size is used, it can slip out of place and contraceptive failure can result. Fitting of the diaphragm by a physician is thus important. Since the size normally changes following childbirth, remeasurement should be made at the appropriate time.

The effectiveness of the diaphragm probably approaches that of the condom. The problem is, again, one of 100 percent use. The woman must use the device during every act of coitus. Failures (conceptions) occurring in women who use this method are often due to lack of use based on the supposition that a particular part of the cycle was "safe." Retention of the device for the prescribed length of time following coitus is also essential (about eight hours is usually recommended).

Spermicidal Agents Alone

Special spermicidal jellies and creams have been developed for use alone. They have a different consistency from those used with diaphragms, and have a much stronger spermicidal action. Spermicides have been incorporated into vaginal suppositories, foaming tablets, and foams packaged under gas pressure. All these preparations are harmless and can be bought without a prescription. Since no medical instruction or physical examination is needed for the use of these agents, they are particularly useful to women who for one reason or another find it embarrassing to ask a physician for a contraceptive prescription or who have no other method available.

Intrauterine Devices (IUDs)

Many years ago it was discovered that any foreign object in the uterus will prevent pregnancy. The first such devices, called Grafenburg rings, after their inventor, were made of silver or platinum, but the more recent versions are stainless steel or polyethylene plastic and, instead of being ringlike, may be of a variety of shapes, such as spirals, loops, or bows (Fig. 8–1).

The earlier devices became discredited because they seemed to cause excessive menstrual bleeding and pelvic pain. In addition, a number of women would expel them. The more recent devices appear to be tolerated much better, especially by women who have had a previous pregnancy.

Several explanations have been advanced to explain how IUDs exert their effect. One is that its presence increases the intensity of the peristaltic waves which pass through the uterine tubes and uterus (part of the mechanism which transports the ovum to the uterus). A fertilized ovum, then, would reach the uterus before it had developed sufficiently to implant in the uterine wall, and would simply pass on through the cervix into the vagina. It has also been suggested that the IUD causes the release of toxic substances from the uterine epithelium that destroy the fertilized egg. Because of these interpretations of IUD effects, they are frequently regarded as abortifacients rather than contraceptives.

The IUDs are not infallible. Over a period of a year about 3 percent of women using this technique will become pregnant, either because they expelled it or in spite of its presence. This puts it in about the same group of effectiveness as condoms and vaginal diaphragms.

However, two characteristics of the IUD make it particularly useful in birth control programs in underdeveloped countries: it is cheap, and once inserted, it is effective as long as it remains in place. Many American physicians, however, recommend that patients fitted with an IUD see their doctor for a follow-up examination at least once a year. There is no scientific evidence to support the claim that the IUD causes uterine cancer.

Contraceptive Pills

Contraceptive pills modify the menstrual cycle by substituting synthetic hormones for those normally produced by the ovary. You will recall from Chapter 5 that estrogen and progesterone not only regulate the growth and functional activity of the uterus, but also react with the anterior pituitary gland to suppress the production of gonadotropic hormones (FSH and LH).

If synthetic estrogen-like and progesterone-like hormones[1] are taken early in the cycle, two important physiological responses can be expected: (1) the endometrium will undergo its proliferation and secretory phases and (2) production of FSH and LH by the anterior pituitary will be suppressed. During the early part of the cycle the estrogen-like substance will stimulate endometrial proliferation, and during the latter part of the cycle the progesterone-like substance will stimulate the endometrium to functional activity, the secretory phase. The secretory phase is maintained as long as progesterone (or its equivalent) is present. When its source, the corpus luteum, dries up, or in this case, when the woman stops taking the pills, progesterone stimulation ceases, and menstruation begins. This means that the uterus undergoes a cyclical change similar to that of the normal cycle.

What is happening in the *ovary* of the woman taking the pills? The artificial hormones suppress the production or release (or both) of the gonadotropic hormones of the anterior pituitary. This means that insufficient FSH is present to stimulate maturation of a Graafian follicle. It is as if a Graafian follicle were already present and producing estrogen, and suppressing FSH release. In any case, no follicle is ripened, and no ovum is available for release,

[1] These hormones are often from curious sources; the progesterone-like hormone is derived from an extract from the root of the Mexican giant yam.

so no ovulation occurs. The cycle, then, is an *anovulatory cycle*, that is, a cycle without ovulation.

The usual procedure is to take one pill each day starting with the fifth day of the menstrual cycle (that is, after four days of menstruation). One pill is taken each day for 20 days; the woman then stops taking the pills—no pill on the 25th day—and a few days later menstruation begins. (Some preparations, as in those illustrated in Figure 8–2, utilize 21 pills—pills are then taken on days 5 through 25 inclusive.) This "reaction" period is usually about four days (25th, 26th, 27th, and 28th days) so that a 28-day cycle is established. (Occasionally, menstruation does not occur after the woman stops taking the pills, in which case she starts a new cycle of 20 pills after a period of one week after her last pill.)

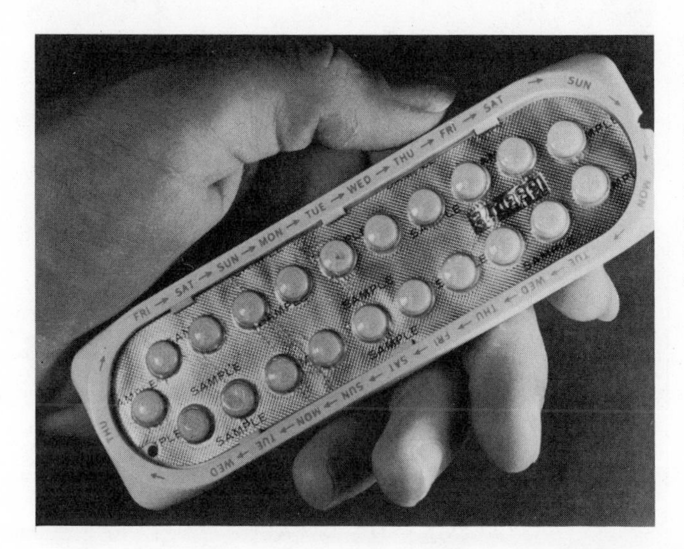

FIG. 8–2 Contraceptive pills. Twenty pills are taken on the fifth through twenty-fourth days. Some brands, including the sequential types, utilize 21 pills, as shown here. All are designed to produce a 28-day cycle.

So-called sequential contraceptive pills have been introduced recently. These pills attempt to duplicate more exactly the natural hormone rhythm of the ovary. Of these pills, usually 21 in number, the first 15 or 16 contain only the estrogen-like hormone, and the last 5 or 6 contain both the estrogen- and progesterone-like hormones. The effects of progesterone stimulation are therefore restricted to the latter part of the cycle.

Contraceptive pills are generally considered to be 100 percent efficient when taken daily as prescribed. Pregnancies among women taking the pills have resulted only when 36 hours or more had lapsed between pills.

Authorities are in agreement that oral contraceptives make up the most effective approach to birth control ever offered. The steadily increasing rate of use in this and other countries indicates a high degree of acceptance by women. It has been estimated that over 7 million in the United States were

TABLE 8–3 Relative Effectiveness of Contraceptive Methods

Group	*Method*	*Effectiveness (No. of likely pregnancies among 100 women using this method for one year)*	*Possible Side Effects*	*Medical Considerations*	*Cost*
Group I—Most effective	Sterilization	0.003	None	Requires surgery	$50–250
	Oral contraceptives	0.1	Nausea, edema, breast changes, thrombophlebitis may occur in susceptible individuals	Supervision by physician is highly recommended	$2.50 per month and physician's fee
	Intrauterine devices	2.7 first year 2.0 second year 1.0 subsequent years	Irregular bleeding; discomfort in first months	Must be inserted by physician; checkup recommended, yearly	Small physician's fee
Group II—Highly effective	Vaginal diaphragm with spermicidal jelly	2.6 (used 100% of time) 14 (used with omissions)	Jelly may cause minor irritations	Proper size required and measurement must be made by physician	$5.00 plus physician's fee
	Condom	2.6 (used 100% of time) 14 (used with omissions)	None	None	10¢–75¢ each
	Coitus interruptus (withdrawal)	18	?	None	None

TABLE 8–3 (continued)

Group	*Method*	*Effectiveness (No. of likely pregnancies among 100 women using this method for one year)*	*Possible Side Effects*	*Medical Considerations*	*Cost*
Group II—Highly effective	Jelly, cream, and foams (spermicidal chemicals) without diaphragm	20	May cause minor irritation	None	$1.00–3.00 per month
Group III—Less effective	Rhythm (use of thermometer increases effectiveness)	24	None	Advice of a physician may be helpful	None
Group IV—Least effective	Douche	31	None	None	$5.00 for douche apparatus
	Breast feeding	No exact data, but highly effective during early months of lactation; ovulation eventually returns even during lactation	None	None	None

using the pill in 1967, roughly 15 percent of all women between the ages of 15 and 45 (Meeker).

The side effects reported for oral contraceptives include just about everything women patients ever complain about. Most seem to be minor. Thrombophlebitis (discussed with diseases of the venous system, Chapter 24) has aroused the most controversy. Two British physicians, Inman and Vessey, reported in 1968 that the mortality rate from thrombophlebitis in women using the pill was about eight times that of nonusers who are not pregnant. Of every 100,000 women taking the pill, 1.3 die each year from thrombophlebitis. This is almost the same number of deaths from the disease as occurs among pregnant women. A risk exists, then, that women taking the pill will die from thrombophlebitis with about the same likelihood that they would if they were pregnant.

The question then becomes: What is the risk of pregnancy-related deaths among those who elect other means of contraception—all of which are less efficient than the pill? One should first consider that the total potential hazard from *all* pregnancy-related deaths, is 22.8 per 100,000 women! If one considers that the average failure rate from other conventional means of contraception is about 10 percent, this means that 10,000 pregnancies per year will occur among 100,000 women owing to failures. Of these 10,000, 2.3 (equivalent to 22.8 per 100,000) will die from one of the various complications of pregnancy. So, statistically, the risk of pregnancy-related death is still greater, 2.3 of 100,000, among those who do not take the pill but use other forms of contraception, than it is among those who do take the pill, 1.3 per 100,000.

Many other problems, mostly minor, but some of major importance, frequently accompany the use of the pills, especially during the first few months. Medical supervision is therefore essential, and the pills should be taken only when prescribed and only by the person for whom they are prescribed.

No drug is 100 percent safe. The risks involved must be weighed against the hazards of the condition that is being treated. So it is with contraception, that to the hazards of pregnancy must be added the psychologic and sociologic problems related to the constant fear of pregnancy, or to the prospect of raising an unwanted child.

Sterilization

The surest, and probably the safest, method of birth control is sterilization. In the long run it is probably also the cheapest. In the male, sterilization is accomplished by cutting the sperm ducts (vasa deferentia). After separation, each end is folded back upon itself and tied. The operation is termed a *vasectomy,* meaning removal of the vas deferens although a more appropriate term would be *vasotomy,* meaning simply a cutting of the vas deferens. With the pathway to the outside blocked, the sperms are retained in the testis, epididymus, and vas deferens. Here, they disintegrate and are reabsorbed. It should be emphasized that virility, libido, and ejaculation remain completely unchanged by the operation. The operation involves little pain, and the whole procedure takes less than 30 minutes, and is usually done in the doctor's office.

The comparable operation in the female is a *tubal ligation*—the cutting and tying off of the uterine (Fallopian) tubes. This operation involves major surgery—as does any operation that requires the entering of the peritoneal cavity. So a regular operating room is needed, and a hospital stay is involved.

Recently, attempts have been made to simplify the operation. A flexible instrument is passed through the vagina, uterus, and into each uterine tube. The tip of the instrument electrically "cauterizes" the interior of the tube. Scar tissue forms, and seals the passageway. Again, as with the sperms in the male, the ova disintegrate without any effect on the woman. Where this procedure is used, hospitalization is not required.

The principal objection people express toward sterilization is that it is irrevocable. Recently, however, success at reconnecting the cut ends of the vas deferens has been about 80 percent [Calderone]. Obviously, reconnection is a much more sophisticated operation than is separation, and is quite expensive. Reconnection of the uterine tubes is not often attempted, and is, apparently, rarely successful.

What is needed, of course, is a simple, easily reversible procedure. One approach receiving considerable attention is the implantation of a non-toxic, insoluble material into the sperm ducts or uterine tubes that would obstruct the passageway, but which would be easily removable.

Until such a technique becomes available, sterilization will continue to be regarded as frequently irrevocable. But for those men and women who can reconcile themselves to the loss of their reproductive function, sterilization is a logical and practical method of birth control.

REVIEW QUESTIONS

1. Discuss the difficulties that may arise in attempting to predict the time of ovulation at the beginning of a menstrual cycle.
2. How is the regularity of the menstrual cycle related to the effectiveness of the rhythm method of birth control?
3. Prepare a table listing the advantages and disadvantages of the principal contraceptive methods.
4. How would you view sterilization as a method of birth control for yourself or your spouse?

REFERENCES

DE BETHUNE, A. J., "Child spacing: The mathematical probabilities," *Science*, **142:** 1629–1634 (1963).

CALDERONE, M. S., *Manual of Family Planning and Contraceptive Practice*, 2nd ed. Baltimore: Williams & Wilkins, 1970.

HIMES, N. E., *Medical History of Contraception.* New York: Gamut Press, 1963.

INMAN, W. H. W., and M. P. VESSEY, "Investigation of death from pulmonary, coronary, and cerebral thrombosis and embolism in women of childbearing age," *British Medical Journal*, **2:** 193–199 (1968).

MEEKER, C. I., "Use of drugs and intrauterine devices for birth control," *New England Journal of Medicine*, **280:** 1058–1060 (1969).

Additional Readings

BIRMINGHAM, W. (ed.), *What Modern Catholics Think about Birth Control.* New York: Signet Books, 1964.

DEMAREST, R. J., and J. J. SCIARRA, *Conception, Birth, and Contraception—A Visual Presentation.* New York: McGraw-Hill, 1969.

FERBER, A. S., C. TIETZE, AND S. LEWITT, "Men with vasectomies: A study of medical, sexual, and psychosocial changes," *Psychosomatic Medicine*, **29:** 354–365 (1967).

HARDIN, G., *Birth Control.* New York: Pegasus, 1970.

HARDIN, G. (ed.), *Population, Evolution, and Birth Control*, 2nd ed. San Francisco: W. H. Freeman, 1969.

HARTMAN, C. G., *Science and the Safe Period.* Baltimore: Williams & Wilkins, 1963.

NOONAN, J. T., JR., *Contraception.* Mentor-Omega Books, 1967.

PEEL, J., and M. POTTS, *Textbook of Contraceptive Practice.* London: Cambridge University Press, 1969.

PISANI, B. J., J. RIVKIND, and R. M. KRISTAL, *The Rhythm Method of Birth Control.* New York: Simon and Schuster, 1967.

SCHMIDT, S. S., "Vasectomy: Indications, technique, and reversibility," *Fertility and Sterility*, **19:** 192–196 (1968).

TIETZE, C., "Contraception with intrauterine devices, 1959–1966," *American Journal of Obstetrics and Gynecology*, **96:** 1043–1051 (1966).

9
PREGNANCY AND BIRTH

From a biological standpoint maternity is the fulfillment of the woman's physiological destiny. However, in spite of the naturalness of the process of pregnancy and childbirth, it cannot be left to the woman alone. Unattended childbirth is regarded as a desperate circumstance by even the most primitive societies. Furthermore, modern medicine views prenatal and postnatal care as indispensable if the health of the mother and child is to be reasonably guaranteed.

Beyond the obvious desirability of proper medical care is the need for education. The expectant mother who is handicapped by ignorance and deluded by folklore that lead to an unhealthy attitude toward pregnancy and an exaggerated fear of childbirth can do unnecessary damage to both herself and to her unborn child.

Conception and Early Development

The life of a new individual begins with the union of two cells—the sperm from the father and the ovum or egg from the mother. This union, known as fertilization, must take place in the upper part of the uterine tube if development is to continue normally beyond the first few days.

Although a single cell results from the fertilization it is a transitory stage since a rapid series of cell divisions, called *cleavage,* commences almost immediately. The result is a closely packed group of cells called a *morula* (Fig. 9–1). By the fourth day the morula has reached the uterus and there develops into a more complicated structure called the *blastocyst.* Only the small inner part of the blastocyst is destined to form the embryo proper. The outer parts develop into the *fetal membranes,* an outer one called the *chorion,* and an inner one called the *amnion.* By the sixth day of development the chorion has developed finger-like outgrowths called *chorionic villi.* The chorionic villi appear to act as digestive agents which enable the embryo to burrow deeply into the endometrium. The tissues of the endometrium and the chorionic villi of the embryo develop together into the *placenta,* which is, then, a composite structure composed of both fetal and adjoining maternal parts. The *umbilical* cord contains blood vessels which carry blood from the embryo to the placenta and back to the embryo.

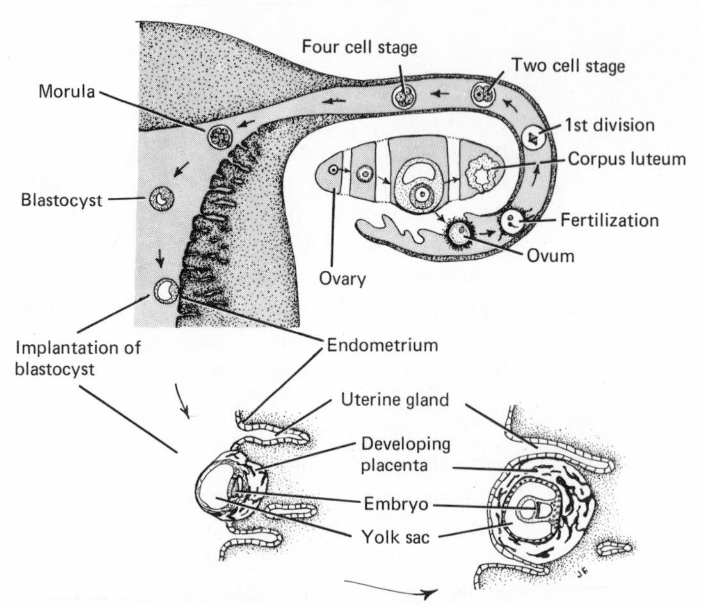

FIG. 9-1 Early stages in human development. Fertilization is believed to occur within 12 hours of ovulation. Other ages are: 4-cell stage, about 40 hours; morula (12 to 16 cells), about 3 days; early blastocyst, about 4½ days; beginning of implantation, about 6 days; a partially embedded blastocyst, about 7½ days. At this latter stage the pregnant woman would be in approximately the 22nd day of her menstrual cycle, and would be unaware of her pregnancy.

The placenta is to the unborn child the functional equivalent of lungs, kidneys, and intestine. However, the blood of the embryo does not normally mix with the blood of the mother. A thin membrane covering the chorionic villi separates the fetal capillaries from the maternal blood, and through this membrane there is rapid exchange of oxygen, carbon dioxide, food, and metabolic wastes (Fig. 9–2). Nerves do not pass through the placenta so that sensory impressions of the mother have no way of reaching the fetus. Hormones, however, do cross so that the emotional state of the mother can to some degree be passed to the fetus.

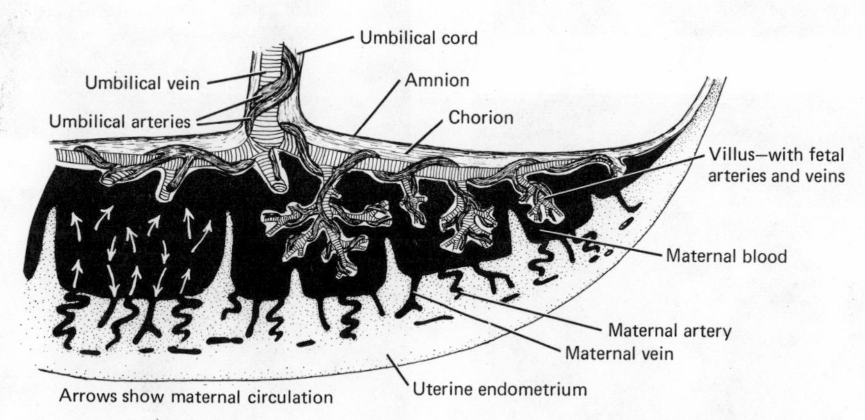

FIG. 9–2 The structure of the human placenta late in pregnancy. The umbilical artery and vein are connected within each villus by a network of capillaries (not shown) covered by a thin layer of embryonic tissue that separates the capillaries from the maternal blood.

The second fetal membrane, the amnion, grows out around the embryo and eventually forms a sac that fills the space between the embryo and the chorion, the amnionic membrane fusing with the chorion. The fluid within the amniotic cavity, the amniotic fluid, provides a protective cushion for the embryo.

The original embryonic part of the blastocyst begins as a flat, somewhat elongated disc of cells which folds, differentially thickens, bends, creases, and develops outgrowths to form the rudiments of the body's organs. The embryo grows rapidly during the second month of life and by the eighth week has taken on a very definite human appearance. By this time, as shown in Figure 9–3, he has a large head with a well-formed face, hands and feet with fingers and toes, a well-developed heart and circulatory system, and he is about 1 inch in length.

By the end of the third month the embryo is about 3 inches in length and development of practically all the organs is complete. After this time the process is principally one of growth in size. The early period of approximately

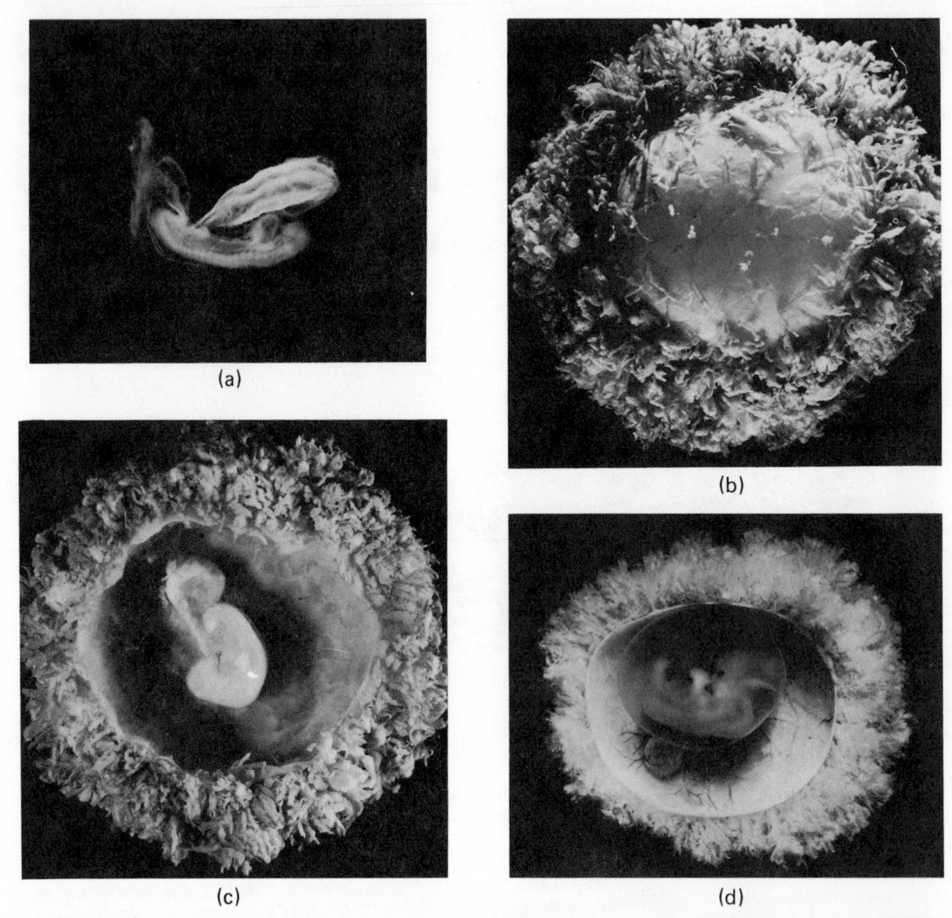

(a) (b) (c) (d)

FIG. 9–3 The following series of photographs of human embryos were made available through the courtesy of the Department of Embryology, Carnegie Institution of Washington. All the ages are given from the time of ovulation. The menstrual age would be, then, about 14 days longer. (a) A human embryo of about 23 days. The embryo is at right and measures 2.5 millimeters. The fetal membranes have been laid out to the left. (b) A 28-day embryo enclosed within the intact fetal membranes. The fingerlike chorionic villi have been separated away from the maternal tissues. (c) The chorion of the 28-day embryo has been opened. The yolk sac lies to the upper left of the embryo. (d) The chorion has been opened to show an embryo of about 40 days. The embryo is enclosed within the amnion. (e) An embryo of 30 days, 7.3 millimeters in length. Limb buds are well-developed. The developing heart lies between the head and the belly. A short tail is present. (f) An embryo of 34 days, 11.6 millimeters in length. The developing eye is now more apparent and the limb buds have become segmented. (g) An embryo of 40 days, 19.0 millimeters in length. The fingers and toes as well as the external parts of the ear have begun to develop. Note that at this time the mother might only just have confirmed the fact that she is pregnant. (h) An embryo of 44 days, 23

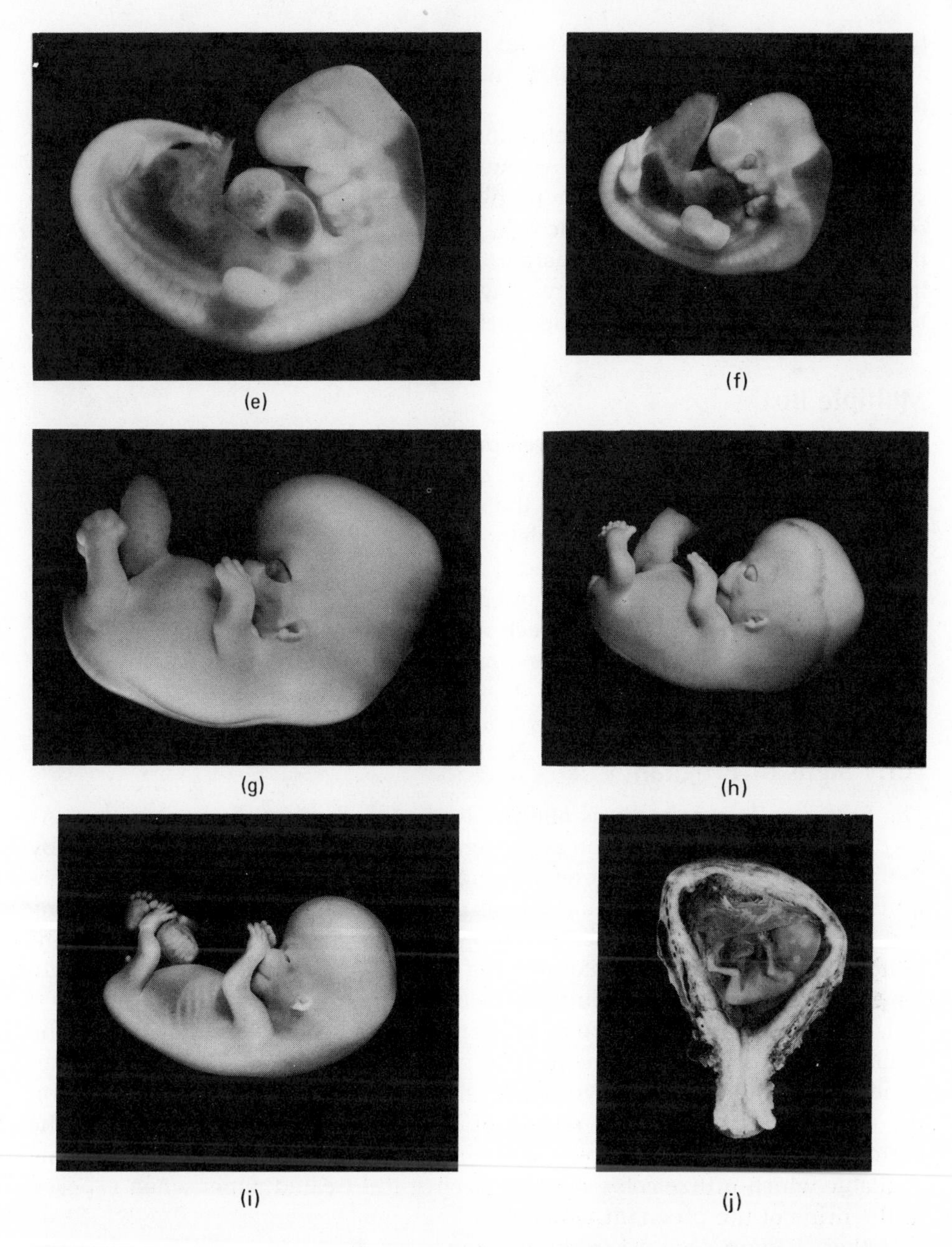

millimeters in length. (i) An embryo of 56 days, 37.0 millimeters in length. (j) Uterus containing embryo of 70 days, 58 millimeters in length. The placenta is located in the upper part of the uterus. The amnion has been cut and folded out of the way.

two months when the embryo is taking form is called the *embryonic period,* and the remaining seven months, principally a time of growth in size, is called the *fetal period.*

Note in Figure 9–1 that the fetus does not develop in the cavity of the uterus as is often believed, but in the endometrial wall. As development proceeds the superficial layer of endometrium (between the chorion and uterine cavity) becomes thinner, loses its chorionic villi, fuses with the chorion and amnion, and the three layers collectively separate the fetus and amniotic cavity from the much compressed uterine cavity. The placenta gradually becomes restricted to the wall opposite to that of the uterine cavity.

Multiple Births

During the very early stages of development, from the two-cell stage on through the blastocyst stage, a separation of embryonic cells may occur, and two or more embryos can develop from the original fertilized ovum. These become identical twins (triplets, quadruplets, or quintuplets) and will be alike in all characteristics determined by heredity. If two or more Graafian follicles mature simultaneously, more than one ovum may be released at the time of ovulation, each may become fertilized, and each will develop in the uterus independently. These are fraternal twins and are no more alike than brothers or sisters borne from separate pregnancies.

Early Signs of Pregnancy

The first sign of pregnancy is almost always the absence of menstruation. As discussed earlier menstruation is prevented by the secretion of a hormone by the placenta, the chorionic gonadotropic hormone, which maintains the functional state of the endometrium. The excretion of this hormone into the urine makes possible the various tests for pregnancy such as the A-Z or Aschheim-Zondek test. When urine containing the hormone is injected into an immature female mammal such as a rabbit, the ovaries respond by becoming swollen and inflamed and by the production of corpora lutea as if overstimulated by the animal's own pituitary hormone. The more recently developed frog test depends upon a similar effect which the chorionic gonadotropic hormone has on the testis, causing a sudden release of sperms which can be detected in the surrounding water a few hours after injection. Even simpler tests are now available which utilize color changes in chemical treated paper when exposed to the urine of the pregnant woman.

Other changes obvious to the mother occur during the first few weeks of pregnancy are, for the most part, due to the extended and increased production of progesterone and estrogen by the corpus luteum. The breasts enlarge and the blood vessels near the surface become more prominent. The nipples and the surrounding areola become more darkly pigmented. The fluid retention typical of the premenstrual period is often even more pronounced during these early stages of pregnancy. Some women suffer from nausea and vomiting during the early part of the day, the so-called morning sickness. It usually disappears after the second month of pregnancy.

None of these changes is, in itself, an infallible sign of pregnancy. However, along with a hormonal pregnancy test and examination of the vaginal entrance (which becomes bluish in color due to congestion of blood) and changes in the firmness of the lower part of the uterus, the physician is usually able to make a positive diagnosis. However, all these signs may be indefinite, and a positive diagnosis may have to await the appearance of more obvious changes which occur during the third or fourth month.

Late Developments in Normal Pregnancy

The fetal heartbeat is detectable by the physician during the fifth month, and active movements of the fetus begin at about the same time. From the third month on the uterus can be felt through the abdominal wall as it steadily enlarges. By the end of the fourth month the uterus reaches to halfway between the pubic symphysis and umbilicus; by the end of the sixth month the uterus reaches to the level of umbilicus (Fig. 9–4). By the ninth month the uterus reaches the lower end of the sternum, but during the last few weeks of pregnancy the fetus settles into the pelvis and the uterus falls a bit.

As pregnancy progresses the fetus makes itself more and more evident, not only through the visibly swollen abdomen, but also through the physical effects on the crowded abdominal viscera and through the increased physiological demands made upon the mother.

This means that varicose veins present before pregnancy become exaggerated, and hemorrhoids become aggravated or may develop for the first time. In both conditions the cause is slower return of blood to the heart because of pressure upon abdominal veins. Neither condition generally becomes serious in women who are in good physical condition and continue a program of regular exercise during pregnancy.

The physiological demands of pregnancy are principally those caused by the rapidly growing fetus which not only increases the mother's dietary requirements, but also increases the load on the mother's kidneys, and to a lesser extent, the load on her lungs and cardiovascular system.

The greater need for all *essential* dietary substances is obvious. However, if the diet is well-balanced and includes ample meat and other sources of protein, fresh vegetables and fruit, the increased demands can usually be met by additional milk (which provides the required calcium) and a supplementary source of iron. The dietary requirements during pregnancy are summarized in Table 17–1.

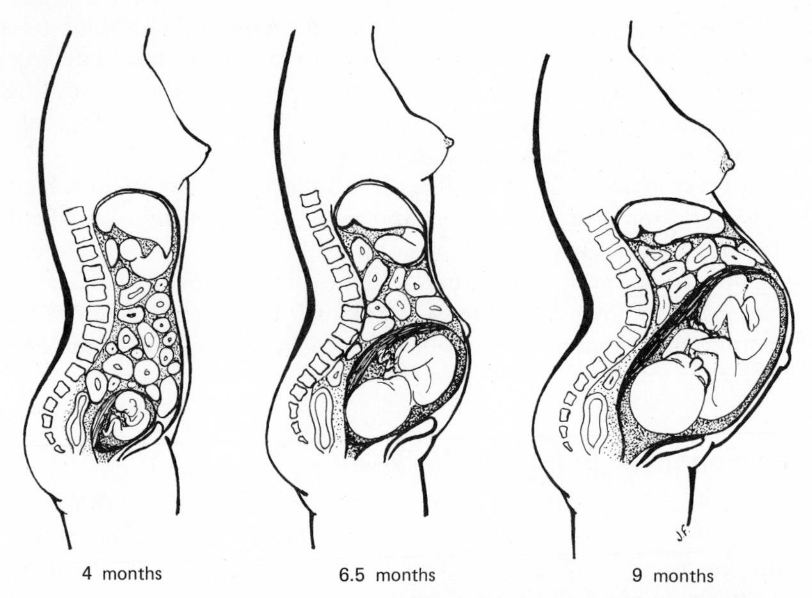

FIG. 9–4 Changes in the mother's body during three stages of pregnancy. In addition to the enlargement of the abdomen note the changes in the curvature of the spine, the enlargement of the breasts, and the crowding of the viscera.

The course of pregnancy should result in a total weight gain of about 20 pounds. Restriction of weight gain to this level may require very careful dietary control and discipline, but is a highly desirable objective because many complications both major and minor are related to excessive weight.

Prenatal Activities and Hygiene

Although pregnancy is most often completed without more than minor difficulties, prenatal care by a physician should be considered essential. Regular examinations will usually ensure that no unexpected complications arise, and if they do, a physician who has seen the patient throughout pregnancy is well equipped to cope with such problems; he knows the patient's medical history, her physical condition, and many other aspects of her life. Although many generalizations can be made about diet and exercise, only the physician is able to specifically prescribe an individual patient's needs which will most reasonably promise a successful pregnancy.

However, among the many generalizations which can be made about pregnancy are the following—given more in the interest of correcting ignorance or myth than in prescribing a regimen of activity.

EXERCISE. The medical attitude toward exercise has tended to become more permissive. About the only limitation on exercise is the caution against excessive fatigue. In fact, it is particularly important that young women who have been active athletes continue such activities in order to prevent excessive weight gain from physiological sluggishness. Obviously, athletics which carry a risk of bodily injury should be strictly avoided.

EMPLOYMENT. Here again the emphasis is on the avoidance of fatigue. When adequate rest is available most jobs can be continued up to six or eight weeks prior to the expected delivery time.

SMOKING. Mothers who smoke during pregnancy tend to bear smaller infants than nonsmokers, and the relationship appears to be in direct proportion to the number of cigarettes smoked. Since nicotine in the blood passes through the placenta, its effects such as increased heart rate and constriction of peripheral arteries would be passed on to the fetus. There is reason to believe that the irritability and nervousness of sudden withdrawal from cigarette smoking in the adult would be simulated by restlessness in the newborn infant when it is separated from the mother.

Whether a woman with an entrenched cigarette habit should attempt to quit smoking at a time when dietary restrictions might become necessary, and when emotional problems might develop from the process, is debatable. Certainly, a pregnant woman should not quit smoking solely as a self-sacrificing gesture to her unborn child, but in respect to the future health of her own body.

ALCOHOL. In spite of much research, the use of alcohol in moderation has not been shown to be deleterious to either mother or fetus, or to affect the course of pregnancy or labor.

TEETH. Calcium is *not* withdrawn from the teeth of the mother in order to meet the calcium demands of the fetus. However, there is some evidence that metabolic changes in the mother, perhaps changes in the nature of the salivary secretions, do encourage a greater growth of acid-producing bacteria. This necessitates greater care in dental hygiene, particularly the removal of food particles from between the teeth.

COITUS. Despite common misconceptions to the contrary, coitus is not discouraged as long as no discomfort is experienced. The notion that a miscarriage is likely if coitus occurs at the time of an expected menstrual period is a myth.

Drugs in Pregnancy

In 1962 the drug thalidomide, commonly prescribed in some foreign countries for the discomforts of early pregnancy, was found to be a teratogen, that is, a substance capable of causing malformations in the developing embryo. Some 10,000 babies born to mothers who took thalidomide early in pregnancy suffered from *phocomelia* or "seal limbs"—shortened, flipperlike arms and legs.

Before this time relatively little concern was given to drugs prescribed to pregnant women. It had been generally believed that most drugs failed to pass through the placenta. The thalidomide disaster awakened the medical profession to the possible serious consequences to the unborn child of a number of commonly prescribed drugs. The result has been that most physicians today assume that when they prescribe a drug to a pregnant woman that the drug and its metabolic derivates will also circulate in the fetus.

The effects of thalidomide emphasized the critical nature of the embryonic period, since it is at this time that the basic structures of the body are laid down. The effects of a teratogen during this period vary according to the developmental processes going on when the teratogen is acting (Langman). For example, if the pregnant woman took thalidomide between days 34 and 38 (counting day 1 as the first day of the last menstrual period), the fetus suffered deformity of the ears, paralysis of the facial nerve (which supplies the muscles of facial expression), and duplication of the thumbs. If she took thalidomide between days 39 and 44, the fetus showed shortening of the arms. If she took the drug between days 42 and 45, the result was shortening of the legs.

Several commonly used drugs and chemicals with demonstrable or suspected effects on the fetus are worthy of note. For example, large amounts of aspirin taken by the mother have been associated with excessive bleeding in the infant. Phenacetin, frequently used with aspirin in pain-killing preparations, may cause anemia and kidney damage to the fetus. Antihistamines are under suspicion as possible teratogens. Vitamins appear to present special problems since many people think that excessive vitamins are essential to maternal health. Excessive vitamin D in the mother's diet can result in infants with hypercalcemia (high blood calcium). These infants show constipation, vomiting, growth failure, and mental retardation, the same characteristics associated with vitamin D toxicity from overdosage following birth. There is also suspicion that excessive levels of ascorbic acid and vitamin B_6 in the maternal diet result in an adjustment by the fetal metabolism to a higher requirement so that the individual goes through life with an abnormally high demand for these vitamins.

A few simple common-sense rules in regard to drug use during pregnancy can be offered.

1. Drugs should be taken during pregnancy only when there is a medical need, and should be taken under the exact conditions prescribed by the physician. (Obviously, the failure to take necessary drugs may adversely affect the baby's health as much as unnecessary drugs. A diabetic, for example, requires proper drugs during pregnancy as during other times.)
2. Women who are pregnant or even suspect that they *may* be pregnant should so inform their physician when he is prescribing a drug.

3. Mothers who breast-feed their infants should continue to exercise caution since many drugs taken by the mother are secreted in the milk.

Disorders of Pregnancy

Any serious infectious disease is a threat to the health of both the mother and fetus, and should be brought to the attention of a physician. However, certain virus infections such as influenza, measles, and mumps, and particularly German measles, are capable of causing damage to the developing fetus, especially during the embryonic period of development. Over 50 percent of babies borne to mothers who had German measles early in pregnancy suffer from congenital deformities such as heart defects, blindness, deafness, and mental retardation. Any woman contemplating motherhood who has not had German measles should deliberately expose herself to the disease in order to avoid the risks to the unborn. Women who have not had regular measles or mumps should request immunizations against these diseases as well as against influenza if they have reason to believe that they are pregnant.

Both syphilis and gonorrhea are dangerous to the infant—syphilis through the production of the congenital form of the disease in the fetus, and gonorrhea through exposure to the organism at the time of birth. Although the early part of pregnancy, usually extended to include the first three months (trimester) probably presents the period of greatest danger, hazards exist for the remaining six months. Differentiation of the external genitalia occurs during the middle trimester, and hormones resembling those produced by the sex organs and the adrenal cortex (the so-called steroids) can sometimes produce a masculinizing effect on the female fetus.

During the latter part of pregnancy there is a danger that drugs may be carried in the blood of the fetus into the postnatal period when the newborn must detoxify and excrete the chemicals derived from the drug independently of the mother. The newborn is sometimes unable to accomplish the necessary chemical conversions and some types of blood and liver disorders of infants have been traced to such circumstances.

Eclampsia

Eclampsia or toxemia of pregnancy is a serious and poorly understood disease characterized by excessive weight gain, hypertension, and albuminuria. The cause is unknown. Some 10 percent of cases prove fatal for the mother (about one quarter of all maternal deaths), and about 50 percent are fatal to the fetus.

Eclampsia is some 15 times more common among teenage first pregnancies than among women who have their first child in their twenties. It also seems to occur much more commonly in women who are resentful and unhappy during their pregnancy. In one study it was found that more than 77 percent of those who developed eclampsia had an unhealthy attitude toward their pregnancy.

Ectopic Pregnancy

Ectopic pregnancy simply refers to a pregnancy at a site outside the uterus. The usual site of ectopic pregnancies is a uterine tube, but various locations within the peritoneal cavity such as the abdominal wall, a segment of intestine, or the outer surface of the bladder can serve as a site for placenta formation. Since the placenta is induced by the blastocyst, by its effect on the maternal tissues, the process is the same whether it occurs at the normal location in the uterus or elsewhere. The uterus, however, is obviously designed for the purpose, and other organs are not. For example, the wall of the uterine tube or intestine is too thin and compact in structure, and, as a consequence, the growth of the placenta causes a rupture of the maternal tissue and severe hemorrhage is likely to occur.

Ectopic pregnancy is a serious hazard to the life of the mother, and since spontaneous abortion is almost always inevitable, the surgical removal of the placenta and embryo is considered essential.

Genetic Defects and Pregnancy

Parents of a defective child usually want to know if they should have more children. If the defect is known to have been caused by an outside agent, such as thalidomide, the family physician is usually able to explain the risks and advise accordingly. If, however, the defect may possibly have had a genetic basis, a professional genetic counselor will usually have to be consulted. Fortunately, techniques for estimating the risks of the birth of future abnormal children are gradually being improved. In more than 70 disorders it is now possible to detect the carrier condition in one or both of the unaffected parents by biochemical tests [Levitan and Montagu]. When the tests are completed, the genetic counselor attempts to present a balanced and dispassionate statement. He will not advise for or against further child-bearing. This is a decision that is left to those who have sought the counseling.

Recently, techniques have been developed for testing the genetic status of the fetus *in utero* [Milunsky et al.]. This is done by obtaining a sample of the amniotic fluid surrounding the fetus by the process of *amniocentesis* (usually by passing a long hypodermic needle through the abdominal and uterine walls, with little risk to either mother or fetus). The amniotic fluid can be examined for many factors—enzymes, amino acids, metabolic waste products, and the like—but especially for the fetal cells that have been shed. The fetal cells can then be examined or cultured to obtain much important information. For example, Down's syndrome (Chapter 3) can be identified, as well as a number of other defects due to abnormal chromosome number. Sometimes therapeutic abortion is suggested. But in all cases, the decision to initiate such action is made by the parents.

Genetic counselors will frequently suggest other advisors, such as ministers or psychiatrists. Geneticists do not know the answers to many of the questions they are asked, and they feel they know very little of what needs to be known. But it is certain that tragedies could have been avoided for innumerable human beings had genetic counseling been available to them.

The book by Levitan and Montague contains a list of genetic centers in the United States and Canada, including those that provide genetic counseling.

Gestation and Labor

The average gestation period for the human is 280 days. This is long in comparison to smaller mammals, such as the mouse (21 days), rabbit (35 days), or dog (60 days), but is about the same as that for several larger mammals, such as cattle. The gestation period is much longer in those species where the newborn is developed almost to the point of self-sufficiency, for example, the horse (about 350 days), fur seal (about 340 days) and elephant (about 640 days).

The human gestation period is often listed as 270 to 295 days, and pregnancies terminating within these limits are considered normal. The approximate delivery date is calculated by adding seven days to the date of the last menstrual period and then counting back three months. For example, if the first day of the last menstrual period was January 7, the approximate date of delivery would be October 14.

Labor

Labor or parturition is the process by which the products of pregnancy are expelled by the mother. Normally it occurs when the fetus is sufficiently mature to cope with the external environment but not so large as to cause difficulties during labor. Exactly how the process is initiated is not clearly understood, but in some way a change occurs in the random and painless contractions of the uterus (which persist all through pregnancy) so that they become strong and effectively coordinated, the cervix is dilated, and the fetus and placenta are efficiently expelled. It is probable that the placenta itself is largely responsible for initiating labor by decreasing the amount of progesterone production which apparently is necessary to maintain tolerance of the placenta by the uterus; but other factors are undoubtedly involved.

The *first stage* of labor begins with the first true labor pain or with the breaking of the fetal membranes and the escape of amniotic fluid. The labor pains are caused by the strong, coordinated contraction of the smooth muscle of the uterus. The contractions have the effect of thickening the upper part of the uterine wall and thinning the lower cervical part. This forces the fetus down into the cervix and toward the vagina (Fig. 9–5). It is doubtful that the muscle contractions themselves are the cause of the pain. More likely the pain is caused by stretching of the perineal structures (principally the wall of the vagina and the area between the vagina and anus) by the downward expulsive force of the uterine contractions.

At first the uterine contractions are brief and far apart, but gradually they become strong and persist longer. The head of the fetus is normally down, and as the thinning of the cervix proceeds, the cervix is more and more dilated until its opening reaches a diameter sufficient to pass the fetal head.

The *second stage of labor* begins when the fetal head passes through the cervix. The continuation of the uterine contractions pushes the fetus into the

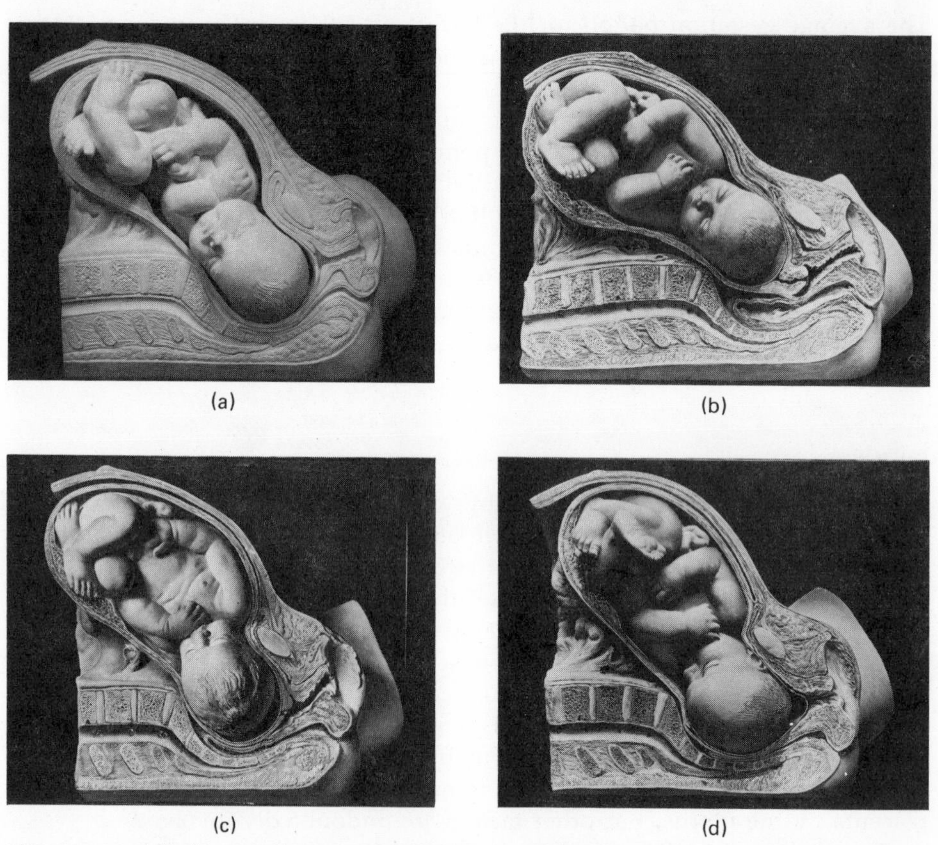

FIG. 9–5 Models showing the events in childbirth. (Maternity Center Association)

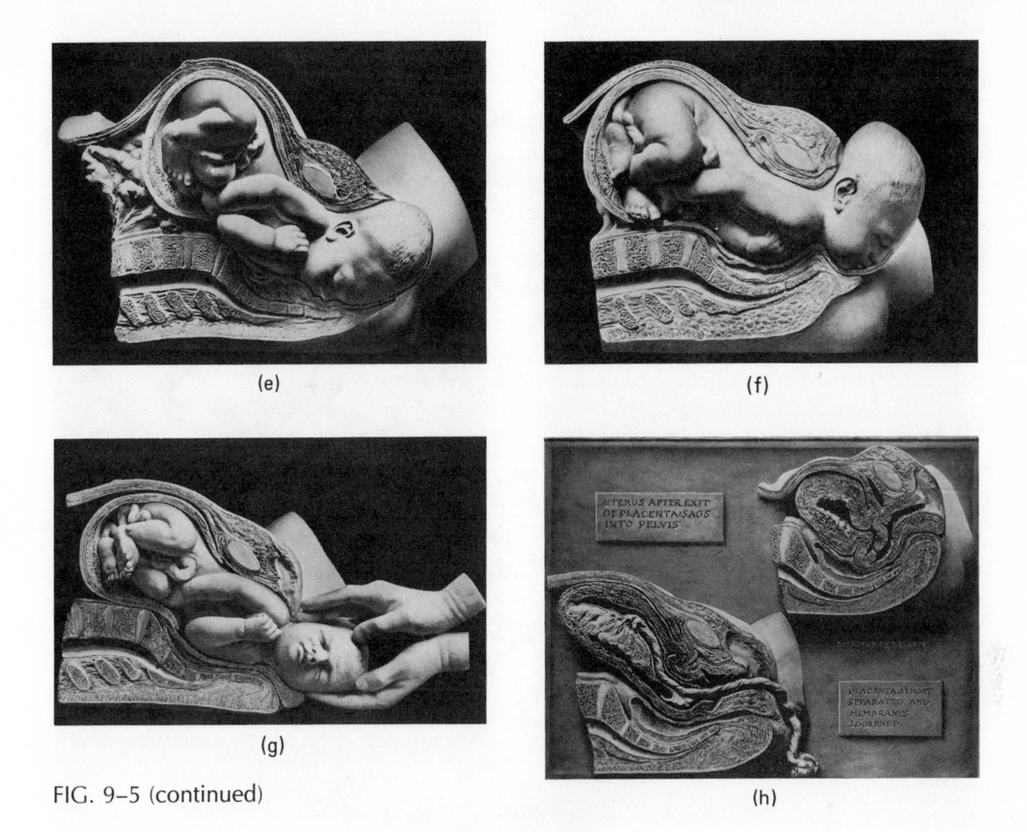

(e) (f) (g) (h)

FIG. 9–5 (continued)

vaginal passage. The complete delivery is much facilitated by the mother "bearing down," that is, by holding her breath and contracting the muscles of the abdominal wall as in the process of defecation. When these abdominal contractions are absent or inadequate the second stage is frequently prolonged and the use of obstetrical forceps is often required.

The delivery of the baby is sometimes complicated by a so-called breech-presentation in which the legs or buttocks of the fetus are delivered first. Such conditions often require great skill on the part of the physician if the fetus is to survive.

The *third stage of labor* is the birth of the placenta. After the birth of the baby there is a short period during which the mother experiences no labor pains, but after a few minutes uterine contractions begin again and continue until the placenta becomes separated from the uterine wall and is expelled from the uterus, as the "after-birth" (Fig. 9–6). The separation of the placenta results in the loss of blood associated with childbirth. The amount of blood lost is usually a pint or less and is easily tolerated.

FIG. 9–6 The human infant at birth, with the placenta and membranes. This illustration is from the anatomical plates of Julius Casserius, published by Adrianus Spigelius in 1626. (New York Academy of Medicine)

Natural Childbirth

The pains of childbirth have probably been the stock and store of intimate conversation among women since speech began, and it should be little wonder that many young women approach childbirth in great fear. Women often wonder why, in the age of great medical advances, an effective anesthesia for completely painless childbirth is not available. The problem, of course, lies in the fact that the obstetrician has two patients to consider while the surgeon has but one. Anesthesia given to the mother traverses the placenta and may jeopardize the initiation of breathing by the infant following birth; it may even prove lethal to the fetus before birth if the anesthesia reduces even slightly the oxygen level in the maternal blood. Also, the voluntary contractions of the abdominal wall, so important in labor, do not occur if the mother is under anesthesia. In spite of these hazards, the skilled use of some anesthesia is often considered beneficial to both mother and baby, especially during the late stages of labor.

However, a good deal of debate exists on how much pain should accompany a normal and uncomplicated birth. How much of the anguish is brought on by emotional factors and the attitude of the woman toward her pregnancy? The late British obstetrician, Grantley Dick Read, asked the questions: "Is labor easy because a woman is calm, or is she calm because her labor is easy?" and "Is a woman pained and frightened because her labor is difficult, or is her labor difficult and painful because she is frightened?" Read believed that fear in some way acts as the chief pain-producing agent in an otherwise normal labor. Also it quite likely exerts undesirable effects on uterine contractions and cervical dilation.

Read and other obstetricians developed the method of "natural childbirth" in which the pregnant woman is taught the basic facts of labor in order to eliminate fear and is trained in exercises which develop the capacity for greater muscle relaxation and control. Proponents of natural childbirth emphasize that their patients suffer pain, but not to the extent suffered by patients without the program of instruction. In addition, the regular methods of anesthesia are available to the patient if she decides they are needed or if the physician believes them desirable.

Caesarean Section

Caesarean section is the delivery of the infant through incisions in the abdominal and uterine walls. In spite of a long and often colorful history, caesarean births were exceedingly hazardous to the mother up until the twentieth century. Fewer than half the mothers survived the operation even late in the nineteenth century.

Even today caesarean birth is considered a major surgical procedure, and is undertaken only when normal birth is considered impossible or dangerous to the mother or fetus. The most common cause for caesarean section is an inadequate birth canal due to small pelvic measurements. Also, caesarean delivery is often utilized when the placenta is located in a lower portion of the uterus (*placenta previa*) so that dilation of the cervix involves premature delivery of the placenta and severe hemorrhage. Fortunately, placenta previa is no longer a serious hazard to the life of either the mother or fetus when competent prenatal care is available.

Breast-Feeding of the Infant

The ideal food for the newborn infant is the milk of the mother, and all mothers should be encouraged to nurse their child unless medical reasons indicate otherwise.

The breasts are prepared for milk production during pregnancy by the action of estrogen and progesterone. Estrogen promotes the development of the duct system and progesterone of the milk-secreting glands, but neither causes secretion of milk. Indeed, it is believed that both these hormones inhibit the secretion of the *lactogenic* (milk-stimulating) *hormone* of the pituitary gland. After parturition, estrogen and progesterone levels fall, and the inhibition is removed. The lactogenic hormone is released and stimulation of actual milk production and secretion is begun. During the first two days a thin, yellowish fluid called *colostrum* is secreted from the nipple, and this is followed on the third or fourth day by milk. Usually the supply of milk appears inadequate, but almost always becomes sufficient if suckling is continued. Nursing also exerts an early beneficial effect on the mother by stimulating the release of a hormone (*oxytocin*) by the posterior part of the pituitary gland. The uterus responds to this hormone by a more rapid return to the nonpregnant state.

If the secretion of milk is to continue, the milk must be removed from the breasts. When breast-feeding is not desired, the "drying-up of the milk" is often hastened by the use of binders and ice packs. Usually the symptoms of milk production will disappear in two or three days.

Prematurity, Perinatal Death, and Abortion

Prematurity and Neonatal Disease

A premature infant is one born at a stage compatible with independent life but with less than the chance of survival of a full-term infant. Infants that weigh less than about two pounds (a fetal age of about 6 months) are unlikely to survive even with the most expert care. More often than not the causes of prematurity are unknown. However, maternal disorders such as heart disease and hypertension are among the causes.

Prematurity accounts for more than half of all deaths during the neonatal period. These infants die of a variety of causes, but one of the most common is the *respiratory distress syndrome* or *hyaline membrane disease.* This disease is estimated to occur in over half of premature infants weighing less than five pounds at birth, but also has a relatively high incidence in full-term caesarean births. The name hyaline membrane disease refers to the fact that the air sac walls become thickened and glassy appearing. The cause of the respiratory distress syndrome is obscure, but it is obviously related to lungs insufficiently developed to take over normal respiratory function following birth.

In about 1 percent of all births the infant is affected by *erythroblastosis fetalis,* the disease which results from Rh incompatibility of the mother and fetus. If a person is Rh^+, it means that an antigenic factor is present in the red blood cells; an Rh^- person does not have the factor. If the mother is Rh^- and the father Rh^+, the fetus may inherit (but not necessarily) the Rh^+ characteristic. When this occurs there is the possibility that a few red blood cells of the fetus may escape through the placenta and enter the maternal blood stream. If this happens, the mother's tissues respond by producing antibodies against the Rh antigen. If these antibodies pass through the placenta into the fetal blood, an immune reaction occurs which causes destruction of the fetal red blood cells resulting in a severe fetal anemia.

The production of antibodies by the mother is not usually sufficient during the first pregnancy to cause anemia. But the antigenic (Rh) factor remains in the maternal blood and antibody production continues. Subsequent pregnancies run increasingly higher risks of severe anemia due to Rh incompatibility.

The usual treatment of the disease has been by the complete exchange of fetal blood for Rh^- blood in what is called exchange transfusion. The rationale is to introduce red blood cells which will not be affected by the maternal antibodies present in the fetus. During the following months the maternal antibodies gradually disappear; the infant eventually replaces the transfused blood by his own, and these new red blood cells appear to be little affected.

Recently, however, a protective vaccine has been developed. It had been demonstrated long ago that if a specific antibody is introduced into the blood, it will inhibit production of that antibody by the body even when the antigen is present. Using this observation as a basis for their studies, two unallied groups of investigators, one in this country and one in England [Freda and Gorman; Finn et al., respectively] injected large amounts of anti-Rh anitbody into Rh^- women who had just given birth to Rh^+ babies. In one study [Freda and Gorman] Rh antibodies did not develop in 48 women given the Rh antiserum, while 7 of 52 women not treated had produced Rh antibodies by six months postpartum. The combined results of the many medical centers that have carried out trials with the anti-Rh serum both in this country and abroad are truly impressive [Hamilton]. Of 559 control cases, 75 (or 13 percent) showed Rh sensitization (presence of Rh antibodies) six months postpartum and later; whereas, of 628 women treated with the anti-Rh serum, only one developed antibodies.

There seems to be no doubt about the vaccine's effectiveness, but some problems are posed by the method. For example, the anti-Rh serum is obtained from Rh^- male volunteers who have been injected with the Rh antigen. Since Rh^- individuals amount to only about 15 percent of the population, some difficulty is involved in enrolling over 1000 regular donors with high levels of Rh antibody accessible to production facilities. With time, however, this and other problems will be solved, and this anemia shall become, along with pernicious anemia, a problem of identification and treatment rather than of basic and applied research.

Other blood group factors, such as A, B, O factors, may also lead to erythroblastosis fetalis, but these cases are less common and are usually much milder.

Abortion

Abortion is the termination of pregnancy at any time before the fetus has reached sufficient development for survival. Abortion is divided medically into two main forms: *spontaneous abortion* in which the termination of pregnancy is through natural causes; and *induced abortion*. Induced abortion may be *therapeutic* which means that it is performed by a physician and with legal justification; or *criminal* which means that it is performed without legal justification by someone who may or may not be medically qualified. To many laymen abortion means criminal abortion, and when spontaneous abortion is meant the term "miscarriage" is used, although the latter term is not generally used medically.

If abortion is induced under proper hospital conditions during the first three months of pregnancy it is usually safe and uncomplicated. Performed at later times than this, it becomes progressively more complicated and dangerous, and *especially so when self-induced, or when performed in circumstances where proper operating conditions are unavailable.* Unfortunately, state laws governing abortion differ widely. In most instances therapeutic abortion is permitted when the mother's life is in danger. A number of states permit therapeutic abortion when a likelihood of serious congenital disease exists or when conception occurred as a result of rape or incest. A few states permit therapeutic abortion when either the physical or mental health of the mother is in jeopardy, and this leaves the matter largely in the hands of a panel of medical experts. Other states, such as New York, have adopted laws permitting abortion solely upon the decision of the woman and her physician.

As a consequence of the liberalized laws over 500,000 legal abortions were reported in 1971 as compared with 6000 in 1966.

The statistics on criminal abortion have been appalling. The real incidence is, for obvious reasons, unknown, but authorities estimate that before liberalization of abortion laws between 600,000 and 2,000,000 illegal abortions occurred each year in the United States! The Kinsey Sex Research Institute [Gebhart et al.] gave the following figures: 88 to 95 percent of all premarital conceptions terminate with induced abortion; 22 percent of all married women have had at least one induced abortion; 87 percent of all abortions are undertaken by physicians, while 8 percent are self-induced. In Central American countries the frequency of abortion has been estimated to approach 50 percent of all live births. The magnitude of the problem is illustrated by the fact that, before the advent of legalized abortion in New York, *nearly half the maternal deaths in New York City were the result of criminal abortion* (Gebhart et al.)!

The liberalization of abortion laws would undoubtedly tend to reduce the maternal mortality rate since more abortions would be performed in hospitals. But even with the most permissive laws, such as exist in Japan and Sweden, illegal abortions continue to flourish.

The enormity of the problem means that a deep conflict exists between our legal and moral concepts and the *de facto* conditions. The most obvious solution appears to be the introduction of universal sex education which teaches how conception can be prevented. Until this aim is realized, the hysterical reaction of the unwed pregnant girl and the hopeless dilemma confronting the married woman unable to provide for another child will continue to be plagues on our society.

REVIEW QUESTIONS

1. What is the function of the placenta?
2. In pregnancy, the absence of menstruation and the commonly used lab tests for pregnancy both depend upon the hormone chorionic gonadtropin. Explain.
3. What are some of the common problems that women are likely to face during pregnancy?
4. Discuss some of the reasons that make prenatal medical care important.
5. Discuss some hazards of excessive drug use during pregnancy.
6. What kind of assistance is available to couples who believe that they may carry the genes for hereditary disorders?
7. Describe the stages of labor.
8. Explain the nature of the vaccine given to Rh– women who give birth to Rh+ babies.
9. Under what circumstances do you believe that therapeutic abortion should be performed?

REFERENCES

FREDA, V. J., and J. G. GORMAN, "Antepartum management of Rh hemolytic disease," *Bulletin of the Sloan Hospital for Women*, **8:** 147–158 (1961).

FINN, R. et al., "Experimental studies on the prevention of Rh haemolytic disease," *British Medical Journal*, **1:** 1486–1490 (1961).

GEBHART, P. H. et al., *Pregnancy, Birth, and Abortion.* New York: Harper & Brothers, 1958.

HAMILTON, E. G., "Prevention of Rh isoimmunization by injection of anti-D antibody," *Obstetrics and Gynecology*, **30:** 812–815 (1967).

LANGMAN, J., *Medical Embryology*, 2nd ed. Baltimore: Williams & Wilkins, 1969.

LEVITAN, M. and A. MONTAGU, *Textbook of Human Genetics.* New York: Oxford University Press, 1971.

MILUNSKY, J. W. et al., "Prenatal genetic diagnosis," *New England Journal of Medicine*, **283:** 1370–1382; 1441–1447; and 1498–1504 (1970).

Additional Readings

BARNES, A. C., *Intrauterine Development.* Philadelphia: Lea & Febiger, 1968.

BECK, M. et al., "Abortion: A national public and mental health problem—past, present, and proposed research," *American Journal of Public Health,* **59:** 2131–2143 (1969).

Consumer Reports, Legal abortion: how safe? how available? how costly?" pp. 466–470, July 1972.

EDWARDS, R. G., and R. E. FOWLER, "Human embryos in the laboratory," *Scientific American,* **223:** 44–54 (1970).

FISHBEIN, M. (ed.), *Birth Defects.* Philadelphia: J. B. Lippincott, 1963.

FRIEDMAN, T., "Prenatal diagnosis of genetic disease," *Scientific American,* **225:** 34–42 (1971).

GREENHILL, J. P., *Obstetrics,* 3rd ed. Philadelphia: W. B. Saunders, 1965.

LEDERBURG, J., "A geneticist looks at contraception and abortion," *Annals of Internal Medicine,* **67:** supplement 7, 25–27 (1967).

NEWTON, N., and M. NEWTON, "Psychologic aspects of lactation," *New England Journal of Medicine,* **277:** 1179–1187 (1967).

RUGH, R., and L. R. SHETTLES, *From Conception to Birth.* New York: Harper and Row, 1971.

SLOAN, R., "The unwanted pregnancy," *New England Journal of Medicine,* **280:** 1206–1213 (1969).

SPOTNITZ, H., and L. FREEMAN, *How to be Happy Though Pregnant.* New York: Coward-McCann, 1969.

TIETZE, C., and S. LEWIT, "Abortion," *Scientific American,* **220:** 21–27 (January 1969).

WEISER, E., *Pregnancy: Conception and Heredity.* Waltham, Mass.: Blaisdell, 1965.

10
VENEREAL DISEASES

Venereal diseases are a group of infectious and contagious diseases which have their common denominator in the fact that they are usually contracted during sexual contact. The two of greatest importance, syphilis and gonorrhea, as well as some others, are caused by bacterial organisms; however, protozoa and viruses are responsible for a few.

In a monogamous society in which sexual contacts are supposedly restricted to man and wife, the contraction of a venereal disease is generally taken as evidence of immoral behavior, and a stigma is placed upon the diseased individuals. So it is not surprising that in most societies infected individuals are secretive about their affliction, and that the treatment and control of the diseases is consequently difficult. Successful control of venereal diseases can come only through education of the public with frank discussion of the social aspects of the diseases.

The incidence of venereal diseases in the United States is rising steadily. New cases of gonorrhea or syphilis develop about every 30 seconds—roughly 3000 each day, or over a million new cases a year. Over a quarter of these new infections are found in persons under 22 years of age [American Social Health Association and National Communicable Disease Center].

Gonorrhea accounts for almost all of this increase. While the incidence of syphilis has remained about steady in recent years, gonorrhea increased 35 percent over the six-year period from 1963 to 1969. For the year ending June 1971 the incidence of gonorrhea jumped 15 percent over the 1969 rate. The American Social Health Association estimates that one out of every 50 persons contracts gonorrhea during the late teens (ages 15–19). Dr. Walter Smartt, chief of the Los Angeles County Venereal Disease Control Division, is quoted by *Newsweek* (January 24, 1972) as claiming that "the probability that a person will acquire VD by the time he is 25 is about 50 percent."

Gonorrhea

Gonorrhea is the most common and most widespread venereal disease, and is caused by the diplococcus bacterium *Neisseria gonorrhoeae* (Fig. 10–1). The symptoms of the disease appear three to eight days after invasion by the organism. In the male the lining of the urethra is almost always infected, and its inflammation results in a burning sensation during urination (Fig. 10–1). Pus may also be exuded from the urethral orifice. In untreated cases the infection may spread to the prostate, bladder, sperm duct, or epididymus. Inflammation

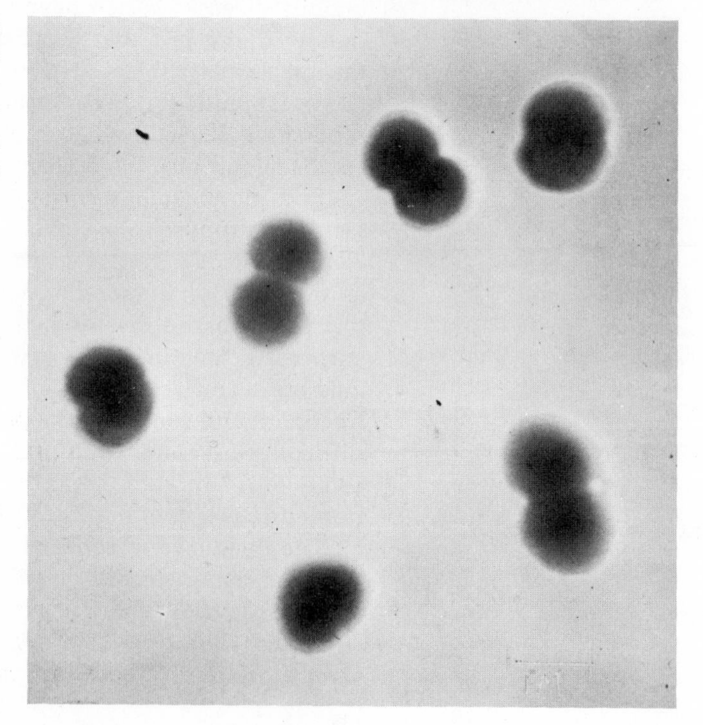

FIG. 10–1 An electron micrograph of *Neisseria gonorrhoeae*, the bacterium responsible for gonorrhea. (American Society of Microbiology)

of the epididymus or sperm ducts can lead to sterility due to closure of the genital passages.

In a female the symptoms are not as clear-cut, with the result that an estimated nine out of ten women with gonorrhea are unaware of their infection. She may feel no discomfort and be unaware of the disease. If the urethra is affected there is a burning sensation during urination. Sometimes there is a discharge of pus from the vagina, and occasionally mild irritation of the mucous membranes of the vulva. During pregnancy and after delivery, the gonorrhea infection may be spread to the endometrium and on to the uterine tubes. Severe and painful inflammation results, and sterility through closure of the tubes may occur. If the infection spreads to the peritoneal cavity (through the openings of the uterine tubes), peritonitis is likely, and death may follow. In either sex the gonococci may enter the bloodstream and infect other parts of the body, notably the heart lining (gonorrheal endocarditis) and joints (gonorrheal arthritis).

Newborn infants may be infected by contact with discharge from the mother during the birth process. The most severe manifestation is infection of the conjunctiva of the eye which may lead to permanent damage to the cornea and blindness (gonorrheal ophthalmia neonatorum). The application of dilute silver nitrate solution to the eye of the newborn, required by law in most states, has all but eliminated this once common cause of blindness.

During World War II and shortly thereafter the treatment of gonorrhea with penicillin gave such dramatic cures, it was believed the disease was coming under control. By 1957 the incidence of gonorrhea reached an all-time low. But since that time its incidence has steadily increased, in marked contrast to most nonvenereal, infectious diseases. The reversal of the trend is attributed to several factors. Most important is the development of several strains of gonococcus which are resistant to penicillin (although they are effectively treated with penicillin in combination with certain other antibiotics). Then there is the difficulty of recognizing the disease in patients, particularly women, who show minimal signs of the disease yet continue to spread the infection. The control of gonorrhea will probably have to wait until an effective blood test is developed which will identify carriers of the disease. Another factor is the lack of funds needed to implement the VD control programs of existing or formative agencies, especially on the massive scale required to prevent or contain epidemics.

Syphilis

Although outstripped by gonorrhea as to incidence (gonorrhea is more than 10 times as common) syphilis makes up for this lack of quantity by its catastrophic effects. It was long believed that the disease had been introduced into Italy from the New World by Columbus' sailors, because syphilis first appeared in its most virulent form in Italy (during the French siege of Naples) in the late

1490s. The present consensus, however, is that the disease can be traced back to biblical times, and for some unknown reason became particularly malignant late in the fifteenth century [Rosebury]. In any case, syphilis occurred in epidemic proportions during the fifteenth century, spread to a large extent by mercenary armies. In France it was known as the Italian disease, in Russia as the Polish disease, and in Turkey as the French disease. Later it was known as the Great Pox, and finally in 1530, named *Syphilis,* after the "hero" in a poem by Fracastorius.

The spirochete *Treponema pallidum* (a member of the Spirella group of bacteria) is the causative organism of syphilis (Fig. 10–2). It is a fragile organism which requires a moist, warm environment, and has not been successfully grown in culture, making research on the organism very difficult. The organism can invade any moist mucous membrane, but since the disease is usually acquired through coitus, the most common site of initial infection is in the genital area.

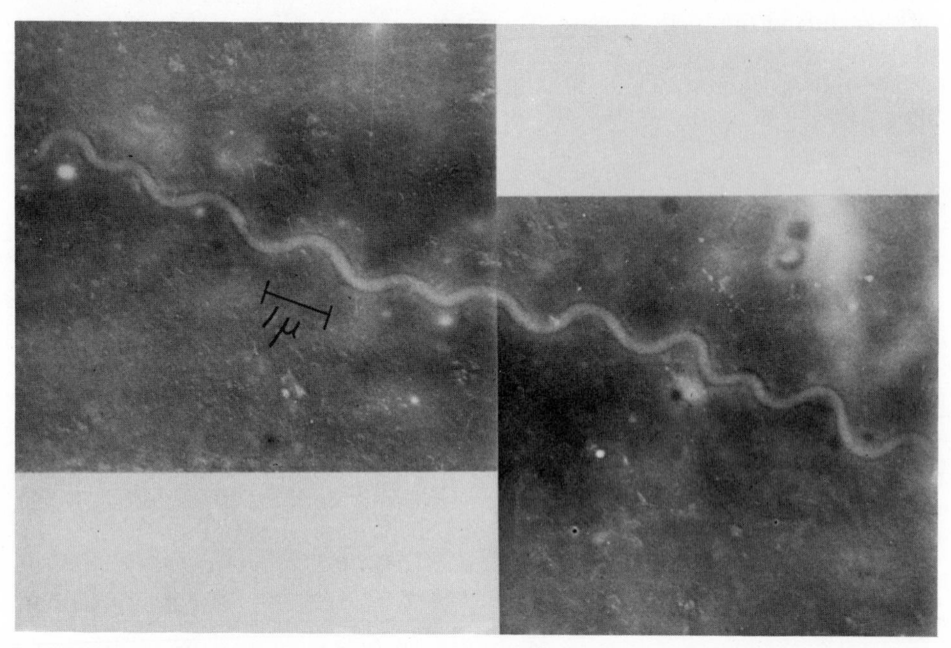

FIG. 10–2 An electron micrograph of *Treponema pallidum,* the bacterium responsible for syphilis. (American Society for Microbiology)

Primary Stage

The first manifestation of the disease is the appearance of a lesion called a chancre (shan'ker), and the enlargement and inflammation of neighboring lymph nodes. The chancre (Fig. 10–3) may be anything between a small, red painless swelling to a deep ulcer surrounded by a margin of hard tissue. After four to six weeks the chancre heals even without treatment and leaves but a barely perceptible scar. The infected individual may be lured into the notion that he has had a "false alarm."

The only sure diagnosis of the disease during this primary stage is by the identification of the spirochete in material removed from the chancre. The blood tests for syphilis (Wasserman and Kahn tests) are usually not positive during this primary stage.

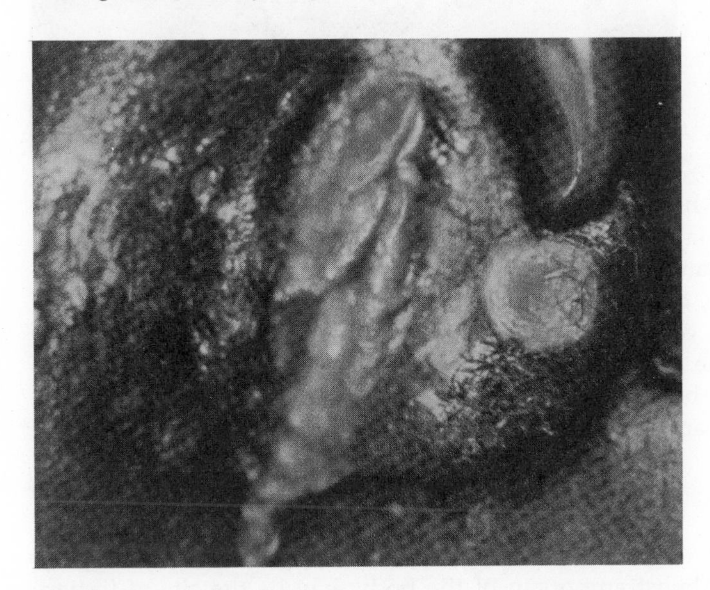

FIG. 10–3 A chancre of primary syphilis on the major labium of the female genitalia. (Armed Forces Institute of Pathology)

Secondary Stage

The lesions of the secondary stage develop 2 to 10 weeks after first appearance of the chancre, and so the chancre may be healed in the meanwhile. Also, the secondary stage may occur without a prior primary stage, or the chancre, if present, may have gone unnoticed. In any case, the secondary stage is characterized by headaches and body aches, and by the appearance over large areas of the body of red spots (Fig. 10–4) or swellings which tend to ulcerate. Ulcers develop in the mucous membranes of the mouth, vulva, vagina, and rectum, and large numbers of spirochetes may be found in these lesions. Warty growths frequently appear around the anus and genital organs, and some patients additionally suffer sore throats, gastrointestinal upsets, or loss of hair. The secondary stage of syphilis is the most highly contagious period of the disease.

All the lesions of the skin and mucous membranes clear up spontaneously

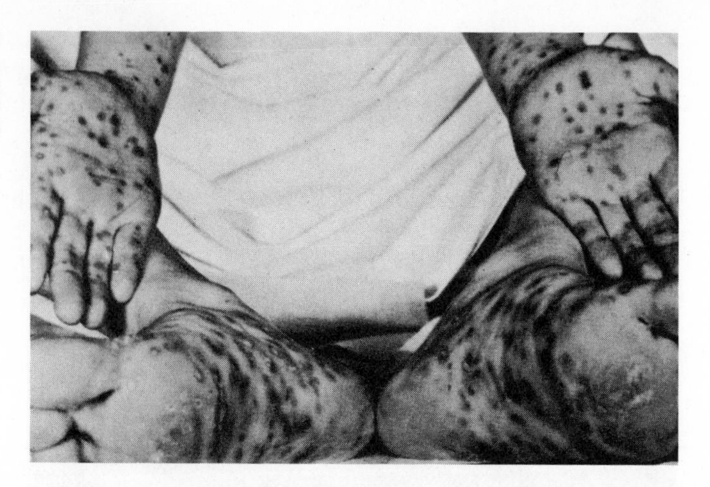

FIG. 10–4 The skin rash of secondary syphilis in a child with congenital syphilis. (Armed Forces Institute of Pathology)

within a few weeks whether treatment is taken or not. However, the symptoms of the second stage may recur on and off for up to three years. During the secondary stage the organism releases several protein substances into the blood. Because one of these, called *reagin,* has the property to react immunologically with certain proteins of beef heart—a little understood immune reaction—we have the basis for the Wasserman and other blood tests for syphilis. There is no evidence that the body manufactures effective antibodies against the bacterium *Treponema* or any of its metabolic products such as reagin.

Tertiary Stage

The lesions of the tertiary or last stage of syphilis appear from 8 to 25 years after the infection. Almost any organ of the body may be the site of tertiary lesions, but the large arteries, the brain and spinal cord, and the skeleton appear to be most commonly involved.

The aorta is commonly affected, sometimes becoming narrowed through thickening of the leaflets of the aortic semilunar valves, or by the development of an *aneurysm.* An aneurysm is a weakening and ballooning out of the wall of an artery, and the possibility of rupture and massive internal hemorrhage is ever present. The effects on the brain and spinal cord may lead to visual loss, deafness, paralysis, and mental degeneration. Syphilis accounts for some 33,000 patients in mental institutions in the United States! [Rosebury]

Congenital Syphilis

Syphilis may be transmitted by the infected mother to the fetus, particularly during the first two years of infection. If the fetus survives, the course of the disease in the infant and child is similar to that of secondary and tertiary stages in the adult. The incidence of congenital syphilis has declined over the past 20 years, in large part due to legal requirements of blood (Wasserman) tests for marriage licenses.

Control of Syphilis

Almost all early infectious stages of syphilis are curable with penicillin. In spite of such remarkably effective therapy, the incidence of syphilis is only slightly lower than it was a decade ago.

Other Venereal Diseases

Chancroid has been called the "third venereal disease" and occurs throughout the world, particularly in warmer climates, and accounts for about a third of all venereal disease. The causative organism is a bacillus. As the name indicates, ulcers similar in appearance to those of syphilis appear in the genital area. Sulfonamides and antibiotics are effective in the treatment of the disease.

Granuloma inguinale is a disease of low contagiousness found commonly in many parts of the world where soap and water and strict morals are not considered essentials of life. The disease organism is a bacillus which causes skin eruptions similar to those of the second stage of syphilis. Ulcers sometimes develop on the external genitalia which heal into scars which frequently mutilate the scrotum or vulva. Granuloma inguinale is effectively treated with certain antibiotics.

Lymphogranuloma venereum is a venereal disease caused by a rickettsia-like virus. The disease is widespread in South America and in southern parts of the United States. The disease primarily infects the lymph nodes, particularly those of the inguinal region and in the anal area. It is a serious disease and may lead to considerable injury to the external genitalia and rectum. Unlike most viruses this particular organism is sensitive to certain antibiotics.

Epidemiology of Venereal Disease

Since no vaccine is available which will prevent either gonorrhea or syphilis, and since neither infection confers immunity (repeated infections are common), the control of these diseases must come through use of prophylactic methods and epidemiological control measures. The only prophylactic device generally available is the condom, but it seems to be little used among the young and poor. Epidemiological control is expensive, time-consuming, and inefficient, but even with all its shortcomings is presently the most effective approach. It involves, in its basic features, the systematic identification and treatment of infected individuals and their sexual contacts so that transmission to others is prevented. For epidemiological control to work, physicians must

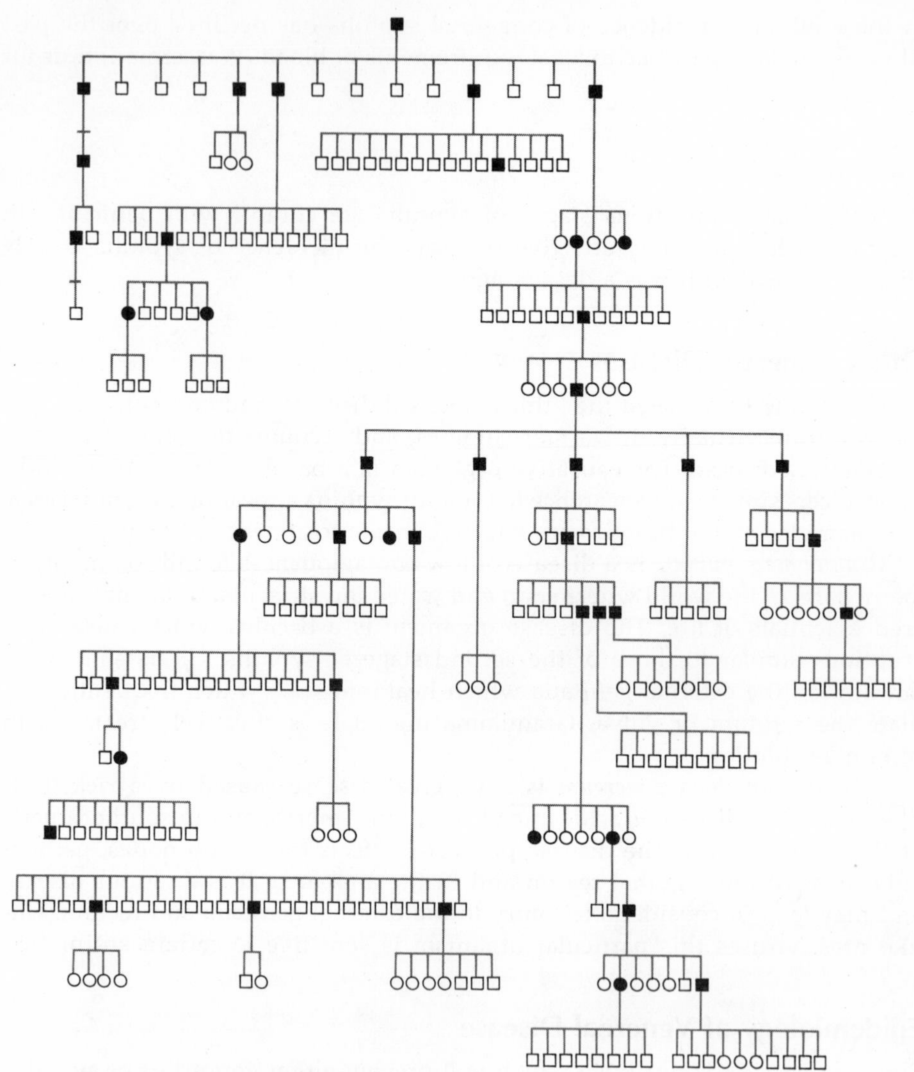

■ MALE—With infectious syphilis
□ MALE—Exposed to syphilis but not infected
● FEMALE—With infectious syphilis
○ FEMALE—Exposed to syphilis but not infected

FIG. 10–5 This chart shows the growth of one large syphilis chain in a midwestern urban area in 1965—from the time one man introduced it early in 1965 until health officials finally wiped it out in November of 1965. The black squares in the chart indicate those who had infectious syphilis. By following the black squares you can see how the disease spread. The white squares indicate persons who were in contact with syphilis but did not become infected. In all, 274 persons were exposed through the man at the top and 42 were infected with syphilis. (American Social Health Association)

report all cases of venereal disease to local public health officials. The patient is then interviewed by a public health investigator who attempts to identify and interview all the patient's sexual contacts. Contacts are then examined for the presence of the disease and, if necessary, treated. Such interviews can reveal widespread chains of infection (Fig. 10–5).

Weak links in the program limit its effectiveness, however. The stigma associated with syphilis constitutes a powerful restraint to cooperation with public health officials. The patient is understandably reluctant to disclose to these investigators the names of his (or her) recent sexual partners. Also, the family physician is under strong pressure from his patients to withhold information. According to the American Social Health Association (Social Health News, January 1969) only 11 percent of all cases of venereal disease are reported to public health agencies, even though law requires such reports in all states. When diagnostic laboratories are legally required to report positive tests for venereal disease, as in California, control programs become more effective. A shortage of trained public-health case workers is also a problem. There are only 800 in the entire country—less than 16 per state! Also, low salaries generally paid often mean poorly trained or understaffed personnel.

Epidemiological control of gonorrhea presents special problems. The incubation period (time between exposure to the disease and the first appearance of symptoms, at which time it can be passed to others) is so short—three to eight days—that caseworkers have little time to trace contacts before they have passed the disease to others. This, coupled with the difficulty of diagnosing the disease in women, has made the epidemiological control ineffective. Until a simple screening test for gonorrhea is developed control will probably continue to be ineffective.

The eradication of venereal disease must wait, like trichinosis, the resolution of an enlightened public. Only when venereal disease is regarded as an intolerable and needless tragedy of society, rather than as a stigma of social misconduct, will eradication be possible.

REVIEW QUESTIONS

1. What are some of the consequences of untreated gonorrhea?
2. What are some explanations of the increase in the incidence of gonorrhea?
3. Briefly explain the three stages of syphilis.
4. Children born to mothers with gonorrhea or with syphilis may develop the disease, but they contract the diseases in different ways. Explain.

REFERENCES

AMERICAN SOCIAL HEALTH ASSOCIATION. Venereal disease rates and projections are reported by ASHA in its *Social Health News.* Current issues should be consulted for up-to-date compilations and projections.

NATIONAL COMMUNICABLE DISEASE CENTER. Cases of venereal disease are reported to the Public Health Service by State and Territorial health departments for each quarter and published in *VD Statistical Letter.*

ROSEBURY, T., *Morals and Microbes: The Strange Story of Venereal Disease.* New York: Viking Press, 1971.

Additional Readings

CONSUMER REPORTS, "Venereal Disease," pp. 118–123, February 1970.

LUCAS, J. B. et al., "Diagnosis and treatment of Gonorrhea in the female," *New England Journal of Medicine,* **276,** 1454–1459 (1967).

NEUMANN, H. H., and J. M. BAECKER, "Treatment of Gonorrhea," *Journal of the American Medical Association,* **219:** 471–474 (1972).

SPARLING, P. F., "Diagnosis and treatment of syphilis," *New England Journal of Medicine,* **284:** 642–653 (1971).

PART THREE
Etiology of Disease

11
THE NATURE AND CAUSES OF DISEASE

You readily understand disease in terms of how you feel (your symptoms), and in some cases, by obvious changes in the appearance of your body. You feel discomfort, and the tasks of everyday existence become difficult or impossible. To the physician disease means the patient's symptoms *and* the findings of his physical examination which suggest to him structural changes or *lesions* in the body, either gross or microscopic. The symptoms may have an obvious relationship to the lesion, as in the case with lower abdominal pain and acute appendicitis, or the lesion may be present without symptoms, as is the case with most early cancers. In addition, it is now beginning to be suspected that biochemical lesions may exist without apparent structural change, and account for *functional disease* in contrast to *organic disease* in which a visible cause can be found. When the complex chemistry of cellular processes is fully understood the distinction between organic and functional disease will likely disappear.

The analysis of the causative factors in the development of a disease is termed *etiology.* Until the middle of the nineteenth century, disease had been regarded as resulting from a lack of harmony among the four humours, or an imbalance between the humours and the patient's environment. The four humours were blood, phlegm, black bile, and yellow bile. Each humour became associated with an organ: blood with the liver, black bile with the spleen, phlegm with the lungs, and yellow bile (choler) with the gall bladder. In this day and age, the humours persist only in the form of adjectives of temperament: sanguine, melancholy, phlegmatic, and choleric (respectively). Through the research of Rudolf Virchow, Robert Koch, and Louis Pasteur, the concepts of disease were revolutionized. Their laboratory experiments revealed that a disease could be produced at will by introducing a single factor, a virulent microorganism, into a healthy animal.

These observations led to the doctrine of *specific etiology*. Today a great number of well-defined disease states can be ascribed to specific causative factors. A broad classification of such identified states are those caused by: invading microorganisms, hormone deficiencies, dietary deficiencies, hereditary defects, radiation, extreme temperatures, and physical trauma. However, despite intense efforts, the causes of cancer, of arteriosclerosis, of most mental disorders, and of other great classes of disease remain undiscovered. The general medical view now is that the search for *the* cause of any of these diseases is futile, because they are probably conditions which are the result of a variety of circumstances.

In addition, the doctrine of specific etiology is now somewhat modified even in regard to infectious disease. Exposure to many of the most virulent microorganisms frequently causes *no* disease! For example, the incidence of infestation by the tuberculosis organism is far greater than the incidence of clinical tuberculosis; the cholera organism can be ingested in enormous numbers and persists in the intestinal tract with but few indications of disease; bronchitis is associated with the activities of viruses and bacteria ubiquitous in all human communities. The etiology of these and many other diseases is not as simple as had been supposed. The causation must include other factors such as prolonged stress, varying degrees of starvation, heredity, and environment. These and other factors are referred to as *predisposing conditions* (Chapter 3).

Diseases caused by hormone deficiencies, dietary deficiencies, hereditary defects, radiation, extreme temperatures, and physical trauma are discussed at length elsewhere in the book in relation to the physiological processes which are affected. However, it would seem appropriate to give some special attention in this chapter to the disease-causing organisms which are responsible for infectious disease.

Infectious diseases are simply those which are caused by an invading organism or its products. Organisms which cause disease are said to be *pathogenic*. The pathogenic organisms of certain infectious diseases can be transmitted from one sick person to another, in which case the disease is said to be *con-*

tagious. For example, measles, chickenpox, and colds are infectious and contagious. Many other infectious diseases are transmitted through drinking water, food, air, disease-carrying animals, or other means, and are not contagious; exposure to infected persons carries no hazard. Examples of this type would be malaria, which is transmitted only through the bite of an infected mosquito, and tetanus, which is acquired through contamination by cuts or other injuries.

In the rest of this chapter we shall discuss, in general terms, the kinds of pathogenic organisms. A number of specific diseases caused by bacteria and viruses are discussed in detail, in relation to physiological effects on the body, elsewhere in the book, making further discussion here unnecessary. However, extended discussions are presented for the principal diseases caused by pathogenic organisms other than viruses and bacteria, since, with a few exceptions, these are not referred to again.

General Classification of Pathogenic Organisms

Bacteria

Bacteria are probably the single greatest cause of disease. Even so, the bacteria that have acquired the ability to live and multiply in the human body amount to only a small fraction of all bacteria. In fact, the vast majority of bacteria live only on dead material, and are called *saprophytes*. Those that do exist on living material—and bacteria are able to infest practically all living creatures—are called *parasites*. Some bacteria can exist as both saprophytes and as parasites. For example, the typhoid organism can live on decaying matter, and then shift to a parasitic existence if the opportunity arises. The syphilis organism, on the other hand, has lost the ability to live as a saprophyte, and for this reason must be passed from one person to another.

Biologists generally agree that bacteria are most closely related to the primitive one-celled plants. Even though bacteria are also one-celled they have a simpler organization. Their cytoplasm has no true mitochondria and no endoplasmic reticulum. Chromatin never seems to take on the appearance of chromosomes, and no distinct nucleus is present.

Most bacteria are classified in one of three types: *bacilli* or rod forms; *cocci* or spherical forms; and *spirella* or curved forms. All types reproduce simply by dividing in two (although sexual reproduction too has now been demonstrated in many bacteria). Under optimum conditions division may occur as often as once every 20 minutes. This means many billions can arise from a single cell in less than one day's time.

The bacilli constitute the largest group, and account for the greatest number of human bacterial diseases, including botulism (food poisoning), conjunctivitis ("pink-eye"), diphtheria, dysentery, leprosy, plague, tetanus, tuberculosis, and whooping cough. Some of the bacilli possess flagellae—whiplike tails—which gives them motility. The bacilli also include a number of forms which can form spores—dormant stages which can resist both very high and very low temperatures, as well as desiccation.

The cocci include a number of groups which tend to form characteristic collections after cell division, and the appearance of these groups, such as pairs, strings, chains, forms the basis for classification. The *diplococci* appear as two cells linked together, and include the organism which causes gonorrhea. Some other forms have chains and are called *streptococci* (Fig. 11–1). The "strep" sore throat and certain types of food poisoning are caused by streptococci. *Staphylococci* are those which collect in bunches, and are infectious organisms in pimples, boils, and eyelid styes. The cocci do not include motile forms, and they probably do not form spores.

The spirella usually have flagellae, and do not form spores. They are responsible for the diseases of cholera and syphilis.

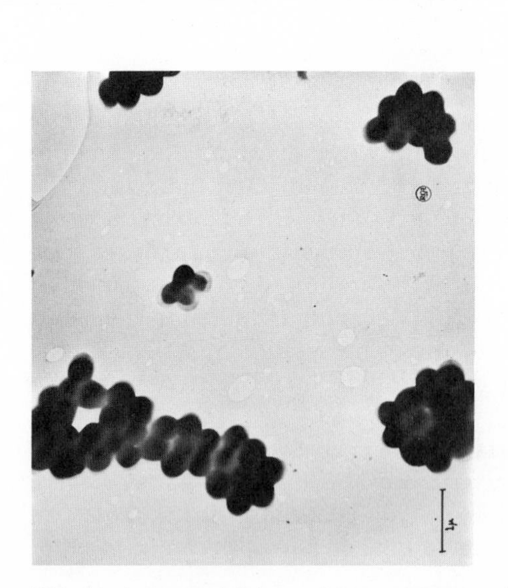

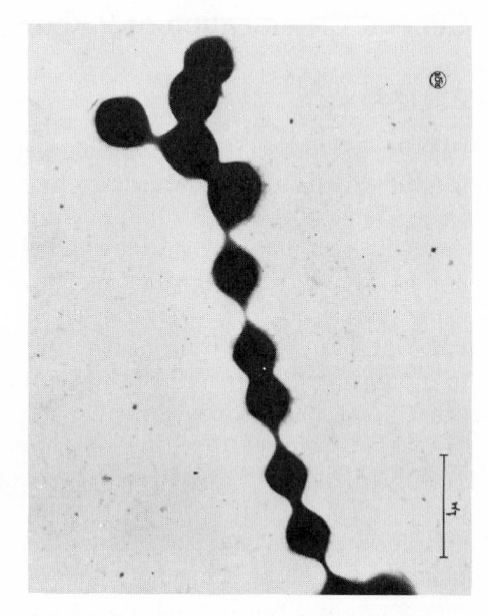

FIG. 11–1 On the left is an electron micrograph of *Staphylococcus aureus,* the common bacterium responsible for pimples, boils, and abscesses. Staphylococci are among the most widely distributed of disease causing bacteria, and are usually present in the respiratory passages where they generally do no harm. Penetration into deeper tissues through hair follicles or ducts of sweat glands brings about the typical localized infection. On the right is *Streptococcus pyogenes.* The streptococci are highly invasive bacteria which spread rapidly all over the body from a local infection. Secondary effects of untreated streptococcus infections often result in valvular heart disease and kidney damage. (American Society for Microbiology)

The great number of diseases caused by bacteria often overshadow the fact that the great majority of these organisms are beneficial, and, indeed, indispensable. They are responsible for the decay of dead plant and animal remains, and without them the essential materials of life would soon be bound up in the corpses of the dead. Soil bacteria also make possible the continuous reuse of nitrogen by converting the nitrogen-containing animal wastes into the nitrates usable by plants. Other bacteria are able to convert the molecular nitrogen of the atmosphere to nitrates.

A number of chemicals (antiseptics) can effectively kill bacteria, but all are exceedingly toxic to the cells of the body. Specific substances which would kill bacteria without damaging effects to the body's cells were rare until antibiotics were discovered. Modern antibiotics, such as penicillin and tetracycline, are substances extracted from soil fungi which can kill bacteria very rapidly in laboratory cultures. In the living body the antibiotics generally act by preventing the bacteria from dividing which gives the defensive mechanisms of the body a chance to overcome the invasion. *Bactericidal drugs* kill bacteria (a few antibiotics are bactericidal); most antibiotics are *bacteriostatic,* meaning that they inhibit or retard growth of bacteria.

When antibiotics were first available, early in World War II, they were not only scarce, but were used before extensive preliminary studies were possible. For these reasons doses we now know to be inadequate were used (50,000 unit dosages compared to present dosages of 600,000 units or more in the case of penicillin). This is one of the explanations given for the fact that some groups of bacteria were able to develop *antibiotic-resistant strains;* occasionally, resistant mutant bacterial cells could flourish and multiply without competition and become prevalent. This seems to be particularly true of some staphylococci. More recently, the gonorrhea organism has developed a penicillin-resistant strain, but apparently from causes other than underdosage.

Fungi

Fungi, yeasts, and molds are plants which lack chlorophyll, and therefore cannot manufacture their own food. This means that they are either saprophytes or parasites living on other plants or animals. Fungi may exist as single cells, but more often they form filaments called *hyphae.* Fungi are spread to new locations by releasing *spores* (seed-like cells) which are generally windblown. They are so light and are produced in such prodigious numbers that they are present almost everywhere.

Fortunately, only a few cause disease. And since they require an environment of high humidity and warmth, their growth is more favored in tropical climates, and as expected, tropical countries have a higher incidence of fungus diseases.

The fungi responsible for most skin infections are the *dermatomycoses.*

Those that cause common ringworm are spread from animals (usually pets) or from other humans. Those that cause athlete's foot are usually spread from person to person in showers or locker room facilities. Once established, they are very hard to eradicate. More generalized fungal infections, sometimes attacking the lungs or even the meninges of the brain, usually come from the soil or vegetation.

We have already mentioned that almost all antibiotics are extracted from soil fungi. It should not be surprising, then, that antibiotics are of little benefit against fungal disease. In fact, the use of antibiotics frequently destroys a number of harmless bacteria which normally function to restrain the growth of fungi. The increased use of antibiotics has, accordingly, given rise to an increased incidence of fungal disease. This is one of the reasons it is unwise to insist that your physician prescribe an antibiotic for you when he is unsure of the cause of your symptoms. Generally speaking, the physician prescribes an antibiotic only when he is reasonably certain that a bacterial infection exists. The use of cortisone and certain other steroids also seems to favor fungal infections for unknown reasons.

Viruses

A good deal of argument exists as to whether viruses are forms of life or not. They do not fulfill all the traditional criteria of living things, but these criteria were established before viruses were understood. The virus consists of a core of nucleic acid, either DNA or RNA, covered with a protein. However, unlike bacteria and other primitive forms of life, they lack enzymes, and are therefore incapable of independent metabolism. In order to reproduce—which, in this case, means the manufacture of viral nucleic acid and protein—they must enter a living cell and use that cell's machinery for this purpose. Obviously, viruses are always parasitic, and never saprophytic. Gel-like cultures containing essential nutrients are adequate to grow most bacteria; viruses require cultures of living cells.

Although viruses are too small to be seen with the ordinary light microscope, they were discovered almost 100 years ago by a Dutch scientist named Beijerinck, who showed that a disease of tobacco plants could be transmitted from plant to plant by a fluid filtered through pores too fine for bacteria to penetrate. For this reason Beijerinck regarded the disease as being due to an infectious liquid, and he named the fluid *virus,* from the Latin word meaning poison. (Actually the word "virus" was used in much the same way in the fifteenth to eighteenth centuries when it meant the agent in any infectious disease.)

Since structural characteristics of viruses can only be determined by examination with an electron microscope—an expensive and time-consuming process—classification is based principally on the characteristics of the infection. For example, tobacco mosaic virus (Beijerinck's virus) causes a mottling of tobacco leaves; the bacteriophage viruses seemingly devour bacteria. In humans, the disease-causing viruses are named according to the signs and symptoms of the disease. Those which caused skin disfigurement were poxes. Originally there were four poxes: Great Pox (syphilis), small pox, cow pox, and chicken pox (so-named because the chicken is thought of as harmless). Syphilis is now known to be caused by a bacterium (the spirella *Treponema*), and is no longer referred to as a pox. Several other virus diseases also show skin involvement: measles (rubeola); German measles (rubella); herpes simplex, the virus which causes "cold sores" and "fever blisters"; and herpes zoster, a virus which infects the sensory nerves. The common cold, influenza, poliomyelitis, rabies, yellow fever, and mumps are also viral diseases, and most of these are discussed elsewhere (Fig. 11–2).

In Chapter 19 we shall discuss the role of immunity in the body's defensive mechanisms against both viruses and bacteria. Recently, however, another agent produced by the body, specific against viruses, has been discovered. The agent is a protein named *interferon,* and seems to act as a virus antibiotic. Unlike immune antibodies, which are specific against each infectious agent, interferon is the same no matter what the nature of the infectious virus. Apparently, interferon somehow changes the host cell so that it will no longer function to reproduce the viral nucleic acid and protein. Our insufficient knowledge of interferon has resulted in only a limited clinical application, but its future use-

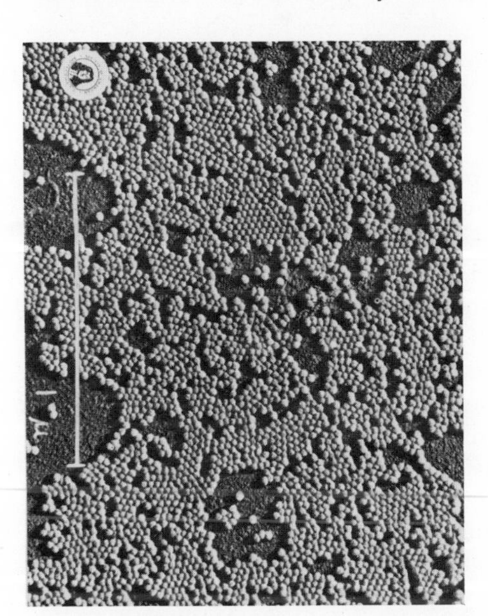

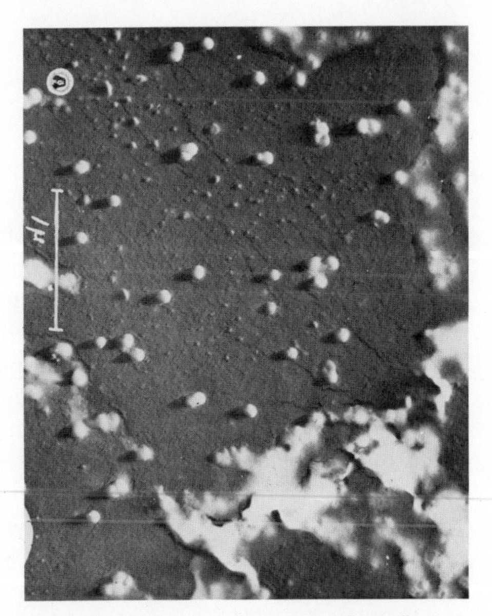

FIG. 11–2 *Left:* An electron micrograph of the poliomyelitis virus; *right:* one of the influenza viruses. (American Society for Microbiology)

fulness could be as great as that of antibiotics. In general, however, the development of antiviral drugs has been slow. Antibiotics are notably ineffectual against viruses, safely lodged as they are in the interior of the cell.

Rickettsiae

The rickettsiae are organisms which seem to be on the borderline between viruses and bacteria. They are large enough to be visible with the light microscope, yet, like viruses, they lack enzymes and must therefore utilize the metabolic machinery of living cells. All the rickettsiae apparently infect some arthropod—louse, tick or mite—and it is from these animals that the rickettsial diseases are spread to man.

Fewer than a dozen diseases in man are caused by rickettsiae. The two important ones are Rocky Mountain Spotted Fever, spread by infected ticks, and typhus, usually transmitted by the bite or feces of infected body lice and preventable by vaccination. Typhus is one of that handful of diseases that have had great historic significance, causing the defeat of armies and contributing to the fall of empires. One of the great books in the history of social aspects of medicine is Hans Zinsser's entertaining *Rats, Lice and History* which relates the history of typhus.

Several of the great research workers in typhus have died of the disease including H. T. Ricketts after whom this group of organisms is named.

Animal Parasites

Animals which have adapted themselves to living on or in the body of another living organism—the host—are animal parasites. Animal parasites do not necessarily produce disease in the host; on the contrary, it is in the interest of the parasite that the host remain healthy. Those parasites that do cause disease are believed to be animals that have only recently acquired the parasitic mode of life and have not yet established a metabolism in balance with that of the host. These infections are not unusual in certain localities of the United States, but fortunately chances of infection can be reduced almost to nil by following the simple basic habits of good sanitation.

Most animal parasites have developed a very curious life history or life cycle in which part of their life is spent in one animal and part in another. Both hosts are almost always essential for the continued existence of the parasite. Early development from the fertilized egg to some intermediate stage, usually called a *larva,* occurs in one host, while maturation to the fully adult form occurs in another host. Occasionally, we find an animal parasite that requires three, or even four different hosts. In any case, the typical life cycle involves production of young in one animal and their growth in another. Obviously this involves a considerable survival risk, and in order to ensure continuation of the life cycle, eggs are produced in enormous numbers. For example, the infamous beef tape worm typically lays a million or more eggs a day. To cut the risks of survival even further, some animal parasites have evolved a life cycle in which the young develop from *un*fertilized eggs (certainly an observation which gives all males a pause for reflection).

Protozoa

The protozoa are one-celled animals which are considerably advanced over the bacteria. Some species are associated in clusters, but the cells have little or no dependency on each other and are not organized into tissues. However, these organisms are far from simple; each protozoan carries on all the fundamental functions common to higher animals, and some forms are astonishingly complex in structure.

The Malarial Parasite

By far the most important protozoan parasite is *Plasmodium* (Fig. 11–3), which causes the disease malaria—one of the great scourges of mankind. Malaria has a history of political significance second only to that of typhus, and persists today as the most important worldwide medical problem. About 2 million human deaths each year are attributable directly or indirectly to malaria. In many parts of the world virtually 100 percent of the population is infected with

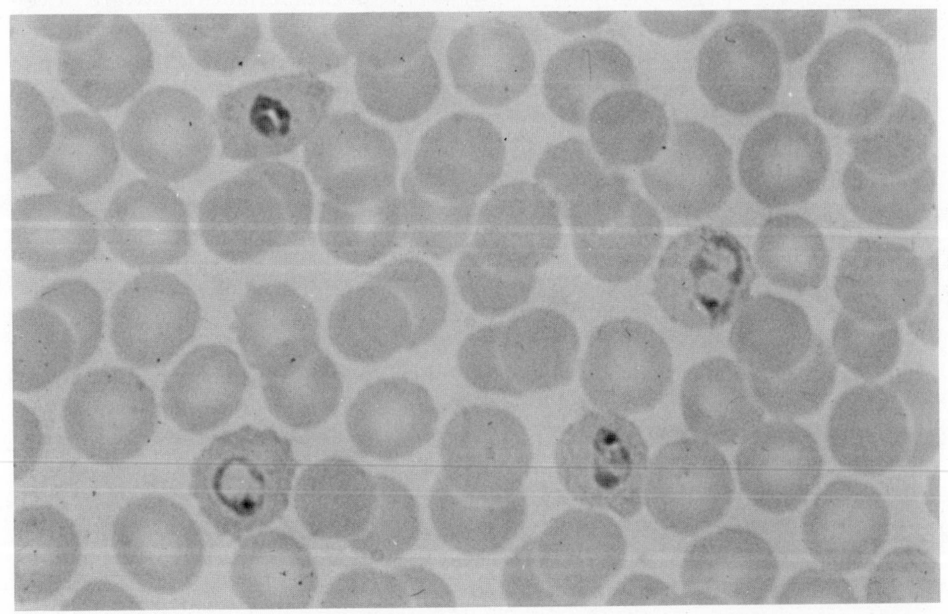

FIG. 11–3 The malarial parasite (*Plasmodium vivax*) in a red blood cell. (Dr. H. Zaiman)

the parasite from the age of a few weeks to the time of death. Among such populations it is not even thought of as a disease since everyone experiences the symptoms—none ever knowing what freedom from disease really means.

The plasmodium of malaria has part of its life cycle in man and part in the *Anopheles* mosquito (Fig. 11–3). When the infected mosquito draws blood from her human host, she injects the parasites into the bloodstream by which they reach and settle in the liver. After a period of time in the liver each parasite enters a red blood cell and multiplies by simple cell division. The new parasites finally burst the blood cell membrane and are liberated into the plasma. Each attacks a new red blood cell and the multiplication process is repeated. Billions of parasites are formed and billions of red blood cells are destroyed. The new generations of parasites develop in phase with one another, requiring either 48 hours (the common form) or 72 hours. This massive release of parasites into the plasma causes a chill or shivering followed by a sudden high fever. If untreated the disease will usually subside within 30 days, but generally recurs. Relapses can suddenly occur as much as 25 years after the initial attack was presumed cured.

For over 150 years the classical treatment for malaria was quinine, a drug extracted from the bark of the cinchona tree (native to South America, but cultivated principally in the Dutch East Indies). Quinine is a drug toxic to nucleic acid metabolism, but with effects much less severe on human cells than on the plasmodium parasite. During World War II, when the Japanese had control of the quinine production, a number of synthetic antimalarial drugs were developed, and these now have largely replaced quinine. Particularly important has been the introduction of quinacrine and chloroquine which give protection to uninfected individuals during exposure to the disease. However, the reduction in recent years in the incidence of malaria has been accomplished primarily by control of the *Anopheles* mosquito through the use of insecticides and the destruction of breeding areas.

African Sleeping Sickness

African sleeping sickness is another protozoan disease transmitted by an insect, in this case the tsetse fly. The tsetse fly occurs only in tropical Africa, so the disease is restricted to that region.

The organism, *Trypanosoma,* lives in the blood, causing fever, weakness, and emaciation. Only late in the disease is the central nervous system invaded, and the lethargy and coma characteristic of sleeping sickness develop. No effective treatment is available and the disease may be fatal.

Flatworms

The flatworms include three major groups: the free-living forms of which the familiar planaria is an example and which are of no particular medical concern; the *trematodes* or *flukes;* and the *cestodes* or *tapeworms.*

Trematodes

The trematodes or flukes are leaf-shaped parasites which attach themselves to the host by means of a suckerlike mouth. All those which infect man, such as *Schistosoma,* require another host, a snail.

The Schistosoma group of organisms is responsible for the disease *schistosomiasis* (Fig. 11–4). The larval stage is produced in a snail and released into the water. The organism penetrates the skin of bathers, washerwomen, and irrigators, and eventually becomes lodged in the liver, lung, or wall of the bladder where it grows to maturity. The adult organism produces eggs in the infected man which are passed out either in the urine or feces depending upon the particular species and the site of the infection. The eggs in turn hatch in the water, and the young larvae seek out a particular snail which once again serves as the intermediate host.

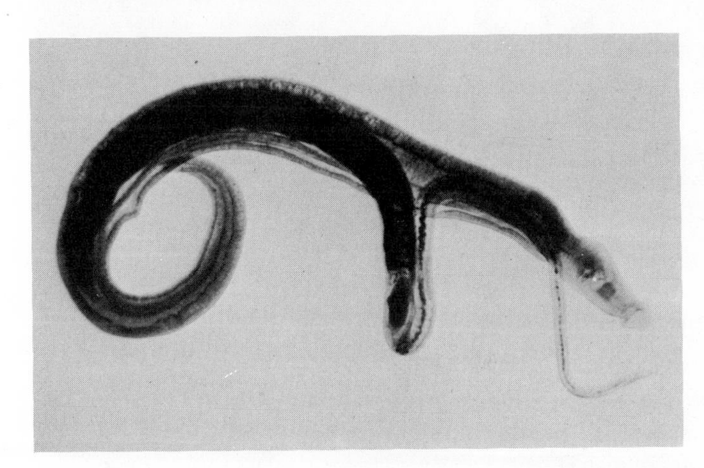

FIG. 11–4 *Schistosoma japonicum.* Schistosomiasis is a disease common in tropical countries that causes enlargement of the liver and spleen. (Dr. H. Zaiman)

The flukes cause widespread disease and disability in Egypt and other parts of North Africa. The contamination of practically all bodies of water with human wastes, and the propensity for outdoor bathing as part of religious practice in these areas makes the disease exceedingly difficult to control.

Certain other flukes, for example, the Asiatic liver fluke, have a part of their life cycles in fish, and the infection is spread to man by the common practice in Asia of eating raw fish. Fortunately, none of the flukes are important causes of disease in North America.

Cestodes or Tapeworms

The tapeworms, as their name implies, are very long, narrow, and flat. The head is very small and equipped with hooks and/or suckers which serve to attach the parasite to its host. The body is composed of many rectangular segments which are continually produced at the head end and broken off at the posterior end where they have matured into bags of mature eggs (Fig. 11–5).

The three common tapeworms in this country have the adult stage in man, and the intermediate or encysted stage in cattle, swine, or fish. This is fortunate since the adult stage is quite treatable while the encysted stage is not.

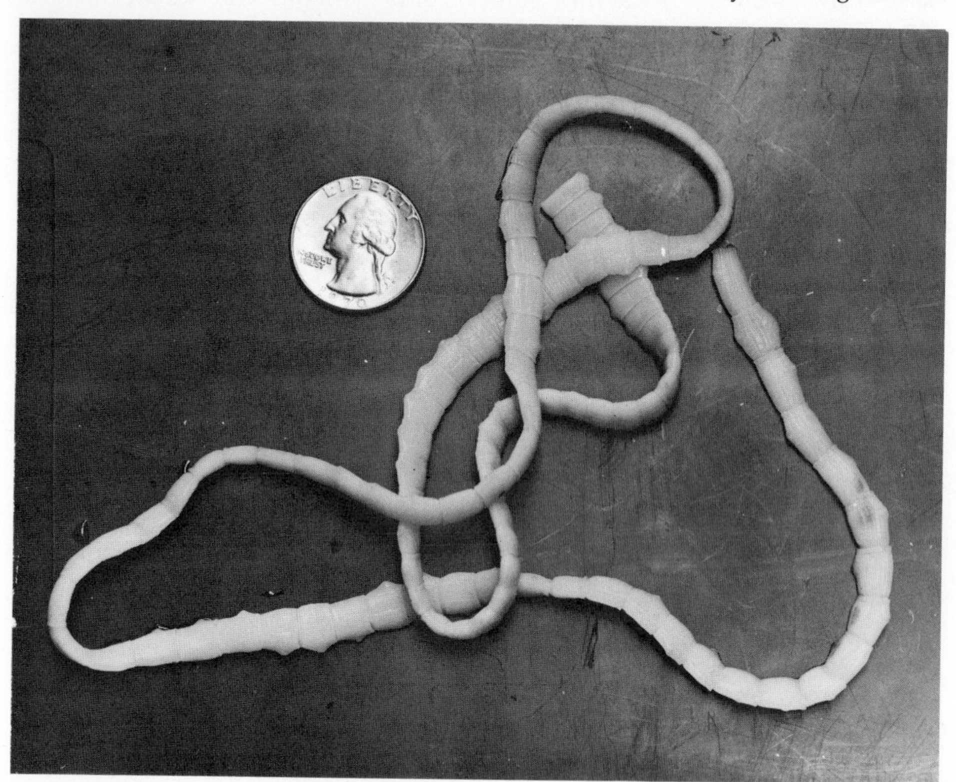

FIG. 11–5 The beef tapeworm of man (*Taenia saginata*). Shown is a short segment of a 6 to 8-foot worm passed in the feces over a period of three days. The infestation was picked up in Guatamala from eating uncooked beef. (Dr. Harry Huizinga)

In all of these forms the head of the adult tapeworm attaches to the lining of the intestines, and the body, 10 to 30 feet in length, is suspended in the intestinal contents from which it absorbs its necessary nutrients. The egg-laden segments are passed out in the feces, and must be picked up and eaten by the particular intermediate host. The eggs hatch in the intestine and the larvae migrate to the muscles where they encyst. Man is infected when he eats inadequately cooked pork, beef, or fish. Infestation is determined by examining the stool for ripe segments of the worm, or for the presence of eggs, in the case of the fish tapeworm.

A fourth tapeworm, the hydatid tapeworm, encysts in man and sheep, and passes its adult stage in the dog or other canines such as a fox. The dog becomes infected by eating the flesh of sheep. Man becomes infected when eating uncooked vegetables contaminated by fecal material of the dog. The disease is quite serious in man. The cysts seem to locate principally in the liver, and may become quite large and cause death. The dog tapeworm is principally a problem in those parts of the world where sheep-herding is prevalent.

Nematodes or Roundworms

The nematode worms appear superficially like the true worms (such as the earthworm) but belong to a quite different group of the animal kingdom. Most are free-living, but a great number parasitize fish and other aquatic organisms. The few which parasitize man cause great misery and debility, and remain a serious problem in the poor and underdeveloped areas of the world.

Hookworm

Hookworm is one of the commonest diseases of the world, probably numbering some 100 million cases. The small worm, about one inch in length, possesses four teeth or hooks which serve to attach it to the intestinal wall where it hangs sucking the blood of the host.

The young larva lives in the soil, and infects its host by burrowing through the skin of the feet of those who go barefoot. It enters a vein and is carried in the bloodstream through the heart to the lungs where it breaks through into the air passages. The upward movement of the mucus through the bronchial system carries the larva to the pharynx where it is swallowed, and it finally ends its long journey when it reaches the small intestine. By this time the worm has grown to the adult stage, and uses its hooks to attach to the intestinal wall. Here they may remain indefinitely, laying eggs which are carried out in the feces. In a heavy infestation of these worms the continuous blood loss and iron depletion can cause severe anemia and almost disabling weakness. Fortunately, hookworm can be readily diagnosed by a physician and there are several effective remedies that he can administer.

Since no host other than man is required, continual reinfection is likely, particularly when there is no proper disposal of sewage.

Ascaris

Ascaris is a much larger roundworm, sometimes up to 10 inches in length (Fig. 11–6), which has a life cycle very similar to that of the hookworm. But in this case the eggs must be ingested, and this usually occurs when children put objects contaminated with eggs in their mouths.

The larvae hatch in the intestines, burrow through the intestinal wall to reach the bloodstream, and undertake the journey described for the hookworm. By the time the worms return to the intestines they are adults. Although many are lost in the feces most remain in the intestines long enough to produce prodigious numbers of eggs.

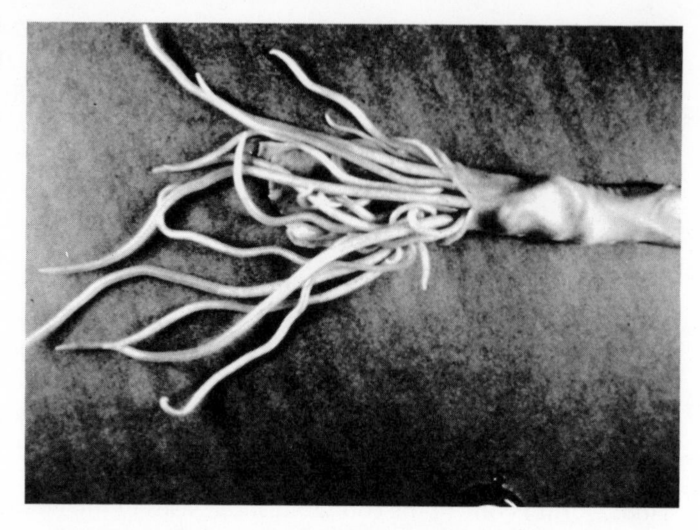

FIG. 11–6 Segment of small intestine from a pig containing numerous *Ascaris* worms. The same species infects man. (Dr. H. Zaiman)

Symptoms of ascaris infection vary with the stage of the infection. During migration through the body they may cause fever, headache, coughing, and muscular pain. If the worms become lodged in the heart or brain they can cause death. Adult worms in the intestines frequently cause no symptoms. As in the case of hookworm, prevention of the disease involves proper sewage disposal; and diagnosis is rather simple with effective treatment administered by the physician.

Trichinella

The roundworm *Trichinella,* or trichina as it is sometimes called, causes the serious disease *trichinosis. Trichinella* infects many mammals, but seems to have a predilection for swine and man. Upon being eaten the larvae hatch in the intestines, reach adulthood, reproduce young which enter the bloodstream, and finally settle in voluntary muscles (strangely enough, if they are carried to other tissues they simply die). If the infected animal is a pig, and his muscle is eaten without proper cooking, the cycle occurs in man. Since man is not usually eaten, the cycle would normally end here. However, the disease is perpetuated in swine by the practice among hog raisers of feeding garbage, some of which seems inevitably to contain trichinous meat.

In the muscle the tiny larvae are coiled within a gelatinous capsule called a cyst (Fig 11–7), and as many as 1200 cysts have been counted in a single gram of human muscle. When the muscle is eaten it is the encapsulated larvae which hatch in the intestines.

Swine seem to have a high degree of tolerance for the disease, but man can be easily overwhelmed. In one case a man suffered an acute attack of the disease as a result of simply eating a piece of bread buttered with a knife that had been used to slice an infested sausage. The hog from which the sausage had been made had appeared to be in excellent health.

Trichinosis is exceedingly painful and debilitating, and in some outbreaks as many as a third of the victims have died, although the over-all mortality of those treated for the disease is about 6 percent. Even when death does not occur complete recovery is impossible since no drug is available which will kill the encysted larvae.

Although no particular treatment is very satisfactory, prevention is quite simple. Thorough cooking will make even the most heavily infested pork quite harmless. Also, refrigeration below 5° F for at least 20 days kill the larvae, as will proper methods of salting, smoking, and pickling. Despite the availability of simple preventive measures trichinosis is a major health problem. On the basis of autopsy studies it is estimated that 20 percent of the American population has some evidence of trichinosis. Although most states have laws prohibiting the feeding of uncooked garbage to hogs, the enforcement of these laws is difficult, and consequently ineffective. Federal laws require that all spiced varieties of pork which are often eaten raw—frankfurters, salami, prosciutto, and the like—be refrigerated sufficiently to kill *Trichinella*. However, these laws apply only to those meat packers who sell their products over state lines, and at least one-third of the pork in this country is processed and

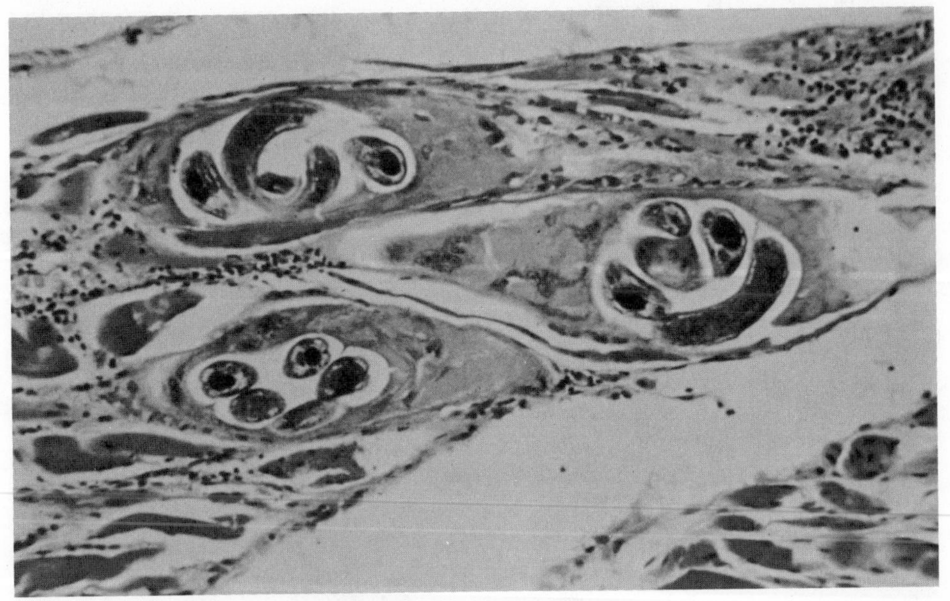

FIG. 11–7 A microscopic section of human muscle infected with the round worm parasite *Trichinella spiralis*. The disease trichinosis is contracted by eating inadequately cooked pork. (Dr. H. Zaiman)

sold locally. Since about 2 percent of the hogs slaughtered in this country are trichinous, the danger is obvious.

The high estimated incidence of the disease is not consistent with the number of reported cases simply because it is only seldom recognized. The signs and symptoms of trichinosis may be mistaken for arthritis, food poisoning, mumps, asthma, tuberculosis, nephritis, appendicitis, malaria, tetanus, syphilis, and cholera, among others. The diagnostic failures are not of great consequence in themselves, since no effective treatment is available, yet they do serve to mask the seriousness of the problem. We would all be wise to thoroughly cook all fresh pork, and to restrict our purchases of spiced meats to those processed by meat packers certified by the United States Department of Agriculture.

Filaria

Filariasis is a roundworm infection of tropical countries transmitted by its intermediate host, one of several mosquitos. The disease so well illustrates the function of the lymphatic system that it is discussed at length in Chapter 21.

Arthropods

The arthropods, or animals with jointed legs, are the insects, and their relatives. The parasitic arthropods are external parasites, that is, they infest the skin or hair and never enter the body.

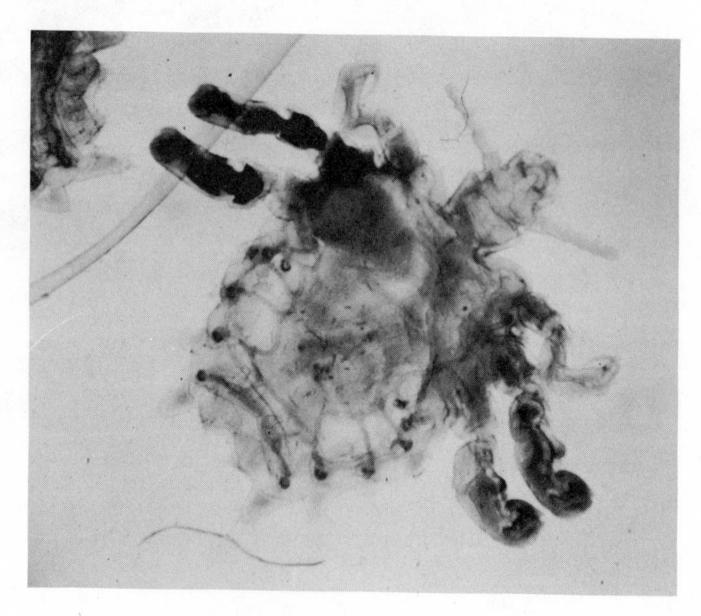

FIG. 11–8 The crab louse (*Phthirus pubis*) infects the pubic hair. (Dr. H. Zaiman)

Insect Parasites

The common mosquito is, of course, an external parasite which lives on the blood of its host. We have already discussed the role of certain species of mosquitos in the spread of malaria and of the tsetse fly in the spread of African sleeping sickness. Insects are responsible for the spread of a number of other diseases as well. However, the common species of North America are principally a nuisance rather than vectors of serious disease. The itching of the bites, which may be very intense to the host, is due to an allergic reaction to the insect saliva. Fortunately, a number of mosquito repellents are available which are quite effective.

The body louse, head louse, and pubic or crab louse (Fig. 11–8) are all insects that live on the skin and which feed on blood. Treatment is usually by dusting with an insecticide followed by the application of a disinfectant lotion.

Arachnid Parasites

The arachnids (spiders and their relatives) comprise the largest number of parasitic arthropods, and include chiggers and ticks.

Scabies is a disease caused by a mite no more than 0.5 millimeters in length. The disease is transmitted by intimate contact with an infected individual. The adult female burrows into the skin of the axillae or between the fingers or toes, and lays her eggs at the end of her excavation. The eggs hatch, and the young excavate new burrows. Intense itching develops a week or more after infection, and the effects of the scratching may be more serious than the work of the mites. Scabies is rare wherever even rudimentary efforts at personal cleanliness are found. Epidemics seem to break out where there is disruption of a culture through war or other disaster.

Chiggers (also mites) are quite small, seldom exceeding 0.5 millimeters in length, and are parasitic only during the larval stage. The young larvae are so small that they easily penetrate clothing; they use their hooked appendages to attach to the skin. The chigger usually feeds by entering a hair follicle ("skin pore"), where it injects a fluid which liquefies the surrounding tissue which then becomes hardened to form a tube. The liquefied tissues are eaten by the chigger, and as feeding continues the tube becomes lengthened and more tissues are dissolved. When the chigger finally releases itself, the tube is left in the tissue and is responsible for the intense itching. Contrary to popular notion, the chigger does not enter deeply into the skin nor does it lay eggs within the skin. The common insect repellents are also generally effective against ticks and chiggers.

Ticks (Fig. 11–9) are larger arachnids which usually attach to the skin for several days or weeks, increasing enormously in size as they become engorged with blood. Ticks should be removed by grasping them firmly and pulling them off. (Much misinformation is written about tick removal.)

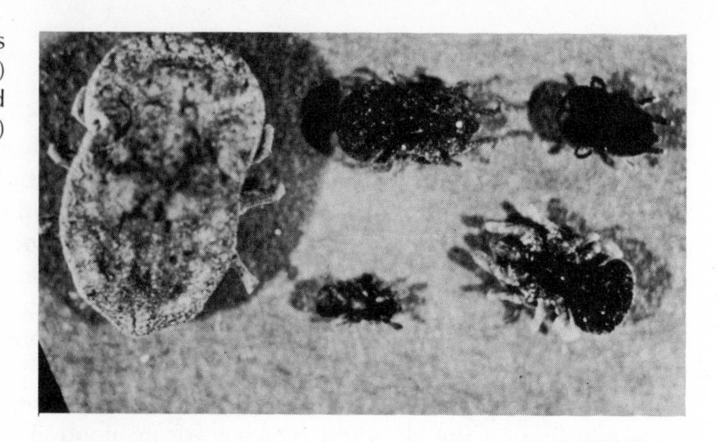

FIG. 11–9 Soft ticks (*Ornithodoros hermsi*) shown in unfed and fed states. (Dr. H. Zaiman)

REVIEW QUESTIONS

1. What is meant by specific etiology of disease?
2. How can it be explained that many pathogenic organisms may be present in or on the body and cause no disease?
3. What are the three basic types of bacteria? Give examples of diseases caused by each type.
4. How do antibiotic drugs combat bacterial diseases?
5. What are some diseases caused by fungi?
6. How do viruses infect the body?
7. How could trichinosis be eradicated?

REFERENCE

ZINSSER, HANS, *Rats, Lice, and History*. Boston: Little, Brown, 1935. (Paperback)

Additional Readings

BURNET, F. MACFARLANE, *Natural History of Infectious Disease*. London: Cambridge University Press, 1962.

DUBOS, RENÉ, *Mirage of Health*. New York: Anchor Books, Doubleday, 1961. (Paperback)

DUBOS, RENÉ, *So Human An Animal*. New York: Charles Scribner, 1968. (Paperback)

FIENNES, RICHARD, *Man, Nature, and Disease*. New York: New American Library, 1964. (Paperback)

HYMAN, HAROLD T., *Your Complete Home Medical Reference Book*. New York: Avon Books, 1963. (Paperback)

GRISSOM, DEWARD K., *Communicable Diseases*. Dubuque, Iowa: William C. Brown, 1971. (Paperback)

ROSEBURY, THEODOR, *Life on Man*. New York: Viking Press, 1969. (Paperback)

ROUECHÉ, BERTON, *Curiosities of Medicine*. New York: Berkley Medallion Book, 1964. (Paperback)

ROUECHÉ, BERTON, *Eleven Blue Men.* New York: Berkley Medallion Book, 1966. (Paperback)
ROUECHÉ, BERTON, *The Incurable Wound.* New York: Berkley Medallion Book, 1964. (Paperback)
ROUECHÉ, BERTON, *A Man Named Hoffman.* New York: Berkley Medallion Book, 1966. (Paperback)

12
TUMORS AND CANCER

Cancers and tumors of various sorts can affect all parts of the body and, for this reason, require some general remarks. They are also of enormous concern since tumors are the second most common source of death in this country.

Many people are reluctant even to discuss cancer and do so only with distaste. Granting that the subject is indeed an unpleasant one, and that death from cancer is sure to horrify anyone who has watched its advance, mere revulsion is futile. Resolute and well-informed confrontation of unpleasant facts has resulted in conquest of many menaces from smallpox to malaria; the need for a similar public attitude toward cancer is evident.

Younger people frequently feel that the topic is remote from them. Cancer, they often believe, is primarily a disease of later life, and cases in younger people are tragic exceptions. But in this belief they are only partly right: Some forms of cancer occur predominantly in youth, or even childhood, and these are on the increase. For that matter, all forms of cancer are growing more common in the early decades of life, probably due to the by-products of modern living ranging from air pollution to X-rays.

What Are Tumors and Cancers?

The term "tumor" is variously defined. In its widest sense, it means any swelling or growth, including such things as tubercular masses, goiters, and so on. More strictly it refers to growths produced by uncontrolled multiplication of body cells. The second definition is the one we shall follow.

Tumors in this sense are also often called *neoplasms,* meaning new-formations. The word *cancer* is used loosely for all such growths but should be reserved for *malignant* forms, those which are dangerous to life. This distinction helps to emphasize that many tumors are relatively harmless or *benign,* so that a suspicion or diagnosis of tumor is not necessarily a matter of alarm. The distinction between malignant and benign tumors is of the very greatest significance (Fig. 12–1).

Tumors are further classified into a wide variety of types. Distinctions are often of great importance in medicine, since a tumor of one type may be quite harmless, whereas one of a closely similar type may be highly dangerous. For this reason biopsies (removal of living samples of tissue) are often performed so that the tumor can be carefully studied before treatment is decided on. In one case a simple removal would be adequate, whereas in another a radical, widespread operation would be necessary to eradicate all traces of the growth. Thus adequate classification is a major medical need.

Its details, however, need not concern the nonprofessional reader. Suffice it to say that almost every organ and tissue can give rise to various types of tumors, some benign and some malignant. These tumors are named after the organ or tissues (though not always by the common popular name), usually with the ending *-oma.* For example, hepatoma is a tumor of the liver (Greek *hepar:* liver); fibroma, of fibrous tissue; meningioma, of the meninges; and so on. The list, with many fine distinctions, is quite endless.

Two rather general terms, *carcinoma* and *sarcoma,* are in common usage. Carcinomas are typically derived from surface tissues such as skin, lining of the stomach, covering of the uterine cervix, and so on. Sarcomas are typically derived from internal connective tissues which permeate all parts of the body in various forms. Both carcinomas and sarcomas are malignant in various degrees, sarcomas usually more so.

In sum, tumors, or even cancers alone, are not really *a* disease. They are a group of diseases as different as are measles, smallpox, and tuberculosis. For this reason, a general "cancer cure" is as unlikely as a general cure for infections would have been a century ago. A particular cure or preventive method will probably have to be worked out for each form.

Causes of Tumors

We do not yet fully understand the causes of tumors. In view of the preceding paragraph, one might expect that causes would be varied and numerous. In

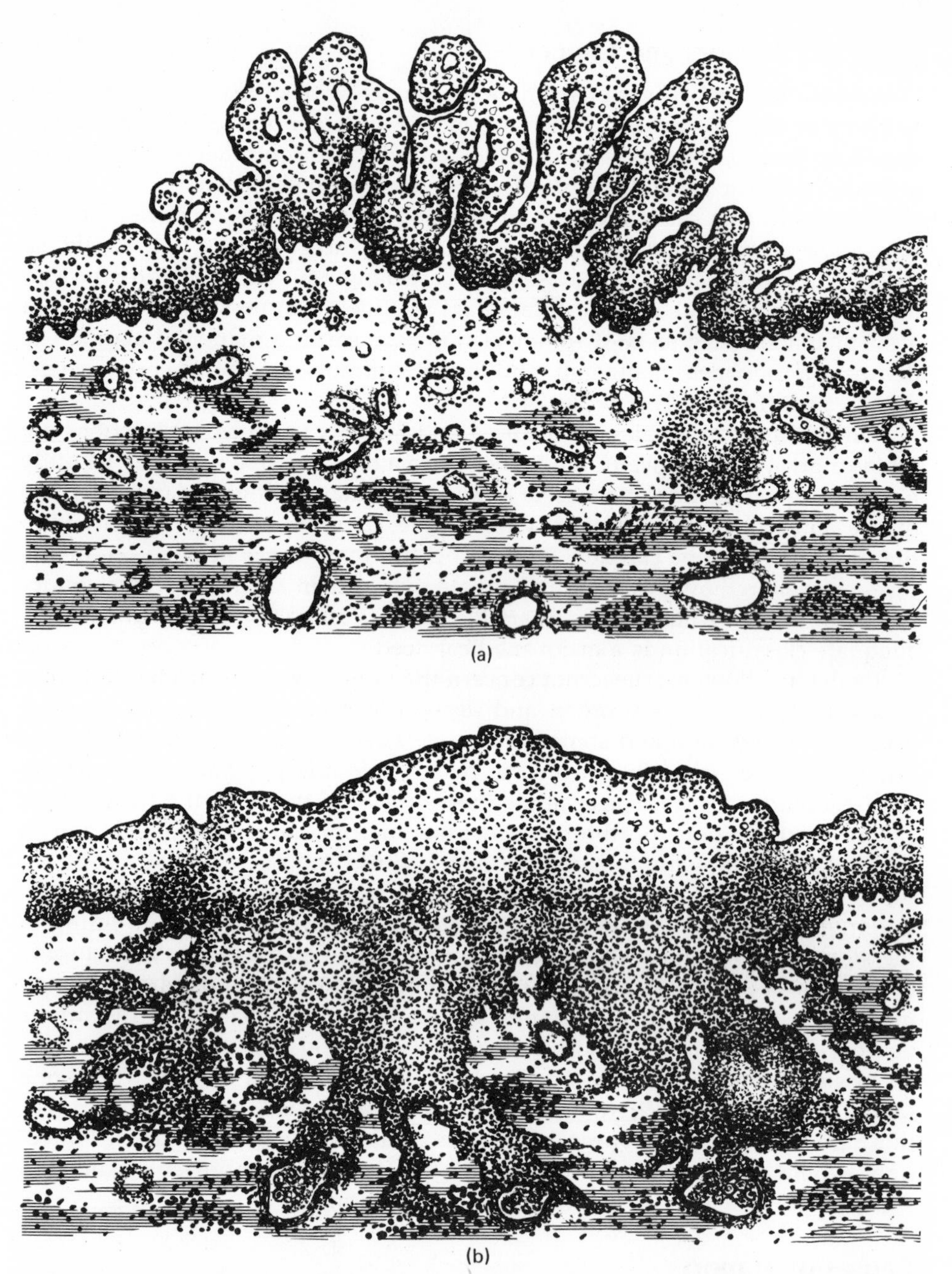

FIG. 12--1 (a) Benign tumor (papilloma). (b) Malignant tumor (epidermoid carcinoma). The principal difference seen here is the manner in which the malignant tumor penetrates the surrounding tissues like the claws of a crab—hence the term *cancer,* meaning crab.

fact, we find this to be so, with unknown causes still to be discovered in many cases.

In general, true neoplasms are due to disorder of the cellular genetic mechanism. The latter, we may recall, is centered in the chromosomes, made up of genes, in the nucleus of every cell. When a cell divides, as most body cell-types are continually doing even in an adult, the chromosomes not only divide evenly to distribute genetic material equally to each daughter cell, but they regulate the process of division itself. They perform this duty accurately and efficiently in perhaps 99.999 percent of cases; but the 0.001 case may be disastrous. It can result in abnormal cells that will continue to divide beyond control of the interlocking body chemical systems that normally limit cell proliferation. Whereas normal skin cells, for example, multiply just enough to replenish lost surface cells or to repair a cut, abnormal skin cells, even without injury, may produce an ever-increasing growth that has no definite limits at all. Indeed under the microscope, tumor cells often show abnormal chromosome patterns, evidence of abnormally frequent cell division. In some cases, these characteristics are hardly noticeable; in others, they are very pronounced.

Certain influences have been shown to disturb the chromosome mechanism and thus to produce forms of cancer (Fig. 12–2). Some chemicals, called *carcinogens,* especially some derived from or resembling coal tar, cause cancers when persistently applied to the skin of experimental animals—and presumably would do so in humans too. Possibly similar chemicals, if eaten, produce some gastric and intestinal cancers; according to a number of doctors and research workers [Ryser; MacMahon], overcooked and, especially, charred foods contain large quantities of carcinogens and could well be a cause of cancer in the digestive system. Viruses are very similar to genes and often act like abnormal genes using up cell material for their own purposes; certain types of virus, replacing normal genes and perverting rather than destroying body cells, are most probably the cause of some types of cancer [Allen]. Continual irritation by hot, though not burning, objects has been given as a cause for other types. Radiation by X-ray, radioactive substances, or cosmic rays, is a known cause for yet other cancers [Epstein].

No cause at all can as yet be plausibly suggested for many cancers. Indeed, spontaneous change in genes and chromosomes is likely a factor in many types. Such changes are like a warped form of the mutations that produce ordinary variants in living bodies.

But with or without external cause, genetic change seems to be the common basis of all neoplasms.

Behavior of Tumors

The vast variety of neoplasms results in a wide range of virulence. A few are quite harmless and can be tolerated almost indefinitely without danger. Others

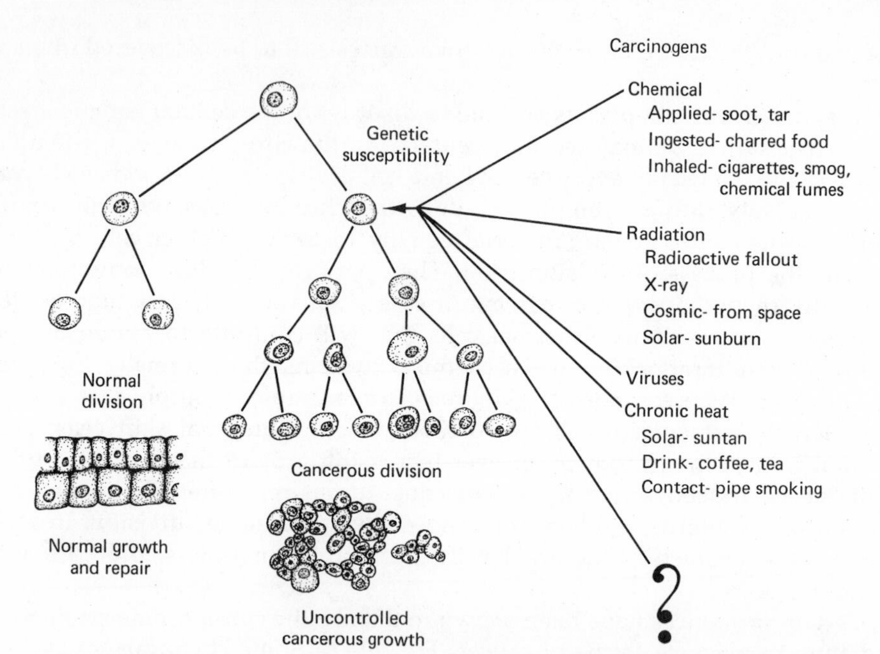

FIG. 12–2 The causes of cancer. The basic cause of cancers is disorder of the genetic code in the chromosomes. This disorder may be due to an inherited or mutational susceptibility; usually this susceptibility does not express itself unless triggered by environmental forces. Nonsusceptible cells, too, may become disordered if subjected to sufficient environmental insult.

Many chemicals, experimental and industrial, charred food, fumes, and atmospheric pollution, and so on, are known to cause cancer. Viruses are closely related to disordered genes, and almost certainly cause some forms of cancer. Radiation of many kinds causes mutations, some of which are cancerous, though some radiations, in controlled and directed dosage, may kill cancer cells. Chronic exposure to heat is linked to forms of cancer: People constantly exposed to sunlight, such as sailors and farmers, are prone to skin cancer; habitual use of scalding drinks has been linked to cancer of the esophagus; and pipe stems of materials and designs that allow overheating are often associated with cancer of the lip.

However, a certain percentage of cancers cannot be traced to any definite cause. Many of these are probably due to heredity or spontaneous mutation.

are so dangerous that they can cause death within months or even weeks of their symptoms. Most are spread along the whole spectrum of virulence between these extremes. They are more or less benign or malignant.

The word "benign" means kindly, mild, harmless. Technically, a benign tumor is one that grows in a single place and does not disperse or ramify diffusely. In a typical case it is compact and well outlined and so can be removed wholly and more or less easily by surgical methods. However, these features of a "benign" tumor must often be qualified.

In any case benign tumors are benign in only a relative sense. No unrestrained, abnormal growth can be entirely harmless in the body. Sometimes patients come for medical help with benign growths as big as grapefruit which must have been progressing for many years and causing increased disfigurement, discomfort, and disability. At last the patient begins to feel graver effects and rouses from his negligence to seek help. Such neglect seems incredible.

Technically, benign neoplasms may compress blood vessels, nerves, digestive tract, and so on, to the great detriment or even death of their hosts. Then too, they may adhere to or envelop vital structures without actually invading them, and so become difficult or impossible to remove wholly, in which case they will grow again from remnants. Such threats can be judged only by a trained expert.

So-called brain tumors are especially dangerous even when benign (Fig. 12–3). These neoplasms are better called "intracranial tumors" because many of them arise not from the brain itself but from meninges and other accessory tissues, and the term "intracranial" points up an important fact. The cranial

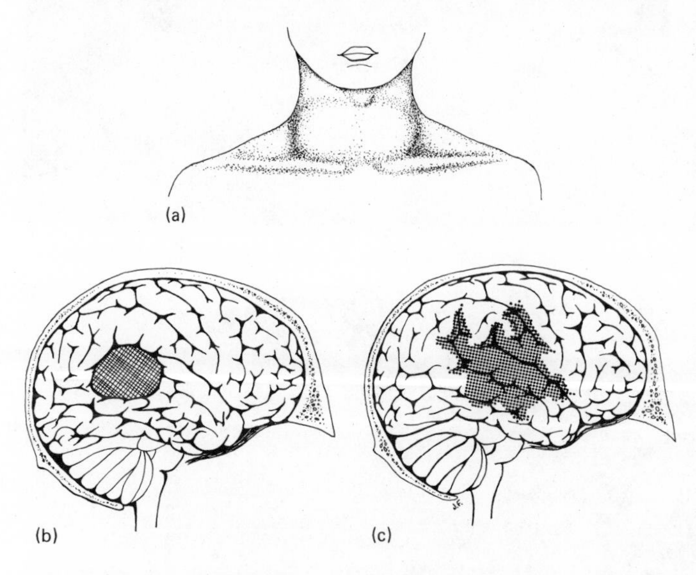

FIG. 12–3 The effects of tumor masses. Tumors are divided into two groups: benign, those that simply grow as a compact mass; and malignant, those that throw off colonies, metastases, to lodge in other parts of the body and/or that invade their surroundings diffusely and, so, are extremely difficult to remove entirely.

However, very few tumors are harmless if allowed to grow unrestricted. The "benign" tumor in (a) will progressively obstruct swallowing and breathing, reduce circulation to the head and brain, and even impinge on the nerves that pass through the neck. The tumor in (b), although easy to remove entirely by surgery, can grow only at the expense of other structures in the tightly closed cranial cavity, especially the brain, and, so, will ultimately be fatal unless removed. The malignant tumor in (c) has the deadly properties of that in (b) and, in addition, is almost impossible to remove entirely. Any remaining fragment will renew the growth.

The only wise treatment for any tumor is early recognition and prompt removal by surgery or destruction by other means.

FIG. 12–4 Stages in the development of bronchial cancer. In (a) normal epithelium is shown. The free edge is composed of the hairlike cilia that move mucus upward to the pharynx. The cilia are parts of the tall columnar cells. Smaller basal cells are arranged along the base of the columnar cells. Deep to the epithelium is seen connective tissue and blood vessels. In (b) the basal cells have undergone enormous multiplication to form a thick layer of cells resembling skin. In (c) the basal cells have become malignant and have invaded the underlying connective tissue. (Dr. Oscar Auerbach)

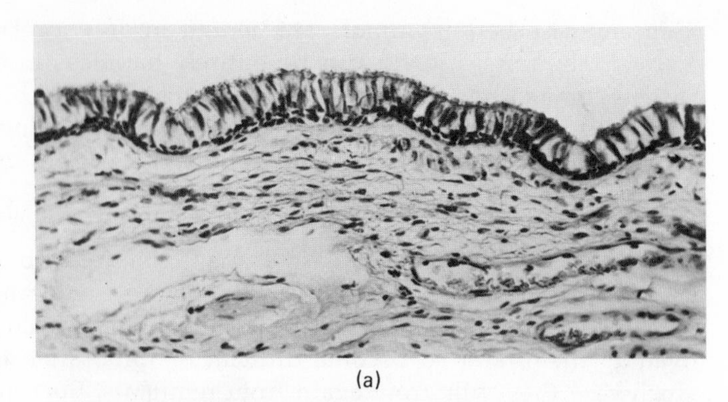

(a)

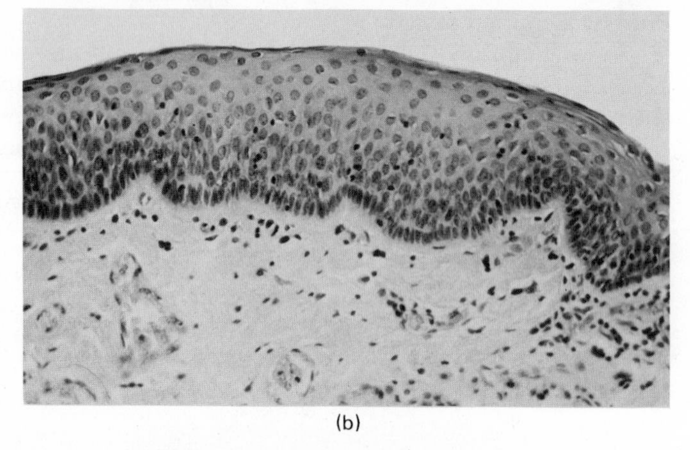

(b)

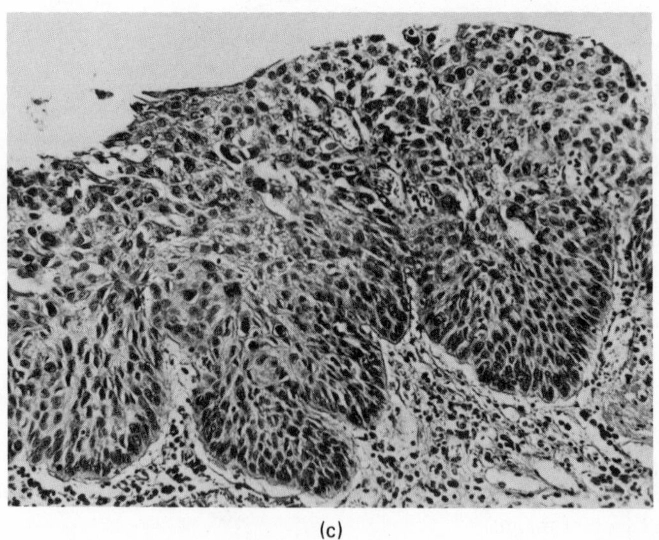

(c)

cavity, in which the brain lies, is a tightly closed space, so that any expanding disorder—hemorrhage, abscess, or tumor—grows at the expense of brain tissue with disastrous results. Thus no intracranial tumor is benign in the sense of harmlessness; all are very harmful, and even fatal if untreated.

Finally, benign tumors may turn malignant without warning. Thus a mass that has been tolerated for years, or even dismissed by a physician as benign, may suddenly change its nature. True, this transformation is not very common; but abnormal benign tissue does seem more prone than normal tissue to turn malignant. The risk of leaving a benign growth undisturbed, as balanced against the risk of operation can, again, be wisely judged only by an expert.

Malignancy itself, of course, is a very different matter. "Malignant" means evil, vicious, destructive; and a malignant neoplasm is always just exactly that. Such a neoplasm is what we properly call a cancer, carcinoma, or sarcoma. Malignancy expresses itself in various ways.

The growth may be invasive. That is, it may not remain in one place but may spread diffusely through its surroundings. In fact, *cancer* comes from the Latin word for crab, because of a fancied resemblance between such spreading neoplasms and a crab with its sprawling legs. The invasive growth may penetrate and mingle with healthy tissue and destroy it (Fig. 12–4). Such growth finally becomes too widespread to be removed except by also removing so much normal material that the patient would die. Thus an invasive type of cancer must be detected and removed *before* it spreads beyond control.

The Lymphatic System and Cancer

The other aspect of malignancy is *metastasis* or "seeding." It entails breaking of bits of the tumor, sometimes just a few cells, and their migration to other parts of the body (Fig. 12–5). When a localized cancerous growth metastasizes it is most often by way of the lymphatic circulation.

Solid cancerous growths, which are what we are concerned with here in contrast to cancers of the blood, differ from normal tissue in several ways. Our concern here is with their tendency to lose the organizational pattern of the parent tissue. For example, pancreatic cells may become cancerous, and in doing so retain many characteristics of normal pancreatic cells, at least to the point that they are usually recognizable as pancreatic cells, yet fail to become organized into the normal grapelike clusters of secretory cells and tubes which are typical of functional cells. Instead they divide rapidly and at random, in a quite chaotic pattern, with little or no adhesion with their neighbors. Such malignant cells break away from the normal tissue, continue their rapid division, and appear in irregular clumps scattered among the lymphatic capillaries. It is natural to expect that many isolated cancerous cells would be picked up by the lymphatic capillaries in the same manner the capillaries pick up extracellular debris such as bacteria or dead white cells (see Chapter 21).

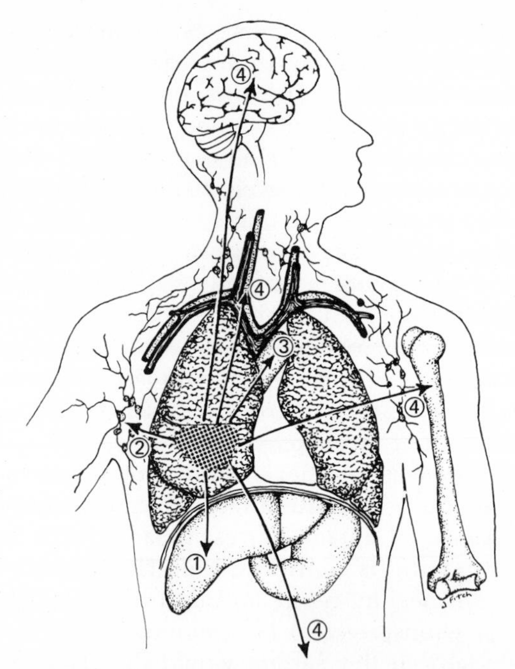

FIG. 12–5 How cancer spreads. Here a cancer of the lung is spreading in four ways as seen in an actual case history. (1) By direct invasion: Any neighboring organ may be invaded by the growth. In this case it is the diaphragm and liver, but lung cancer can also invade the heart, trachea, or blood vessels. (2) By the lymphatic system: Fine lymphatic vessels form a web throughout the body and can conduct cancer cells (metastasis) anywhere. (From the lung, direct conduction to the throat, esophagus and upper abdomen can occur.) (3) By the veins: Veins draining the lung to the left atrium offer a pathway into the general circulation. (4) By arteries to the general circulation: Here, arteries carried metastasizing cells to the brain, bone, and abdominal viscera.

In this case the symptoms that caused the patient to see his doctor were brought on by the tumors in the brain and liver. Upon questioning he revealed that he had suffered from a chronic cough that began almost a year earlier. Examination and treatment at that time might have saved his life.

The cancerous cells are moved along to the lymph nodes where they are destroyed in large numbers by the phagocytic cells in the nodes. It is probable that many cancerous growths end here: If only a few such cells are formed and break cleanly away from the parent tissue to enter the lymphatic vessels, the cancer could be destroyed without the individual ever being aware of the condition. But if the cancer continues to grow, the lymph nodes draining the lymphatic vessels of the region become overwhelmed and the cancerous cells continue growth within the node, and the daughter cells are passed out the efferent lymphatic vessels to the next nodes in the chain.

Each set of lymph nodes, in order, becomes overwhelmed. Since all lymph fluid is finally returned to the blood by emptying into large veins in the neck, the cancer cells will enter the venous circulation. Once in the bloodstream the body's principal means of defense are lost. The malignant cells are carried to the heart and then into the pulmonary circulation. It is not surprising that the lungs are the most common site of metastasis—they contain the first capillary system through which the cancerous cells must pass. They become lodged in the capillaries, break through the walls as they continue to divide, and form new colonies among the air sacs and bronchial tubes. Many of these cancerous cells, of course, are returned to the heart through the pulmonary vein, and may be carried to other parts of the body.

Such is the pernicious consequence of metastatic cancer. With this realization we should little wonder that surgeons go to such exhaustive efforts to search out and remove all lymph nodes which drain a cancerous area. If just one cancer-laden node goes undiscovered, it can serve as the source of later metastasis.

The breast, for example, has a rich system of lymphatic channels which drain into nodes located in the axilla. The surgeon must not only cleanly remove the cancerous breast, he also must reflect the skin from the front of the shoulder and from the axilla, and meticulously search out and remove the nodes lying in the subcutaneous tissue and along the large arteries and veins. It is this part of the operation which requires such perseverance and skill; the removal of the breast itself is a much simpler matter.

Symptoms and Signs

The first necessity for checking cancer is an ability to recognize it in time. This would be true even if we had universally effective cures available. The necessity is even greater in our present state of helplessness against advanced cancer.

The public itself is largely responsible for such recognition. A yearly check-up is very important for allowing the physician to detect early symptoms; but even then, he must rely largely on intelligent reports and questions from his patient. Furthermore, the patient must be faithful in seeking the check-up. And finally, since a year is a long time and some forms of cancer can begin and progress disastrously between two checkups, the patient should be able to distinguish between really significant conditions and trifles exaggerated by fear or ignorance.

For the purposes of the widest possible public dissemination, the American Cancer Society has summarized the clues to early cancer detection as the "Seven Danger Signals of Cancer." These seven signs and/or symptoms have been italicized in our discussion.

The first significant fact is strangely a negative one: Cancer is *not* always, or even usually, painful during its early stages; it may progress beyond control before it begins to give pain. Thus the attitude, "It doesn't hurt, so it can't be serious'' is one of dangerous ignorance. Of course pain, *along with* other symptoms, may be significant. But it is not the first thing to be expected with cancer.

Perhaps the clearest warning of possible cancer is a persistent growth of any sort. *Any lump anywhere that continues and enlarges* should be suspected, observed, and when its permanence is reasonably certain, taken promptly for expert examination. *Surface growths or sores that ulcerate or discharge and fail to heal* and *warts or moles that suddenly begin to enlarge or otherwise change* are particularly significant. Internal lumps are not so readily examined by a layman, but may be noted and monitored to see if they enlarge. In particular, a chronically enlarging, hardening, or lumpy organ in the abdomen is suspicious. Many sensible women choose a definite, and not easily forgotten, time every month to examine their breasts for lumps, the breasts being a common and dangerous site of cancer in the female (Fig. 12–6). A neoplasm *is* a ''growth'' and often a feelable mass.

This does not mean that one should become hypochondriac or overly anxious about ''lumps.'' Lumps of many sorts occur in the human body, most of them relatively harmless cysts, scars, inflammations from bacteria, and so on. Sometimes these other forms are worth reporting on their own account; but they are not cancers. The chance of even a persistent lump being a neoplasm is perhaps one in ten; the chance of a neoplasm being malignant is perhaps one in five. Thus a constant apprehension about any and every mildly suspicious knob is unwarranted, and may even become a form of neurosis. Careful attention to any lump out of the ordinary is something else.

Chronically swollen lymph nodes may also be the cause of suspicious lumps. These little organs, often incorrectly called ''lymph glands,'' are widely scattered throughout the body (Chapter 21). In most places, they are not accessible even to the expert except by surgery; but those in the armpit, groin, and neck can be felt by anyone. Normally, they are not easy to find except by trained fingers; but they swell and grow hard, and are quite easy to feel in various conditions, especially infections. Thus swollen nodes are no more a certain indication of cancer than are lumps in general. They can be ruled out of this role by the presence of inflammation, infected wounds, and such, in the limb or adjacent areas. If they arise and persist without infection, persist long beyond infection, or especially, if they coincide with nonhealing lumps and ulcers in the area, they are more suspect.

A *persistent cough or hoarseness* can be indicative of cancer of the larynx, trachea, bronchi, or lungs, especially among cigarette smokers. The pathological changes cigarette smoking imposes on the respiratory passages and lungs, and their relation to cancer are discussed in Chapter 28. It is sufficient to say here that these changes would be evident to the patient only through the distresses they cause him: shortness of breath, excessive or persistent coughing, and changes in the voice.

Among the most dangerous of cancer signals is *unusual bleeding or discharge from any body opening.* This is easy to detect in the mouth or nose: vomiting or

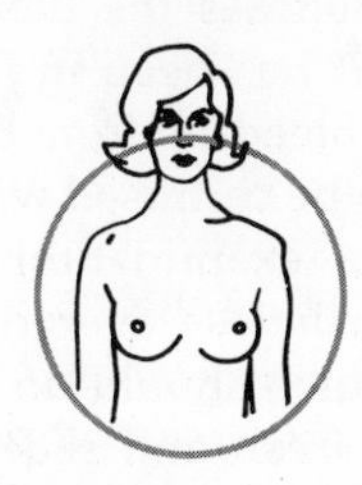

1. Sit or stand in front of your mirror, arms relaxed at your sides, and look for any changes in size, shape and contour. Also look for puckering or dimpling of the skin and changes on the surface of the nipples. Gently press each nipple to see if any discharge occurs.

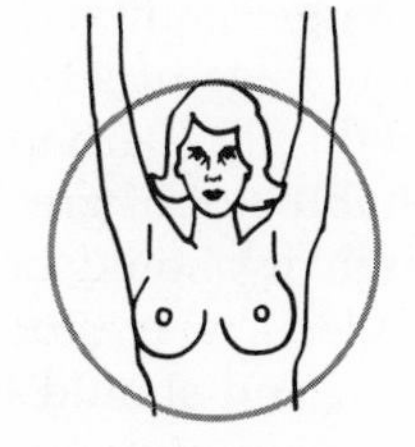

2. Raise both arms over your head, and look for exactly the same things. Note differences since you last examined your breasts.

3. From here on you will be trying to find a lump or thickening. Lie down on your bed, put a pillow or a bath towel under your left shoulder, and your left hand under your head. With the fingers of your right hand held together flat, press gently against the breast with small circular motions to feel the inner, upper portion of your left breast, starting at your breastbone and going outward toward the nipple line. Also feel the area around the nipple.

4. With the same gentle pressure, feel the low inner part of your breast. Incidentally, in this area you will feel a ridge of firm tissue. Don't be alarmed. This is normal.

5. Now bring your left arm down to your side and, still using the flat part of the fingers of your right hand, feel under your left armpit.

6. Use the same gentle pressure to feel the upper, outer portion of your left breast from the nipple line to where your arm is resting.

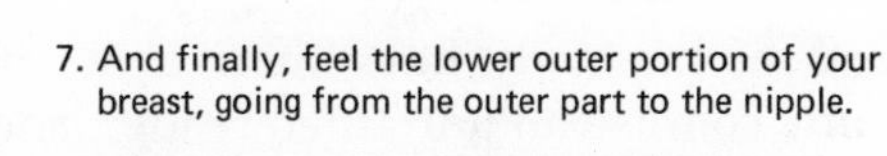

7. And finally, feel the lower outer portion of your breast, going from the outer part to the nipple.

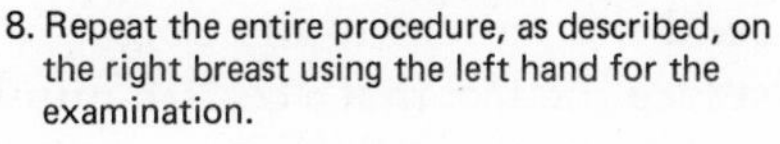

8. Repeat the entire procedure, as described, on the right breast using the left hand for the examination.

Your own doctor may want you to use a slightly different method of examination. Ask him to teach you that method.

Examine your breasts every month, about one week after each menstrual period. Be sure to continue these checkups after your time of menopause.

If you find a lump or thickening, leave it alone until you see your doctor. Don't be frightened. Most breast lumps or changes are not cancer, but only your doctor can tell.

FIG. 12–6 Breast self-examination (American Cancer Society)

coughing up blood is sure to draw attention to itself. Sometimes the blood looks unfamiliar, for example, rather like coffee-grounds if it has been in the stomach for a time, and such forms should arouse suspicion. Bleeding from the rectum or urogenital tract is more likely to escape notice or to be confused with menstrual discharges. Further, most people feel distaste for examination of discharges and incline to leave this for specialists to do. Nevertheless, everyone past the age of 40 and anyone at any age with other symptoms should do so periodically, perhaps once a month. Blood in the feces, if fresh and red, is probably not of great concern and is likely due only to bleeding hemorrhoids, though it should be reported if it continues; clotted blood turning black in so-called "tarry stools" is of graver import, though by no means a proof of cancer, and should be reported at once. Irregular menstrual flow, especially if continuous or frequent, is also suspicious. Blood in the urine usually escapes ordinary inspection unless it is so copious as to produce an odd color; this latter should certainly be reported if it continues. In all these, and other rarer conditions, bleeding is even more significant if combined with obstruction or palpable lumps.

Obstruction is the most serious symptom of neoplasm whether benign or malignant. *Difficulty in swallowing or severe prolonged constipation or indigestion* should never be left "to get better" for long, and most especially not if they are accompanied by bleeding. *Any persistent changes in bowel or bladder habits* are noteworthy, particularly since they may be related to cancers of the colon or prostate gland—two rather common but curable malignancies.

Symptoms of intracranial tumors are peculiar and very important. As noted above, *any* expanding growth in the cranial cavity is dangerous. The first indication of such a growth is most often a convulsion like those of epilepsy. Epileptic convulsions seldom begin in later life, so that such an event in an older person strongly suggests tumor. Headaches of increasing violence and continuity are also common. Nausea, vomiting, giddiness, and other symptoms follow. All these symptoms may be due to other causes; but the decision should be left to a qualified specialist as soon as possible.

Tumors in other parts of the body have their own special signs. But a catalog of these would only confuse an untrained reader, and most of them are sufficiently disturbing to send the patient to his doctor in time.

To be realistic, one must face the fact that a certain number of cancers do not declare themselves to simple observation till they are well advanced. Even many of these can be detected by a thorough physical examination, which emphasizes the need for the annual check-up. Many special tests are available, some highly reliable and some increasingly helpful. For example, a simple microscopic study of a vaginal smear can detect the otherwise insidious beginnings of all-too-common uterine cancers (Chapter 5). Indeed, this test could and should be routine for all adult women. In such ways, "undetectable" cancers can be brought to light with increasing frequency.

Treatment of Tumors

Many people still believe that tumors are essentially incurable. This belief is handed down from earlier times when it was practically true, but it is far from true today. Even without a "cancer cure" along the lines of the antibiotics for infection, the majority of tumor patients today could be cured by various means *if their condition were detected and treated in time.*

If a tumor has been detected, by the symptoms described in the last section or otherwise, what can be done about it? In general, and in descending order of importance today, there are three things that can be done: surgery, irradiation, and chemical (drug) therapy. Sometimes two or all three of these are combined.

Surgery is by far the most common, and still the most effective treatment for a majority of tumor types. If a tumor is removed, *all* removed, it is finally and completely cured. Even incomplete removal is very effective in many benign, slow-growing forms, and gives relief for many years and is renewable for extension of the relief. The obstacles to complete removal have already been indicated: benign adhesion to or entanglement with vital structures, malignant invasion of vital structures, and metastases. All these obstacles are due to delay in detection, and to growth because of extended delay.

Irradiation is growing in effectiveness and frequency of use (Fig. 12–7). It is based on the fact that some tumor cells are more easily killed by certain radiations than are normal cells; so that just the right dose kills the one but not the other. Many ingenious and powerful methods of applying radiations have been devised and are being added to yearly, and precision of technique is steadily increasing. The value of irradiation, however, varies according to the type of tumor, being great for some but negligible for others. Sometimes this therapy can be used to slow down growth of a tumor so that this does not become serious for a long time or even during a normal lifetime. Often when surgery cannot deal with parts of a tumor, irradiation will destroy or slow the growth of the remnants. Undoubtedly great further advances are possible in irradiation therapy for tumors.

Chemotherapy would be the ideal treatment for tumors. A drug that, like radiation, killed tumor cells and spared normal cells would solve the problem completely and finally. If carried by the bloodstream, it could seek out and kill abnormal cells wherever hidden. Thus it could deal even with invasions or metastases. Unfortunately, such therapy is still only a hope. There are drugs that are known to slow down, or even temporarily halt and reverse, the growth of some tumors. But the relief, so far, is almost always only temporary and is often at the cost of side-effects. In general, the drugs do affect tumor cells more strongly, but still affect normal cells strongly enough to prohibit prolonged use.

Nevertheless, everyone should promote drug therapy to the limit of his power. A life prolonged may be extended to the time when a real cure becomes available. More certainly, anyone taking drug therapy is helping the battle for

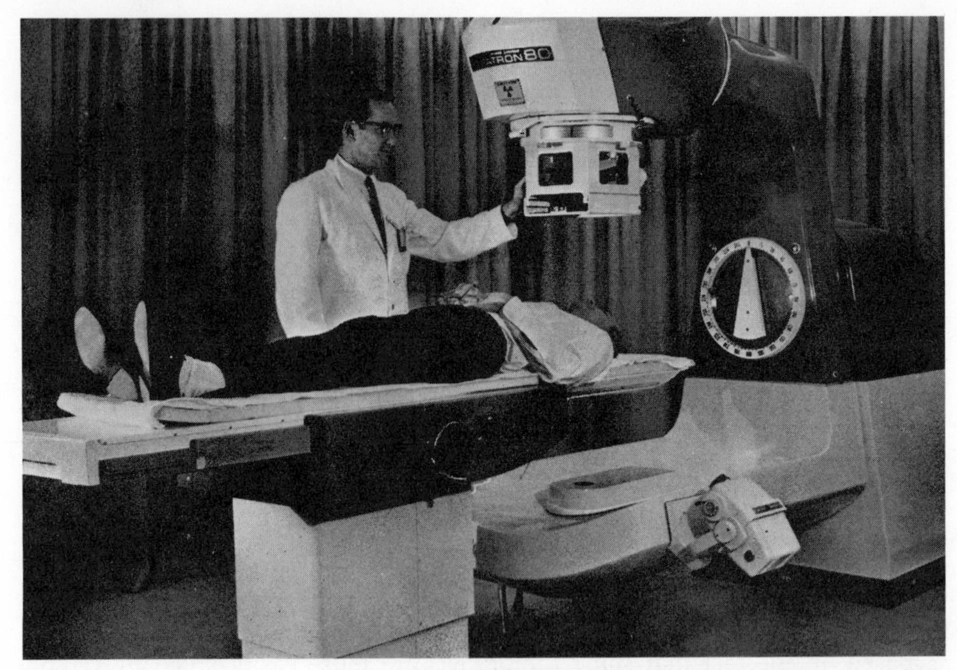

FIG. 12–7 Sixty percent of cancer patients are treated with radiation from machines similar to the one shown here (National Institute of Health)

others, since we cannot tell for sure the effects of a drug on man from the effects on laboratory animals. Hundreds and thousands of human subjects must be tested to discover how effective a drug is and to suggest how it might be improved. Such a life seems to us to be far better than an otherwise fatalistic surrender.

The Citizen and Cancer

The final conquest of cancer, as can be seen, must come from highly technical work. Indeed, this is the reason why "cancer cure" has lagged so far behind control of the major infectious diseases (Fig. 12–8). Hence, effective action may seem to be far beyond your competence and that of the man in the street.

You and that man, however, can do three highly effective things to reduce cancer disability and death. Two of these concern you and your circle, and the third may affect even the public domain of research and treatment.

First, you can avoid and persuade others to avoid known causes of cancer such as cigarette smoking and unnecessary X-rays or other radiation. A sound climate of public opinion and education could probably eliminate a large proportion of the annual deaths from cancer. Apathy, ignorance, and lack of self-control are responsible for those deaths.

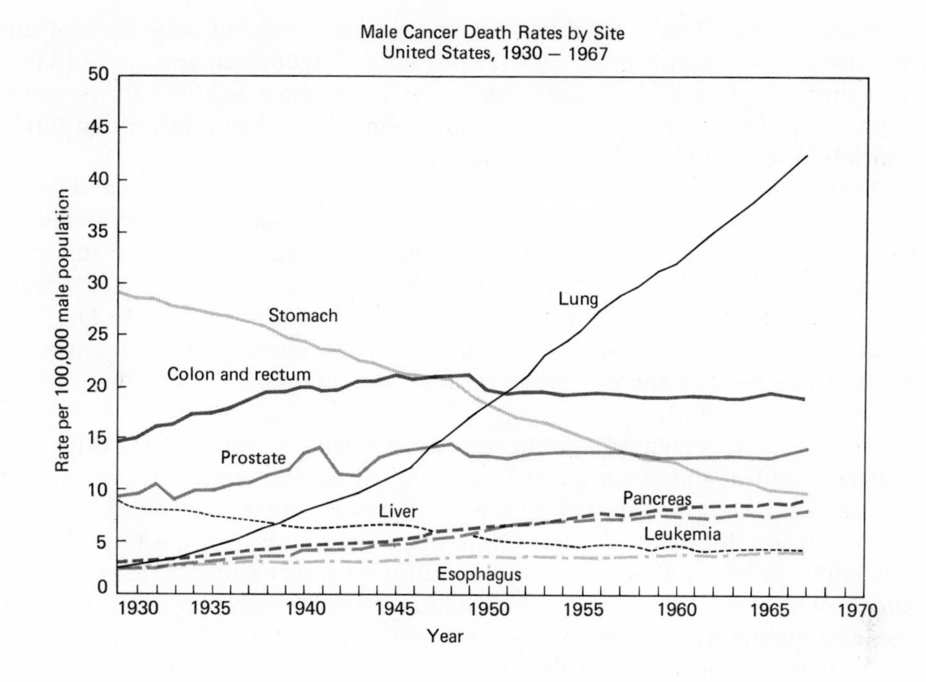

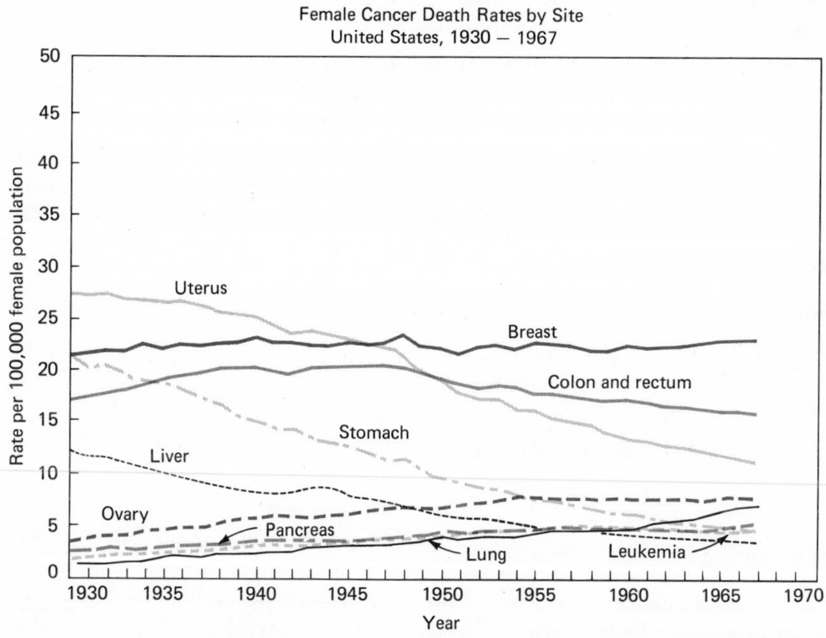

FIG. 12–8 Cancer death rates by site, 1930–1967. (Vital Statistics of the United States. U.S. Public Health Service)

Second, you should resolve, and encourage others, to take prompt and resolute action when tumors *are first suspected.* Anyone can and should know the symptoms, general and particular, of possible cancer as given above and as often presented by public-service advertisements. Yet possibly three out of four adults do not know these vital facts.

Even more deplorable is vacillation when symptoms *are* known and recognized. To a person in the healing professions, such delay is almost incredible in view of the penalties entailed. What could be more childish than the quite common attitude: "If I went to the doctor, he might find I had a cancer"? Then, too, many people have the "mañana" attitude of procrastination. In fact, they would rather risk infiltration or metastasis than inconvenience themselves. Yet any day may be the day when a tumor, easily operable until then, passes the point of no return!

Third, an informed lay public can exert effective influence in cancer research. It can abandon its attitude of indolent confidence in everything that claims to be scientific, and adopt one of critical impatience; it can discuss the issue with its doctors and other trained personnel to learn expert opinions; it can know as much about the administrators of funds and policies as it does about politicians, athletes, and entertainers, and equally question how well the administrators and researchers are performing; and it can be as vocal, through press, elected officials, and personal communication as it is about other issues of vital public concern. In these ways it might well stimulate progress to achieve success years earlier.

Finally, the refrain sounded throughout this chapter should be reemphasized in conclusion: Death and disability from cancer could be halved, or more, by simple public knowledge of symptoms and by readiness to act on those symptoms through the seeking and following of professional advice. Here is a matter where every individual must largely be his own guardian.

REVIEW QUESTIONS

1. How do benign and malignant tumors differ?
2. Discuss the causes of cancer.
3. What is meant by malignancy?
4. How are lymph nodes related to metastasis?
5. List the seven danger signals of cancer.
6. How is cancer treated?

REFERENCES

ALLEN, DAVID W., and PHILLIP COLE, "Viruses and human cancer," *New England Journal of Medicine,* **286:** 70–82 (1972).

AMERICAN CANCER SOCIETY, *Cancer Facts and Figures.* 1969.

EPSTEIN, SAMUEL, "Chemical hazards in the human environment," *Ca—A Cancer Journal for Clinicians,* **19:** 276–281 (September–October 1969).

MACMAHON, BRIAN, "Epidemiologic aspects of cancer," *Ca—A Cancer Journal for Clinicians,* **19:** 27–35 (January–February 1969).

RYSER, HUGUES J-P., "Chemical carcinogenesis," *New England Journal of Medicine,* **285:** 721–734 (1971).

Additional Readings

ACKERMAN, LAUREN V., "Some thoughts on food and cancer," *Nutrition Today,* **7:** 2–8 (1972).

BENSON, RALPH C., "Cancer of the female genital tract," *Ca—A Cancer Journal for Clinicians,* **18:** 2–12 (January–February 1968).

MCGRADY, PAT, *The Savage Cell.* New York: Macmillan, 1967.

MOORE, FRANCIS D. el al., "Carcinoma of the breast," *New England Journal of Medicine,* **277:** 293–296; 343–349; 411–416; and 460–468 (1967).

WHITE, RAYMOND L., "A life with cancer: A physician's personal experiences," *Ca—A Cancer Journal for Clinicians,*185–187 (May–June 1969).

PART FOUR
NUTRITION AND METABOLISM

13
THE NATURE OF FOODS

It is not especially difficult to determine the chemicals forming the human body. In fact, the basic elements are well known. But such information tells us little about our food requirements since only a few of the chemical elements are of use to us in their uncombined form, and none of them provide us with energy. Whence, then, comes the chemical energy used to operate our bodies?

The story can be briefly summarized.

Calories

The energy content of food is measured in terms of the amount of heat which can be produced when it is burned. The calorie is a measure of heat. One kilocalorie is the amount of heat required to raise the temperature of 1 kilogram of water 1° C (this is approximately the amount of heat given off by a medium size candle in about 30 seconds). In nutrition, we usually use the capitalized form *Calorie* to mean a kilocalorie which is equal to 1000 calories.

The Food and Nutrition Board of the National

Research Council states that a man between the ages of 25 and 40, weighing 154 pounds, 5-feet 9-inches tall, and engaged in moderate activity, requires about 2900 kilocalories daily; a woman between the ages of 25 and 45, weighing 128 pounds and 5-feet 4-inches tall requires about 2100 kilocalories daily. These recommendations can serve as the basis for a quick rule-of-thumb method of calculating very approximate individual caloric requirements:

Men:	Moderate activity	$\frac{2900}{154}$	=	19 kilocalories per pound
	Heavy work	3542	=	23 kilocalories per pound
	Sedentary activity	2156	=	14 kilocalories per pound
Women:	Moderate activity	$\frac{2100}{128}$	=	16 kilocalories per pound
	Heavy work	2560	=	20 kilocalories per pound
	Sedentary activity	1536	=	12 kilocalories per pound

To find your approximate caloric requirement simply determine your desirable weight (see Table 13–1) and multiply by the kilocalories per pound as determined above. For example, a woman age 20, 5 feet 5 inches tall and with a medium frame should weigh about 126 pounds. If she is engaged in moderate activity her total caloric requirement would be $126 \times 16 = 2016$ kilocalories per day. A man 6 feet 2 inches tall with a heavy frame who does heavy work would require 4209 kilocalories (183×23).

TABLE 13–1 Desirable Weights for Height

Height[a] *in Inches*	*Weight in Pounds*[a] *Men*	*Women*
60		109 ± 9[b]
62		115 ± 9
64	133 ± 11	122 ± 10
66	142 ± 12	129 ± 10
68	151 ± 14	136 ± 10
70	159 ± 14	144 ± 11
72	167 ± 15	152 ± 12
74	175 ± 15	

[a] Heights and weights are without shoes and other clothing.

[b] Desirable weight for a small-framed woman at this height would be approximately 109 pounds minus 9 pounds, or a total of 100 pounds; for an average-framed woman, 109 pounds; for a large-framed woman 109 pounds plus 9 pounds, or a total of 118 pounds.

SOURCE: Food and Nutrition Board, National Research Council.

One might wonder why women require fewer calories per pound than do men. The reason can probably be attributed to the fact that a woman of desirable weight has a higher proportion of fat to total body weight than does her male counterpart. And fat is largely inert in calorie-burning activity.

Note that these caloric requirements assume at least some degree of physical activity. However, a certain number of calories are expended even at complete rest in order to maintain what are called vegetative functions—activities such as breathing, circulation of blood, and maintenance of body temperature. This minimal level of energy expenditure is termed the *basal metabolic rate* or BMR and represents the amount of heat generated by the body surface under so-called basal conditions: (1) lying comfortably with eyes closed, (2) 12 hours after eating, (3) after a restful night's sleep, (4) after 30 minutes of inactivity, and (5) in a comfortable temperature. The heat generated results from the oxidation of food, and is most conveniently measured by the rate of oxygen consumption. The oxygen consumption, in turn, can be converted to kilocalories expended per hour. For example, the basal rate of energy expenditure for a 25 year old 154 pound male is about 55 kilocalories per hour, or approximately 1320 kilocalories per day. This means that a moderate level of physical activity in a 25-year-old 154 pound male increases the daily caloric requirement by about 1580 kilocalories per day (2900 − 1320). The basal rate of metabolism for the 25-year-old 128 pound female is about 36 kilocalories per hour, or about 864 kilocalories per day; moderate activity requires 1236 kilocalories per day (2100 − 864). When a clinical laboratory determination is made of the basal metabolic rate, it is usually reported as a plus or minus percent. For example, +25 percent means a BMR 25 percent higher than the normal average rate for a person of the same sex, age, and weight. However, in physiology the BMR is more conveniently expressed in kilocalories liberated per hour per square meter of body surface or, more simply, as total kilocalories liberated per hour.

Our basal rate of energy expenditure declines with age, so that a male 5 feet 9 inches tall and weighing 154 pounds at age 30 expends about 48 kilocalories per day less than at age 20; at age 40, about 120 kilocalories per day less; at age 50, about 192 kilocalories per day less; and at age 60, about 240 kilocalories per day less. Thus a person with a constant level of caloric intake would tend to store calories (put on fat) at a constant level of physical activity (Chapter 17). *Clinically,* however, the BMR would remain the same *if* it is reported as a percentage above or below normal for a given age, sex, and weight. When expressed as kilocalories liberated per hour it would be seen to decline with age.

The BMR is usually increased in hyperthyroidism (Chapter 32) and in fever; it is usually depressed in hypothyroidism, and in people with poorly developed muscles and consequent lack of stamina.

In the American diet about 45 percent of the energy is derived from carbohydrates, 40 percent from fats, and about 15 percent from proteins. This is typical of an affluent society since fat and protein-containing foods are generally a good deal more expensive. In poorer areas of the world the fats and proteins of the diet fall to as little as half these values, and the carbohydrates provide approximately 80 percent of the energy. It is obviously much less expensive to eat cereal grains directly rather than feed them to animals and eat the animals. But it is also duller, and all peoples eat foods rich in fat and protein when these are available.

Energy Content of Foods

The amount of energy provided to the body by 1 gram of carbohydrate is 4.1 kilocalories; 1 gram of protein also provides 4.1 kilocalories; and 1 gram of fat provides 9.3 kilocalories. Thus fats, by weight, have over twice the caloric value of either carbohydrates or proteins. And although a person thinks that his fat intake is small in any given meal, it can be deceptively high, because, in addition to their high caloric value, fats are usually eaten in pure form while proteins and carbohydrates are generally diluted with water or indigestible substances. For example, in a serving of potatoes and gravy, the gravy will probably contain more calories than the potatoes. The starch in the potatoes contains less than one-half the calories of fat per ounce or gram, and the potato is only about one-sixth starch, the rest being water. The gravy is, by weight, largely fat. This fact should make dieters reflect upon their eating habits.

Carbohydrates

Carbohydrates are organic foods containing carbon, hydrogen, and oxygen with the hydrogen and oxygen always occurring in the same ratio as in water, that is, two hydrogen atoms to one oxygen atom.

Forms of Carbohydrates

The simplest form of a carbohydrate is termed a *monosaccharide* or simple sugar, and a familiar example is glucose (grape sugar or dextrose). All the more complex carbohydrates can be broken down to monosaccharides, and for this reason the monosaccharide is referred to as the basic unit of carbohydrates.

A carbohydrate composed of two monosaccharide units is called a *disaccharide* or double sugar. Examples of common disaccharides are *sucrose* or table sugar and *lactose*, the sugar of milk. Lactose is almost tasteless, so that milk is not sweet.

When many monosaccharide units are combined together a *polysaccharide* results. Most polysaccharides are high molecular-weight molecules with more than 100 monosaccharide units. Probably the most common polysaccharide is *cellulose*, which forms the tough framework of plants and from which cotton and linen are made. The other common polysaccharide of plants is *starch*. And although both cellulose and starch are made up entirely of glucose units, one (starch) is entirely digestible by humans, while the other (cellulose) is not

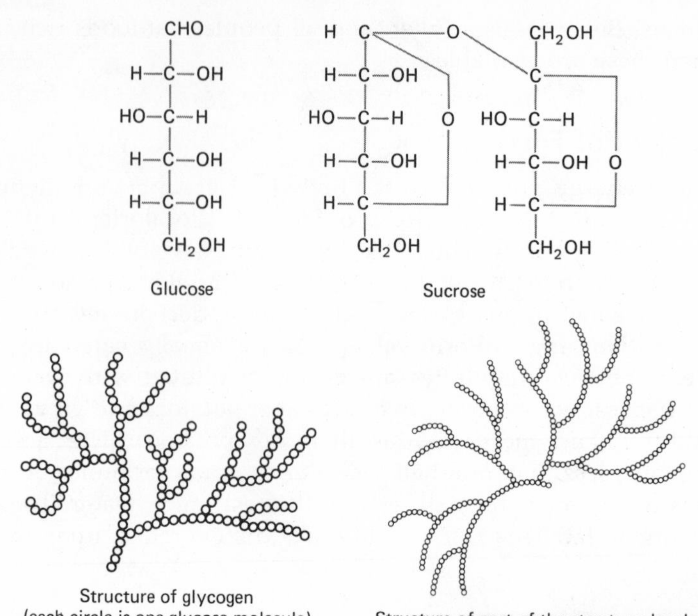

FIG. 13–1 The chemical structures of glucose, sucrose, and starch.

(Fig. 13–1). This is the reason that the potato, with its high starch content, is food, while grass is not. (Incidentally, cattle are able to digest the cellulose of grass only because of the presence in the gut of certain bacteria which carry on the digestion for them; certain protozoa serve the same function in the intestine of the termite.) The principal polysaccharide of animals is *glycogen* or "animal starch". The storage form of glucose, glycogen, is found in large amounts in the liver and in muscles. But in man not more than 1 percent of the total body composition, including the glucose in the blood, is composed of carbohydrate.

The Importance of Carbohydrates in the Diet

The digestion or breakdown of carbohydrates reduces polysaccharides and disaccharides to monosaccharides, and it is only in this simple state that they are absorbed. The sugars are picked up by the blood as it courses through the intestinal wall and delivered directly to the liver. The level of sugar in the circulating blood will rise after a meal. Nevertheless the greater part of the sugar is picked up by the liver which converts it into glycogen (Fig. 13–2). The liver then proceeds to reconvert the glycogen back to glucose as needed, and to dole it out to the blood. The liver maintains the level of glucose in the blood simply by drawing upon its stores of glycogen. Although sugars other than glucose are metabolized by the liver and fed into the blood, glucose is by far the most important and of most general significance.

Since glucose is the body's principal source of energy it is constantly withdrawn from the blood by all the cells of the body. Our muscles, which account for most of our energy expenditure, remove the glucose from the blood and build up their own storehouse of glycogen. However, this glycogen is reserved for the muscle in which it is stored, and muscle is not a general storage depot for glucose as is the liver.

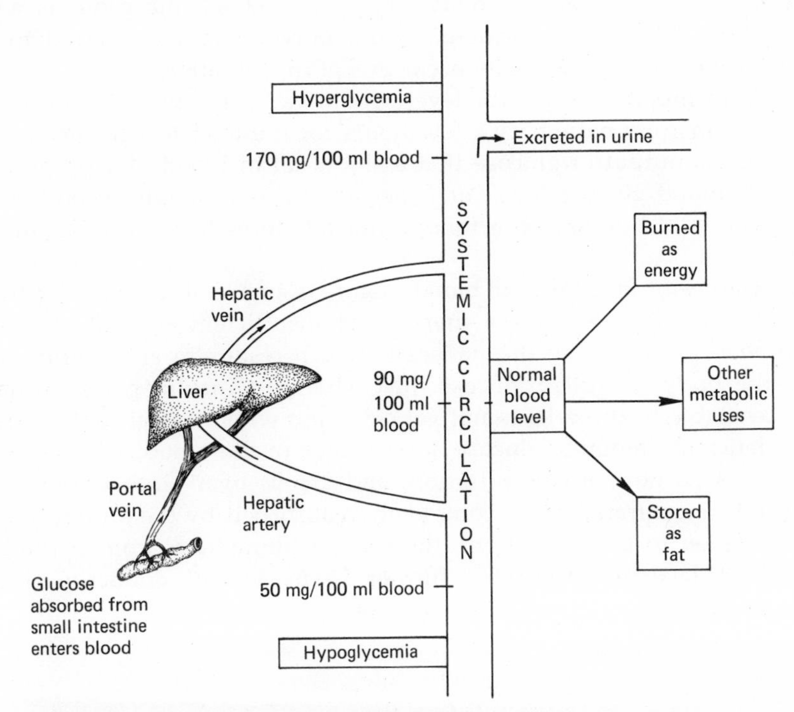

FIG. 13–2 The absorption and fate of glucose. Glucose is the principal breakdown product of carbohydrates of the food. After absorption into the blood, glucose is carried to the liver by way of the portal vein. The liver manufactures glycogen from much of this glucose, but high intake of sugar or metabolic or endocrine disturbances, such as diabetes mellitus, can bring âbout abnormally high levels of glucose leaving the liver in the hepatic vein. If the amount of glucose exceeds 170 milligrams per 100 milliliters of blood (hyperglycemia), this excess will be lost in the urine. Normally, the liver will alternate between the chemical conversions glycogen → glucose, and glucose → glycogen in coordination with energy requirements and metabolic needs to maintain a relative constancy of glucose in the blood. When the liver has stored its maximum amount of glycogen, the excess glucose is removed from the blood and converted to fat. Hypoglycemia occurs when glucose falls below about 50 milligrams per 100 milliliters of blood. This could come about as a result of an excessive intake of insulin, just as hyperglycemia can result from a deficiency of insulin (see Chapter 32).

Since the liver contains only enough glycogen to maintain blood glucose levels for about 24 hours, liver glycogen should be regarded as only a temporary energy source. When these glycogen stores begin to be depleted, the body draws more and more upon fat for energy. Fat molecules are carried to the liver and converted to glycogen to help maintain a sufficient level of glucose in the blood. Fats, broken down to glycerol and fatty acids, are also used directly for energy. When the body's fat is exhausted, the proteins will be removed from the tissues, principally the muscles, and converted to liver glycogen. This does not normally occur except in starvation.

All this to maintain a proper level of glucose in the blood! And we only have to refrain from eating for a few hours for the liver to become so parsimonious in its output of glucose that a slight fall in blood glucose develops. This fall in blood glucose level, or *hypoglycemia,* is probably responsible for the beginnings of hunger, even though our fat stores have only begun to be tapped.

Hyperglycemia, an excess of blood sugar, is one of the characteristics of *diabetes mellitus.* In this disease there is an insufficient amount of the hormone *insulin,* produced by the pancreas, which results in an inability of the cells of the body to utilize glucose. Just what part insulin plays in making glucose available to the cells is not yet clear, but when it is absent or present only in deficient amounts, glucose accumulates in the blood, may reach four or five times its normal concentration, and "spills over" into the urine (becoming too concentrated to be completely reabsorbed by the kidneys). When insulin is given to such a patient, there is an immediate drop in the blood glucose level. Diabetes mellitus is further discussed with endocrine diseases in Chapter 32.

Fats

Fats are composed of carbon, oxygen, and hydrogen, but in a different proportion from that of carbohydrates. When fats are broken down, they yield *fatty acids* and *glycerol;* and they are usually absorbed from the intestine in this form, though some partially digested fat is also absorbed.

Forms of Fat

Saturated fats are those whose fatty acids contain all the hydrogen atoms they can hold. If the fatty acids are able to take on more hydrogen atoms, the fat is said to be *unsaturated* (Fig. 13–3). Fat sources high in unsaturated fatty acids are fish oils, shellfish, corn oil, and cottonseed oil. Our principal dietary source of fats high in saturated fatty acids are pork, beef, and dairy products. The commercial process of hydrogenation converts unsaturated fats to saturated fats simply by the addition of hydrogen. We can recognize the difference

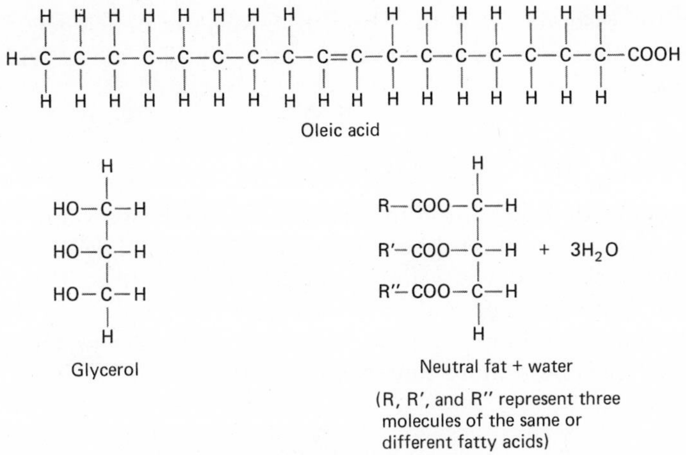

```
   H  H  H  H  H  H  H  H        H  H  H  H  H  H  H
   |  |  |  |  |  |  |  |        |  |  |  |  |  |  |
H—C—C—C—C—C—C—C—C—C=C—C—C—C—C—C—C—C—COOH
   |  |  |  |  |  |  |  |  |  |  |  |  |  |  |  |  |
   H  H  H  H  H  H  H  H  H  H  H  H  H  H  H  H  H
```

Oleic acid

```
       H                          H
       |                          |
  HO—C—H               R—COO—C—H
       |                          |
  HO—C—H               R'—COO—C—H   +   3H2O
       |                          |
  HO—C—H               R''—COO—C—H
       |                          |
       H                          H
```

Glycerol

Neutral fat + water

(R, R', and R'' represent three molecules of the same or different fatty acids)

FIG. 13–3 The chemical structure of oleic acid (a common fatty acid); glycerol; and of a neutral fat.

between the two by the general tendency of unsaturated fats to be liquid at room temperature, and their greater liability to become rancid. Much has been done in recent years to modify the latter characteristic, so that the unsaturated fats (cooking oils) are becoming more popular.

The Importance of Fats in the Diet

After absorption fats are utilized: (1) as a constituent of protoplasm; (2) as a product of certain glands such as the sebaceous glands; (3) as an energy source; (4) as stored body fat; (5) in very special structures such as nerve sheaths; and (6) by conversion to special substances such as cholesterol (which may also be taken in directly). The amount of fat required in the diet for the manufacture of protoplasm, and for the production of glandular secretions, is so modest that no American diet would be deficient. Some stored body fat is significant as insulation against cold (skinny Eskimos are hard to visualize), and the kidneys and other organs are protected from shock by the fat deposited around them. Again, modest requirements are needed for most of us, and the bulk of the fat in our diet is used immediately for energy or stored later for such use. Consequently, the amount of fat we require in our diet is dependent upon our energy needs and upon the amount of carbohydrate in the diet. Most nutritionists suggest that fat should provide, optimally, only about 25 percent of our energy foods instead of the 40 percent which is typical of the American diet.

No deficiency syndrome is known to occur in adult humans resulting from a lack of essential fatty acid (that is, those fatty acids which the body requires but cannot synthesize). However, it is generally assumed that unsaturated fatty acids are essential, and the absence of specific deficiency disorders is due to the fact that both unsaturated and saturated fats are so widespread in common foods.

Fat, Cholesterol, and Atherosclerosis

There is evidence that diets high in saturated fats give rise to high cholesterol levels in the blood. And it has been observed that these high cholesterol levels are frequently associated with atherosclerotic heart disease. Thus a relationship between saturated fats and atherosclerosis has been suggested. But the subject is a complicated one, and the evidence can be interpreted in different ways. The matter is more fully considered in Chapter 23.

Proteins

Proteins differ from the carbohydrates and fats by the presence of nitrogen in addition to carbon, oxygen, and hydrogen (Fig. 13–4). The basic unit of protein is the *amino acid,* and there are 20 different amino acids which occur in most animal proteins. These are linked together in long chains to form protein macro(super)-molecules. Since a typical protein has several hundred amino acids arranged in a strict order, the number of different kinds of proteins possible becomes almost infinite. Proteins are by far the most complex chemical substances known. They are the substance of life and are responsible for the unique characteristics of living things.

Even though animals are able to assemble a vast variety of proteins from a store of amino acids, they are unable to manufacture the amino acids themselves. However, if we have a certain ten of the amino acids we are able to build the other ten from them. For example, the liver is able to convert the amino acid phenylalanine into tyrosine, so that tyrosine is not required in the diet as long as phenylalanine is present. But since the body is unable to

FIG. 13–4 The chemical structure of four amino acids.

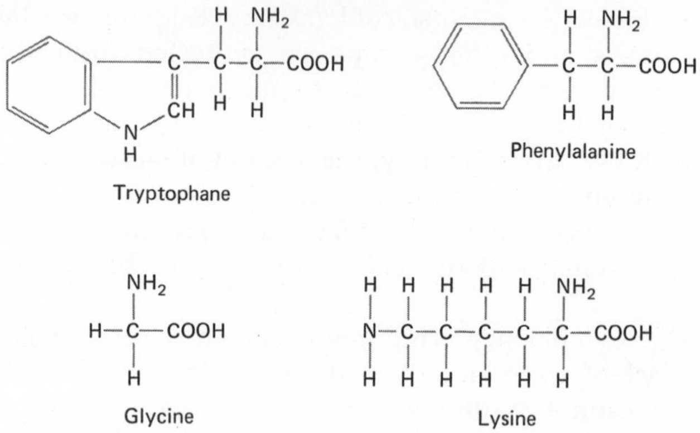

Tryptophane

Phenylalanine

Glycine

Lysine

manufacture phenylalanine from any other substance, this amino acid is one of the essential amino acids.

The digestion of proteins involves their breakdown into smaller units called *polypeptides,* then into two amino acid units called *dipeptides,* and finally into the amino acids themselves. However, proteins are not synthesized in such a step by step manner. Instead, all the amino acids of a given protein are assembled together in a single reaction. In other words, if any one amino acid is deficient, we do not get part of a protein, we get no protein at all. This emphasizes the importance of having "complete" proteins in the diet, that is, proteins which contain all 20 of the amino acids or at least the 10 essential ones. Gelatin is an incomplete protein and although it is a good accessory food, it would not support protein requirements by itself. Most plant proteins are incomplete although a few have the 10 essential amino acids. However, the common proteins from animal sources, for example, those found in eggs, milk, meat and fish, are complete.

Yet another problem in our protein metabolism is the fact that we do not store excess amino acids. Those amino acids in the blood in excess of our immediate protein needs are removed by the liver, and are processed partly into an energy source, and partly into the waste product, urea. For this reason the expensive, high protein breakfast cereals are of value only if we do not have otherwise adequate protein in the diet.

These three major ingredients of diet provide all our energy. Proteins and, to a lesser degree, fat and carbohydrate build and replace components of living tissue. But they do not provide certain small, but very essential "spare parts" for the body. For such needs, we must have other, small, but likewise essential, ingredients in our diet. These ingredients, the vitamins, we shall next discuss.

REVIEW QUESTIONS

1. What is a calorie?
2. How can you determine your caloric requirements?
3. What are some of the forms of carbohydrates in your food?
4. How is the level of sugar in your blood controlled?
5. Why is the possible number of different proteins so much greater than the possible number of polysaccharides or fats?

REFERENCES

U.S. Department of Agriculture, *Food—The Yearbook of Agriculture, 1959.*

The Heinz Handbook of Nutrition. New York: McGraw-Hill, 1959.

Wilson, E. D. et al., *Principles of Nutrition.* New York: John Wiley, 1959.

Wohl, Michael G., and Robert S. Goodhart, *Modern Nutrition in Health and Disease,* 4th ed. Philadelphia: Lea & Febiger, 1968.

Additional Readings

CAHILL, GEORGE F., JR., "Starvation in man," *New England Journal of Medicine,* **282:** 668–675 (1970).

DEUTSCH, RONALD M., *The Nuts among the Berries: An Exposé of America's Food Fads.* New York: Ballantine, 1961. (Paperback)

LEVERTON, RUTH M., *Food Becomes You.* Iowa State University Press, 1960.

MAY, JACQUES M., and HOYT LEMONS, "The ecology of malnutrition," *Journal of the American Medical Association,* **207:** 2401–2405 (1969).

MCHENRY, E. W., *Foods without Fads.* Philadelphia: J. B. Lippincott, 1960.

14
VITAMINS AND MINERALS

The Nature of Vitamins

Late in the last century the chemists finally developed methods for the purification of the carbohydrates, fats, and proteins found in foods. Yet when animals were fed these purified preparations, along with water and minerals, they promptly died. If small amounts of yeast and certain fats such as those found in butter and egg yolk were included, the animals (rats) remained healthy. There began the long and arduous task of purifying and identifying substances necessary for life, evidently hidden in the supplementary foods. These substances were later named vitamins.

A vitamin, then, is a food substance necessary to the body but in far smaller quantities than those of carbohydrates, fats, and proteins. These vitamins are chemicals that, like the essential amino acids, cannot be manufactured by the body itself. Today, we know about a dozen vitamins with maybe half a dozen more compounds claimed as vitamins by some investigators but not generally accepted as such.

At first, vitamins were named simply by letters

TABLE 14–1 Recommended Dietary Allowances

	Age[b] (years)	Weight		Height				Fat-Soluble Vitamins			Water-Soluble Vitamins							Minerals				
	From Up	to (kg)	(lb)	cm	(in.)	kcal	Protein (gm)	Vitamin A Activity (IU)	Vitamin D (IU)	Vitamin E Activity (IU)	Ascorbic Acid (mg)	Folacin[c] (mg)	Niacin (mg equiv)[d]	Riboflavin (mg)	Thiamin (mg)	Vitamin B_6 (mg)	Vitamin B_{12} (μg)	Calcium (g)	Phosphorus (g)	Iodine (μg)	Iron (mg)	Magnesium (mg)
Infants	0–1/6	4	9	55	22	kg × 120	kg × 2.2[e]	1,500	400	5	35	0.05	5	0.4	0.2	0.2	1.0	0.4	0.2	25	6	40
	1/6–1/2	7	15	63	25	kg × 110	kg × 2.0[e]	1,500	400	5	35	0.05	7	0.5	0.4	0.3	1.5	0.5	0.4	40	10	60
	1/2–1	9	20	72	28	kg × 100	kg × 1.8[e]	1,500	400	5	35	0.1	8	0.6	0.5	0.4	2.0	0.6	0.5	45	15	70
Children	1–2	12	26	81	32	1,100	25	2,000	400	10	40	0.1	8	0.6	0.6	0.5	2.0	0.7	0.7	55	15	100
	2–3	14	31	91	36	1,250	25	2,000	400	10	40	0.2	8	0.7	0.6	0.6	2.5	0.8	0.8	60	15	150
	3–4	16	35	100	39	1,400	30	2,500	400	10	40	0.2	9	0.8	0.7	0.7	3	0.8	0.8	70	10	200
	4–6	19	42	110	43	1,600	30	2,500	400	10	40	0.2	11	0.9	0.8	0.9	4	0.8	0.8	80	10	200
	6–8	23	51	121	48	2,000	35	3,500	400	15	40	0.2	13	1.1	1.0	1.0	4	0.9	0.9	100	10	250
	8–10	28	62	131	52	2,200	40	3,500	400	15	40	0.3	15	1.2	1.1	1.2	5	1.0	1.0	110	10	250
Males	10–12	35	77	140	55	2,500	45	4,500	400	20	40	0.4	17	1.3	1.3	1.4	5	1.2	1.2	125	10	300
	12–14	43	95	151	59	2,700	50	5,000	400	20	45	0.4	18	1.4	1.4	1.6	5	1.4	1.4	135	18	350
	14–18	59	130	170	67	3,000	60	5,000	400	25	55	0.4	20	1.5	1.5	1.8	5	1.4	1.4	150	18	400
	18–22	67	147	175	69	2,800	60	5,000	400	30	60	0.4	18	1.6	1.4	2.0	5	0.8	0.8	140	10	400
	22–35	70	154	175	69	2,800	65	5,000	–	30	60	0.4	18	1.7	1.4	2.0	5	0.8	0.8	140	10	350
	35–55	70	154	173	68	2,600	65	5,000	—	30	60	0.4	17	1.7	1.3	2.0	5	0.8	0.8	125	10	350
	55–75+	70	154	171	67	2,400	65	5,000	—	30	60	0.4	14	1.7	1.2	2.0	6	0.8	0.8	110	10	350
Females	10–12	35	77	142	56	2,250	50	4,500	400	20	40	0.4	15	1.3	1.1	1.4	5	1.2	1.2	110	18	300
	12–14	44	97	154	61	2,300	50	5,000	400	20	45	0.4	15	1.4	1.2	1.6	5	1.3	1.3	115	18	350
	14–16	52	114	157	62	2,400	55	5,000	400	25	50	0.4	16	1.4	1.2	1.8	5	1.3	1.3	120	18	350
	16–18	54	119	160	63	2,300	55	5,000	400	25	50	0.4	15	1.5	1.2	2.0	5	1.3	1.3	115	18	350
	18–22	58	128	163	64	2,000	55	5,000	400	25	55	0.4	13	1.5	1.0	2.0	5	0.8	0.8	100	18	350
	22–35	58	128	163	64	2,000	55	5,000	—	25	55	0.4	13	1.5	1.0	2.0	5	0.8	0.8	100	18	300
	35–55	58	128	160	63	1,850	55	5,000	—	25	55	0.4	13	1.5	1.0	2.0	5	0.8	0.8	90	18	300
	55–75+	58	128	157	62	1,700	55	5,000	—	25	55	0.4	13	1.5	1.0	2.0	6	0.8	0.8	80	10	300
Pregnancy						+200	65	6,000	400	30	60	0.8	15	1.8	+0.1	2.5	8	+0.4	+0.4	125	18	450
Lactation						+1,000	75	8,000	400	30	60	0.5	20	2.0	+0.5	2.5	6	+0.5	+0.5	150	18	450

[a] The allowance levels are intended to cover individual variations among most normal persons as they live in the United States under usual environmental stresses. The recommended allowances can be attained with a variety of common foods, providing other nutrients for which human requirements have been less well defined. See text for more-detailed discussion of allowances and of nutrients not tabulated.

[b] Entries on lines for age range 22–35 years represent the reference man and woman at age 22. All other entries represent allowances for the midpoint of the specified age range.

[c] The folacin allowances refer to dietary sources as determined by *Lactobacillus casei* assay. Pure forms of folacin may be effective in doses less than 1/4 of the RDA.

[d] Niacin equivalents include dietary sources of the vitamin itself plus 1 mg equivalent for each 60 mg of dietary tryptophan.

[e] Assumes protein equivalent to human milk. For proteins not 100 percent utilized factors should be increased proportionately.

of the alphabet, but as some of them, notably "vitamin B," proved to be mixtures of several distinct vitamins, and others, notably A and D, were found to exist in several variant forms, alphabet names became complicated by subscripts such as B_{12}, and so on. Furthermore chemists deciphered the true nature of each vitamin and named them accordingly, so that you will see names such as "ascorbic acid—vitamin C," or "riboflavin—vitamin B_2" on bottles of medicine or pills.

Vitamins are classified as fat- or water-soluble. Fat-soluble vitamins, such as A and D, are found in egg yolk, butter, meats, and richly in fish liver oils. Water-soluble vitamins are found in yeast (B group), fresh fruits (C), greens (folic acid), and so on. This classification, though useful, is very rough and does not correspond to any essential common quality of the group but only to their associations in food. Each vitamin has its own individual nature. Table 14–1 summarizes the recommended daily allowances for calories, protein, minerals, and vitamins. The most important vitamins are as follows.

Fat-Soluble Vitamins

Vitamin A

This is a product of animal metabolism, and, as mentioned, is most abundantly found in butter, egg yolks, and fish liver (cod and halibut) oils. However, it is doubtful that any animal synthesizes vitamin A without the precursor substance carotene derived in the diet from plants. Carotene is the yellow pigment which gives the characteristic color to carrots, sweet potatoes, and rutabagas, but it is also found in green vegetables where the yellow color is masked by the green, and presumably in the plankton eaten by fish. We can satisfy our vitamin A requirement then by either taking in vitamin A directly from its animal sources or by eating carotene-containing plants, in which case the liver makes the necessary conversion.

The main function of this vitamin seems to be the maintenance of normal epithelium of the skin and eye and the linings of our digestive and respiratory passages. Animals with a severe vitamin A deficiency show a dry and brittle skin through which bacteria may pass and cause infection. For this reason vitamin A has been called the "anti-infection vitamin." But since a dull razor blade can nick the skin and provide a similar entrance, a sharp razor might be

Table 14-1 is taken from the Food and Nutrition Board, National Academy of Sciences—National Research Council. As the Board states in their discussion on the purposes and intended uses of their recommendations, "the allowances are designed to afford a margin sufficiently above average physiological requirements to cover variations among practically all individuals in the general population. The allowances provide a buffer against increased needs during common stresses and permit full realization of growth and productive potential, but they are not necessarily adequate to meet the additional requirements of persons depleted by disease, traumatic stresses, or prior dietary inadequacies.

called an "anti-infection razor blade" with just as much justification. In fact, studies of children in orphanages, in which half the children received only normal dietary vitamin A while the other half received supplementary allowances, have disproved any special anti-infection properties. About the same number of children in each group came down with colds, influenza, sore throats, and middle ear infections. In other words, a small amount of vitamin A is essential for normal skin, and a larger amount will not give supernormal skin or any other protection against infection.

Two principal A-deficiency states are found in humans, and both are related to the eye. One of these is *xerophthalmia* in which the cornea (the transparent front "window" of the eyeball) becomes dry and eventually opaque, causing blindness. The other is *nyctalopia* or nightblindness in which the individual is unable to see in dim light. The retina of the eye incorporates vitamin A into the compound called visual purple; when light strikes visual purple, the compound is broken down, and when this happens, the sensory cells of the retina are stimulated and impulses are sent to the brain, giving the sensation of light. Normally, the visual purple is quickly rebuilt, but if a vitamin A deficiency exists the resynthesis is slowed down, and the lack of sufficient visual purple results in the nightblindness.

Xerophthalmia is practically unknown in this country, and the incidence of nightblindness has not been determined. It is well known, of course, that fighter pilots during World War II were given supplementary doses of vitamin A as protection against nightblindness. If they had more acute vision during their nocturnal excursions than their comrades in the bombers, it has gone unreported.

The Food and Nutrition Board of the National Research Council has proposed a daily allowance of 5000 international units of this vitamin. This is double the minimum daily allowance, and is probably justified in view of the fact that there is considerable individual variation in our vitamin requirements. For a general idea of how much 5000 units amounts to, there is approximately this much vitamin A in, collectively, four cups of milk, one egg, and one-third serving of spinach. Some vitamin A is stored in the liver when an excess is present in the diet, and this provides an added margin of safety.

A daily intake of vitamin A in excess of 50,000 units over long periods of time can give rise to symptoms of toxicity, such as loss of hair, bone and joint pains, and loss of appetite. This would most likely occur in those people who feel the necessity of taking concentrated vitamin supplements, but a few cases have occurred in people who had eaten polar bear liver which, we must assume, is rich in vitamin A.

Vitamin D

Vitamin D is another vitamin which is available as vitamin D itself, and as a precursor substance, *ergosterol,* a cholesterol-like substance found in plants.

The dietary ergosterol is deposited in the skin and on exposure of the skin to the ultraviolet rays of the sun (or to an ultraviolet lamp) is converted to a form of vitamin D called *calciferol.* Although the vitamin D found in animal livers, and particularly in fish liver oils, is more active than calciferol, both meet our vitamin D requirements.

This is the vitamin which is most likely to be deficient in the diet, hence the practice of giving children a daily tablespoon or capsule of cod liver oil. Since ergosterol is a natural constituent of milk, the irradiation of milk with ultraviolet light results in a ready source of calciferol. The National Food and Nutrition Board recommends a daily allowance of 400 international units of vitamin D for pregnant and lactating women and for children, and this is almost exactly the amount of vitamin D found in a quart of irradiated (fortified) milk. No other adult requirements are given.

The effects of vitamin D deficiency are all related directly to defective bone formation, and the deficiency disease is known as *rickets.* The defective bones lack adequate calcium, and the presence of vitamin D is essential for the proper absorption of calcium from the intestines. A deficiency of calcium in the diet has exactly the same effects. It is convenient that a quart of milk provides not only the vitamin D but also most of our daily requirement of calcium.

Rickets was widespread in Northern Europe during the sixteenth and seventeenth centuries. It seemed to occur particularly in children of the more privileged social classes. The fewer hours of daylight in the northern latitudes, the tendency for well-to-do children of the day to wear clothing which shielded all but their hands and faces from the sun, and the general feeling that the outdoors was unhealthy were contributing factors to the high incidence of rickets. Many of the artists of the period faithfully portrayed such children with their bowlegs, enlarged joints, and protruding breastbones, and may even have regarded them as normal. Some portraits by Breughel and Vermeer and the sculptures of della Robbia are examples.

Excessive vitamin D in the diet can be dangerous since it causes an overload of calcium in the blood, a condition referred to as *hypercalcemia.* Infants are usually the victims, and show loss of appetite and retarded growth. The condition has most often developed in cases in which a vitamin D supplement such as viosterol has been prescribed. One drop of viosterol will supply 400 units of vitamin D (a very adequate allowance for an infant); yet a mother believing that a little more "would not hurt" or simply misunderstanding the directions may give the infant a teaspoonful. Such daily overdoses over an extended period would probably cause permanent disability.

Vitamin E

Vitamin E was discovered in 1922, but only recently has there been assurance of its need in man. Apparently, it functions in some certain tissues by combining with oxygen and thereby protecting certain easily oxidizable substances, such as vitamin D and carotene, from destruction.

A deficiency of vitamin E in rats causes sterility in the males and absorption of the embryos in the pregnant females. For this reason it has been called the "antisterility vitamin," and this is an apt term when one is concerned with fertility in rats. No human deficiency disease is known for this vitamin, but it has been shown to be of benefit in the treatment of anemia associated with low protein diets. Presumably, vitamin E is so widespread in foods that deficiency symptoms just do not develop. In spite of this it is usually included in multiple vitamin supplements.

Vitamin K

Vitamin K is required by the liver for the formation of prothrombin, a protein of the blood essential for blood clotting. In fact its name comes from the German *Koagulation*. Although vitamin K is widely distributed in vegetables, and is also found in egg yolk and meat, our requirements for vitamin K are easily satisfied by its production by bacteria normally resident in our intestine. Since vitamin K is soluble in fats, and is absorbed along with fats, its absorption is dependent upon the presence of bile salts which break up fat globules (emulsification). If liver disease is present, or if gallstones block the bile duct, there can develop a bile deficiency and a consequent defect in vitamin K absorption. Whereas all the fat-soluble vitamins become deficient when there is interference with fat absorption, no others show such a dramatic effect as does the clotting disability associated with vitamin K deficiency. Such problems would develop only in severe disease states, and a vitamin K deficiency is of no concern to a healthy person. In fact, the National Food and Nutrition Board makes no recommendation for daily allowances of vitamin K. However, since newborn babies lack intestinal bacteria, and may not acquire them quickly under sterile hospital conditions, they may lack vitamin K and so be subject to uncontrolled bleeding unless the vitamin is supplied.

Water-Soluble Vitamins

Thiamin

Thiamin was the first of the B-vitamins to be discovered and is still often called B_1. In the 1890 a Dutch physician named Eijkman, working in Java, noted that chickens fed on a diet of polished rice developed a paralysis. When the chickens were fed the cheaper, unhusked rice, they recovered. The Dutchman claimed the chickens were suffering from *beriberi*, a disease then prevalent among the natives in the East Indies, and suggested that the human disease might be cured by something in the rice husks. Later a Pole named Funk, working in the Lister Institute, isolated the substance of the rice husk and referred to it as a "vitamin." Funk was later to isolate the second B vitamin, B_2 (which later turned out to be itself a mixture of several compounds—niacin, riboflavin, and others) from the rice husks, and suggested that scurvy and rickets might also be diseases due to dietary deficiencies.

It is ironic that beriberi is still a disease of serious concern in rice-eating countries. Public health authorities in Taiwan recently attempted to combat the disease by adding a small percentage of unpolished rice kernels to the rice supplies distributed by the government. When results were disappointing they came to discover that the Formosans were carefully picking out the brown, unhusked kernels as contaminants.

Beriberi is principally a disease of the nervous system. It results in loss of sensation, muscular weakness, and eventually paralysis; it can also affect the heart. A milder thiamin deficiency, found in this country, results in chronic fatigue, loss of appetite, and personality changes such as moodiness and depression. Massive doses of vitamin B-complex are sometimes prescribed for these symptoms; but the treatment is effective only if the symptoms are really due to thiamine deficiency. Many other disorders can cause very similar symptoms.

Thiamine is widely distributed in foods, but plentiful in only a few. Particularly good sources are pork and liver. The recommended daily allowance is about 1.4 milligrams, and could be obtained from a total of two servings of meat, a bowl of cereal, three glasses of milk, and two vegetables.

Niacin (Nicotinic Acid)

This is one of the "B_2" group. The requirement of niacin in the diet can be partly met by the intake of increased levels of the amino acid *tryptophan* which can serve as a precursor. This has complicated the establishment of a minimum daily requirement, which has been set, rather arbitrarily, in niacin equivalents per kilogram of body weight. The figure given in Table 14–1 of recommended dietary allowances are considered to be at least 50 percent greater than the minimum daily allowance. Since animal foods are sources of both niacin and tryptophan, it is difficult to see how niacin deficiencies could arise in the general population, at least in this country.

A prolonged deficiency of niacin results in pellagra. The disease is characterized by inflammation of the skin and the membranes lining the digestive tract. The tongue becomes swollen and sore, irritations in the mouth develop, and there is a chronic diarrhea. Psychological changes such as irritability, anxiety, and depression are usually present.

Pellagra was at one time rather widespread in the southern parts of the country where many people lived on a diet almost exclusively of corn. Pellagra is uncommon where cereals other than corn make up a part of the diet. Corn has relatively low concentrations of both niacin and tryptophan, but no less than certain other cereals such as white rice or oatmeal. So it may be that part of the niacin and tryptophan is unavailable; or possibly some other factor, as yet undiscovered, is involved. Much remains to be learned about this complex disease.

Riboflavin

Riboflavin, another "B_2" vitamin, was discovered before a deficiency disease for it was recognized. And the symptoms of riboflavin deficiency are so similar to those of pellagra that patients are simply treated for both. Since corn is also deficient in riboflavin, pellagra victims generally have a riboflavin deficiency too.

The daily recommended allowance for riboflavin has been set at 1.7 milligrams for young adult males with the allowance almost as high for 15 to 18-year-old boys (probably due to the accelerated muscular development during this period). The allowances for women vary from 1.3 to 1.8 milligrams, depending upon age, pregnancy, and lactation. These allowances are regarded as at least double the *minimum* daily allowance.

Riboflavin is found to some extent in practically all foods. The best sources are lean meat, liver, dairy products and eggs; in other words, about the same sources as niacin. Since World War II, bread, flour, and corn meal have been enriched by the addition of all three of the foregoing primary B vitamins, and this has virtually eliminated dietary deficiencies of these vitamins.

Other B-Vitamins

Vitamin B_6 (pyridoxine), pantothenic acid, biotin, folacin, vitamin B_{12}, and choline are also included in what is referred to as the vitamin-B complex. No daily recommendations have been made for some of these vitamins, but diseases due to dietary deficiencies are rare since the recommended allowance is so easily met in common foods. Some peculiar situations can arise: biotin can be deficient in those who eat large quantities of raw egg white (a protein in the egg white combines with the vitamin and prevents its absorption); B_{12} intake can be deficient in a strict vegetarian diet, or in persons with pernicious anemia who are unable to absorb the vitamin (see Chapter 18).

Ascorbic Acid (Vitamin C)

Over 200 years ago it was known that the disease *scurvy* could be cured or prevented by eating fresh oranges or lemons (Fig. 14–1). In fact, British sailors were long required by law to take a daily ration of lemon or lime juice, whence their nickname "limey." This mild jeer was compensated manyfold by their

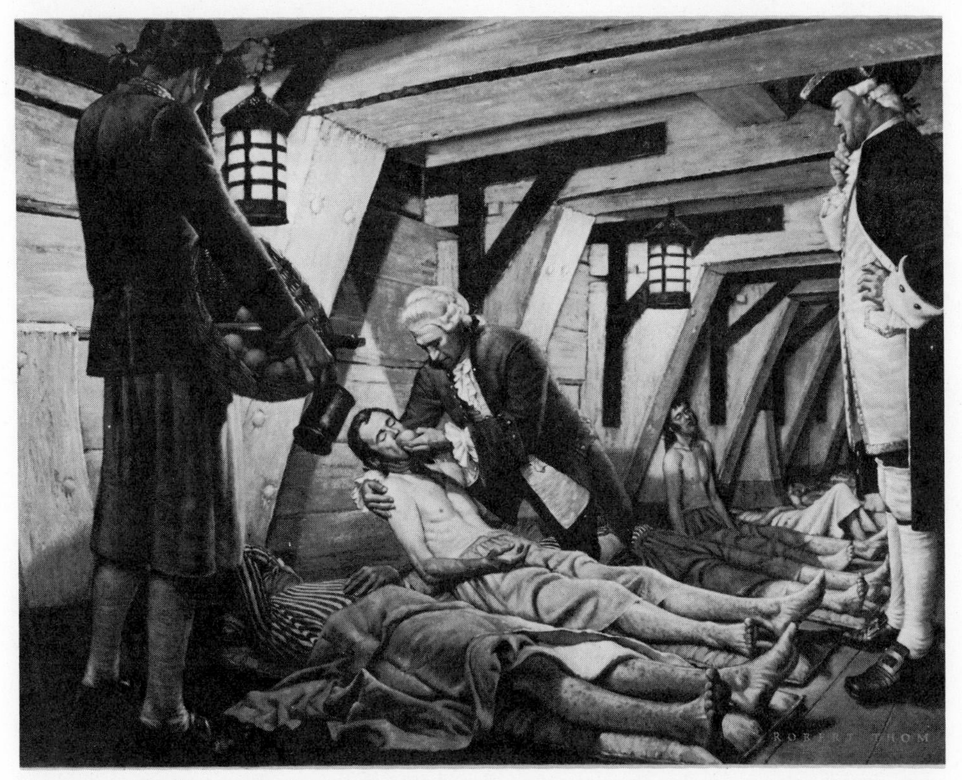

FIG. 14–1 The first use of fresh fruits for the treatment of scurvy occurred in the British navy in 1754. Here Dr. James Lind, a surgeon in the Royal Navy, is shown feeding lemons to sailors severely ill with scurvy. (Parke-Davis)

freedom from the scurvy that scourged other crews on long voyages without fresh fruit. But it was not until 1933 that ascorbic (Greek for nonscurvy) acid was isolated and identified as the effective agent. Much of the delay was due to the fact that rats and mice do not require ascorbic acid in the diet. The guinea pig does, and studies on scurvy are responsible for the introduction of the guinea pig as an experimental animal.

Important food sources of ascorbic acid, in addition to citrus fruits, are onions, cabbage, broccoli, cauliflower, tomatoes, apples, and grapes. Since much ascorbic acid is destroyed by cooking, at least some of these foods should be eaten fresh.

Ascorbic acid is required by the body for the development of the fibers of connective tissue. These collagen fibers, formed from the protein collagen, provide the strength of such connective tissues as tendons and ligaments, and the dermis of the skin. The failure of these connective tissues causes the loosening of teeth and the poor healing of wounds associated with scurvy. Persistent open sores, bleeding and infected gums, and swollen joints are characteristic.

Scurvy reached its greatest incidence in this country during the Western migrations of the last half of the nineteenth century. It is estimated that at least 10,000 men died of scurvy in the California gold rush, half of them in 1848–1849.

The disease is rare in the Western world at the present time, but occasionally is seen in children. However, vitamin C is not stored in the body as are some other vitamins, so that an inadequate ration, while averting frank scurvy, may permit a smoldering form resulting in general poor health. Since overdoses of vitamin C are simply excreted without doing any harm, one can safely err on the side of taking too much rather than too little.

Minerals

Numerous minerals have been shown, by experimental studies, to be dietary requirements in laboratory animals, and therefore are presumably required in man. Only three of these, however, may become nutritionally deficient when the dietary intake is sufficient to meet the requirements for calories and protein. These three are calcium, iron, and iodine.

Calcium

Most of the body's calcium (about 99 percent) is found in the bones and teeth, but the small amount in the other tissues serves in such vital functions as the transmission of nervous impulses, muscular contraction, and blood clotting. When a deficiency of calcium occurs, these functions are given priority, and the needed calcium is withdrawn from the bones. For this reason skeletal defects become the major symptom of calcium deficiency and are most likely to develop in children where bone growth necessitates high calcium intake. The resulting skeletal deformities, known as rickets, have been discussed in relation to vitamin D requirements.

The hormone produced by the parathyroid glands (parathormone) controls the level of calcium in the blood, and if a deficiency of the hormone exists, blood calcium levels fall (hypocalcemia) and the nerves and muscles of the body become so irritable that they respond to the slightest stimulus with localized spasms or generalized convulsions. The condition, called *tetany,* can also be produced by a deficiency of calcium or vitamin D in the diet.

Tetany sometimes accompanies rickets in children, and in women during lactation or in pregnancy when excessive demands for calcium by the fetus cause maternal deficiency. Depending upon the cause of the disturbance, tetany is treated by administration of vitamin D (viosterol), a calcium-rich diet, calcium injections, or injection of parathormone.

The recommended daily allowance for calcium has been set at 0.8 gram for adults, and up to 1.4 grams for growing children. The high level of calcium in milk (about 1.1 grams per quart) makes it our most important calcium source, but there are also substantial amounts in eggs and green vegetables. And few foods are without at least some calcium. However, it is difficult, if not impossible, to obtain the recommended daily allowance of calcium if dairy products of some sort are not included in the diet.

Iron

Iron is required for the manufacture of hemoglobin which is responsible for the transport of oxygen by the red blood cell. A deficiency of iron can give rise to the reduced oxygen-carrying capacity of blood, or in clinical terms, a state of "*nutritional anemia.*"

The newborn infant has more hemoglobin than he requires. During intrauterine life, relatively greater amounts of hemoglobin and greater numbers of red blood cells are needed to compensate for the less abundant oxygen supply provided by the mother's blood. During the first few weeks after birth more red blood cells are destroyed than are manufactured, and the iron is recovered and stored for later hemoglobin production. Even though milk has practically no iron, the nursing infant gets along quite well for some months on an iron-free diet. After this period the need for iron increases with growth and becomes pronounced during puberty.

The daily physiological requirement for iron (that is, the iron the body uses in its metabolism) has been calculated on the basis of growth demands and on daily loss from the body, which is a different matter from the nutritional requirement which cannot be so easily determined. One reason for this is that much of the iron in foods is in combination with certain proteins from which it cannot be removed by digestive processes. Also, ferrous iron is absorbed much more readily than ferric iron; and a number of iron salts are insoluble, making them unavailable for absorption. Some people simply seem to absorb iron more easily than others, and, logically, those with iron-absorbing difficulties tend to be the ones who develop iron-deficiency anemia.

The female has greater iron requirements than the male because of loss of iron in the menstrual blood. An unsupplemented diet does not generally contain sufficient iron to meet this loss during the period of menstruation. Demands for iron are even greater during the last three months of pregnancy, and an unsupplemented diet during this particular period would rarely be adequate.

The National Food and Nutrition Board has recommended a daily dietary allowance of 10 milligrams of iron for men and 18 milligrams for all women from age 10 to 55 including pregnant women and nursing mothers. These allowances were intentionally set at liberal levels to offset the unpredictable nature of iron absorption, and can be considered quite adequate. Sources of iron include red meat (which owes much of its color to iron compounds), kidneys, liver, and whole grains; but iron is present in at least small amounts in most natural foods.

Iron deficiency has been claimed to be the most common of all dietary deficiency diseases in this country. It has been estimated that as many as 40 percent of all pregnant women develop iron deficiency anemia. Not all anemias can be attributed to insufficient absorbable iron in the diet; abnormal loss of blood from any cause though slight, can give rise to anemia.

There seems to be little doubt that many of the improvements people experience when taking vitamin supplements can be credited to the small amount of iron usually included in such preparations. The tonics for "tired blood" we hear promoted on television, costing four to five dollars per monthly supply, are no bargain since iron supplements as pills would not usually exceed one dollar per month.

Iodine

Iodine forms part of the thyroxin molecule, the hormone of the thyroid gland. A deficiency of iodine results not only in a reduced thyroxin production, sometimes with metabolic effects, but also in a reaction by the thyroid gland itself. The gland may enlarge to many times its normal size in an effort to capture all available iodine. The resulting bulge on the front of the neck is termed a simple goiter or "colloid goiter" (Fig. 14–2).

Soils in the Great Lakes area and, to some extent, in the Northwest are deficient in iodine, and goiters were quite common in these areas a number of years ago. Particularly afflicted were teenage girls, who presumably have high thyroxin requirements during puberty. The addition of iodine to table salt has virtually eliminated any dietary deficiencies of iodine. Natural sources of the element are most seafoods, especially shellfish. It is interesting to note that the ancient Greeks prescribed eating sponges for goiter because sponges look like goiters and "like cures like." Of course, the sponge diet would often effect a cure because of its iodine content and "prove" that the theory was correct.

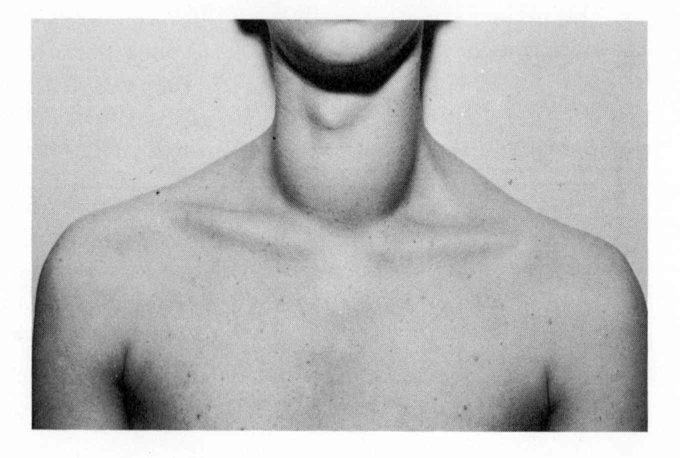

FIG. 14–2 Simple goiter is an enlargement of the thyroid gland that tries to compensate for the lack of iodine in the diet.

Water in Nutrition

The thirst mechanism is extremely sensitive to small fluid deficiencies; thus if we are not thirsty, we probably do not need water. This sensitive mechanism is very important for life. A healthy person may go for many days without eating, and suffer no ill effects—Mahatma Ghandi did so repeatedly in "hunger strikes." But lack of water can cause death in a few days even under average conditions.

Under comfortable temperature and humidity conditions the normal adult loses about 2½ quarts of water a day: about 1 pint each through the skin (in perspiration) and expired air, about 1½ quarts in the urine, with small amounts lost in the feces (Guyton). If the water intake is in excess of that required to replace this loss, the volume of urine increases. In normal persons no benefit is derived from such increased urine production.

During hot weather the daily loss of water through increased perspiration can amount to many quarts, and this obviously must be accompanied by an equivalent water intake. Now, all body fluids and excretions contain sodium (not discussed in this section on minerals, since it is amply provided in any normal diet) which is very essential for life, and loss of sodium in sweating has been looked on as dangerous. However, the loss of sodium during even excessive sweating has been generally exaggerated earning little justification for the common practice of taking salt tablets during summer exercise. Experiments to induce a sodium deficiency in humans by sodium-free diets, accompanied by profuse sweating in heat chambers, have been unsuccessful. After initial losses of sodium and chloride, the urine and sweat become virtually free of these substances. The body will presumably give up its excess salt and no more. An exception, of course, are diseased conditions such as Addison's disease in which the insufficiency of adrenal secretion results in excessive sodium loss. Secretion of the hormone aldosterone by the adrenal cortex seems to ensure adequate sodium retention except under extraordinary conditions. A low salt diet along with profuse occupational sweating over long periods of time, as in the case of furnace stokers (the subject of Haldane's original study where all this began), can probably lead to a sodium deficiency, though even this has now been questioned. People in many parts of the earth have very low salt intake with little observable effect. However, the taking of a few salt tablets during hot summer weather is certainly harmless unless it causes nausea.

REVIEW QUESTIONS

1. What is a vitamin?
2. What is the view of professional nutritionists toward vitamin supplements for normal healthy people?
3. Which vitamins may be toxic if taken in excessive amounts?

4. One vitamin is not necessarily required in the diet. Explain.
5. Most of the vitamins required by people are also required by most laboratory animals. There is one exception. Explain.
6. Does a person suffering from iron deficiency necessarily benefit by taking iron supplements in the diet? Explain.

REFERENCES

GUYTON, A. C., *Textbook of Medical Physiology,* 4th ed. Philadelphia: W. B. Saunders, 1970.

FOOD AND NUTRITION BOARD, NATIONAL RESEARCH COUNCIL, *Recommended Dietary Allowances,* 7th ed. New York: National Academy of Sciences, 1968.

Additional Readings

LOCKE, DAVID N., *The Agents of Life.* New York: Crown, 1969.

MARK, JOHN, *The Vitamins in Health and Disease.* Boston: Little, Brown, 1969.

15
THE ANATOMY OF DIGESTION

Digestion refers to the processes by which food is broken down to a state suitable for assimilation into the body. The food must then be absorbed, further processed by body chemistry, and distributed. This covers a great deal of anatomical territory, and a reasonable explanation of all the processes requires not only some understanding of the chemical nature of foods but also of the basic features of the organs involved.

The Digestive System

The *alimentary canal* includes the mouth, pharynx, esophagus, stomach, small and large intestines, rectum, and anus. It is about 18 feet long in life (in the cadaver it is longer than this due to the loss of muscle tonus). We can look upon this long tube as a continuation of the outside covering of the body. In fact, food contained within this tube is not strictly within the body. This concept is the essence of digestion, for most of the food contained within the gut (a quite respectable term in the singular) must be physically and chemically reduced to a simpler form before it is absorbed. This absorption involves passing through the wall of the gut and entering the bloodstream—after which it is really within the body. The alimentary canal can be conveniently divided into upper (mouth through esophagus) and lower (stomach through anus) parts.

The canal has several important *glandular accessories:* salivary glands, liver, pancreas, and lesser (but collectively important) glands. These accessory glands contribute in various ways to the digestion of food; together with the canal they form the *digestive system.*

The Mouth

We are fairly well acquainted with the structures of the mouth since we can see many of them in the mirror when we brush our teeth. As shown in Figure 15–1, they can be identified as follows: Above the soft palate is the posterior portion of the nasal cavity. The anterior and posterior fauces (openings at back of the mouth) pass downward to the base of the tongue and to the wall of the pharynx. Between the palatal arches (or pillars as they are sometimes called) framing the fauces, nestle the palatine tonsils.

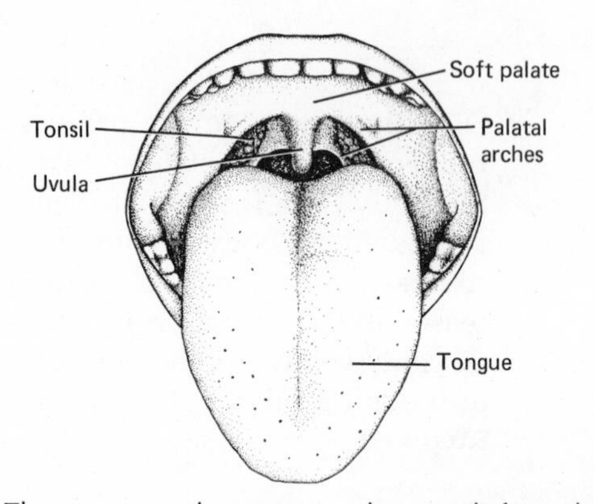

FIG. 15–1 The structures that separate the mouth from the pharynx.

The *tongue* is a highly mobile mass of muscle and, besides its importance in speech, it has several digestive functions: (1) it adroitly manages food during mastication (how rarely one bites one's tongue, and how skillfully one can detect and isolate, say, a hair or seed in the food); (2) it takes part in swallowing; and (3) it is the site of the taste buds which can guard the digestive system by alerting the presence of crude poisons (but these detect only sweet, sour, salty, and bitter, many "tastes" being really smells).

The saliva which moistens our mouth lining, and our food when we eat, comes from the *salivary glands.* (Fig. 15–2). There are three pairs of these glands.

1. The *parotid* is the large gland lying in front of the ear; its swelled condition in mumps brings it into prominence. The secretion of the parotid is carried

by way of a duct which opens into the mouth opposite the upper second molar (the last tooth in the upper jaw if you've lost your wisdom teeth). A flap of skin guards the opening and is sometimes slightly torn when we accidentally bite our cheek. This torn flap can be very irritating but will heal rapidly if left alone.

2. The *submandibular glands* lie just under the skin under cover of the lower jaw. They can be palpated as easily movable, almond-shaped masses. The submandibular glands also can become swollen during mumps.
3. The numerous smaller *sublingual glands* are buried in the root of the tongue. Masses of small glands line the palate and cheeks.

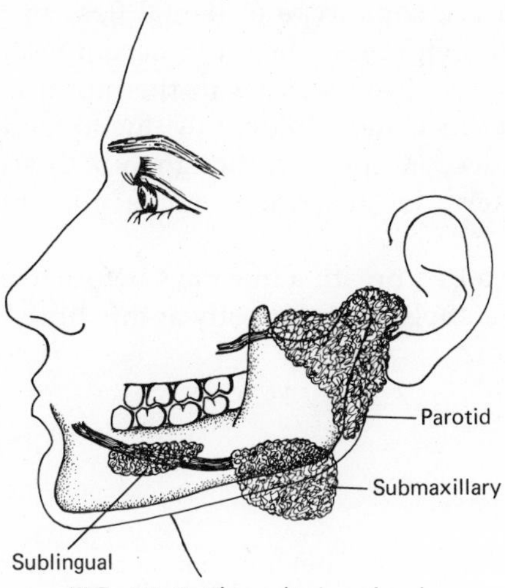

FIG. 15–2 The salivary glands.

Hygiene of the Mouth

The presence of an odor in the mouth has been named halitosis by the advertising industry. This can be due to a number of causes, some of them far from the mouth, including: decaying food particles between the teeth or, sometimes, in a fold at the back of the tongue; dry mucous membranes due to mouth-breathing; diseased tonsils; diseases of the nose, sinuses, the bronchial system, or the lung tissue (the latter sometimes with an odor like boiling cabbage); diseases of the gastrointestinal tract or liver (the latter sometimes with a strong fecal odor); ingested substances, such as onions or alcohol, absorbed by the blood and excreted into the air from the lungs or in the saliva; heavy smoking; and other less common sources.

In all these conditions the bad breath is only a symptom of the cause. Thus antiseptic or aromatic mouth washes do little more than temporarily mask it.

The most common source of halitosis is food particles decomposed by bacteria normally present in the mouth and pharynx. Antiseptic mouth washes, even if they act as claimed, would destroy only the bacteria that they reach, and this would not include those located above the soft palate in the nasopharynx, in the crevices of the tonsils, or in the upper air passages. A short time is all that would be required to recontaminate the entire area.

Obviously, the best treatment of chronic bad breath is elimination of the cause. One can much more easily remove food particles than bacteria, and your anxiety can be relieved by simply brushing your teeth after meals, or at least rinsing out your mouth with water, preferably salted. Brushing up or down, parallel with the crevices of the teeth and including the gums is obviously more effective than side brushing which sweeps particles into the crevices. It is also very helpful to use some type of dental floss to remove food particles from between the teeth before they begin to decompose.

A bad taste in the mouth on waking in the morning *is* usually associated with a bad breath and this is most frequently due to food particles that remain in the mouth during sleep. Generally, though, a bad taste is not indicative of halitosis and, conversely, its absence is not a guarantee of an inoffensive breath.

Some women have a bad breath a few days before menstruation. This must be attributed to the metabolism of the body at this time, and little can be done except to mask the odor.

Teeth

Importance of Dentistry

The miseries and general poor health that accompany diseased teeth and gums are a great and common human affliction, and good teeth are evidently essential not only to masticate food but also for proper speech and normal appearance.

The prevalence of poor teeth in children, as well as in adults, is illustrated by the fact that many parents with growing families have yearly dental expenses exceeding those for medical and surgical care. The propensity for bad teeth and the financial burden it brings, plus an ever increasing shortage of dentists, poses a problem of growing social importance.

The Nature of Teeth

Figure 15–3 shows the structure of a typical tooth. The part of the tooth that projects above the gum, the *crown,* is covered by *enamel,* the hardest substance produced by animal tissues. This outer layer is continuous with a layer below the gum line called the *cementum.* The cementum is bound to the bony socket and gum by the *periodontal membrane*—really a mass of tough fibers. *Dentine* lies deep to the enamel and has a hardness more closely resembling that of bone. A *pulp cavity* and one or more *root canals* form the core of the tooth and

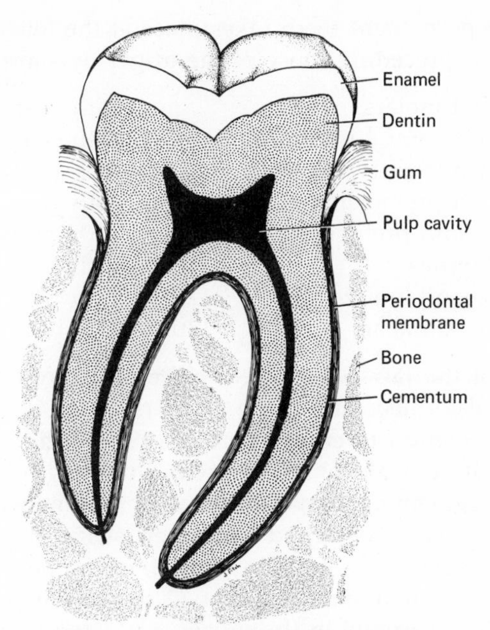

FIG. 15–3 The structure of a tooth.

contain nerves and blood vessels embedded in a fatty pulp. Nerves penetrate through the dentine layer but probably do not enter the enamel. This accounts for the fact that teeth with shallow cavities, restricted to the enamel, are sometimes more sensitive to the dentist's drill than are the deeper cavities which penetrate into the dentine and destroy the nerve endings.

Two sets of teeth are produced. The first or *deciduous* set (baby teeth, milk teeth, primary teeth) is composed of two incisors (cutters), one canine (dog-type), and two molars (grinders), on each side of each jaw, thus 20 teeth in all.

The times of eruption of these teeth are subject to considerable variation. Usually the eruption begins about the 6th month after birth and is completed at about the 25th month. The following are the usual times of eruption:

Lower central incisors	6 to 9 months
Upper incisors	8 to 10 months
Lower lateral incisors and first molars	15 to 21 months
Canines	16 to 20 months
Second molars	20 to 24 months

In general, the 1-year-old child should have 6 teeth; at 1½ years, 12 teeth; at 2 years, 16 teeth; and at 2½ years, 20 teeth.

In the *permanent* set of teeth, the so-called molars of the deciduous set are replaced by the bicuspids or premolars; the three permanent molars emerge in successive positions to the rear. Thus this set comprises 32 teeth in all: two incisors, one canine, two premolars, and three molars in each half of each jaw.

The eruption of the permanent teeth takes place at the following periods, the teeth of the lower jaw preceding those of the upper by short intervals:

First molars	6th year
Two central incisors	7th year
Two lateral incisors	8th year
First premolars	9th year
Second premolars	10th year
Canines	11th to 12th year
Second molars	12th to 13th year
Third molars	17th to 25th year

The radiogram of the jaws of a 5-year-old child shows the crowns of the permanent teeth well-formed within the gums (Fig. 15–4). As the permanent tooth moves into a position to erupt, the root of the deciduous tooth is gradually resorbed until it loses all connection with the bony socket, being held in place simply by a fragment of the periodontal membrane where it attaches to the gum. Hence it is easily detached by finger, string (whether attached to the proverbial door or not), or simply by growth of the permanent tooth.

Evolutionists believe that man has lost 12 teeth (four incisors and eight bicuspids) and he is apparently in the process of losing four molars since the third molars or "wisdom" teeth fail to develop in many people. In many of us, when these teeth do come in they are undersized or, because of the lack room

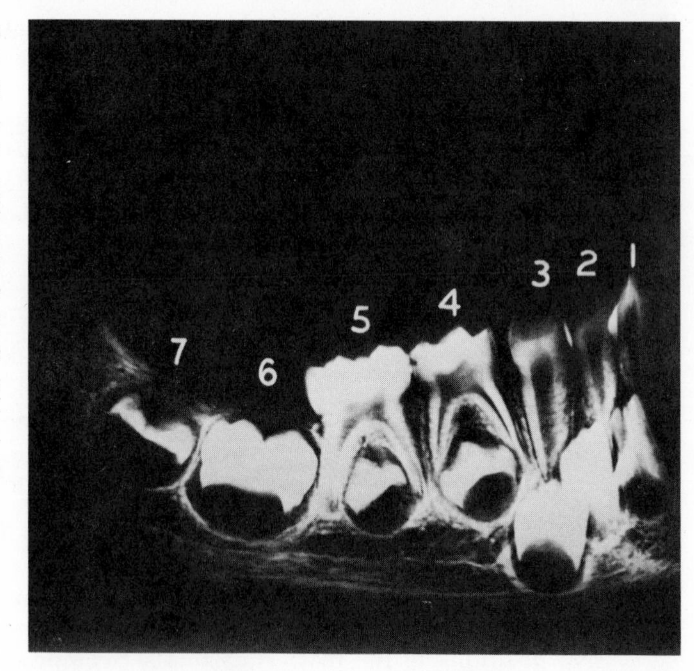

FIG. 15–4 X-ray of child's lower jaw. Teeth numbered 1 to 5 are the primary teeth (1 and 2, the incisors; 3, the canine; and 4 and 5, the molars). Permanent teeth are developing in the jaw: incisors and canines to replace their primary counterparts; two premolars to replace the primary molars; and, 6 and 7, the first two permanent molars. (American Dental Association)

in the jaw, they grow in at an angle toward the second molars and become "impacted."

Dental Caries

Three factors are apparently involved in dental decay (caries), and probably all three must be present or decay will not occur. These factors are a susceptible tooth surface, acid-producing bacteria, and carbohydrates upon which the bacteria grow. The last two are closely related because the bacteria produce acid from the fermentable carbohydrate (mainly sugar). This acid, when mixed with food debris, saliva, and other substances in the mouth, produces dental plaque. Dental plaque is probably only incidental to tooth decay, but as a mechanical factor it serves to hold the acid against the tooth surface. The enamel of a susceptible tooth will erode in the presence of this acid (Fig. 15–5), and any treatment that prevents formation of the acid, or neutralizes it after it is formed, will prevent decay.

However, this is not easily accomplished. As stated earlier, the bacteria cannot be thoroughly removed, and any effect of so-called antiseptic mouthwashes or toothpaste is transitory. Even if your mouth could be cleansed of all bacteria, this might do more harm than good, since there is some indication that mouth bacteria serve to control certain harmful fungi and protozoa which can infect the mouth and throat.

The removal of the carbohydrate-containing food debris is a more practical approach. As long as carbohydrate is present in the mouth, acid will be produced. This is the reason dentists stress the importance of brushing the teeth or, at least rinsing the mouth after eating or even drinking a sweet drink. Particularly harmful are foods containing a high proportion of ordinary table sugar (sucrose) since it is easily fermentable into acid, and frequently, by way of candy, cake, and other sweet foods, forms a sticky paste which adheres to the tooth surface. Also harmful, from the standpoint of acid production, is food taken between meals and as bedtime snacks without immediate brushing.

Dental plaque seems to form more rapidly in caries-susceptible persons than in those who are caries-resistant. This guilt by association is probably justified, and the elimination of plaque is certainly desirable. Dental plaque can best be minimized by daily brushing, by water or peroxide rinses, by the use of dental floss or tape, and by dental-metal, wooden, or water-pick instruments for the removal of food particles packed between the teeth. Cleaning of the teeth by a dentist or dental hygienist involves the removal of plaque which has accumulated and should be accomplished in at least yearly or twice yearly intervals. All care, both personal and professional, is essential to dental health.

Why some teeth are more susceptible to decay than others is not clear.

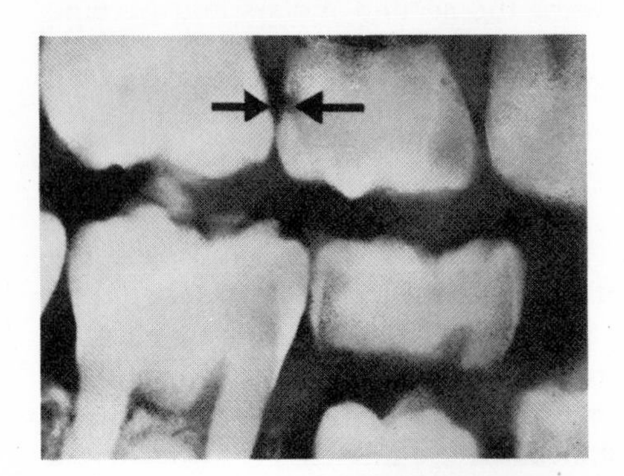

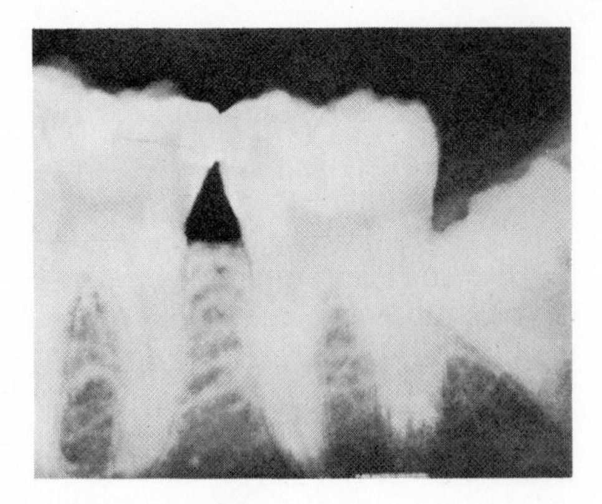

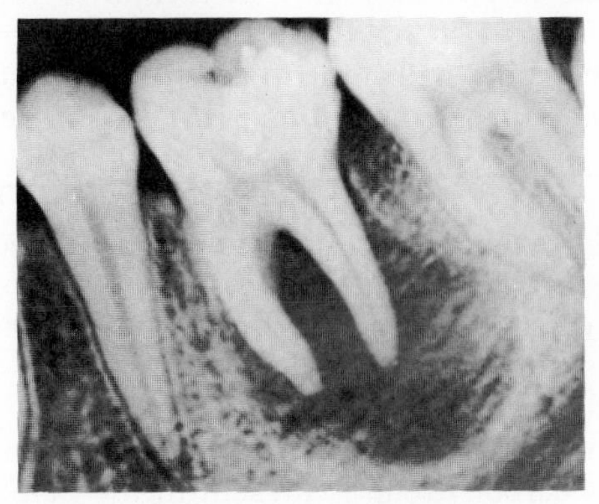

FIG. 15–5 Dental problems revealed by X-ray photographs. *Left:* A cavity in a deciduous tooth that has infected a permanent tooth. *Right:* An impacted third molar which will have to be removed surgically. *Bottom:* An abscess surrounding the root (dark area). (American Dental Association)

Inheritance is undoubtedly an important factor, and one which we can do little about. For instance, large teeth seem to be more prone to decay, particularly large teeth with furrows or deep pits. The fact that large teeth are usually closely spaced is also a contributing factor. Such teeth require more careful attention.

Fluoridation

Of very practical significance is the fact that you can increase caries-resistance when fluoride is included in the diet; and the one practical, safe, and effective method to accomplish this is to add fluoride to the community water supply. Communities across the country have repeatedly demonstrated that children who drink fluoridated water for the first eight years of life have less than 65 percent of dental caries that are found in those who do not drink fluoridated water [McClure]. When the child is supplied with fluoridated water later in life, the proctection is less but still significant. In fact the efficacy of fluorides in preventing tooth decay is no longer denied even by those who oppose fluoridation.

A secondary benefit from reduced caries formation is protection from later orthodontic problems. Early loss, from decay, of the deciduous molars or the first permanent molars is a leading cause of malocclusion (bad bite).

Why do only about one-third of all Americans have the benefit of fluoridated water? Because when fluoridation is proposed in a community it often arouses very bitter opposition. Opponents frequently claim that fluorides are poisonous. Indeed they are—and so too are most standard remedies, such as aspirin, when taken in overdoses. But the dosage recommended, one part per million, entails absolutely no danger. In fact, some communities have water supplies with naturally occurring fluorides up to the level of eight parts per million. Studies have shown that people in these communities live as long and suffer no greater propensity for human ills than other people. They are peculiar only in the fact that they have fewer dental caries and, if the level of fluoride exceeds 1.5 parts per million, in the possible appearance of brown mottling on the teeth. Mottling has never been observed when the level of fluoride is below this figure.

Opponents frequently allege some undesirable effect on old people such as weakening of their bones or circulatory disability, yet such claims lack evidence. In fact there is some indication that fluoride may be helpful in *retarding* osteoporosis, a condition in which the bones of older people become porous and weak. Thus we must conclude that resistance to fluoridation is largely emotional and misinformed.

Community fluoridation is an example of sensible public health practice: It is inexpensive; it is highly effective; it has no proven, or likely, undesirable effects; and it reaches practically every member of the community. If your local community does not have fluoridated water, do your part to see that such a program is adopted.

In the absence of fluoridated water, what can be done? Fluoride-containing drops can be obtained from your druggist and added to drinking water according to directions. Fluoride can also be applied by the dentist; typical treatment is the application of stannous fluoride at every regular visit to the dentist until the age of 15. Such treatment will significantly reduce caries formation. If none of these treatments is available to you, the use of fluoride-containing toothpaste gives some measurable protection, as the advertising industry has so diligently informed us. Any of these measures is far cheaper and less trouble than even just half an average dental bill!

Tooth Transplantation

In a few cases enterprising dentists have transplanted natural or ivory teeth into empty sockets. These transplants are often sloughed off, as is typical of most homografts (one person to another) or xenografts (one species to another, as with ivory).

However, autoplastic transplants (transplants of tissue from one part of a person's body to another part) are almost always successful if done under proper surgical conditions. This makes possible the transplantation of an impacted third molar into the socket of a lost first molar, for example. If the transplanted tooth develops a normal periodontal membrane and blood supply, the graft becomes permanent and can be treated as any other tooth. This technique is still in its infancy, and not too often applicable; but it opens interesting prospects.

Orthodontics

When we think of orthodontics, we think of children of junior or senior high-school age (Fig. 15–6). However, orthodontics are now doing much to realign the teeth of older people, even those of middle age. Many of these people have had a malocclusion most of their lives, and others have gradually developed one through loss of teeth. Orthodontists are usually able to fit these adults with appliances which are inconspicuous or which need be worn only at night.

Gingivitis

Gingivitis refers to gums which are swollen and red and which tend to bleed easily. The condition is usually not painful. It is most often caused by the accumulation of tartar or calculus (the hard calcified deposits that form on the surface of the teeth as distinguished from the softer dental plaque), but also may be caused by malocclusion and certain metabolic disturbances. The net effect of chronic gingivitis is frequently a gradual recession of the gums until the cementum is exposed resulting in the loss of the tooth. Pyorrhea is simply gingivitis in which pus is present, indicating infection.

Gingivitis can be prevented by careful dental hygiene and regular visits to the dentist. The eating of foods requiring thorough chewing also seems to help in the prevention of gingivitis.

Pharynx and Esophagus

The pharynx is a common passageway for inhaled air and ingested food. During swallowing the larynx moves upwards against the epiglottis which closes its opening (touch your "Adam's apple" when you swallow and note

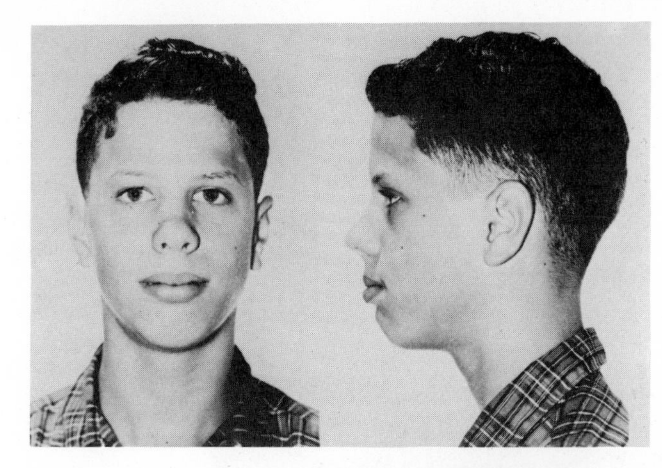

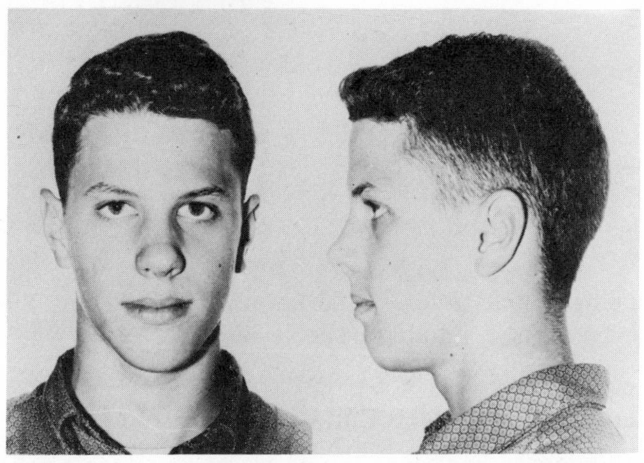

FIG. 15–6 An orthodontist can do a great deal to correct irregular teeth. Corrective treatment not only enables proper chewing of food but also improves the appearance of the face. Notice the improvement of the boy's profile on the right after orthodontic treatment. (Unitek Corp.)

how it bobs up). This prevents food and liquids from going down "the wrong way," that is, into the highly sensitive larynx and trachea. Such an accident though generally minor and taken care of by coughing reflexes can occasionally be serious; material inhaled into the bronchial system can cause real trouble. Hence one should never try to give fluids by mouth to an unconscious or semiconscious person whose swallowing and coughing reflexes are sure to be functioning badly.

The pharyngeal muscles contract so as to propel the food into the esophagus (Fig. 15–7). A wave of muscular contraction called *peristalsis* passes down the walls of the esophagus carrying the food along ahead of it and delivering it to the stomach about an inch below the diaphragm. At the upper end of the esophagus, swallowing can be controlled by the will; but at the lower end, involuntary control takes over to persist through the rest of the canal. Hence the common experience of feeling, say, an aspirin tablet stick in the esophagus and

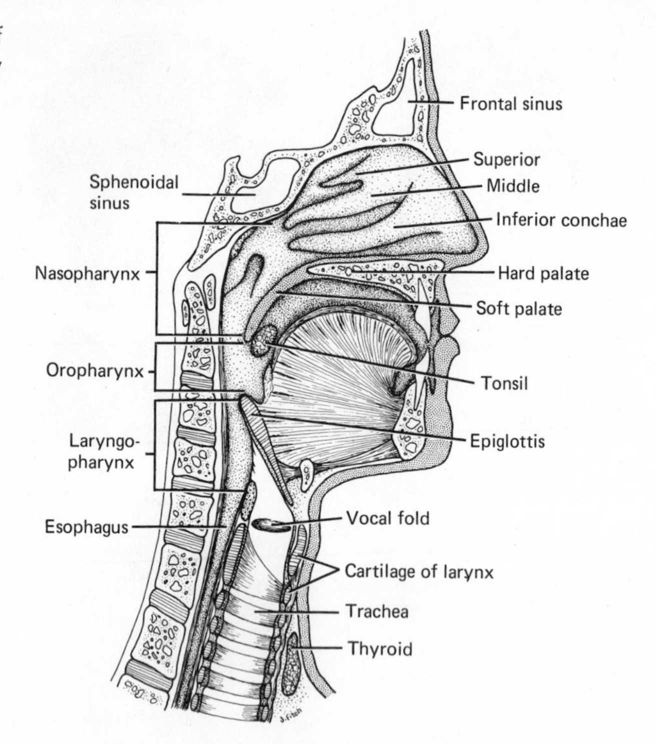

FIG. 15–7 The upper parts of the digestive and respiratory systems.

being unable to force it farther until it dissolves. Therefore irritant substances, even aspirin, should not be swallowed dry.

Lower Alimentary Canal

This is also known as the gastrointestinal tract or GI system. It is subdivided into stomach, small intestine, and large intestine, each with its own special functions. It is supplemented by the liver and pancreas, the great digestive glands.

The *stomach,* in the living person, is not the bloated sac usually pictured in most texts, but is rather J-shaped in outline and measures about 10 inches in length with a maximum breadth of about 4½ inches (Fig. 15–8). Most of the stomach lies on the left side of the upper abdomen under protection of the lower ribs; but its position can vary widely and still be perfectly normal. The capacity of the stomach is about one quart.

The lining of the stomach has a honeycomb appearance. The minute glands that form the lining produce digestive juices (see below). The lower curved part of the J leads into the narrow *pylorus* which connects with the first part of the small intestine.

The *small intestine* is subdivided into three successive parts not very dis-

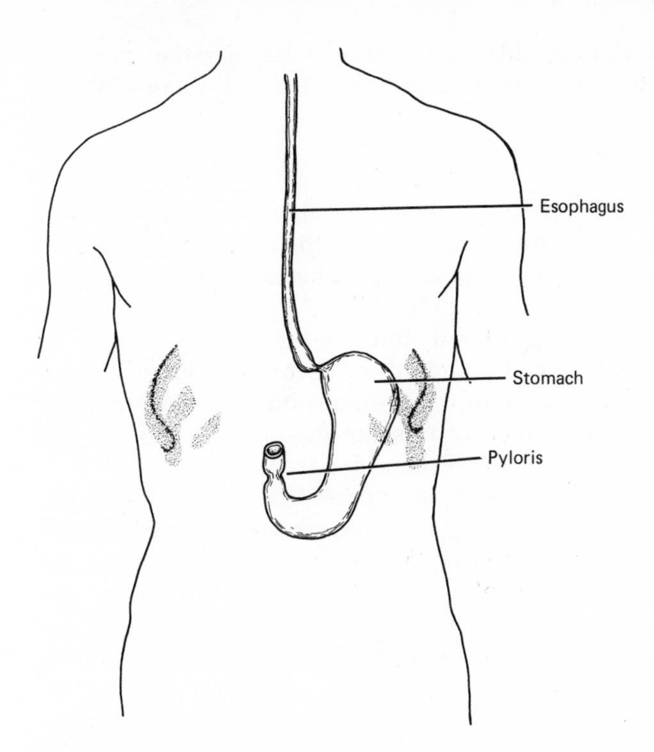

FIG. 15–8 The form and position of the stomach.

tinctive for our purposes. The *duodenum* lies in the form of a C just below the stomach (Fig. 15–9). Within the curve of the C it receives the common bile duct which drains the liver, gall bladder, and pancreas. The duodenum is tightly bound to the back wall of the abdomen by the overlying *peritoneum,* the membranous lining of the abdominal cavity. The duodenum leads into the more loosely attached remainder of the small intestine.

This remainder is in two successive parts called the *jejunum* and *ileum,* and is about 12 feet long in life. Although it is mobile, it cannot be unwound and laid out on the operating table as is often thought; it too is bound, although loosely, to the body wall by a sheet of peritoneum.

The small intestine ends by passing via the *ileo-cecal valve* into the side of the *cecum* which forms the beginning of the *colon* (or *large intestine*).

The *large intestine* is also subdivided (Fig. 15–10). The cecum lies in the lower right part of the abdomen, and its principal distinction is its *appendix,* properly called the vermiform appendix because of its resemblance to a large worm.

The major part of the large intestine forms an inverted and shallow M which more or less "frames" the small intestine; it is subdivided into *ascending, transverse,* and *descending* portions. In the lower left part of the abdomen it forms an S-shaped curve which terminates in the pelvis as the *rectum.* Although this last name suggests a straight tube, the rectum takes a rather twisting course before ending in the short anal canal and its opening the *anus*.

The Great Digestive Glands

The *liver* is the largest gland in the body, weighing about four pounds. Just as the stomach lies under cover of the lower left ribs, the liver occupies essentially the same position on the right. Its upper border lies almost at the level of the nipple. Its secretion to the intestine is called *bile*.

The system of ducts which transport and store the bile is called the *biliary system* (Fig. 15–9). Much of the bile is poured directly into the duodenum by way of the hepatic ducts which lead into the common bile duct. A branch of the common bile duct, the cystic duct, carries the excess bile into the gall bladder where it is concentrated and stored (Fig. 15–11).

Although the secretion of bile is the most apparent sign of liver activity, numerous other liver functions, some 500 in number, are vital to health. These include the synthesis of certain blood-clotting factors; the synthesis of albumins and globulins of the blood; the storage of fat-soluble vitamins; the storage and liberation of the body's sugar reserves; the maintenance of the proper acid-base balance of the body; the detoxification of ingested poisons as well as those

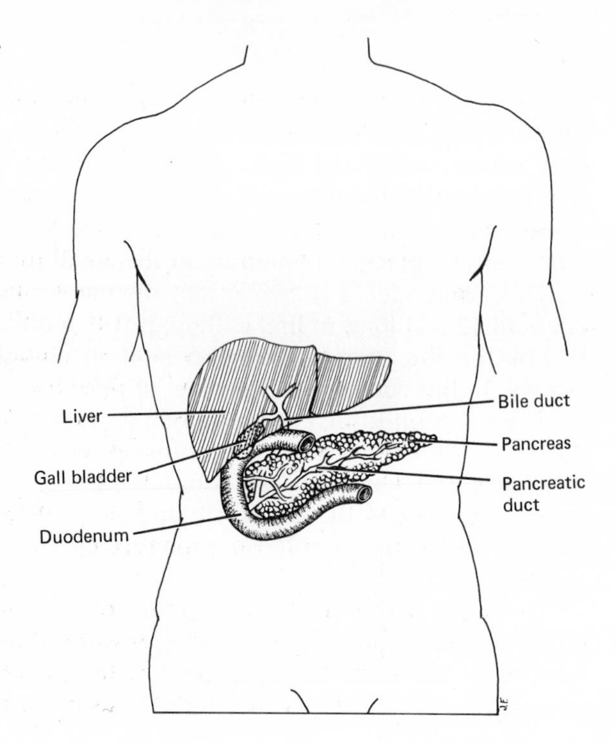

FIG. 15–9 The biliary system, pancreas, and duodenum.

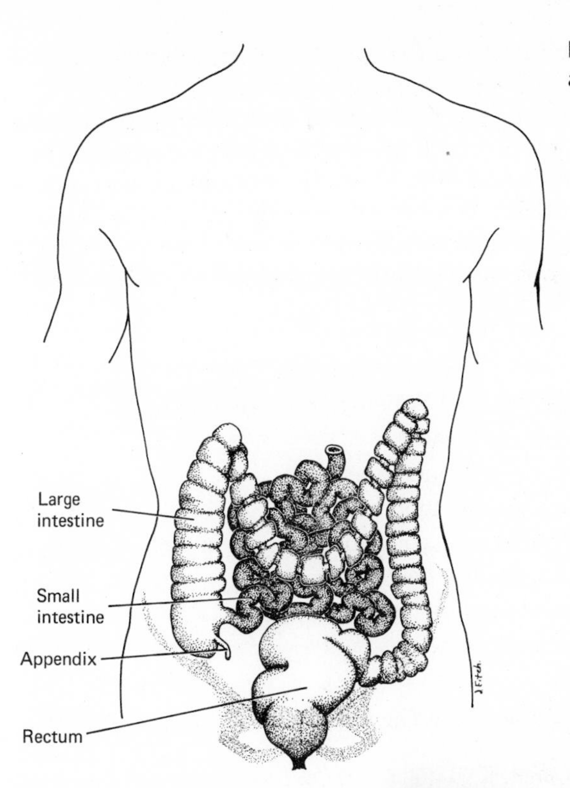

FIG. 15–10 The lower parts of the alimentary canal.

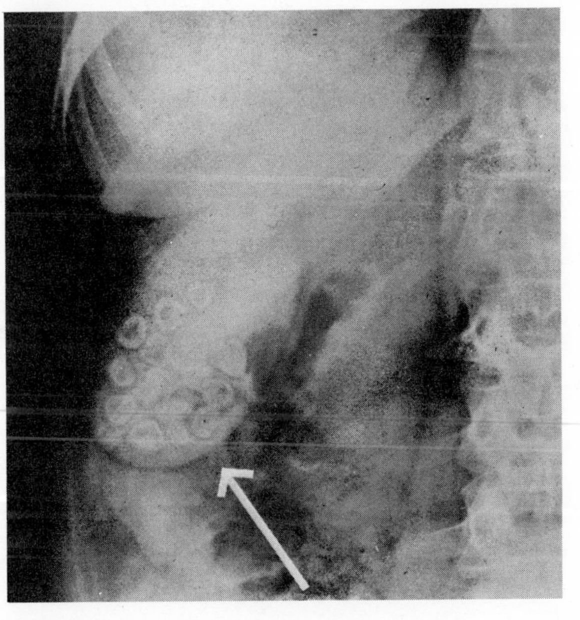

FIG. 15–11 Gallstones (arrow) in a dye-filled gall bladder and bile duct. (Dr. J. Edward Tether and Jerry E. Harrity)

which are normal metabolic products; and the conversion of sugars and amino acids to fats. Thus it is perhaps the most versatile organ of the body.

The *pancreas* lies flattened against the back wall of the abdomen behind the stomach. It produces digestive juices which are emptied into the common bile duct for final passage to the duodenum (Fig. 15–9). In addition, certain cells of the pancreas produce insulin which is released into the blood (an internal secretion or *hormone*). Insulin is of vital importance to sugar metabolism and will be dealt with in connection with the organs of internal secretion Chapter 32.

REVIEW QUESTIONS

1. Why is food inside the intestine regarded as being "outside" the body?
2. What are the common causes of bad breath?
3. How do the baby teeth differ from permanent teeth?
4. What are the factors that are involved in tooth decay?
5. How would you describe the position in your abdomen of the following structures: liver, stomach, duodenum, cecum, pelvic colon?
6. What are some of the functions of the liver? of the pancreas?

REFERENCES

GLICKMAN, IRVING, "Periodontal disease," *New England Journal of Medicine*, **284**:1071–1077 (1971).

LAUTERSTEIN, AUBREY M., and THOMAS K. BARBER, *Teeth, Their Forms and Functions*. Boston: Heath, 1965.

MCCLURE, F. J. (ed.), *Fluoride Drinking Waters*, National Institute of Dental Research, 1962.

MOORREES, COENRAAD F. A., "Orthodontics," *New England Journal of Medicine*, **279**:689–694, (1968).

Additional Readings

DAWSON, HELEN L., *Basic Human Anatomy*. New York: Appleton-Century Crofts, 1966.

GARDNER, W. D., and W. A. OSBURN, *Structure of the Human Body*. Philadelphia: W. B. Saunders, 1967.

NIZEL, ABRAHAM E. (ed.), *The Science of Nutrition and Its Application in Clinical Dentistry*, 2nd ed. Philadelphia: W. B. Saunders, 1966.

16
THE PHYSIOLOGY OF DIGESTION

The previous chapters have laid the groundwork for our final considerations—the processes by which the foods are made available to the cells of the body. These processes are most conveniently described by simply following the course of events set in motion by eating.

The soaking of food with saliva during chewing primarily serves a lubricating function. Anyone who attempts to swallow dry food knows this. But an enzyme present in saliva (ptyalin) also serves to begin the digestion of polysaccharides. Such digestion can be easily demonstrated in the laboratory by collecting saliva and by allowing it to act on a previously boiled solution of powdered starch. The deep blue color reaction resulting from the addition of iodine disappears as the starch is digested. Or you can detect the change more easily by chewing unsalted crackers for some time and noting the emergence of a sweetish taste. (Hence, particles of crackers, bread, and the like, left between the teeth, can be turned into bacteria-supporting sugar). However, the enzyme responsible, ptyalin, is very sensitive to acids, so that what little digestion it does achieve must occur in the mouth before the food reaches the acid stomach. Chewing each morsel of food 50 to 100 times might allow time for considerable carbohydrate digestion, but is hardly to be recommended. The principal function of saliva, then, is to moisten and lubricate the food mass preliminary to swallowing.

Gastric Secretions

For practical purposes we can say that digestion begins in the stomach. Parts of a meal may remain in the stomach for as long as four hours. During this time the food mass, now referred to as chyme, is churned about by muscular contraction and mixed with the gastric juices. These juices are produced by tiny glands lining the stomach and include (1) hydrochloric acid in dilute solution, which provides a favorable environment for (2) the enzymes, digestive chemicals that break down the food-chemicals into smaller more easily absorbed forms.

However, the food is not physically crushed by the stomach, as is frequently imagined. In fact, as was long ago demonstrated, fragile, hollow wooden balls can be passed through the length of the alimentary canal without being crushed. The churning action of the stomach wall serves rather to ensure thorough saturation of the chyme with the stomach acid and enzymes. Fats and carbohydrates are fairly stable in acid and are little affected by gastric digestion. But the principal enzyme of the stomach, pepsin, acts in the acid medium to initiate the breakdown of protein. Since most animal and much plant tissue is essentially protein, its digestion results in a general disintegration of most foods.

You may be interested to note that proteins are normally quite resistant to chemical breakdown outside the body. For instance, the digestion of egg albumin (egg white) into its constituent amino acids requires boiling for 24 hours in a concentrated acid. An equivalent digestion is accomplished in the stomach and intestine in four or less hours, with a very much weaker acid and at considerably cooler temperature, by the aid of the enzyme pepsin, and the protein digesting enzymes of the pancreas and intestinal wall greatly accelerate chemical activity; in fact, this is their essential function throughout the body.

Control of Gastric Secretion

About 800 milliliters of gastric juice are secreted before, during, and after a typical meal. Approximately 200 milliliters of this are under nervous control, and is called the *psychic phase* of gastric secretion. Such nervous control of digestive processes was first demonstrated by Pavlov in his research on "conditioned reflexes." Pavlov found that a dog, conditioned to the ringing of a bell before presentation of food, will produce saliva upon hearing the bell, even if he is *not* given food. We now know that the stomach responds in a similar manner. The nervous stimulation of gastric secretion is a function of the *vagus nerve,* which carries impulses from the brain to the stomach and to much of the small and large intestines. Much that we know about the control of gastric secretion is the result of research on persons who, through surgical necessity, have permanent openings through the abdominal wall into their stomach (Fig. 16–1).

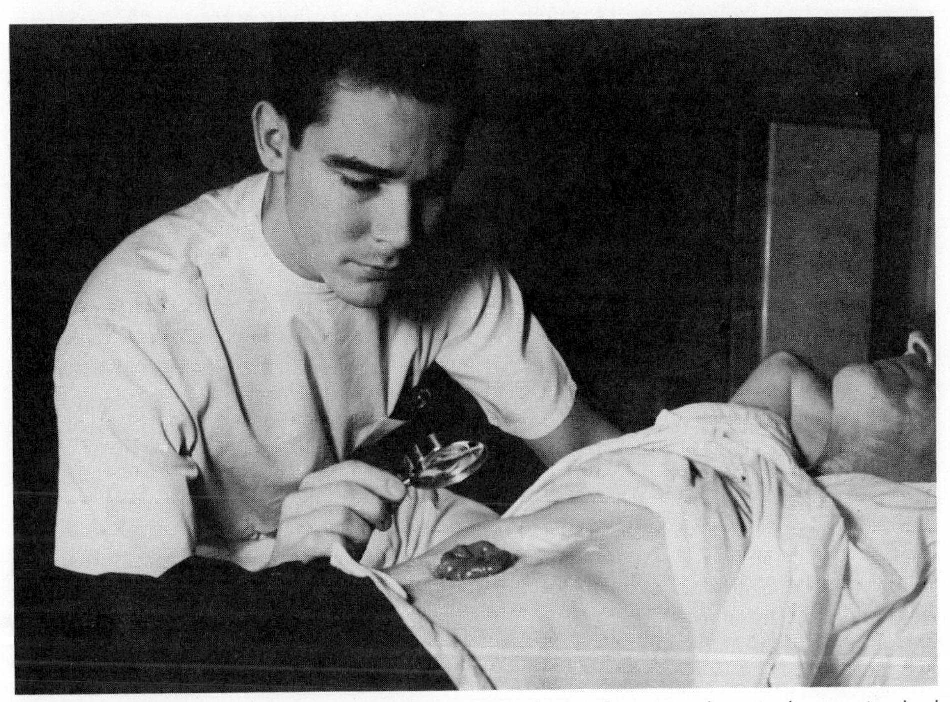

FIG. 16–1 Tom was a patient at New York Hospital who, because of surgical necessity, had a communication between the interior of his stomach and the outside of his abdomen. By means of such patients as Tom, doctors have been able to study the effects of emotions on gastric digestion. (Dr. Harold G. Wolff and Dr. Stewart Wolf)

We are all aware that our appetite is much affected by our "state of mind," that a state of excitement or depression will cause us to lose all interest in food. This is a normal physiological response and is referred to as the *sympathetic response.* This sympathetic response will be discussed at greater length in Chapter 33 but for our purposes here it is sufficient to say that readying of the body for active and alert behavior is incompatible with normal digestive processes and, as a consequence, the brain suppresses such a state.

The significance is obvious. Meals should be approached in a relaxed and calm state of mind. If an upsetting argument arises during a meal (unfortunately, a convenient time to thrash out family problems), there will be a suppression of gastric secretion along with the loss of appetite. No wonder that "indigestion" sometimes follows such meals, since the amount of gastric juice may have been reduced by as much as 20 percent. (Some experiments have suggested that gastric secretion may be totally suppressed.)

The other 600 milliliters of the gastric secretion is under hormonal control, and is called the *gastric phase.* Simply, the presence of food in the stomach causes certain cells in the stomach lining to release a hormone called gastrin. Gastrin is carried by the bloodstream to all parts of the body, but has its effect only on the digestive glands of the stomach, causing them to release their secretions. This seems to be rather a roundabout way of doing things, but that is the nature of hormones: They are carried to all the cells of the body even though they may act on relatively few.

The existence of gastrin was demonstrated by the following experiment. A piece of a dog's stomach was transplanted to a site on the outside of the dog's body where it could be observed. Such a transplant will develop a blood supply and heal normally, so that the dog has a piece of "outside stomach." When food was placed in the intact stomach, the transplanted piece began to secrete digestive juices. Since the transplant had no nerve supply, the stimulus must have come from the blood. Although this experiment was first done more than 50 years ago, it was not until 1964 that the stomach cells that produce gastrin were identified and the hormone purified [McGuigan].

Some foods more effectively promote gastric secretion than others. Partially broken-down proteins, which would be found in thoroughly cooked foods, are very effective, hence starting a meal with soup. Dilute alcohol also is effectual, so that the habit of a predinner cocktail has some measure of justification. It is interesting to note that alcohol is one of the exceptions to the rule that food is not absorbed from the stomach.

Intestinal Digestion

The chyme, in a semiliquid state, is passed into the duodenum. Here it is set upon by a great variety of enzymes. From the pancreas come enzymes that continue the digestion of protein and others that commence the digestion of carbohydrates and fats. In addition, the pancreas releases large amounts of sodium bicarbonate which neutralizes the acid-soaked chyme. Glands lining the wall of all parts of the small intestine produce the enzymes that are responsible for the terminal stages of chemical breakdown which result in the release of the basic food units: amino acids, simple sugars, fatty acids, glycerol, and a mixture of other molecules liberated in the process. These substances "enter" the body by passing through the wall of the small intestine into the bloodstream.

The glands of the pancreas and small intestine are also under hormonal control, that is, the presence of food in the duodenum causes the release of hormones by cells in the intestinal wall, which in turn effect the release of the pancreatic and intestinal juices. The gall bladder is emptied by a similar mechanism. In this case the presence of fat in the duodenum is the trigger for the release of the hormone, and the hormone (cholecystokinin: gall bladder activator) causes the wall of the gall bladder to contract. The nervous system apparently is of less importance in the control of these activities.

Role of the Large Intestine

The chyme is moved along the gut by slow peristaltic waves of contraction, and when these waves reach the end of the ileum they pass through the ileo-cecal valve, relax the valve, and push the chyme through. Physiologists now generally believe that the ileo-cecal valve serves principally to prevent a backup of the contents of the large intestine into the small intestine rather than to hold the chyme in the ileum.

When the chyme enters the large intestine, it has been pretty well depleted of usable food substances. It then consists largely of indigestible residues and water, and is referred to as *feces.* Bacteria normally residing in the large intestine act on these residues and the products of this bacterial action are responsible for the unpleasant odor of the feces.

As noted in Chapter 14, certain vitamins (K, biotin, folic acid, pantothenic acid) are produced by these bacteria and absorbed into the body, so that the daily requirements of such vitamins in the diet are indefinite. Use of some antibiotics may greatly depress the numbers of these useful bacteria, cut down production of the vitamins, and thus call for vitamin supplements to the diet until the normal bacterial population is restored. As also noted in Chapter 14, in newborn babies kept in a sterile environment the production of vitamin K, in particular, may be deficient; and since this vitamin is needed for proper blood clotting in case of an injury, it must be supplied in the diet until K-producing bacteria develop. Thus the large intestine and its bacteria play a useful role in nutrition.

The feces is gradually dehydrated as it passes through the large intestine. Normally by the time the feces reaches the transverse colon it has the consistency of soft mush, and in the descending colon it is a soft solid. The slower the passage through the colon, because of slower or weaker peristaltic waves, the greater is the degree of dehydration and the consequent possibility of constipation. Conversely, rapid or strong peristaltic waves, which can arise through irritation of the colon or in certain anxiety states, prevents proper water absorption, and diarrhea can result. Thus the colon plays a major role in the water balance of the body.

The extent of the fluid resorption by the colon is surprising. Large quantities of fluids are poured daily into the gut; figures which might be taken as typical for a 150 pound man are: a quart each of saliva and pancreatic juices; 2½ quarts of gastric juice; 2 quarts of intestinal juices; and ½ quart of bile, or a total of 7 quarts! Yet the normal daily loss of water by way of the feces is less than half a cup; and when we consider that the total blood volume of this 150 pound man is only about 5 quarts, we begin to appreciate the delicate balance maintained between secretion and absorption within the digestive tract. If this is upset by severe diarrhea or persistent vomiting, vigorous steps must be taken to replace the lost fluid by mouth, or by vein if necessary.

Gastro-Colic Reflex and Defecation

Strong peristaltic waves occur in the large intestine only about three or four times a day. They seem to arise following meals when the stomach is contracting, and this correlation is referred to by physiologists as the *gastro-colic reflex.* These contractions bring the feces down into the rectum and result in the desire to defecate. The common habit of emptying the bowels an hour or so after breakfast or after the evening meal takes advantage of this reflex. This desire can be suppressed and, after a short period of discomfort, generally disappears. However, feces retained in the rectum tend to become excessively dry, and the subsequent bowel movement may be difficult or even painful. In general, the dictates of nature should be followed if these are not unduly inconvenient.

Constipation and Laxatives

We are besieged with advice from the purveyors of laxatives regarding our bowel habits, even to the extent that the word "regularity" has achieved a rather singular connotation. No one can argue with the virtue of such regularity; the difficulty is in the definition of the word. Many people in perfect health regularly defecate two or three times daily, and others, just as healthy, have only two or three movements weekly. Proper bowel habits cannot, then, be expressed in terms of daily bowel movements; any healthy person's digestive tract will establish its own rhythm if left alone. A trip or a change in your daily routine may upset this rhythm for a few days; however, the return of a normal pattern can be expected without the aid of a laxative. Indeed, laxatives may interfere with normal function and may set up a vicious circle of dependence and further use. Therefore they should be taken only on advice of a physician, not as a "home remedy."

The Nature of Laxatives

Because of the general tendency of the public to resort to laxatives, from either real or fancied need, it might be useful to discuss their nature. The *cathartic drugs*, as they are called by the medical profession, may be classified according to their mechanism of action.

Irritant Cathartics

Irritant cathartics include senna, cascara, phenolphthalein, and castor oil.[1] They all cause a direct irritation to the wall of the colon inducing strong contractions and defecation. One would think that side effects, including the serious discomfort which accompanies the use of phenolphthalein and castor oil, would cause their disappearance from the druggist's shelves, but such is not the case. None of these substances are generally recommended by physicians except for very special purposes.

Bulking Laxatives

The gels such as agar and commercial preparations such as Metamucil, Methocel, Hydrolose, and Carmethose are generally indigestible substances which absorb large quantities of water (up to five times their weight). This increase in bulk distends the colon, stimulating the colon muscles, and results in defecation. The gelatins are probably less objectionable than the other types of laxatives because their method of action resembles the normal stimulus. Bran acts in a similar manner although it is not to be considered a drug.

Saline Cathartics

These are salts which are poorly absorbed by the intestinal tract. Their presence increases the osmotic pressure of the intestinal contents so that water is retained, and a fluid feces results. Examples are magnesium citrate (Milk of Magnesia) and sodium phosphate (effervescent sodium phosphate).

Intestinal Lubricants

Certain vegetable oils and mineral oil are indigestible. When ingested these substances become mixed with the fecal mass and act as mechanical lubricants. The principal objection to these oils is that they can interfere with the absorption of fat-soluble vitamins as well as of calcium and phosphates. When taken in excess they can also result in involuntary leakage of "incontinence."

Most laxatives are taken without medical advice, and with the mistaken impression that they cure constipation. They will not cure constipation since

[1] The seeds of the castor bean plant grown in flower gardens are the source from which castor oil is extracted. But the whole seed includes highly toxic substances which can result in fatal poisoning. Children are usually the victims of such poisoning.

the cause has not been treated, only the symptom. In fact their continued use can transform a mild constipation into a chronic one. They can also cause dehydration, loss of food nutrients, and irritation of the gastrointestinal lining.

Diseases of the Liver and Biliary System

Hepatitis

The term hepatitis means inflammation of the liver, but it is not an inflammation in the usual sense with inflammatory (white blood) cells pouring out of the blood vessels into the infected area. Instead, the reaction is usually one of death or *necrosis* of the liver cells. When damage is slight the dead cells are eventually removed and replaced by normal liver cells. If the damage is severe or prolonged, the dead cells are replaced by fibrous tissue laid down by connective tissue fibroblasts, and the resulting fibrosis is termed *cirrhosis.*

Viral Hepatitis

A number of viral infections of the liver are recognized according to the mode of transmission. "Serum hepatitis" or "transfusion jaundice" is transmitted by injection with virus-contaminated needles or syringes or by transfusion of blood containing the virus. What is commonly called "infectious hepatitis" is contracted from the feces of infected persons. Yellow fever is primarily a virus disease of the liver which is transmitted by the *Aedes* mosquito. Public health measures have greatly reduced the incidence of yellow fever, but the incidence of infectious hepatitis and of serum hepatitis seems to be increasing.

In the common infectious hepatitis, the first symptom is a loss of appetite (anorexia), and even the sight of food may bring on nausea. A jaundice (yellowing) then develops due to a deposition of bile pigments in the skin and in the whites of the eyes. Fever is present in about half the cases. The patient suffers great lassitude and fatigue, sometimes feeling almost well on rising in the morning but becoming completely worn out after an hour or more of ordinary activity. The patient usually recovers in two to six weeks, but may feel a general weakness for several months.

Although complete recovery is usual, some cirrhosis may result, and in occasional cases the liver becomes very soft and bright yellow in color and loses much of its weight. This is known as *acute yellow atrophy,* and ends fatally.

Unfortunately for the community the patient is infectious both before and for a long period after the actual course of the disease. He may unsuspectingly contaminate needles used by doctors and dentists, or even donate blood which contains the virus.

When viral hepatitis appears in a community, great care should be taken to avoid exposure. Particularly thorough washing of the hands following the use of public toilets, and before preparing or handling food is absolutely essential. If possible, all meals eaten away from home should be cooked foods and beverages packed in containers. Gamma globulin injections can offer a temporary immunity (perhaps up to 8 weeks) when exposure is likely.

Toxic Hepatitis

One of the functions of the liver is the detoxification of ingested poisons. Consequently, the poisoning resulting from ingestion of toxic metals such as lead or mercury, or chemicals such as carbon tetrachloride, primarily affects the liver. A necrosis results with symptoms similar to those of infectious hepatitis. Chronic poisoning usually results in cirrhosis.

Deficiency Hepatitis

Protein deficiency in experimental animals causes the liver cells to become distended with fat; this is followed by necrosis and finally cirrhosis. It is probable that the protein-deficiency disease of children, *kwashiorkor* (an African term meaning red boy) is principally a liver disease. In this disease the liver cells show fatty deposits characteristic of other liver degeneration, and, if the patient survives, cirrhosis and mental retardation result. The hair and skin sometimes show a red pigmentation, the cause of which is unknown. Where kwashiorkor is common there is a high incidence of carcinoma of the liver.

Gall Bladder Disease

Cholecystitis means inflammation of the gall bladder. The chief cause of cholecystitis is obstruction of the biliary system, usually by a gallstone. The inflammation is probably due to an irritative action of the bile salts.

Gallstones consist of bile pigments, cholesterol, and calcium which become precipitated in solid form. Gallstone formation usually occurs after age 35, and is more common in females than in males. It is also associated with overweight and with conditions in which there is an elevated blood level of cholesterol and other fatty substances (atherosclerosis and diabetes mellitus). The most typical patient, then, is an obese, middle-aged, diabetic woman who is beginning to show signs of advanced atherosclerosis (Fig 15–11).

The gallstones may be many and small, or single and large. The large stone is least likely to cause problems, but may be the cause of reflex symptoms such as indigestion and intolerance of fatty foods. Small stones are more likely to enter the duct system, become impacted, and cause cholecystitis. When this occurs the usual treatment is *cholecystectomy*, that is, the surgical removal of the gall bladder and the clearing of the biliary ducts of impacted stones.

Disorders of the Lower Alimentary Canal

Gastric and Duodenal Ulcers

The word *ulcer* is used to designate an open sore in the skin or in the alimentary canal. Ulcers in the stomach or duodenum are due to the action of the gastric acid and pepsin on the mucous membrane. Since acid and pepsin digest protein, and since the cells of the mucous membrane are largely protein, this should not be surprising. But normally this does not occur for two reasons. The stomach glands produce a great deal of mucus which provides a protective coating to the mucous membrane; and the gastric juices are usually mixed with food and do not normally become concentrated along the stomach or duodenal lining. The presence of gastric juices in an empty stomach tends to wash away the protective mucus, resulting in the digestion of the stomach wall, or when the juices are passed on undiluted into the duodenum, they attack the duodenal wall (Fig. 16–2). In fact the duodenal wall seems to be more liable to such autodigestion than is the stomach, probably because its lining is not so constructed to resist the onslaught of such concentrated gastric juice. When any such digested area becomes an open wound, it is termed a peptic ulcer; the terms gastric ulcer or duodenal ulcer merely indicate the location.

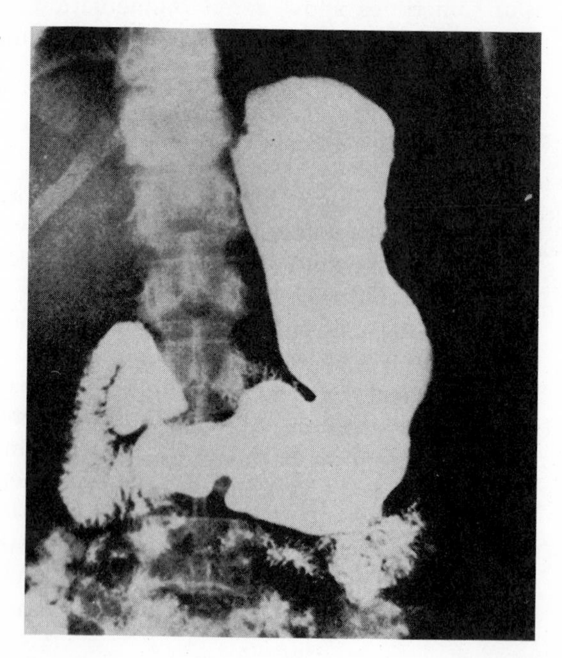

FIG. 16–2 A barium X-ray showing the stomach on the right, and the duodenum beginning at the sharply constricted area, and forming a "C" with its feathery part passing behind the stomach.

The typical ulcer victim usually complains of burning pain in the region of the stomach about an hour after meals. The pain is usually relieved by antacid preparations or by food. If there is bleeding from the ulcer, the stool becomes soft and black, and some symptoms of anemia (blood loss) may be present. These are conditions which obviously justify a physician's examination.

The question is why gastric juice should be secreted into an empty stomach. Recall that the anticipation of food brings about gastric secretion. This is an example of our senses effecting changes in the function of our digestive organs, preparing them, so to speak, for the acceptance of food. However, mental states other than appetite can effect such changes. For example, a high tension state, anxiety, or persistent worry can bring about a discharge of nervous impulses from the vagus nerve, and a consequent release of gastric juices. This close association between certain emotional states and the development of peptic ulcers has led to the recognition of *psychosomatic illness,* that is, disease states brought on by psychological disturbances. There is some evidence of association between peptic ulcers and coronary heart disease. This is not surprising since one of the factors implicated in coronary heart disease is anxiety.

Peptic ulcers seem to be by-products of a highly civilized society. They develop particularly in those people who are continually exposed to the stresses and strains of a complex life. Since it is a type of neurosis, one would think that it could be treated with psychotherapy, but this does not seem to be the case. However, most ulcers heal if the individual is relieved of his strains and tensions, and avoids foods which aggravate the condition (coffee, aspirin, and tobacco appear to be special offenders). The usual ulcer victim has recurrences even after complete healing, and many eventually require surgery.

Vomiting

Vomiting is a protective mechanism in man and animals which enables the stomach and upper part of the small intestine to rid itself of substances detrimental to the organism's well-being. It commonly occurs when the stomach or duodenum is overly sensitive because of infection (as in influenza) or other disorder (peptic ulcer), or as the result of the ingestion of a poison which irritates the stomach or intestinal lining. Nerve impulses are transmitted from the stomach or intestinal wall to the vomiting center of the brain which initiates the motor activities resulting in the expulsion of the stomach and intestinal contents. The sequence of events are an inspiration of air and a contraction of the abdominal muscles; closure of the air passages; relaxation of the sphincter separating the stomach and esophagus; and strong reverse peristaltic contractions upward through the small intestine, stomach, and esophagus.

Since the act of vomiting is under the control of the brain (the vomiting center is in the medulla oblongata, a part of the brainstem; see Chapter 35), vomiting can be induced by any factor acting directly or indirectly upon this center. Such factors include the drugs apomorphine, morphine, copper sulfate, and some forms of digitalis; electric shock; migraine; conditions related to the vestibular apparatus (Chapter 38) such as motion sickness; and even the cerebral cortex as shown by the vomiting induced in response to certain nauseous odors.

As with diarrhea, persistent vomiting can lead to dehydration and require restoration of fluids by intravenous injection. If the vomiting is unusual in any respect, for example, if even a small amount of blood is lost, a physician should be called. Usually, however, the cause of an episode of vomiting is apparent. Moreover, in many instances, such as food poisoning, the relief that follows ridding the body of pernicious substances is ample evidence that the mechanism is basically protective.

"Colitis"

Many persons complain of an irritable lower colon. The name "colitis" has been applied to the condition, and it is inappropriate inasmuch as the irritable colon is neither inflamed or ulcerated. A more or less harmless condition should not be given a name which indicates a serious and even fatal disease. The colon is normal in these cases (Fig. 16–3); the trouble lies in certain emotional states of these persons that cause superfluous nervous discharges to the colon, much as anxiety causes such nervous discharges to the stomach in persons with peptic ulcers (Fig. 16–4).

Most persons with an irritable or "spastic" colon are tense and sensitive. Almost any emotional stress seems to cause pain in the lower colon with the need to go to the toilet to pass a little gas or mucus. Neither gas nor mucus is cause for alarm in themselves, but, if after thorough examination, the physician diagnoses the condition as irritable colon, the patient must adjust to the fact. No "cure" seems to be available—the best which can be expected is prescriptions for drugs which decrease intestinal motility to quiet the "rumbles" and advice on a diet which has less tendency for gas formation. This at least will make the condition less of a social handicap.

Indigestion

The complaint of indigestion is so vague as to have little medical significance. It may be a nausea due to undigested food in the stomach; vomiting; diarrhea; heartburn due to regurgitation of stomach acid into esophagus. Very similar pain to that of indigestion can arise from kidney stones, menstrual cramps, gall bladder disease, or appendicitis. Some people experience an occasional abdominal pain of a given sort which they find to be relieved by aspirin or

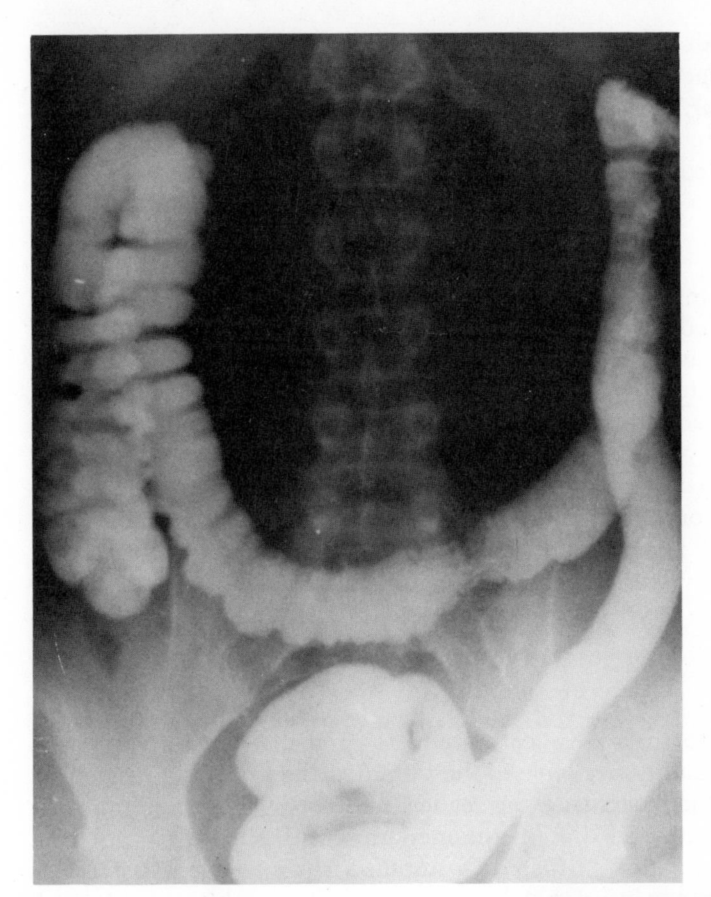

FIG. 16–3 An X-ray photograph of the colon. Identify the ascending, transverse, descending, and pelvic parts.

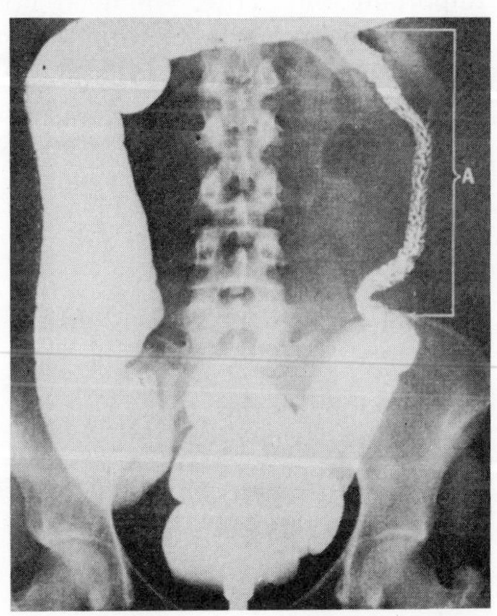

FIG. 16–4 Spasm in the colon, shown in the area marked "A." (Dr. John A. Campbell)

an antacid preparation or some previously prescribed medication, and there is probably little objection to such symptomatic relief. However, any abdominal pain of particular severity, or accompanied by a fever, or which persists or recurs over a long period should be regarded with more concern, and a physician should be consulted.

Appendicitis

Appendicitis is an infectious inflammation of the vermiform appendix. If untreated the diseased appendix may burst and its infective contents freed into the peritoneal cavity. The peritoneum is then subject to the infection, and the resulting peritonitis can cause death.

The condition is accompanied by severe abdominal pain, and the widespread notion that all abdominal pain should be treated with laxatives cannot be severely enough condemned. Severe abdominal pain accompanied by a fever and vomiting or nausea may indicate appendicitis, and a physician should be consulted immediately. Laxatives place further stress upon the delicate tissues already weakened by the inflammation, and a bursting of the appendix can result.

Conclusion

Most people attend to their own eating and to their own digestive upsets, since medicines are readily available to the public. If the foregoing discussion does nothing more than persuade the reader that his digestive system is no simple matter and that even diet, in the modern world, should be based on sound information and not entrusted to impulse, it will have achieved an important aim. True, we need not burden our doctor with every stomach ache and temporary constipation; nor should we become a faddist worrying about every calorie and vitamin. But chronic digestive trouble should indeed be brought to medical attention lest it lead to ulcer or worse. We would not attempt home treatment of a persisting heart or kidney disorder; why, then, disorders of the digestive system?

REVIEW QUESTIONS

1. What is the role of the stomach in the digestive process?
2. What is the function of the large intestine?
3. What is the gastro-colic reflex?
4. How do laxatives work?
5. What is the most common location of a peptic ulcer? What are some of the symptoms of a peptic ulcer?

REFERENCES

McGuigan, J. E., "Gastric mucosal intracellular localization of gastrin by immunofluorescence," *Gastroenterology*, **55:** 315–327 (1968).

Additional Readings

Alvarez, Walter C., *Nervousness, Indigestion, and Pain.* New York: Collier Books, 1962. (Paperback)

Berland, Theodore, and M. A. Speelberg, *Living with Your Ulcer.* New York: St. Martin Press, 1971.

Davenport, H. W., *Physiology of the Digestive Tract.* Chicago: Year Book Publishers, 1961.

Guyton, A. C., *Textbook of Medical Physiology*, 4th ed. Philadelphia: W. B. Saunders, 1970.

Lyght, Charles E. (ed.), *The Merck Manual*, 11th ed., Merck, 1966.

17
DIET, BODY WEIGHT, AND BODY COMPOSITION

The prevalence of obesity in the same world where millions die each year of starvation is indeed a curious predicament. Statistically, out of every nine Americans one is obese. There is a seemingly endless barrage of weight-reduction diets and simple exercise programs for weight loss directed at the American public via popular magazines. Pharmaceutical companies have increased profits by marketing various over-the-counter aids to weight reduction, and enterprising business men and women have capitalized on the public's apparent need for directed and partly passive exercise programs.

At the same time obesity is an increasing concern, we find that among Americans with even better than average incomes, certain vitamin and mineral dietary deficiencies exist. What should we eat? What should we not eat? How much should we eat? How can we prevent unwanted weight gain? The reader should be able to answer these and other important related questions with the application of a few concepts and principles.

Hunger, Appetite, and Satiation

Why do we eat? Why do we get "hungry" when we do? Why do we eventually stop eating and what determines when we are satisfied? The usual answers to such questions are teleological—we eat to survive; we get hungry so that we will eat; we tend to be satisfied when we have eaten enough—but the regulatory mechanisms that account for our eating habits are far from clear.

There is a difference between *hunger* and *appetite. Appetite* is a sensation, usually pleasant, by which we are aware of a desire for food, perhaps even a special kind of food depending on time of day, mood, weather factors, and so on. *Hunger* is an unpleasant sensation by which we are made miserably aware of our need or desire for food. *Satiation* is the sensation by which we feel that hunger or appetite has been satisfied even though more food is still available.

It is clear, assuming adequate food is available, that the interaction of appetite, hunger, and satiety regulates food intake. The mechanisms controlling appetite (largely psychological), hunger (physiological), and satiety (physiological and psychological) are not so clear.

An area in the hypothalamus of the brain plays a major if not dominant role in food intake regulation. There are two pairs of centers in this so-called appestat that work to control food intake much as the thermostat in your house controls heating. The feeding centers basically initiate the desire or need to eat while the satiety centers, either directly or by serving as a brake on the feeding centers, are responsible for terminating food intake. Experimental destruction of the feeding centers in animals leads to anorexia, loss of the desire to eat. That this is not the only factor involved in hunger is demonstrated by the fact that these animals, if force-fed for a short time, can regain the desire to eat. The "braking" or satiety centers are dominant; once these are destroyed, the animal will never learn to control his food intake at a reasonable level and is said to be *hyperphagic*.

A variety of stimuli apparently affects the appestat centers. Any one or a combination of nervous, chemical, and thermal stimuli serve to initiate the desire to eat. Nervous impulses signal that the stomach is empty when hunger contractions of the stomach occur, but an animal with a denervated stomach still feeds and so does a human whose gastric nerves have been cut during stomach surgery. Slight changes in the temperature of the blood flowing through the appestat may affect feeding—temperature fall, and satiety—temperature rise, but the evidence for such a mechanism is not overwhelming.

One aspect of blood glucose level, a chemical stimulus, has the greatest acceptance among physiologists as a mechanism controlling food intake. The absolute blood glucose level cannot be the stimulus or a diabetic person would never be hungry! Dr. Jean Mayer demonstrated that the arterial-venous *difference* in blood glucose, "effective glucose," is high after eating and gradually falls until it is negligible, at which time the subject is hungry. Free-fatty acids, freed and ready to use for energy, rise during fasting, and the level of circulating free-fatty acids is closely associated with hunger. [This accounts for the difficulty of dieting (losing fat) without hunger.]

None of these factors accounts for the *amount* of food we eat, however. Digestion and absorption of food have barely begun when we stop eating! At present there is no adequate explanation for the amount of food we eat at one sitting.

There are two kinds of food intake regulation: short term and long term. The latter is most effective and accurate. Again, we do not know the mechanisms, but physiologists speculate that feedback from the fat stores, either through the nervous system (sensory receptors in the adipose tissue) or by chemoreceptors in the appestat (sensitive to the fat present in the blood), may play a role. The role of the endocrine system cannot be overlooked as a possible contributing factor nor can genetic control of certain kinds be ignored. The most remarkable thing is that man, on the average, regulates his food intake so accurately with little if any conscious effort!

What to Eat

There are three systematic approaches to the selection of foods. One is the standard and well-known "basic four food groups" plan; a second is a guideline based on recommended fat and carbohydrate caloric percentage plus protein intake as based on body weight; the third and probably best approach is an intelligent combination of the first two methods.

The Basic Four Approach

Foods have been divided into four general groups. Simply following the recommended servings from each of these groups supposedly provides the nutrients regarded as essential. The minimum servings from each of the groups would provide a total of around 2000 kilocalories, which is less than that recommended for most adults. The caloric deficiency would normally be made up by desserts or by additional servings of the same groups, and would generally present no problem. A summary of the four food groups is given in Table 17–1.

MILK GROUP. This includes all dairy foods. The equivalent of two cups of milk per day, as recommended for adults, provides almost three-quarters of the daily allowance of calcium, one-half the allowance of vitamin D, and about one-sixth the allowance of vitamin A. Skim milk has the same amount of calcium as whole milk, but the removal of fat eliminates the A and D vitamins. If the skim milk is sold as "fortified," this generally means that these vitamins have been restored. Cheese and ice cream have considerably greater concentrations of calcium and vitamin A, but variable amounts of vitamin D. Note

TABLE 17–1 Recommended Dietary Allowances

Group	*Servings*	*Includes*	
Milk	Some milk daily	Children	3 to 4 cups
		Teen-agers	4 or more cups
		Adults	2 or more cups
		Pregnant women	4 or more cups
		Nursing mothers	6 or more cups
		Cheese and ice cream can replace part of the milk.	
Meat	Two or more	Beef, veal, pork, lamb, poultry, fish, eggs, with dried beans and peas and nuts as alternates.	
Vegetable-fruit	Four or more	A dark green or deep yellow vegetable important for vitamin A—at least every other day. A citrus fruit or other fruit or vegetable important for vitamin C—daily; other fruits and vegetables including potatoes.	
Bread-cereals	Four or more	Whole grain, enriched, restored.	
Other food	Normally included in the daily diet	Butter, margarine, sugars, and unenriched grain products serve to provide the caloric and nutrient allowances.	

that the milk allowance for children and teenagers is at least double that for adults.

MEAT GROUP. The meat group includes not only meat, poultry, eggs, and fish, but also other so-called alternate protein sources such as peanuts and legumes (peas and beans). The two servings suggested in Table 17–1 will provide about one-half of the recommended allowance of protein. This is adequate since the balance of the protein will be provided by the milk and bread groups. Two eggs are regarded as one serving in this group. The meat group also constitutes a major source of vitamin A and iron.

VEGETABLE AND FRUIT GROUP. The recommended serving of one deep yellow or dark green vegetable contributes substantially to the vitamin A requirement. The other specific recommendation in this group is for one citrus fruit to meet the ascorbic acid requirement. Servings of at least two other vegetables help to meet the requirements for the B vitamins, and also contribute calcium and carbohydrate.

BREAD AND CEREALS GROUP. This group includes bread in any shape or form, and made from any grain cereal. It includes macaroni, noodles, spaghetti, and the hot and cold cereals served at breakfast. In this country most of these products are derived from wheat. The four servings recommended provide a cheap source of calories, as well as iron and B vitamins which are added in the enrichment process now required by law in most states.

Professional nutritionists and many others who write for lay magazines seem to get carried away by milligrams, international units, and long lists of special nutrient foods. Yet they end up with menus that are little if any better than those that a housewife could plan on the basis of such a guide as the foregoing. The major disadvantages of this approach to diet is that one can easily follow it and yet take in an excess of total calories or an excess of fats or both.

The Percentage-of-Calories Plan

Even more fundamental than the Basic Four plan discussed above is the classification of foods according to the three major components: carbohydrate, fat, and protein. The percentage-of-calories plan is based upon what nutritional science knows about the body's need for these major foodstuffs.

PROTEIN. The recommended requirement for protein for the mature adult is roughly 0.9 gram of protein per day for every kilogram of body weight (recently reduced from a former level of 1.0 gram). This is about 0.5 gram per day per pound of body weight. The recommended intake is adequate, with a built-in safety margin, to replace the protein lost daily and to maintain tissue protein. The allowance should be *increased* as follows: for children of both sexes, 1.3 gram per pound per day until age 3; 1.1 gram per pound per day for ages 4–6; 1.0 gram per pound per day for ages 7–9; and 0.9 gram per pound per day for ages 10–12; for ages 13–15, females need 0.6 gram per pound per day and males 0.75 gram per pound per day; for ages 16–19, males need 0.6 gram per pound per day; add 8 grams per day during pregnancy and 20 grams per day during lactation; for an adult to add weight, increase to .75 gram per pound per day.

FAT. There are two factors involved in determining optimal fat intake: total fat calories and proportion of fat calories derived from unsaturated fats. The current recommendation is 25 percent of total calories from fat, allowable range is 20 to 30 percent, and 50 to 70 percent of these fat calories in the form of unsaturated fats. Keep two concepts in mind: (1) fats are needed in the diet; (2) an excess of fats, especially saturated fats, may promote obesity and atherosclerotic disease.

CARBOHYDRATE. The carbohydrate intake is actually determined by subtraction; the caloric intake over and above the 0.5 gram protein per pound of body weight per day and the 25 percent fat will naturally be in the form of carbohydrate. Thus the percentage and absolute carbohydrate intake will vary with the total caloric requirement, but will normally range from about 55 to 65 percent of the total caloric intake. For example, two men, each weighing 150 pounds and consuming a 25 percent fat diet, will ideally take in 75 grams of protein (about 300 calories); the inactive one with a 2100 caloric requirement will consume about 525 fat calories (25 percent) and 1275 carbohydrate calories (60 percent); the more active one needing 3000 calories per day would consume 750 fat calories (25 percent) and 1950 carbohydrate calories (63 percent).

Although the carbohydrate percentage is actually determined by the protein and fat intake, of primary concern is the kind of carbohydrate consumed. There is some evidence that sucrose intake should be minimized, although not necessarily eliminated, so that no more than 80 to 100 grams per day are consumed. It can be assumed that sucrose is the only or predominant carbohydrate in table sugar, soft drinks, cakes, pies, ice cream, and fruits (in other words, things that taste sweet) in contrast to vegetables which contain mostly starch.

The chief disadvantage of the percentage-of-calories plan is the slight inconvenience of completing a dietary record of calculations based on your tastes and on the information from a comprehensive caloric table. Once this is done, however, and you reconcile your food preferences with what you learn from your initial calculations, it is not necessary to continue such diet recalls or calculations so long as you adhere to the diet you have determined as nutritionally optimal and of acceptable palatability. The practical use of this approach will be illustrated in connection with the third approach: a combination of the Basic Four and percentage-of-calories methods.

A Practical Combined Approach

It is quite possible, in fact in America it is quite *likely,* that you can use the Basic Four approach to dietary selection and not only overeat but take in the wrong kind of calories. A literal interpretation of the Basic Four nutritional guide would permit a day's dietary intake as follows: two eggs, toast and butter, and orange juice for breakfast; a hamburger-on-bun, french fries, glass of milk, and chocolate cake for lunch; two pork chops, two boiled and buttered potatoes, spinach, two slices of bread with butter, milk, and peaches for dinner. Without even considering the total calories (about 2700) which may or may not be in balance with energy expended, this diet contains better than 50 percent fat!

It is also possible to select from the Basic Four and derive *less* than the recommended daily allowance of certain vitamins.

The most practical approach, then, uses a combination of the Basic Four with your own food preferences as a guideline and then keeping a record of your diet for one week. At that time check to see if your average daily intake of vitamins, minerals, and protein meets the recommended dietary allowances. Then determine if you are taking in the correct proportion of fat, not to exceed 25 to 30 percent of total calories; the minimum proportion of unsaturated fats, at least 50 percent of fat intake; and no more than 90 to 100 grams of sucrose. If your normal intake meets the percent-of-calories guidelines and your weight is acceptable to you, it is likely that your intake, both in terms of quantity and quality, is healthy and nutritious. If necessary, adjust either the caloric intake or the proportion of foods or both to meet the percent-of-calories and Basic Four guidelines. Once you have determined what is a reasonable and nutritious diet for you, you shouldn't need to continue the dietary record.

How Many Calories?

How many calories per day, on the average, should you consume? The answer depends upon whether you are still growing and whether or not you are presently at your desired weight. Unfortunately, this question is of little practical value though it provides the basis for understanding one's caloric requirements. There are, however, two practical approaches to determining your average daily caloric requirement.

The National Food and Nutrition Board estimates that the average adult American male needs about 2900 calories per day, and the female about 2100 calories per day. It is quite obvious that exceptionally large or small and very active or inactive adults will need more or less calories respectively. The average requirement for infants is about 60 calories per pound from 1 to 3 months of age, 55 per pound from 4 to 9 months, and 50 per pound from 10 to 12 months. From 1 to 3 years of age, 1300 calories per day is recommended with a 400 calories per day increase at ages 4, 7, and 10 so that the recommendation for ages 10 to 12 is 2500 calories per day; the recommendation for boys is 3100 for ages 13 to 15 and 3600 for ages 16 to 19; and for girls it is 2600 for ages 13 to 15 and 2400 from 16 to 19. Again, it should be obvious that these are only averages and must be adjusted according to body size and amount of daily physical activity.

If a child is not growing at a normal rate, one can readily check to see whether he has an adequate caloric intake by keeping a dietary record for one week and comparing his intake with the average figures presented above.

The second method of determining average daily caloric requirement involves, first of all, some common sense. With respect to your body weight the two questions you answer are: (1) "What do I want?" and (2) "What is happening?" To illustrate:

> "I want to maintain my present weight." "I *am* maintaining and have maintained my present weight for two years." These answers indicate that your ideal caloric requirement is obviously just what you *are* consuming.
>
> "I want to lose weight because I have some 10 pounds excess fat." "I am maintaining my present weight." If this is so, your caloric intake is too high. Complete a one-week diet recall, determine your average daily intake, then reduce your intake by 500 to 1000 calories per day; that will temporarily be your daily requirement until your desired weight has been attained. At that point your requirement will increase

slightly in order to maintain your new weight and to prevent further weight loss. (The recall step can actually be eliminated if you like; the same thing can be accomplished by simply eliminating 500 calories from your normal intake each day and/or increasing your energy expenditure by 500 calories.)

"I want to gain weight." "I have weighed the same for six years." If such is the case, it is obvious that your current calorie intake is balanced with your energy needs. To gain, simply add 500 to 1000 calories per day, keeping the principles of "what to eat" in mind, until your desired weight has been achieved; then reduce intake appropriately to prevent further unwanted weight gain.

The point is simply this—you can determine your own caloric requirement based on a simple knowledge of what you want (to lose, to gain, or to maintain weight) and what is currently happening to your weight and body composition. You can calculate the actual requirement if you like or simply work with the knowledge that it is acceptable as it is, or that you need to increase or decrease the intake by so many calories per day. In order to decrease or increase, you should pay close attention to the nutritional requirements and not deal simply in caloric increase or decrease alone. Other factors enter into a discussion of losing and gaining weight and will be amplified later in this chapter.

How Often Should One Eat?

There is probably no absolute rule that prescribes the best meal distribution or meal spacing pattern. At one time man was led to believe that snacking all day and during the evening was bad. Certainly if such a habit leads to obesity it is questionable, but what about the snacker who is of normal weight?

To be sure the first consideration is the total daily caloric intake, but of almost equal importance is the distribution of the caloric intake. It is apparently better to consume the calories at several meals rather than to gorge oneself at one or two sittings. Population studies and small animal experiments tend to support the benefits of an even distribution of food intake as opposed to the one or two large meals per day [Fabry].

Rats fed during a three-hour period each day as contrasted with those fed *ad libitum* ate less and were lighter but were actually fatter, had larger stomachs, and gained more than twice as much weight per gram of food consumed than the *ad libitum* feeders! (See Figure 17–1.)

On the average, animals and people who consume their calories in one or two daily sittings ("meal-eating" or in the extreme, "gorging") definitely tend to be more obese. In addition, such humans have a higher average blood cholesterol level.

It is recommended, then, that the calories be spread out over at least three meals daily and that no greater than 40 percent nor less than 25 percent of the

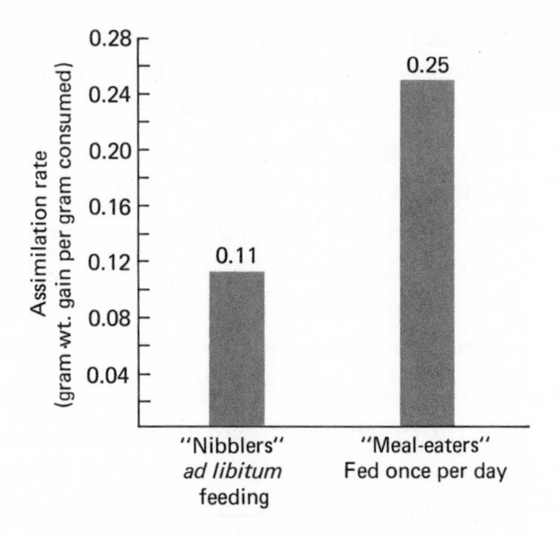

FIG. 17–1 Comparison of the assimilation rate of consumed food in "nibblers" and "meal eaters." The ratio shown is the weight gained per gram of food consumed; there were 16 in each group of male albino rats. (Johnson and Cooper)

daily intake be consumed at any one meal. These figures are arbitrary but do provide for a "nibbling" approach rather than one of "gorging."

How to Gain Weight

It must be obvious that to gain weight caloric intake *must* exceed energy expenditure. First one should be certain that no organic abnormality is either the cause or an associate of the condition of being underweight (anemia, hyperthyroidism, and so on). Second, it should be pointed out that some degree of underweightness, especially after age 20, is no cause for concern, in fact is even desirable. Assuming there is no disorder and you do want to gain weight, the best course of action is probably to eat the normal diet described on pages 240-244, with a slight increase in protein to perhaps 0.75 gram of protein per pound of body weight. Then, in addition to the usual meals that have maintained your present weight, add 300 to 500 calories per day as a "snack." If this snack does not appreciably reduce your daily intake at regular meals, you should certainly gain weight, probably at the rate of some 1 pound per week if the excess is 500 calories per day. High calorie, low bulk preparations are available commercially either as canned liquids or powder to which milk is added. With whole milk, some 350 to 375 calories can be consumed in only 8 ounces of liquid.

Obesity

Though we often confuse the terms, obesity is *not* the same as being overweight. Obesity means excess body fat. Overweightness may be comprised entirely of excess fat, of no excess fat, or of any amount of excess fat between these extremes; but obesity refers only to the excess body fat present. Highly muscled persons may be overweight but not obese, and occasionally, we see underweight people who are actually slightly obese.

Though we do not know what is ideal for a given person, we do know that the average college male is 14 percent body fat and the average college coed is 20 percent body fat. It is also true that these averages increase with age, but there is no logic or evidence to support the *necessity* for any increase in body fat with age. Therefore, it is a fair assumption that anything in excess of 14 percent for men and 20 percent for women probably represents excess body fat.

Determining the Percent of Body Fat and Obesity

Although the standard height-weight tables (Table 13–1) for each sex are useful guides and better than none at all, they leave much to be desired. An actual assessment of percent body fat is preferred. If an appropriate device (skinfold fat calipers[1]) or method (underwater weighing for body density) is not available to you, other guides can be used to supplement the height-weight table.

$$\text{Overweight index} \quad \frac{100 \times \text{body weight (in kilograms)}}{\text{ht (cm)} - 100}$$

Amount over 110 percent is considered "overweight."

THE PINCH TEST. If you can pinch up a skinfold (Fig. 17–2) thicker than 1 inch on the back of the upper arm, abdomen, side of the chest at nipple height, or on the back just below the shoulder blade (scapular area), you are probably obese at least to some extent.

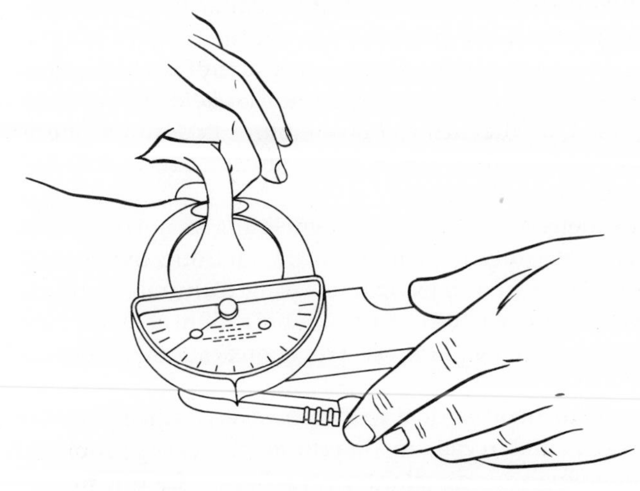

FIG. 17–2 The use of fat calipers for skinfold measures.

[1] The department of physical education or department of nutrition at many universities will have a skinfold caliper. An ordinary caliper will not give reliable and accurate measurements.

Causes of Obesity

Obesity is an excess of a normal constituent of the human body. It is not a matter of whether one has fat in his body but how much. The adipose tissue (fat cells) provide stored energy and help to insulate the body (especially the adipose tissue located in the subcutaneous region). Fat provides over nine kilocalories per gram, whereas glucose provides slightly over four calories per gram; therefore more fuel can be stored in a given space than with glycogen, stored glucose. Most of the excess calories not needed immediately for energy or tissue repair are converted to triglycerides (neutral fats) and then stored as adipose tissue. Some storage of fat is quite normal, indeed very important, but excessive fat storage may be very undesirable.

What then causes excess body fat? The simplest answer is excess calories, that is, more calories consumed than metabolized. No matter what the underlying cause, obesity is directly attributable to storage of calories. An excess of about 3500 calories over any given period of time will mean about one pound of additional body fat and possibly an additional half pound or so in water retention.

Obesity can be classified according to cause: (1) exogenous, where the cause is from "outside" the body and (2) endogenous, from within the body. Although it is estimated that only a small percent of obesity is endogenous in origin (an inherited enzymatic, endocrine, or hypothalamic disorder), it should not be dismissed. For example, Jean Mayer believes that a significant proportion of obesity in this country may be due to an hereditary disorder of fat metabolism, similar to that observed in a strain of hereditarily obese mice. Such mice, if placed on restricted caloric intakes, lose not their stored fat but protein from their muscles. Mayer points out that a measure of consumption of stored fat in a dieting person (or mouse) is the amount of ketones (a waste residue of fatty acids) excreted in the urine. Mayer found that some obese women when placed on low calorie diets fail to show an increase in urinary ketones and complain of muscular weakness.

However, it is generally believed that an overwhelming majority result from exogenous causes. The possible contributors are neurotic overeating (compulsion, boredom, and so on), nonneurotic overeating (cultural), and decreased energy expenditure (forced inactivity or immobilization or voluntary inactivity). There is little question that *all* of these exogenous causes of obesity are widespread in America.

Neurotic overeating may be difficult to deal with. However, a sincere desire plus nutrition education and a diet recall, to determine the exact problem, can go a long way toward effecting sound weight control habits. Caloric modification is mandatory when forced immobilization is necessary. If the problem is *voluntary* inactivity, it must be faced realistically.

Either overeating or inactivity (or both) contribute to the positive caloric

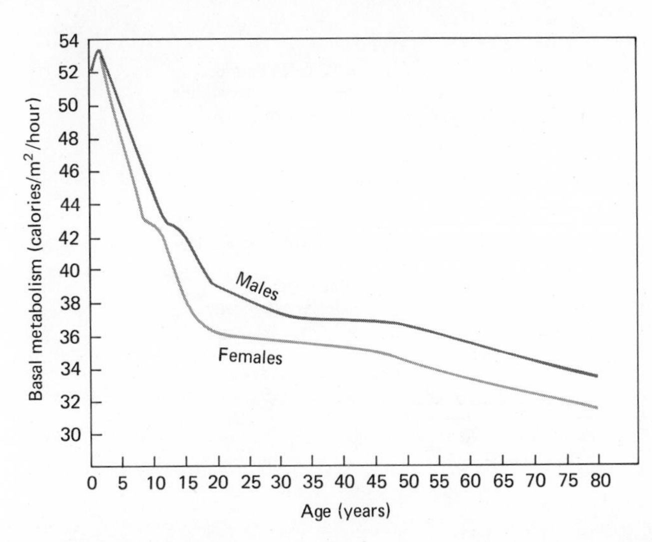

FIG. 17–3 The basal metabolic rate by sex and age. (A. C. Guyton, *Function of the Human Body*. Philadelphia: W. B. Saunders Company, p. 361, 1964.)

balance that leads to obesity. Two practical examples demonstrate the effect such a positive caloric balance can have on fat deposition.

Figure 17–3 illustrates the typical decline in Basal Metabolic Rate (BMR) as we age. The BMR is the metabolic rate at its lowest level, as close to complete relaxation as possible. Though there is a wide variation in BMR for a given age, and men and women differ considerably even when body size is compensated for, it is quite predictable that individuals will nonetheless have a decreasing BMR. The decrease in the postadolescent years approximates only eight calories per day per year; that is, each year one's BMR will probably be some eight calories per day less than the previous year. This seems inconsequential unless some simple arithmetic is employed. Figure 17–4 illustrates the potential weight gain from excess stored fat during a five-year period based on this very subtle eight-calorie-per-day decrease. The excesses are progressively 2900 calories (365 × 8) the first year, 5800 calories the second year, and so on. At this rate a person who eats the same but has the typical slight-but-steady decline in BMR would have an excess of some 43,500 calories over five years. This amounts to about twelve pounds of additional stored fat! By the same token, a young person who exercises vigorously an hour each day in college—let us say he plays recreational basketball expending 400 calories each of five weekdays—would gain some fifteen pounds of excess fat during just a six-month period if he stopped the activity and maintained the same caloric intake (Fig. 17–4). Although these examples seem extreme, they are quite realistic and occur to a greater or lesser degree in a large proporiton of college-age individuals.

Science does not as yet know enough about the causes of obesity, and there may yet be identified some significant endogenous factor. But, we *do* know a good deal about overeating, excess fat intake, and inactivity, all of

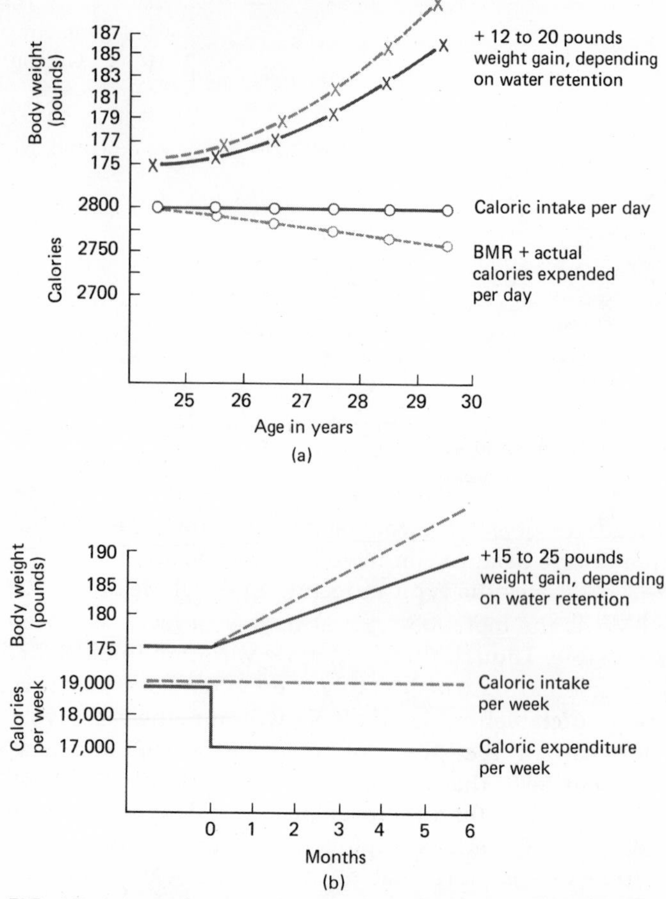

FIG. 17–4 (a) A theoretical weight gain from normal fall in BMR from age 25 to 30. The average fall at this age is only 8 calories per day, but this would cause a weight gain of 12 pounds over five years (even if activities stayed the same), if intake was not reduced. Assume 3500 calories = 1 pound body weight. (b) A theoretical six-month weight gain resulting from a cessation of one-hour recreational basketball, five days per week, and a continuation of the same caloric intake. Assume 3500 calories = 1 pound body weight.

which can and do cause obesity. In most cases, these are conditions over which the individual has some control.

Obesity and Health

Obesity is related to a wide variety of human disorders and diseases. There is not often a single direct cause; however, a predisposition of the person to a number of disease processes seems beyond dispute. For example, obese persons are two-and-a-half times more susceptible to hypertension and twice as likely to have arteriosclerotic heart disease. Death rates for obese men and women are significantly higher than for the nonobese for cardiovascular-renal disorders (including coronary artery disease, cerebral hemmorrhage, and chronic nephritis), liver and gall bladder cancer, appendicitis, and noncarcinogenic gall bladder disease. In addition, death rates for obese men are significantly higher from cirrhosis of the liver, hernia and intestinal obstruction, and automobile accidents, than for nonobese men. For obese women, the death rate for accidents in general is higher than for nonobese women. The most significantly increased death rate for obese men and women is for diabetes; in the obese, the death rate from diabetes is almost four times that of nonobese persons. In general, a 45-year-old man's life expectancy is reduced by 25 percent if he is 25 pounds overweight.

One must wonder, however, what conclusion is to be derived from these facts: obese persons are significantly *less* likely to die from tuberculosis, suicide, and stomach or duodenal ulcer, and obese women are *less* likely to succumb to breast cancer.

These relationships are not well understood, but since obesity so often coexists with serious disease, most medical people agree that obesity is a risk factor to be avoided. Although the basic principles are the same, obesity control, more commonly referred to as "weight control," may involve different goals: (1) the prevention of obesity or weight *maintenance* and (2) reduction of excess body fat, weight *loss.*

Weight Maintenance

Five basic principles, judiciously and faithfully applied, provide the best defense against either short or long range weight gain for the adult.

1. Be certain that the *composition* of your normal diet follows the principles discussed earlier: adequate protein, no more than 25 percent fats with a minimum of saturated fat, minimum sucrose intake, adequate vitamins, minerals, and water.
2. Be certain that your *caloric intake does not exceed your energy requirements;* adjust caloric intake when physical activity is significantly reduced over an extended period of time.
3. *Spread the caloric intake evenly over at least three meals per day.*
4. Include *at least some physical activity each day* throughout your life.
5. *Weigh yourself* on reliable scales at some standardized time of each day. The recommended time is upon arising in the morning. A pound or two fluctuation from day to day is quite normal, but a persistent gain of two or three pounds, if this puts you above your optimal weight, should be dealt with immediately by means of a moderate daily caloric deficit.

Confrontation and positive action are much easier at this point than when the excess reaches greater proportions.

A word of caution to prospective parents: Perhaps the most critical time for applying the foregoing weight maintenance principles is during the early growing years—from birth to age ten—and after puberty. There is some evidence that during the early years, fat cells can multiply, that is, hyperplasia of fat cells has been shown to result from overeating in laboratory animals during the growth years. If this is true for humans, these years are critical in that a lifelong propensity toward obesity may result from overeating when adipose tissue has a hyperplastic potential. Indeed, autopsy studies of adults show that obese persons not only have larger fat cells but more of them. The belief that the fat child is a healthy child is becoming increasingly suspect. It should also be noted that research demonstrates inactivity invariably *precedes* rather than results from obesity in children and teenagers.

Weight Reduction

The approach to weight reduction requires some basic common sense. First, one must be intelligent enough to differentiate between extreme (refractory) obesity requiring the assistance of a physician and the more moderate degrees of obesity that can be dealt with by the individual himself. An arbitrary rule-of-thumb is: If you are under 21 or over 35, and if your excess weight exceeds 110 percent on the overweight index (page 247) or exceeds 20 pounds, a visit to your physician is probably a wise first step. Between the ages of 21 and 35, if you are in good health, an excess of as much as 20 pounds might be intelligently dealt with on your own. Remember that the physician has at his disposal certain methods for weight reduction that you should *not* prescribe for yourself (fasting, for example). On the other hand, the person armed with a sound nutrition education, knowledge of basic weight reduction principles, reasonably good health, and with only moderate obesity can certainly embark on a sound weight reduction program without professional assistance. Physicians would hardly manage to schedule appointments for every moderately obese American.

The principles of weight reduction are identical to those for weight maintenance except that one (proper caloric intake) is handled differently, and some preliminary calculations must be made to determine how much weight *should* be lost. The most desirable method for this determination is the application of the percent body fat estimate. For example, if a male is 18 percent body fat, he has 4 percent *excess* fat (14 percent is average) and thus 4 percent of his weight is excess fat. Let us say he weighs 190 pounds; 4 percent of 190 is 7.6; thus he should set his weight-loss goal at about eight to ten pounds (the extra is to allow for the fact that fat loss is accompanied by loss of supporting water so that a *weight* loss of eight pounds is not eight pounds of fat loss).

If skinfold measures are not available to you, you can use the Overweight Index to estimate an appropriate weight-loss goal. Simply set your goal at a percentage of your weight equivalent to whatever percentage above 100 your overweight index happens to be (for example, overweight index = 107 percent, 0.07 × a 190-pound body weight would be 12.6 pounds or a 13 to 16 pounds weight loss goal, again allowing for supporting water loss).

After first calculating the extent of the excess weight and/or fat, the second approach should be to follow a diet of the proper composition. The diet recommended for weight reduction differs slightly (but in very important ways) from that recommended for weight maintenance. Protein intake should be about 25 to 30 percent of the diet since protein has a high satiety value. Fat intake should also be slightly higher, 30 to 35 percent of calories, but the unsaturated intake should still be from 50 to 65 percent of the total fat intake. These increases in protein and fat intake will proportionately limit and reduce carbohydrate intake to a maximum of 40 percent and will physiologically minimize water retention and fat deposition.

Third, spread the calories over the day.

Fourth, weigh yourself regularly at a standardized time. Do not expect miracles but use the weighings to assure yourself that your caloric deficit is doing what it should for you. There may very well be a week or two delay in noticeable weight loss if there is enough water retention to compensate for the fat actually lost. It may be helpful to use a diuretic prescribed by your physician on a limited basis.

Fifth, regular exercise is an important, and for some, an indispensable, part of the program (see Chapter 31) and can contribute to a caloric deficit as discussed below.

Finally, assuming that action based upon all of the above principles has been taken (especially the realistic weight-loss goal, proper diet composition, and regular exercise), carefully select and determine the routine for insuring a daily caloric deficit. You cannot lose weight without this deficit and it must be persistent, not intermittent.

A daily 500-calorie deficit will result in at least a one-pound fat loss per week (total weight loss can be slightly higher due to supporting water loss). A daily 1000-calorie deficit should mean a little better than two-pounds weight loss per week. However, the total intake should not be less than 1000 calories. The 1000-calorie deficit is the maximum usually recommended for an adult. The deficit can be achieved by caloric reduction, increased energy expenditure, or a combination of both. To reduce caloric intake, simply look up some of the "regulars" in your diet and eliminate enough to total the caloric reduction component of your proposed deficit. Be sure to subject your new proposed diet to the percent-of-calories and Basic Four tests to see that your diet composition is still correct. Obviously, you must be careful not to compensate for the items or proportions eliminated or reduced by taking in more of other foods.

To include an increased energy expenditure component of the caloric deficit, use Table 17–2 as a guide. There are great variations in energy expenditure for a given activity and a given person, but this table will provide a rough estimate of energy expenditure. The activity must be added to your normal routine, of course, if it is to contribute to the deficit.

TABLE 17–2 Approximate Caloric Expenditure for Various Activities

Cal/min/lb of Body Weight	Activity	Cal/hr/lb of Body Weight	Cal/min/lb of Body Weight	Activity	Cal/hr/lb of Body Weight
0.0234	House painting	1.40	0.023	Volleyball	1.40
.026	Carpentry	1.56	.026	Playing pingpong	1.56
.031	Farming, planting, hoeing, raking	1.86	.033	Calisthenics	1.98
			.033	Bicycling on level roads	1.98
.039	Gardening, weeding	2.34	.036	Golfing	2.16
.045	Pick-and-shovel work	2.70	.046	Playing tennis	2.76
.050	Chopping wood	3.00	.047	Playing basketball	2.82
.062	Gardening, digging	3.72	.069	Playing squash	4.14
			.100	Running long distance	6.00
.0078	Sleeping	0.47	.156	Sprinting	–
.0079	Resting in bed	0.47			
.0080	Sitting, normally	0.48		Swimming	
.0080	Sitting, reading	0.48	.032	Breast stroke 20 yd/min	–
.0089	Lying, quietly	0.54	.064	Breast stroke 40 yd/min	–
.0093	Sitting, eating	0.56	.026	Back stroke 25 yd/min	–
.0096	Sitting, playing cards	0.58	.056	Back stroke 40 yd/min	–
.0094	Standing, normally	0.56	.058	Crawl 45 yd/min	–
			.071	Crawl 55 yd/min	–
.011	Classwork, lecture	0.66			
.012	Conversing	0.72	.033	Walking on level	1.98
.012	Sitting, writing	0.72	.093	Running on level (jogging)	5.60
.016	Standing, light activity	0.96			
.020	Driving a car	1.20	.01	Fill-in constant for time unaccounted for (if not completely inactive such as sleeping or resting	0.60
.028	Cleaning windows	1.68			
.024	Sweeping floors	1.40			
.044	Walking downstairs	–			
.116	Walking upstairs	–			
.014	Lecturing	0.84			

EXAMPLE: 150-pound man sitting and reading for 60 min $= 150 \times 0.0080 \times 60 = 72$ calories expended, or 1 hr $= 150 \times 0.48 \times 1 = 72$ calories.

Three examples of determining an estimated caloric deficit are cited in Table 17–3. Since it is assumed your daily physical activity will not be appreciably decreased, you must remember that such a deficit will only be effective if there is no compensatory increases in other foods.

Questionable Self-Prescribed Weight Control Methods

Weight-reduction schemes which ignore the well-established principles of sound nutrition are usually either ineffective or present a potential health

TABLE 17–3 Examples of Determination of Daily Caloric Deficit (for a 240-pound person)

Current intake	3500 cal	3800 cal	4350 cal
Decreased intake	–200 cal	–400 cal	–1000 cal
Increased activity	at 300 cal (1/2 hr walk, 4.47 mph)	at 375 cal (20 min walk, 5.8 mph)	at 300 cal (1/2 hr walk, 4.47 mph)
Total deficit	500 cal	775 cal	1300 cal
New daily intake	3300 cal	3400 cal	3350 cal
Expected loss per week	1 lb	1 1/2 lb	2 1/2 lb

hazard. Remember that the methods employed by physicians that you read about are *not* methods to be self-inflicted inasmuch as you are not qualified to judge the risk factor or to exercise the proper controls. An excellent example is the in-hospital fast for grossly obese persons. A review of the safety and effectiveness of some common self-prescribed weight-loss regimens should be informative and practical.

Aphagic or Semiaphagic Diets

Lay magazines and professional journals alike refer to "starvation diets" when they mean aphagic diets. The totally inappropriate use of the term *starvation* for the purpose of describing a food abstention period in grossly obese people could only be possible in a culture unfamiliar with the act of dying from lack of food. The proper term is *aphagia* (not to eat) and in the adjectival forms, aphagic diet or abstentious diet. Alternatively, the term *fasting* can be used, but it can have a connotation divorced from the intent of weight control. In fact, a person who is starving is rarely aphagic—he is usually eating *something* though grossly deficient calories. And conversely, a person 50 pounds overweight who is aphagic is *not* starving.

There are very definite health hazards associated with aphagia, both during the actual period of abstention and during the period of initial refeeding. Although shorter periods of abstention (say 24 to 48 hours) followed by periods of normal food intake do not subject the person to the same immediate health hazards as total, extended aphagia, the potential for serious degenerative changes that may contribute to poor health and reduced longevity is nonetheless very real. The similarity between aphagia-refeeding and "meal-eating" or "gorging" in regard to caloric intake distribution as discussed earlier must be quite evident.

As strange as it may seem, aphagia is not a very effective means of losing fat and maintaining a new, more desirable body weight. First, aphagia reduces the nonfat stores and tissues of the body equally if not more than it does the *fat stores* [Benoit]. Second, aphagia obviously cannot be continued indefinitely for weight reduction, especially since the body's systems become more effective at absorbing, assimilating, and even storing food with periodic fasting and refeeding [Holeckova; Johnson]. Essentially, remember that aphagia is not *natural,* that it results in no permanent adjustment in the systems of the body and actually tends to ultimately defeat the purpose of weight reduction. One must at some time begin to eat again, and no readjustment of the body processes for normal feeding has had an opportunity to take place. Thus, extended periods or brief periods of aphagia are potentially *very hazardous* and are relatively ineffective for permanent weight control.

DRUGS. Of all the preparations that have been offered to the public as weight control drugs, none has been even reasonably effective except one— the so-called appetite suppressant. Topical anesthetics to "dull the taste buds," diuretics to cause water loss, cathartics and laxatives to hurry the food through the intestines, and so-called lipotropic agents to prevent deposition of fat, have been patently unsuccessful. Furthermore, unregimented and extended use of diuretics or the cathartics and laxatives are dangerous practices.

The anorexigenic drugs (appetite suppressants) have enjoyed some modest success but their mode of action is uncertain, their effectiveness limited, and there are very definite side-effects associated with their use. Nervousness, irritability, insomnia, and elevated heart rate and blood pressure are not at all uncommon when these amphetamine and amphetamine-like drugs are taken. Furthermore, the amounts that can legally be contained in drugs sold over-the-counter (without prescription) is insufficient to cause any significant anorexigenic effect. Several controlled experiments have indicated that the weight loss to be expected from these drugs is minimal (something like four to five pounds in six weeks) and that beyond the four- to six-week period no further weight loss generally occurs.

The safety of certain of the so-called weight reduction drugs, then, when self-prescribed, is highly suspect and they are not particularly effective.

Liquid or Powdered Food Supplements

Several liquids or powders to which milk can be added are available for dieters. Depending on the particular brand and whether fat free, 2 percent, or whole milk is added, anywhere from 250 to 375 calories are available with eight ounces of liquid. These preparations have a reasonably good balance of the major food components plus vitamins and minerals. Although there have been no deleterious health effects noted when such preparations are used on a limited basis, neither is there evidence to recommend their unlimited use over extended periods of time. Exclusive liquid diets of this sort can become extremely boring, and they do not provide normal bulk to stimulate intestinal motility and normal bowel movements. But there is nothing to detract from their occasional use or from their use as a one-meal substitute so long as the intake falls within the caloric requirement and does not throw the component ratios out of balance.

Increased Sweating

Little space needs to be devoted to such pure foolishness as "weight reduction by steam baths" and "weight control by exercising in a rubber suit." Both of these methods are completely ineffective and both are potentially very dangerous. To be sure, some slight increase in metabolism occurs in a steam bath but not enough to be significant. Furthermore, the observed weight loss is misleading, is almost entirely water loss, and will be regained as the body reestablishes its water balance. Overstaying the time limit in a steam room can be dangerous, even fatal. The same is true of exercising in a rubber suit; one man recently died of heat stroke while playing handball in a rubber suit. The very slight increase in metabolism over and above that induced by the exercise itself is hardly worth the risk involved. As in steam room weight loss, most is water loss and will be regained as water balance is reestablished.

Massage

Whether by human hand or machine, exercise participated in by another person or thing is not going to increase *your* energy expenditure. For relaxation, maybe; for weight reduction, preposterous! You might expect about 0.05 ounce of weight loss while standing at a buttocks vibrator for 15 minutes. Over the course of one year, if you did this every day, the weight loss might total about 1.5 pounds!

REVIEW QUESTIONS

1. What is the difference between hunger and appetite?
2. What are the major factors that lead to obesity?
3. How would you plan a program of weight reduction that would provide the most permanent weight loss?
4. Why is "starvation" an ineffective method of weight reduction?
5. Discuss the use of appetite suppressant drugs in weight reduction.

REFERENCES

BENOIT, F. L., "Prolonged fasting: Physiologic undesirability studies," *Medical Tribune*, **6:** 16–17 (1965).

FABRY, P. et al., "The frequency of meals: Its relation to overweight, hyper-cholesterolemia, and decreased glucose tolerance," *Lancet*, **2:** 614–615 (1964).

HOLECKOVA, E., and P. FABRY, "Hyperphagia and gastric hypertrophy in rats adapted to intermittent starvation," *British Journal of Nutrition*, **13:** 260–266 (1959).

JOHNSON, P. B. et al., "Effects of starvation and realimentation," Paper presented at Midwest District AAHPER Research Section, 1964.

JOHNSON, P. B., and J. COOPER, "Comparative effects of 'meal-eating', 'nibbling', and starvation-refeeding in male albino rats," Paper presented at AAHPER Convention, Las Vegas, Nevada, 1967.

MAYER, JEAN, *Overweight: Causes, Cost, and Control.* Englewood Cliffs, N.J.: Prentice-Hall, 1968. (Paperback)

Additional Readings

HAVEL, R. J., "The neurohumoral control of lipid storage, transport, and utilization," *Physiology for Physicians,* **1:** 6–13 (1963).

HODGES, R. E., "Present knowledge of carbohydrates," *Nutrition Reviews,* **24:** 65–68 (1966).

Journal of the American Medical Association (Medical News), "Childhood eating habits may determine obesity," **200:** 31 (May 29, 1967). (For detailed study, see Salans et al., below).

KEYS, A. et al., *The Biology of Human Starvation.* Minneapolis: University of Minneapolis Press, 1950.

MAYER, JEAN, "Exercise and Weight Control," *in Science and Medicine of Exercise and Sports.* New York: Harper and Row, 1960.

MEAD, JAMES F., "Present knowledge of fat," *Nutrition Reviews,* **24:** 33–35 (1966).

NUTRITION REVIEWS, "Effects of meal eating vs. nibbling on body composition," **19:** 9–11 (1961).

SALANS, L. B., J. L.KNITTLE, and J. HIRSCH, "The role of adipose cell size and adipose tissue insulin sensitivity in the carbohydrate intolerance of human obesity," *Journal of Clinical Investigations,* **47:** 153–165 (1968).

WISHNOFSKY, M., "Caloric equivalent of gains or lost weight," *American Journal of Clinical Nutrition,* **6:** 542–548 (1958).

PART FIVE
THE BODY'S FLUIDS

18
BLOOD

The blood and blood vessels serve as the transportation system of the body. They provide a means by which oxygen and nutrients can be distributed to organs and tissues which lie a considerable distance from the heart, lungs, and digestive tract. They also serve as a distributing system for the chemical messengers of the body, the hormones. Other functions are less obvious: Heat is distributed by warming of the blood as it passes through the deeper, centrally located structures and the blood is cooled in its passage through the skin. The rate of this heat dissipation can be regulated by controlling the amount of blood passing through the skin—increased blood flow increases heat loss; decreased blood flow conserves heat—hence the flushing of the skin in warm surroundings and its blanching in the cold. Waste products from the body's metabolism are excreted by the kidneys, but must be transported to the kidneys—this, too, is a function of blood.

In this chapter we shall discuss the nature of blood, its functions, and its disorders. The two following chapters are concerned with subjects closely related to blood—immunity and tissue transplantation. The heart, blood vessels, and the mechanics of circulation we shall consider in later chapters.

The Nature of Blood

If a sample of blood is placed in a glass tube, and the tube placed in a centrifuge, and spun at a high rate, the centrifugal force will divide the blood into its heavier fraction composed chiefly of *blood cells* below, and the lighter, clear, fluid fraction above, the *plasma*. Normally about 45 percent of blood is cells and 55 percent plasma. This determination is called the *hematocrit* (Fig. 18–1).

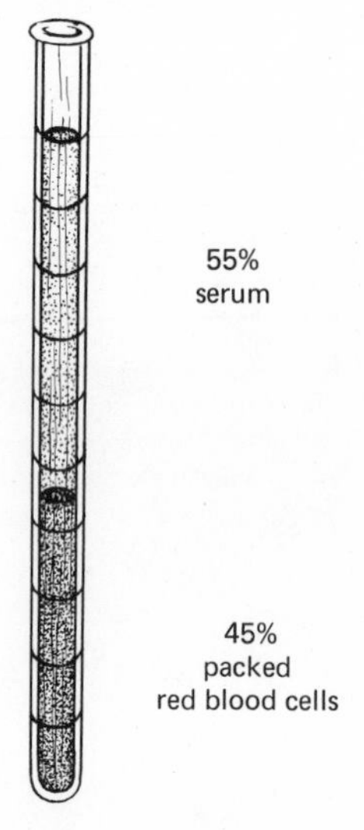

FIG. 18–1 Hematocrit determination. Centrifugation of a sample of blood results in the movement of the heavier cells to the bottom of the tube. The percentage of cells composing the blood (in this case 45) constitutes the hematocrit reading.

The Plasma

Plasma is mostly water (about 92 percent), but it obviously contains all the soluble substances transported by the blood. This includes not only nutrients (glucose, amino acids, fatty acids), but also numerous salts, hormones, vitamins, and diverse proteins. The latter components function in blood clotting, as enzymes, as antibodies, and as regulators of the body's fluid balance. Plasma from which clotting proteins have been removed is called *serum*.

The Red Blood Cells—Erythrocytes

The red blood cells have been reduced to such a degree that they are essentially sacs of hemoglobin. The substance hemoglobin (*heme*, iron; *globin*, a type of protein) functions to transport most of the oxygen and some of the carbon dioxide in the blood. Its oxygen-carrying efficiency is illustrated by the fact that 100 milliliters of plasma alone will dissolve and hold only 0.3 milliliter oxygen, while the hemoglobin of 100 milliliters of blood will hold (combine with) 19.7 milliliters of oxygen. The red cells and their hemoglobin maintain a constant level of oxygen in the plasma, which in turn supplies oxygen to the body tissues. The movements of oxygen and carbon dioxide in and out of the blood are more properly considered in connection with breathing (Chapter 26).

Size, Number, and Shape of Red Blood Cells

Even in the realm of the microscopic, red blood cells must be considered small. They measure 7 microns (7/1000ths of a millimeter) at their greatest dimension. In form they resemble a strongly biconcave lens, that is, they are disc-shaped and thinner in the center than at the edge (Fig. 18–2). This gives them an ideal surface for rapid uptake and release of oxygen.

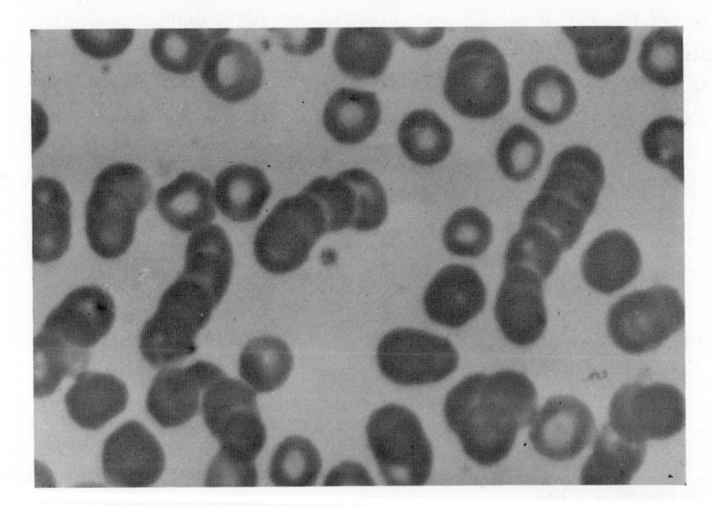

FIG. 18–2 Human red blood cells. Note the "doughnut" shape. (Dr. H. E. Brown)

The number of red blood cells in a cubic millimeter of blood is called the *red cell count.* This is normally about 5,500,000 in the male and about 4,500,000 in the female, which means that one drop of blood from an eyedropper (about 50 cubic millimeters) would contain about 275,000,000 red blood cells in the male and about 250,000,000 in the female. The number in the total circulation of five to seven quarts of blood is astronomical.

Red Blood Cell Production and Lifespan

The lifespan of the red blood cell is relatively brief, probably about 115 days. This means a very high rate of production—in the order of 3 million per second. In the developing embryo the red blood cells are manufactured by the liver and spleen, but after birth this function is taken over by the bone marrow. In the child the cavities of all bones show the red matrix of blood-forming tissue. But gradually the red bone marrow becomes replaced by yellow marrow (fat) until in the adult the red marrow is largely restricted to the ribs, sternum (breastbone), and vertebrae. If a sample of blood-forming tissue is required by a physician he will generally remove it from the sternum by what is known as the "sternal puncture."

Red Blood Cell Destruction

The wear and tear on a red blood cell over the 115 days of its life are considerable. During this time it travels over 700 miles in constant collision with its neighbors, much of the time forced through narrow capillary channels which distort its shape. Eventually, the red cell becomes fragile and is fragmented. The fragments are filtered from the blood by "scavenger" cells (called reticuloendothelial cells) located in the liver. These cells break down the hemoglobin and from it salvage the iron, which is used again in hemoglobin formation, and the protein, whose amino acids can be burned for energy or used again for assembling fresh proteins. The part of the hemoglobin molecule responsible for the red color is excreted in the bile as *bilirubin.* Bilirubin excretion is used as a measure of the rate of red blood cell destruction.

Anemias

An anemia is any condition in which the oxygen-carrying capacity of the blood is less than normal. This could be caused by excessive loss of red blood cells through an abnormal rate of destruction within the body or by hemorrhage. Alternatively, the cause could be a deficient rate of red cell formation, abnormal red cells, or abnormal chemistry such as the lack of iron. In many anemias a combination of these factors is present.

Blood Loss Anemia

After loss of blood from hemorrhage the reduced blood volume causes a drop in blood pressure. The blood pressure is restored to normal in a few hours first by contraction of the spleen, which carries a sizable reserve, and then by movement of body fluids into the blood vessels. (This results in the acute thirst experienced by those who have lost large amounts of blood.) Although the blood volume is now normal, the red cells are more diluted, and anemia will persist until they have been renewed by an accelerated rate of production. This will generally require several weeks. When you serve as a blood donor (usually giving about one pint of blood), the time required for red blood restoration averages about seven weeks, hence the three months lapse required between blood donations. Loss of a pint has no harmful effects, but it does cut down the margin of reserve. Loss of over a quart may produce serious consequences. Included among blood loss anemias are those which arise in women who menstruate excessively or in persons who have bleeding from peptic ulcers.

Iron-Deficiency Anemia

Since iron is part of the hemoglobin molecule, a deficiency of iron in the body would inhibit normal hemoglobin production. These iron deficiencies usually are due to inadequate iron in the diet, but may also be due to an inability to absorb iron properly (Chapter 14). While a minor iron deficiency might be of little consequence in itself, when coupled with excessive menstruation or a bleeding ulcer, it can result in a chronic anemic state.

Pernicious Anemia

Vitamin B_{12} is essential for red blood cell formation. When too little of the vitamin is present in the body, pernicious anemia develops; but this is not generally due to a deficiency of the vitamin in the diet, but rather to an inability to absorb the vitamin. For this reason B_{12} has been called the *extrinsic factor* essential for the prevention of pernicious anemia. An *intrinsic factor* must be present for absorption of B_{12}. The precise identity of the intrinsic factor is still unknown, but we do know that it is elaborated by the stomach of normal people. For example, food known to contain B_{12} provides no benefit for anemic patients; but if it is first fed to a normal person, then removed from his stomach and placed in the stomach of the patient an improvement in the red cell count can be expected. The stomachs of those with pernicious anemia produce no hydrochloric acid; however, the relationship between this achlorhydria and the intrinsic factor is not clear. The term "pernicious" is no longer appropriate since these people are now easily treated by injections of the vitamin, and can expect to live a quite normal life.

Sickle-Cell Anemia

In this type of anemia the red blood cells are exceedingly fragile, rupture easily, and exhibit peculiar shapes, somewhat resembling a sickle as shown in Figure 18–3. The condition is hereditary and is caused by imperfect genetic "coding" (Chapter 3) for the synthesis of hemoglobin. The hemoglobin of the sickle-cells differs from normal hemoglobin by a substitution of the amino acid valine for the amino acid glutamic acid in the globin molecule. This small "error" in synthesis results in a highly fatal disease. Nonafflicted carriers of the gene (heterozygotes) show improved survival in malaria-infested tropics so apparently the gene has had a selective value in such areas. The mutation (spontaneous change in the genes) responsible for the disease must have occurred in Africa, since almost all cases are found among black people in Africa and the United States.

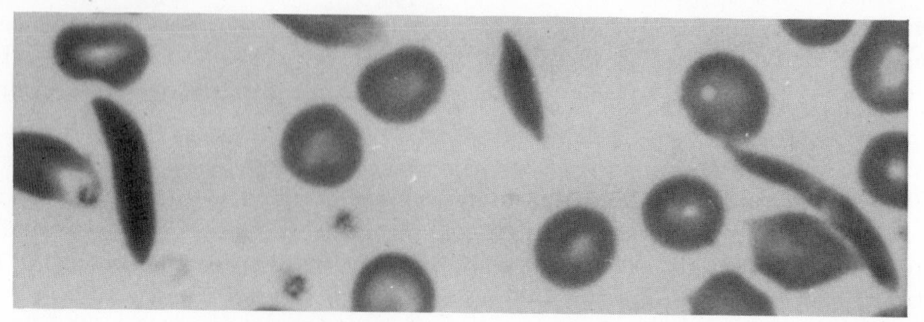

FIG. 18–3 Red blood cells from a patient with sickle-cell anemia. (Dr. D. A. Nelson)

About 10 percent of all black Americans carry the gene for the sickle cell trait; that is, they are heterozygous, with one normal gene and one sickle-cell gene, and are not anemic. About 0.2 percent or one in 500 are homozygous (have two sickle-cell genes) and are anemic.

Aplastic Anemia

If the blood-forming tissues degenerate to the point where red blood cells are no longer manufactured the condition is called aplastic anemia. This type of anemia was practically unknown before the advent of man-made radiation. Aplastic anemia can develop in response to over-exposure to X-rays, radioactive isotopes, or atomic fallout. One of the antibiotics, chloromycetin, is also known to cause aplastic anemia in rare cases. Consequently, the use of this antibiotic is reserved for bacterial infections not effectively treated by other drugs.

The White Blood Cells – Leukocytes

White blood cells number about 7500 per cubic millimeter. If this seems small in comparison to the number of red cells (about 1/750) the figure becomes more impressive when we consider them as an army of some 37,500 million in the circulating blood of an adult!

The majority of white blood cells are manufactured in the bone marrow. These white cells have rather peculiarly lobed nuclei – polymorphonuclear leucocytes, white cells with many shaped nuclei – and have great numbers of large granules, thus the collective name for these cells, *granulocytes.* The nature of the granules are responsible for the names given to these white cells: Those whose granules stain in a neutral manner are *neutrophils* (Fig. 18–4); those whose granules stain red are *eosinophils;* those whose granules stain blue are *basophils.*

The remaining kinds of white cells are without cytoplasmic granules and are produced in lymphatic tissue of such organs as the spleen, lymph nodes,

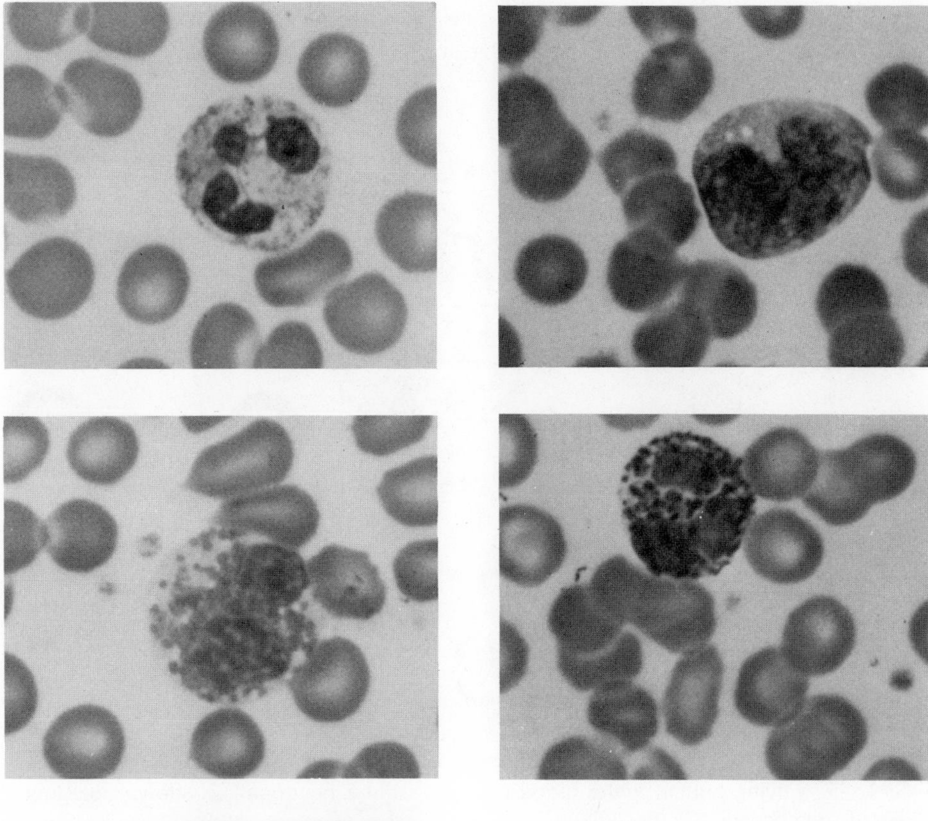

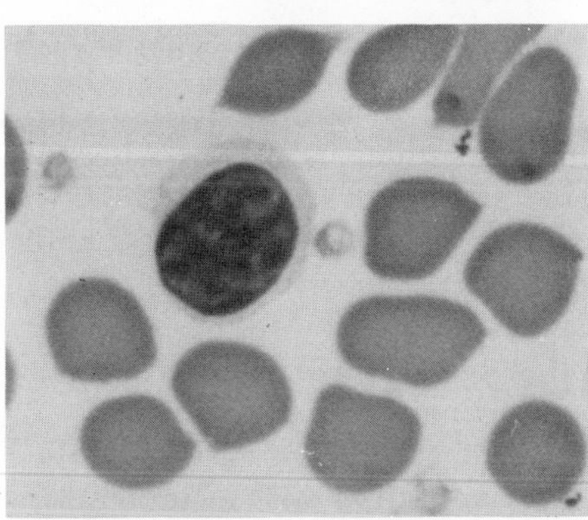

FIG. 18–4 The different types of white blood cells. *Upper left:* a neutrophil; *upper right:* a monocyte; *middle left:* an eosinophil; *middle right:* a basophil; *lower left:* a lymphocyte. (Dr. H. E. Brown)

and tonsils. These cells have a large, almost round nucleus. Some are hardly bigger than red blood cells and are called *lymphocytes;* the others, the largest of the white cells, are called *monocytes.*

Neutrophils are the most common of the white cells (about 65 percent), and

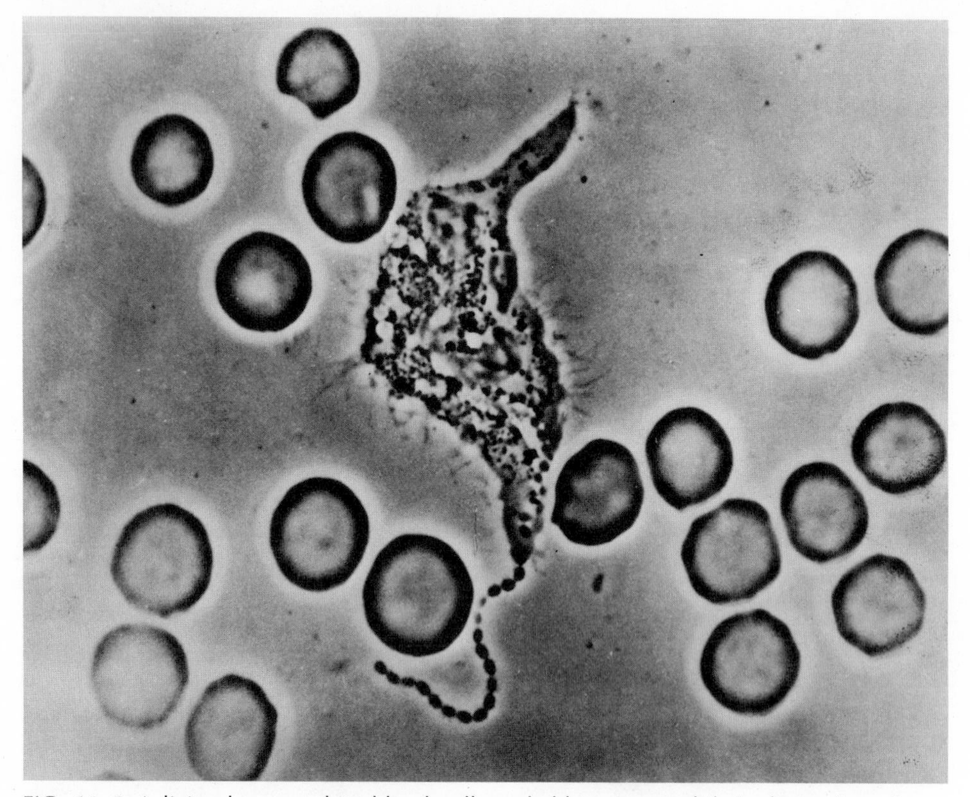

FIG. 18–5 A living human white blood cell, probably a neutrophil, is shown ingesting a chain of streptococci. (Charles Pfizer)

function principally to protect against infection. We can best describe the action of neutrophils by following the sequence of events which occurs when bacteria enter the tissues (Fig. 18–5). The presence of bacteria attracts neutrophils. The mechanism is not clear; perhaps some attracting substance is released by the tissue cells attacked by the bacterial poisons. In any case neutrophils squeeze through minute pores in the capillary walls and migrate toward the site of the infection. Their movement is similar to that of an ameba—they push out extensions of themselves and flow into the extensions. In this way they are able to move through pores in the blood vessel walls too small for a red cell. When they reach the site of the infection they attack the bacteria with a similar technique—extensions of the cell flow out and around the invaders and engulf them in a process called *phagocytosis* (eat-cell) (Fig. 18–6).

The accumulation of these neutrophils, the release of fluid from dilated blood vessels, and the red blood in distended capillaries cause the redness and swelling in the infected area. The pain and tenderness felt are due to the

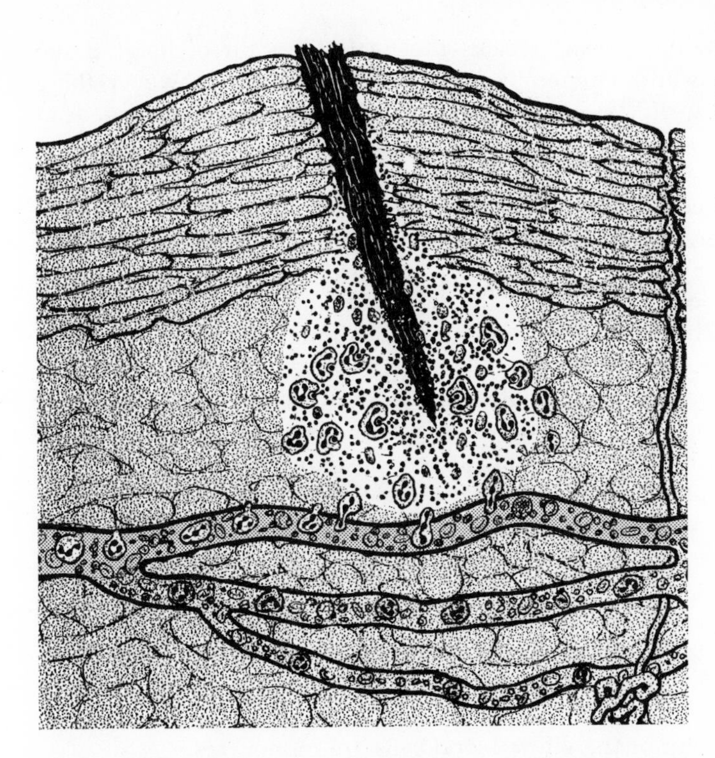

FIG. 18–6 The sequence of events in infection. See text for explanation.

mechanical stretching of the tissues. Collectively, these characteristics define the *inflammation reaction.*

After the neutrophils engulf large numbers of bacteria they themselves may die. If the neutrophils are successful in their struggle, the infected area is walled-off, and the healing process begins. In this walling-off process a cavity is usually formed which contains the debris of battle—fragments of damaged tissue cells, dead neutrophils, and bacteria. This mixture is the *pus.* Pus formation continues until the infection is overcome. Sometimes the pus cavity breaks through the surface of the skin or into a cavity within the body; or the cavity may remain enclosed long after the infection has been suppressed. In this case the debris is gradually removed by lymphatic capillaries (Chapter 21).

The presence of an infection and the accumulation of neutrophils cause a generalized increase in the total number of neutrophils in the blood, a sort of call-to-battle in which reserves are requisitioned from the bone marrow. This may bring about an increase in the total white cell count in the blood to 12–15,000 per cubic millimeter in a few hours. This response is called *neutrophilia.*

Essentially the same course of events occurs wherever there is tissue death. For example, a bruise which is swollen and blue and then becomes yellow has undergone the same process. In this case the neutrophils are acting as scavengers in removing the remains of the dead cells. This is the reason the white cell count will rise following a heart attack—the death of a bit of heart tissue attracts the neutrophils and a generalized neutrophilia results. However, there is usually no "walling off" since there is no infection to spread; the dead neutrophils are eventually removed by the lymphatic vessels.

Eosinophils are less well understood. They normally constitute only about 2 percent of the white cells, and their number seems to increase when the individual exhibits allergic reactions (Chapter 19) or is infected with parasitic worms.

Basophils are the least common of the white cells, numbering less than 1 percent of the total. Their function is unknown, although they may contain a substance, *heparin,* which inhibits blood clotting.

Lymphocytes are second to the neutrophils in number (about 30 percent of the total). Their function in the blood is uncertain: They are believed to live only a few hours after entering the blood stream, have no ameboid movement, and have no power of phagocytosis. Outside the blood, in the lymphatic tissues, they are believed to be responsible for the production of antibodies, and we shall come back to this in our discussion on Immunity and Allergy in the next chapter.

Monocytes accompany neutrophils in their migration to sites of infection. However, the monocytes are slower to respond and require several days before they accumulate in large numbers. For this reason monocytes are more generally associated with rather long-term or chronic infections such as tuberculosis. Normally about 3 percent of the white blood cells are monocytes.

Agranulocytosis

Agranulocytosis (no-granulocyte disease) is the clinical condition in which the bone marrow produces few, if any, white blood cells. Since bacteria, including many responsible for serious disease, are regularly present in the alimentary canal and respiratory passages (as well as in other communication with the environment such as the urethra and vagina), any decrease in the number of neutrophils will be followed by an invasion by these bacteria. About two days after the bone marrow stops making white cells, open sores will begin to appear in the mouth or colon, or the person will develop a serious respiratory infection. With no defense against such invasions the infection will spread into the surrounding tissues and into the blood. Death will usually result in less than a week. This process graphically illustrates the importance of the white cells.

Here, as in aplastic anemia, the condition is most frequently associated with intense X-ray or atomic irradiation. Gamma rays from an atomic explosion are capable of penetrating the walls of buildings, and can cause agranulocytosis and aplastic anemia, even though the individual survives the initial blast.

Leukemia

Leukemia is the uncontrolled production of white blood cells. This is a type of cancer, but whereas most cancers form solid lumps, leukemic cells are dispersed throughout the blood. The result is greatly increased numbers of white cells in the circulating blood.

Since white blood cells have two distinct sites of origin, one might expect two distinct types of leukemia, and this is the case. Leukemias of the bone marrow are called *myelogenous* (marrow-derived) *leukemias* and result in large numbers of abnormal neutrophils, eosinophils, and basophils. Leukemias of the lymphatic tissues are called *lymphogenous* (lymphatic-derived) *leukemias* and result in large numbers of lymphoid (lymphocyte-like) cells.

Leukemias also are referred to as *chronic* or *acute.* In chronic leukemias there is production of large numbers of fairly normal white blood cells which seem to cause very little harm, and the individual may live for many years. The white blood cell count may go extremely high—up to 500,000 per cubic millimeter in the myelogenous type. This eventually results in such a high metabolic drain upon the body that death ensues.

In acute leukemia the dividing cancer cells (called "blast" cells), tend to metastasize or spread from the bone marrow or lymph nodes to the liver, spleen, and other abnormal sites. Although the number of circulating white cells never reaches the high level found in chronic leukemia, there is a rapid growth of the cancer. If untreated, acute leukemia results in death within six months.

All leukemias will eventually invade surrounding tissues. In the case of the bone marrow this invasion can so weaken the bone structure that fractures occur easily. Another common effect of leukemia is anemia, to the extent that the bone marrow devoted to red cell production is taken over by the cancer cells. As in most malignancies the demand for foodstuffs by the growth of cancer cells places such a metabolic drain on the body that there is rapid deterioration of the normal tissues.

To make matters worse these great numbers of leukemic cells do not function as normal leukocytes. The leukemic neutrophils do not combat infections or engulf debris from tissue damage, and leukemic lymphatic cells do not make antibodies. So in spite of great numbers of leukocytes the patient suffers acutely those disabilities from which he is regularly protected by normal leukocytes.

Causes of Leukemia

A number of viruses are known to cause leukemia or leukemia-related diseases in rats and mice; and some evidence now points to viruses as the responsible agent in at least some human leukemias. Radiation also causes leukemia in both animals and man (Fig. 12–1). This may seem strange when we have just said that it causes agranulocytosis, but such is the case; the apparently opposite effects can be accounted for by the type and length of irradiation and other factors. In fact the high incidence of leukemia among radiologists and dentists was the first indication of the danger of overexposure to X-rays. Elevated rates of the disease are consistently found among patients who have been previously irradiated for a variety of conditions. The present, more conservative use of X-rays by the medical and dental professions is an outcome of such findings. An informed person can submit without anxiety to reasonable X-ray procedures, but should question any use that seems excessive.

Treatment of Leukemia

Although no cure is yet available for any form of leukemia, the life expectancy for almost all forms is being extended. Antileukemia drugs are now giving remissions to many children with the disease, and more effective drugs are regularly becoming available. In the area of cancer research where pessimism is the rule, there is remarkable optimism that leukemia will be the first form of cancer to be conquered with drug treatment and within this decade.

Blood Elements and Clotting

Platelets

Platelets are small bits of cells in the circulating blood. They contain a protein substance, *platelet thromboplastin,* which is involved in the clotting of blood. The other clotting factors are found in the plasma and in the tissues outside the blood vessels. The plasma factors include two proteins—*prothrombin* and *fibrinogen;* the tissue factor is a protein called *tissue thromboplastin.* In addition there are some 30 other factors in the plasma and related tissues which affect blood clotting, few of which need concern us.

The normal sequence of events in the formation of a clot following injuries to a blood vessel may be outlined as follows (Fig. 18–7): The ruptured blood vessel and damaged tissue cells bring about a mixing of the plasma with the tissue thromboplastin. In the presence of the tissue thromboplastin, the plasma prothrombin is converted to a new substance, *thrombin.* Thrombin in turn converts fibrinogen into a netlike material called *fibrin* which is the clot. Thus the simple event of blood being brought into contact with tissue fluid automatically triggers a mechanism to plug the opening. The platelets, when freed into the tissue spaces, adhere to cell surfaces and disintegrate, releasing their thromboplastin. The platelet thromboplastin seems to have the same action as tissue thromboplastin in activating prothrombin to thrombin.

The importance of the tissue thromboplastin is evident if you have ever experienced a clean deep cut with a sharp instrument. Such a wound causes

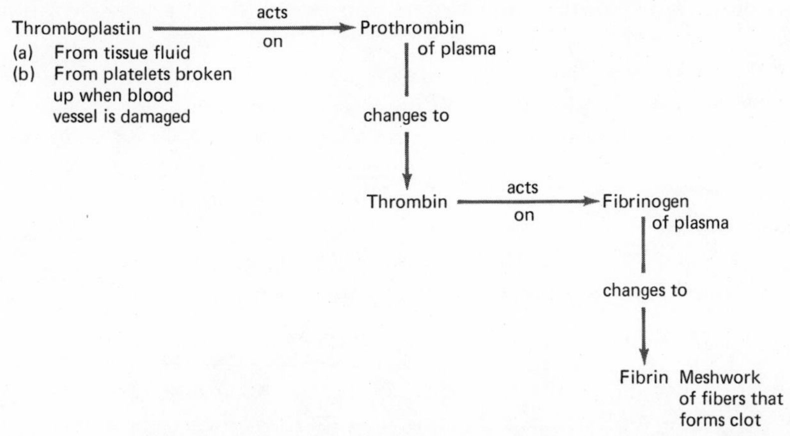

FIG. 18–7 Clotting of blood. Thromboplastin, contained inside the platelets and in the tissue fluid, is normally not in contact with prothrombin. Injury to blood vessels causes the mixing of the prothrombin of the blood plasma with the thromboplastin of the tissue fluid and from ruptured platelets. Thromboplastin causes the prothrombin to be converted to thrombin, which in turn causes the fibrinogen to be precipitated out in the form of the fibrin meshwork.

minimal damage to the tissues and only small amounts of tissue thromboplastin come in contact with the plasma prothrombin, so that thrombin formation proceeds slowly. An abrasive wound, which generally causes considerable tissue damage, releases large amounts of tissue thromboplastin; also, the difference in the available surfaces in the two wounds affects the number of platelets which disintegrate and release thromboplastin. Thus the deep, clean wound bleeds freely, and clotting proceeds slowly.

Clotting within Blood Vessels

The very smooth lining of blood vessels, called the *endothelium,* prevents adherence of the platelets and hence their disintegration. Also, there appears to be an electrically charged film on the inner surface of the endothelium which repels the platelets. In addition certain anticlotting factors, called *anticoagulants,* are present in the blood. One of these factors, *heparin,* blocks the conversion of prothrombin to thrombin. (It is interesting that a similar substance is contained in the saliva of leeches. After rasping an open wound the leech is able to ensure a free flow of blood by inhibiting the normal clotting process.)

However, in certain disease conditions, clots do form within blood vessels. Any roughened surface on the endothelium will cause adherence of platelets and a clot may be initiated. Roughened surfaces can be caused by atherosclerosis, inflammation, or injury. An abnormal clot of this kind is called a *thrombus*

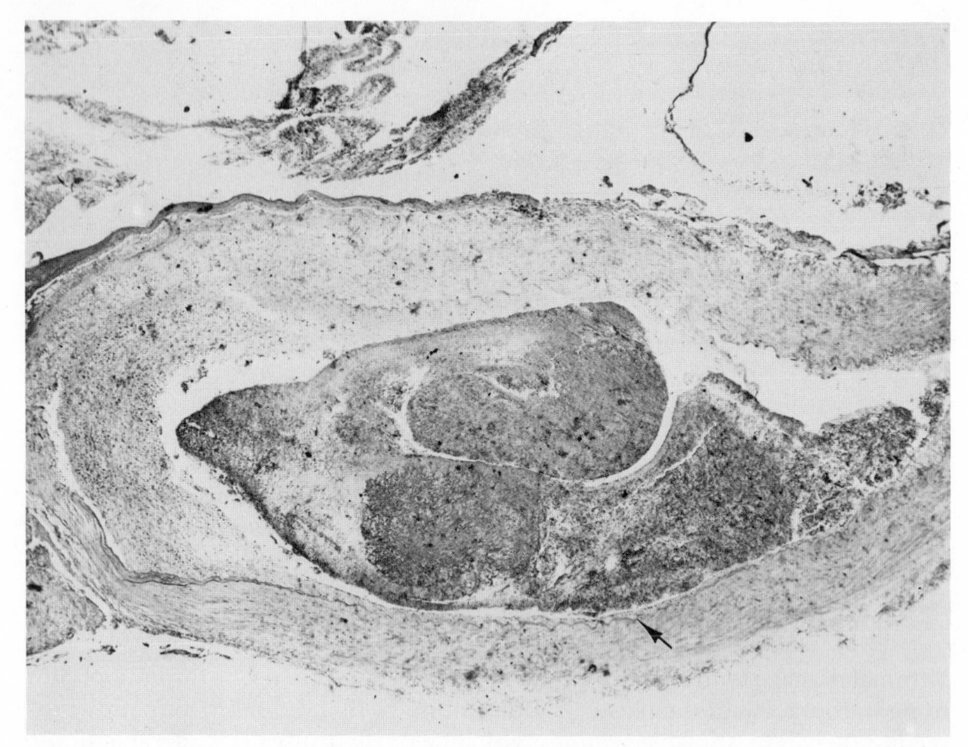

FIG. 18–8 Shown is a blood clot in an artery of the cerebral cortex of the brain. The original lining of the artery can be seen as the wavy line (arrow). The specimen was taken from a man who died in 1900 at the age of 71. (Dr. D. Reed Jensen)

(plural, thrombi). A thrombus may enlarge to the point that it blocks blood flow, or it may break free and be carried along with the blood. These floating clots may lodge in a blood vessel forming a plug in which case it is called an *embolus.* (Any plug which occludes a blood vessel is an embolus, so any foreign body such as a mass of bacteria or a bubble of air would be included in the definition.) Thrombi and emboli are serious disorders of the vascular system, and these are discussed with circulation.

Conditions Related to Excessive Bleeding

We commented earlier upon the role of vitamin K in the synthesis of prothrombin by the liver (Chapter 14). Since vitamin K is continually produced by intestinal bacteria, dietary deficiencies are not likely to develop. In those diseases that may disturb fat absorption (particularly liver disease that interferes with bile production or release) the fat-soluble vitamins may not be properly absorbed. For this reason vitamin K is often given by injection to these patients.

Hemophilia

The hemophiliac suffers from impaired blood clotting, so that simple wounds may bring about severe hemorrhage. Even the extraction of a tooth is a serious undertaking which may require hospitalization.

The blood plasma of the afflicted person lacks a factor essential for the activation of platelet thromboplastin, which means, of course, that clotting is entirely dependent upon the tissue thromboplastin. However, if plasma containing the factor is transfused into the patient, normal clotting will occur, and the bleeding tendency will be temporarily relieved. And this is the procedure followed when a hemophiliac requires surgery. It has recently been found that a rather simple chemical, aminocaproic acid (a six-carbon amino acid), taken by mouth, often gives effective clotting protection to hemophiliacs.

Hemophilia is really a rare disease, occurring in only about one in 3000 persons. However, it has attracted rather special interest because of the fact that it is an inherited disorder which is transmitted by a sex-determining chromosome. Also, the prevalence of hemophilia in the royal families of Europe has lent a romance to the disease that adds general interest. The inheritance of hemophilia is explained in connection with heredity in Chapter 3.

Blood Types and Blood Transfusions

The first attempts to transfuse blood more often than not resulted in the death of the patient. The discovery that different types of blood exist and the characterization of these types have made possible the routine and uneventful blood transfusions with which we are all familiar. A number of blood-type classifications are known, but only two are of particular significance to us. These are the ABO groups and the Rh groups previously discussed in Chapter 3.

ABO Blood Types

Two protein factors which sometimes occur in the cell membranes of red blood cells are known as A and B factors or *agglutinogens.* An individual's red blood cells may contain both of these proteins, one of them, or neither. This makes possible the four ABO blood types: type A in which the red blood cells contain only the A factor; type B in which the red blood cells contain only the B factor; type AB in which the red blood cells contain both the A and B factors; and type O in which both factors are absent from the red blood cells.

Individuals lacking an A or B agglutinogen in the red blood cells have an antagonist or *agglutinin* for the absent agglutinogen in the blood plasma. The agglutinin for the A agglutinogen is called α agglutinin; that for the B agglutinogen is β agglutinin. The agglutinogens and agglutinins present in the blood of the four ABO blood types, as well as the incidence of the types in the American population, are shown in Table 18–1.

TABLE 18–1 ABO Factors in the Red Blood Cells and α and β Factors of the Plasma

Type	Percent American Population	Red Blood Cell Agglutinogen	Plasma Agglutinin
A	41	A	β
B	9	B	α
AB	3	AB	–
O	47	–	αβ

The percentage of each type in the American population is based on a cross section of the population. There are distinctive social difference. The A type comprises about 61 percent of native Hawaiians, whereas B type shows its highest concentration (28 percent) in Chinese. American Indians are almost all type O.

Agglutinins are naturally occurring antibodies against the agglutinogens and will give rise to an immune reaction when in contact with agglutinogens. The general nature of immune reactions is discussed more fully in Chapter 19; for our discussion here we need only note the following. Upon contact with plasma containing the antagonist agglutinins, the red blood cells clump together, a reaction called agglutination. When such clumps of cells reach the narrow capillary channels, the cells tend to be broken up, and the hemoglobin is spilled into the plasma. This plasma hemoglobin can cause considerable damage to the kidneys and kidney failure is the major cause of death following transfusion of incompatible blood.

In a transfusion it is the possible agglutination of the donor's red blood cells which is of greatest concern. For example, the transfusion of a donor's type A cells into a patient with blood of type B would result in the immediate agglutination of the cells as they came in contact with the agglutinins of the patient's plasma. Theoretically, the donor's plasma would also cause agglutination of the patient's red blood cells, but apparently the donor's agglutinins are diluted in the patient's blood to such an extent that the effect on the recipient's (patient's) cells is minor. Since a type O person has no factors in his red blood cells to be agglutinated, he is sometimes referred to as a *universal donor;* and since an AB person has no agglutinins to agglutinate incoming red blood cells, he is referred to as a *universal recipient.* In practice, however, only blood of the same type is transfused.

Because incompatible plasma has little effect on the recipient's cells, it has been useful in emergencies to transfuse plasma as a substitute for whole blood. This restores the blood volume and prevents the onset of circulatory shock (Chapter 24). This was the case during World War II when the use of "pooled plasma" became widespread. The blood from many donors was processed to remove the blood cells, mixed, and then dried. When needed it was added to sterile water and transfused. Since that time a disease known as *serum hepatitis* (p. 230) has been found to be transmitted by this technique. It is a virus disease, and if one of the donors to the pool has the virus in his blood, all those who receive transfusions from the pool will acquire the disease. The recent development of synthetic blood expanders, which have many of the physical characteristics of plasma, has almost eliminated the use of plasma as a transfusion agent. Serum hepatis has recently become a serious problem in ordinary blood transfusions. Because some blood banks rely on paid donors for their reserves and because such blood banks cannot or do not prescreen donors, those who have contracted hepatitis (perhaps through illicit drug use [Chapter 38]) can contaminate the blood supply.

REVIEW QUESTIONS

1. How do each of the several types of anemia cause a reduction in the oxygen-carrying capacity of the blood?
2. What are the principal events in the inflammation reaction?
3. Why does blood not normally clot except following injury to blood vessels?
4. Why is it that the donor's red blood cells (rather than the recipient's) clump together when mismatched blood is transfused?

Additional Readings

Asimov, Isaac, *Bloodstream: River of Life*, rev. ed. Boston: Houghton-Mifflin, 1964.
Ballinger, C. M., (ed.), *Blood Transfusion*. Boston: Little, Brown, 1968.
Hall, C. A. *Blood in Disease*. Philadelphia: J. B. Lippincott, 1968.
Race, R. R., and R. Sanger, *Blood Groups in Man*, 5th ed. Oxford, England: Blackwell Scientific Publications, 1968.

19
IMMUNITY AND ALLERGY

We have already discussed how white blood cells provide us with a means of resisting infection. The production of *antibodies* is another means. Antibodies are proteins that are able to neutralize invading organisms or their toxins to give us *immunity*. However, if we produce antibodies which do damage to our own body's cells, we say that an *allergy* exists. Both instances involve the *immune process.*

IMMUNITY

The Immune Process

Any protein not of our own making may act as an *antigen.* Antigens—proteins of infectious organisms such as bacteria and viruses, or the proteins of animal or plant cells—if they manage to pass through the skin or intestinal wall and enter the blood, bring about a response by particular cells in lymphatic tissues (spleen, lymph nodes, thymus) called *plasma cells.* This response is the production of antibodies which then become a part of the plasma proteins of the blood. (Much evidence suggests that when antibody formation is stimulated, lymphocytes undergo transformation to plasma cells; see Craddock.) Among the plasma proteins are the *globulins*, and if we separate the globulins from the other plasma proteins we also separate out the antibodies. In fact most of the antibodies seem to be among a certain group of the globulins called the *gamma globulins*. Thus a convenient method of isolating the blood's antibodies is of great practical importance, as will be seen later.

Once a plasma cell begins the production of a certain antibody, it apparently is able to continue this production long after the antigen has been destroyed. Even the cell's descendents may continue the antibody production years later. However, the rate of production gradually decreases, and in some cases, re-exposure to the antigen every few years is necessary for continued immunity.

Reactions between Antibodies and Antigens

The manufacture of antibodies in response to a given antigen sets the stage for any later exposure to that antigen. For it is the response of the body to the next encounter with the antigen that is the *immune reaction*. In the immune reaction the antibody combines with the antigen in a way which renders the latter harmless. The antigen-antibody complex, which is essentially the debris of battle, is removed by the scavenging action of the white blood cells.

Acquisition of Immunity

Natural Active Immunity

When an antigenic substance gains entrance to the body, antibodies are formed against it. Foreign proteins, whether free or as parts of bacteria or viruses, may gain entrance in a variety of ways: through the respiratory passage or through breaks in the skin; if they are swallowed they may invade through the intestinal wall. No matter in what way a foreign protein gains entrance, once in the body fluids the plasma cells respond by producing specific antibodies against it. If the antigen is the protein of a disease-causing bacterium or virus, a protective immune reaction will occur when there is a later exposure.

We know that children are generally more susceptible to infectious disease than are adults. This is because the first exposure to pathogenic organisms finds the child without protective antibodies. If he recovers from the effects of the first exposure, he has acquired a *natural active immunity* to the disease through the immune response, and the protection will last as long as the antibodies are present in the body. This is the rest of his life in the case of chickenpox, measles, and whooping cough. Or it may be for only a few weeks or months as seems to be the case with many viruses responsible for colds and influenzas.

TABLE 19–1 Active Immunization Schedule for Children

Diseases		*Material (Antigen)*	*No. Doses*	*Age for Immunization*	*Age for Booster Doses*
Diphtheria Tetanus Pertussis (Whooping cough)		DTP—Triple preparation Diphtheria toxoid Tetanus toxoid Pertussis vaccine Td—Double preparation Tetanus and diphtheria toxoids, adult type	3	DTP 2 months 3 months 4 months	DTP 15-18 months 4-6 years Td 12-14 years every 10 years thereafter
Poliomyelitis (2 schedules)	1	OPV—Oral polio vaccine (Sabin), trivalent vaccine, Types 1, 2, 3	3	OPV trivalent 2 months 4 months 6 months	OPV trivalent 15-18 months 4-6 years
	2	OPV—Oral polio vaccine (Sabin), monovalent vaccine Type1 Type 3 Type 2	1 dose each	OPV monovalent 2 months 4 months 6 months	OPV trivalent 15-18 months 4-6 years
Smallpox		Smallpox vaccine	1	15-18 months	4-6 years 12-14 years every 3-10 years thereafter
Measles or rubeola (3 schedules)	1	Live attenuated vaccine	1	12 months	No recommendation: immunity may be permanent
	2	Same + immune serum globulin	1	12 months	
	3	Live further attenuated vaccine	1	12 months	
German measles or rubella		Live vaccine	1	1 year to puberty	No recommendation: immunity may be permanent
Mumps		Live attenuated vaccine	1	12-14 years	No recommendation; immunity at least 2 years

Adapted from Table 1. *The Red Book,* Report of the Committee on Infectious Diseases, Sixteenth Edition, American Academy of Pediatrics, 1970

Some antibodies appear to be present from the time of birth with no apparent external stimulus for their production. These are called *natural antibodies.* For example, some laboratory animals possess naturally occurring antibodies against such human pathogens as the typhoid fever and cholera organisms. Certain humans seem to possess, from birth, antibodies against the pneumonia and typhoid organisms. However, it is not possible, at present, to determine whether such antibodies are developed as a consequence of the individual's genetic makeup or as a result of the uterine environment, that is, stimulated in the fetus as a consequence of the mother's immunity. Since we cannot make the distinction, it would seem logical to consider that any immunity to disease organisms present without vaccination is natural active immunity.

Active Immunity by Vaccination

Fortunately, it is not necessary to have a disease in order to acquire immunity against it. A technique that introduces the antigen of a pathogenic organism in such a form that antibody production is stimulated in the absence of disease will serve the same purpose as having had the disease. This is the basis for immunization by vaccination. The vaccines used to induce antibody formation may be of a number of different types. (Table 19–1).

Inactivation and Attenuation

Some bacteria and viruses can be killed or otherwise inactivated in a variety of ways that diminish their disease-causing ability while retaining their antigenic properties. Common methods of inactivation are heat, treatment with formalin (Salk polio vaccine) or phenol, or pulverizing them by grinding or by high-frequency sound waves. Vaccines containing organisms that have been inactivated by such physical or chemical means include those for protection against typhoid fever, plague, and cholera.

Attenuation is the diminution of virulence of a pathogenic organism. Many bacteria and viruses lose their disease-causing ability (virulence) to man after repeated growth in other animals or in laboratory culture media (Fig. 19–1). For example, the yellow fever virus was grown in tissue culture for thousands of generations until an adapted strain was found which could not cause serious disease, but which would act as an antigen to induce the formation of antibodies effective against the dangerous form of the virus. Similarly, the rabies vaccine for pets is a live, mutated virus. Sabin (Fig. 19–2) and his coworkers developed nonvirulent strains of the three viruses that cause poliomyelitis.

The term *attenuated virus,* commonly used to describe rabies, yellow fever, Sabin polio vaccines, and that recently developed for mumps [Hilleman],

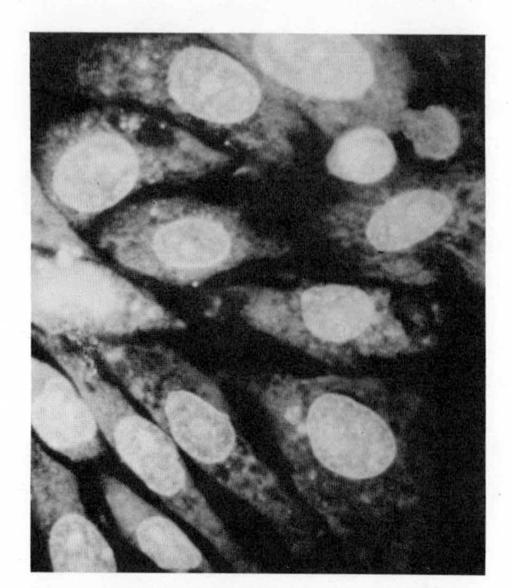

FIG. 19–1 Preparation of measles vaccine. Since viruses will grow only in living cells, large numbers of cells are grown in tissue culture to serve as hosts for measles virus. *Right:* Technician is handling special flasks filled with nutrient fluid in which the cell cultures are grown and inoculated with measles virus. The measles virus, when grown in culture, is attenuated, that is, unable to cause disease symptoms in humans, but able to stimulate antibody production. (Charles Pfizer)

FIG. 19–2 Polio vaccines were developed in large part through the work of Albert Sabin (*left*) and Jonas Salk (*right*). (March of Dimes)

is a correct one according to the usual definition of attenuation as given above. However, it should be understood that these laboratory developed variants are strong and viable in every biological sense—differing from the natural forms only in their inability to cause disease. Thus they are biologically more closely related to the phenomenon of cross-immunity than to the traditional attenuation developed for bacteria.

It is generally considered that live virus vaccines have distinct advantages over killed virus vaccines. Only one injection of the living virus is needed since it will multiply in the body and provide a long-term supply of antigen. Several injections of the killed virus are usually required in order for effective immunity to develop and often the protection is only temporary.

Cross-Immunity

We have implied that antibodies are specific for the antigen which induced their formation. However, immune reactions will also occur with proteins which are chemically similar to the inducing protein. For example, the smallpox virus is very similar, chemically, to the cowpox virus. While smallpox is a disease that may ultimately lead to death—it was responsible for the decimation of several of the Indian tribes of the eastern United States—cowpox is a minor affair that generally causes nothing more than slight fever.

The eighteenth century physician Edward Jenner (Fig. 19–3) noted that dairy maids, who almost inevitably got cowpox, almost never got smallpox even during epidemics. So, immunization with cowpox was successfully tried even before men knew about bacteria or viruses. In short, immunity to one of these diseases gives immunity to the other; the vaccination for smallpox is simply inoculation with the cowpox virus. Since the vaccinated individual will now become immune not only to cowpox but also to smallpox, we say that cross-immunity exists, that is, there is a crossing of the immunity for cowpox to smallpox. At present, smallpox vaccination has been so successful that smallpox is now practically unknown in this country. In fact, more people die from the effects of the vaccination than from smallpox. For this reason it has been suggested that routine vaccination for all children be discontinued [Lane and Millar].

Toxins and Toxoids

In many instances bacteria may grow in the body without apparent harm. The disease processes which develop in the presence of certain bacteria are most often due to the poisonous substances they produce as excretory products of their metabolism—these are referred to as *toxins.*

If the toxins are proteins, they are, of course, antigenic. Some of these toxins can be changed chemically so as to retain their antigenic properties yet lose their poisonous ones. Such altered toxins are called *toxoids.* Examples of

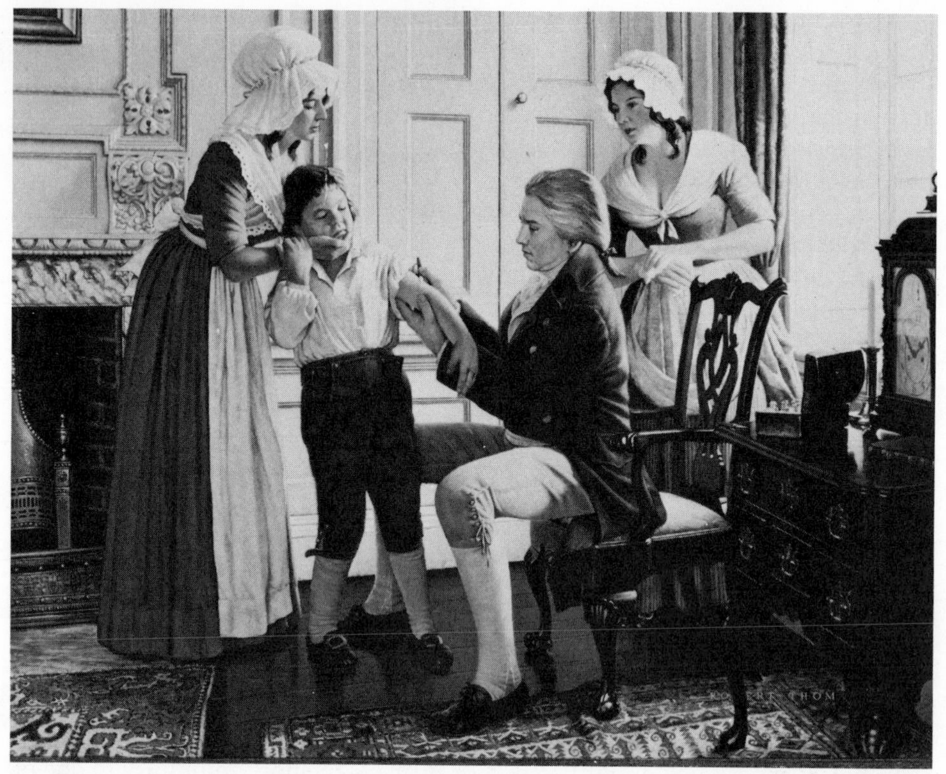

FIG. 19–3 In 1796 Dr. Edward Jenner performed the first vaccination. The cowpox virus is inoculated, and the immunity developed against cowpox also gives immunity to the much more serious smallpox. (Parke-Davis)

toxoids are the vaccines for tetanus, diphtheria, and whooping cough (the three are usually given together).

Passive Immunity

So far we have discussed the acquistion of immunity in terms of the methods which can be used to induce antibody formation. These techniques provide what is called *active immunity*—active in the sense that the body acts in its own behalf.

However, serum of an animal that has been immunized by the injection of an antigen (a disease organism, toxin, or toxoid) contains protective antibodies against the antigen (Fig. 19–4). If this serum, or serum from a person who has recently recovered from the disease, is injected into an individual it will confer passive immunity. The recipient of this *immune serum* is not manufacturing antibodies against the disease, he is merely borrowing those from the immune animal or other person. Since the immunity was passively acquired the person is said to have passive immunity. Antitoxins can be isolated from immune serum and are available for use against tetanus, diphtheria, scarlet fever, botulism (a form of food poisoning), and the venoms of several snakes.

Passive immunity is the only practical means of providing immunity after exposure, since effective levels of active immunity require at least two weeks to develop. However, passive immunization gives protection lasting only a few weeks. Indeed, it is the immune response of the recipient which destroys the introduced antibodies. The antibodies are proteins, hence they are also antigens, and antibodies will be made against them—thus antibody-destroying antibodies. This is the reason subsequent passive immunization against a specific disease is less effective, for the antibodies are quickly destroyed by the recipient's own immune system.

Antibodies are generally held to be identical or closely associated with the gamma globulin of the blood. Gamma globulin from human blood serum has

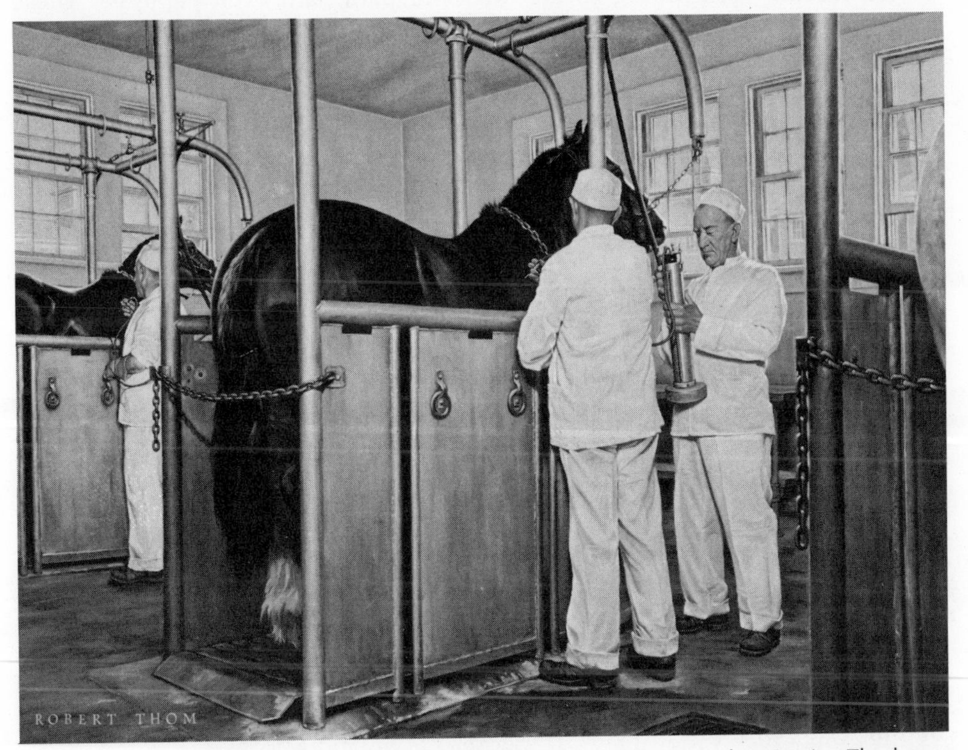

FIG. 19–4 The injection of a horse with toxin for the production of antitoxin. The horse produces antibodies (antitoxin) against the antigen (toxin) and thereby develops immunity to the disease. The horse is then bled periodically and the antitoxin extracted from the blood for use. (Parke-Davis)

been used to reduce the severity of viral diseases, notably measles and infectious hepatitis.

Rabies Vaccines

Treatments for rabies administered to those bitten by rabid animals may be either a vaccine (antigen) or an antiserum (antitoxin). Since the symptoms of the disease do not develop until the virus has reached the central nervous system, there is a long incubation period—if the bite is on the head, the incubation period is 12 to 13 days; on the arm, about 21 to 30 days; and on the leg as long as 64 to 70 days. For this reason both active and passive immunization are possible.

The original *Pasteur treatment* consisted of a series of 14 injections of the virus extracted from the brains and spinal cords of rabbits inoculated with rabies—the live virus preparation becoming a vaccine when killed by treatment with phenol. The bite victims thus vaccinated sometimes suffered severe reactions, including severe muscular pain, and even paralysis. These were the result of an antigen-antibody reaction to the foreign brain material in the vaccine. These reactions can give rise to inflammation and degeneration in the nervous system. In fact as many as 15 percent of all patients receiving the vaccine prepared from nervous tissue develop abnormalities in their brain waves, determined through electroencephalograph readings (EEG).

In 1957 Dr. H. M. Powell, a research virologist, developed a much safer rabies vaccine [see Criep]. Dr. Powell prepared the new vaccine by growing rabies virus in the tissues of a duck embryo. After 14 days growth in the egg, the suspension of rabies virus is extracted and killed by chemical treatment. The duck embryo vaccine (DEV) is dried and administered after dilution with distilled water. Like the Pasteur Vaccine the DEV is also administered in a series of 14 injections. The absence of central-nervous-system effects makes the DEV far superior to the original vaccine, and the bad reputation of rabies injections is no longer deserved—no one need be anxious about pain, stiffness, or possible paralysis.

Passive immunity is provided by injection of immune serum, which is prepared by the inoculation of horses with rabies virus, and the extraction of the immune serum (antibodies) from the horses' blood plasma. When definite exposure to rabies is suspected in a patient, both the vaccine and the antiserum are administered as the immune serum is rarely administered by itself. In any serum of equine origin there is always the danger of an anaphylactic reaction (see discussion below), so doctors are cautious in their use of the immune serum. (The antirabies vaccine used for pets and livestock is an attenuated virus grown in chick embryos. Such a live vaccine will produce immunity of over three years duration in the dog, but state laws usually require yearly vaccinations.)

Tests for Immunity

Often it is desirable to determine the degree of immunity of an individual to a specific infectious agent or toxin of a disease. For example, an individual who has had a disease, and recovered, would possess active immunity to the disease and carry antibodies against the disease. Specific tests using the toxin or other antigen have been developed to detect the presence of such antibodies.

SCHICK TEST. Schick discovered that a small amount of diphtheria toxin injected into the skin of a nonimmune person will cause a redness and swelling of the skin around the site of the injection after 24 to 48 hours. Immune individuals usually give no reaction. This absence of a reaction (or "negative reaction") indicates the presence of the diphtheria antitoxin which presumably combines with the toxin and thereby prevents any skin reaction.

DICK TEST. The Dick test for immunity to scarlet fever is similar to the Schick test. A small amount of scarlet fever toxin is injected into the skin. A positive reaction is an area of redness and swelling at least 1 centimeter in diameter around the site of the injection after about 16 to 24 hours. No reaction (a negative reaction) develops in an immune person.

ALLERGY

The Allergic Reaction

Most people refer to allergies or allergic reactions without a very clear idea of what they mean. You are probably aware that certain foods when eaten or certain substances when applied to the skin will result in hives (urticaria) or even more serious symptoms such as vomiting and severe headache, or that some pollens will cause itching eyes and a stuffy nose. It is now known that these and a number of other conditions are reactions occurring when a person is *mildly* immunized against an antigen. When complete immunity is developed, the allergy disappears. For example, infants are sometimes allergic to egg albumin or meat, and later lose these allergies.

Moreover, it is believed that the antibodies involved in these allergic responses are either attached to cells or are within cells, and are not generally present in significant amounts in the globulin fraction of the blood. What this means is that the immune reaction, when it occurs, has direct effects only on the cells associated with the antibodies. As a result these cells begin to swell and may even rupture. Such damaged cells release certain chemicals which are normally restricted to within the cell. These chemicals include *histamine.*

There is abundant evidence that histamine, or a very similar substance, is responsible for many of the serious reactions in allergy. We may cite the effects of histamine on blood vessels and on the bronchioles. Histamine causes intense dilation of the very small arteries, called arterioles, leading into the capillaries. This has the effect of allowing an abnormally large amount of blood into the capillary beds. This in turn raises the pressure within the capillaries to such an extent that fluid of the blood leaks out into the tissues in greater than normal quantities. The vasodilation and fluid accumulation in the tissues are responsible for the redness and swelling present in sites of allergic reactions. The fluids exuded from the delicate membranes of the eyes and nose of the hay fever victim are also a result of the dilation of arterioles.

Quite the opposite effect occurs in the bronchioles of the lungs leading to the air sacs; they are constricted through the action of histamine on their muscular walls. (The reason that allergic reactions dilate some organs and constrict others can be better understood after discussion of the autonomic nervous system, Chapter 33.) There is evidence that one or more of the other chemicals liberated from affected cells are principally responsible for the bronchiolar reaction. At any rate this generalized bronchiolar constriction causes the asthma which is often one of the serious consequences of allergy.

Types of Allergic Reactions

Local Reactions

Most of the common allergic reactions occur at the site of entry of the antigen (usually called an *allergen*). For example, if you happen to be allergic to a certain kind of soap, a rash will develop at the site of application. In some instances—antibacterial soaps are good examples—the sensitivity is restricted to given areas such as the face or hands. Another example is the *contact dermatitis* which arises in most people upon exposure to poison ivy.

In the typical *hay fever* sufferer the allergens, whether pollens, horse dander, house dust, or whatever, comes in contact with the nasal tissues and eyes, and the reactions are limited to these areas. The asthma that develops from airborne allergens is another type of local reaction. In this case the immune reaction occurs within the bronchiole and the local release of histamine results in a generalized bronchiolar constriction which impedes the movement of air in and out of the air sacs.

Generalized Allergic Reactions

If a severe immune reaction occurs within the blood vessels, or if an intense local reaction releases large amounts of histamine into the blood, then a generalized allergic reaction can result. Such body-wide reaction is termed *anaphylaxis* or *anaphylactic shock.*

Presumably, the reaction involves the destruction or damage of many cells such as those forming the walls of arterioles and those lying in close contact with the blood vessels. The release of large quantities of histamine and the subsequent dilation of arterioles throughout the body brings about a virtual collapse of the circulation. Large quantities of blood accumulate in the capillary beds—fluids escape from the capillaries—and a drastic drop in blood volume is the essential result. The net effect is much the same as that of severe hemorrhage, that is, a failure of blood to return to the heart in sufficient quantity to maintain circulation. Heart failure and death may follow. The circulatory failure may be accompanied by a histamine-induced contraction of the bronchiolar muscles, making breathing difficult. In fact, death in anaphylactic shock may be caused by suffocation.

Anaphylactic shock causes death occasionally following the injection of certain immune sera and antibiotics. For example, some people develop a sensitivity to horse serum, and since horses are used in the production of a number of antitoxins, progressively more severe reactions are liable to occur after each injection. Such a situation could develop, for example, from the injections of serum containing tetanus antitoxin for producing a state of passive immunity. The second or third injection of the serum may cause only a local reaction, but later injections could have a generalized reaction. Many people develop similar sensitivity to penicillin. This is the reason physicians are interested in learning of any previous reaction to this drug. It is also the reason physicians keep a person in the office for some period of time after a penicillin injection, unless they already know he is not penicillin sensitive.

The Basis of Allergy Treatments

Antihistamine Drugs

Since most of the effects of allergy are due to histamine, any agent which blocks either the release or the action of histamine will prevent some of the symptoms. The antihistamine drugs are believed to act by competing with histamine in the chemical reactions which cause the allergic symptoms. This means that the antihistamine drugs react with the cells in the same manner as histamine but produce few or none of the effects of histamine.

Antihistamines are particularly valuable in relieving the effects of histamine in anaphylaxis, urticaria, and hay fever. Several of these drugs have an additional effect of inducing drowsiness in many people—sometimes to the extent of preventing normal activity. Indeed, several of the over-the-counter (nonprescription) "sleeping pills" are in fact antihistamines!

Epinephrine (Adrenalin)

Epinephrine is a hormone produced by the medulla (inner part) of the adrenal gland which has effects almost exactly opposite to those of histamine. When released into the blood this hormone causes constriction of arterioles of the skin and visceral organs, bronchiolar dilation, increased heart rate and hyperglycemia (the release of sugar into the blood), and a generalized excited state. These responses are characteristic of what is called the *sympathetic reaction.* The sympathetic reaction is due in large part to the release of epinephrine by the adrenal gland, but also to the release of an epinephrine-like substance by nerves belonging to the sympathetic division of the nervous system.

Drugs which bring about responses that mimic the normal sympathetic reactions are called *sympathomimetic drugs,* and include not only epinephrine itself, but also ephedrine (a plant derivative) and several synthetic agents. Some of these drugs will cause bronchiolar dilation without concomitant effects on the heart and nervous system, so that relief from the asthma is not necessarily accompanied by palpitations and "jitters." Aerosals of sympathomimetic drugs are now available in pocket units, and are indeed a blessing to asthma victims. These inhalers can frequently give relief from an asthma attack in seconds. However, only an experienced allergist can select the drug most suited to the individual needs. Do not use any potent pharmaceuticals such as the above without competent medical guidance.

Cortisone

Cortisone is also a hormone secreted by the adrenal gland, but is manufactured by the cortex (outer part) of the gland. Cortisone is now synthesized as a pure chemical, and is a valuable drug used widely in medicine. Our concern here is in its ability to disrupt the immunological response, apparently by blocking the interaction between antigen and antibody. When given cortisone, patients with hay fever, poison ivy, or hives will often make a dramatic recovery.

However, cortisone has other effects which caution against its use. It suppresses the movement of white blood cells through blood vessel walls and also seems to inhibit their formation. Hence a person receiving cortisone has decreased resistance to infection. It may also cause changes in facial appearance—a "moonface" condition which results from a reddening and puffing of the skin. Cortisone has other effects which are equally serious, and physicians are reluctant to commence cortisone treatment except where circumstances demand it.

Desensitization

The "shots" which many people take for the treatment of an allergy are attempts to change the mild immune state into a complete one. To do this the offending allergen is collected and purified, and then injected into the allergic person in small but progressively larger and larger doses, usually over a period of many months. Mild immune reactions are expected to occur at first, but as the circulating antibodies are developed, the allergic reactions become weaker and weaker until the person is actually immune rather than allergic to the allergen.

Generally, persons who are allergic to one agent are allergic to others also, and desensitization treatments are most successful in those in which the allergy is due to intense reaction to a few allergens rather than mild reactions to a great number. Even when the nature of the allergy is well understood the effectiveness of the treatment is unpredictable. Some persons are essentially cured of an allergy after a few months treatment, whereas others derive little or no benefit even after several years.

Although medical science has learned a great deal about the immune process and allergies, about active and passive immunizations, and about prevention and treatment of allergies, there is yet a wide world of knowledge, especially concerning allergies, that remains a mystery. It is quite possible that many a person's vague and frustrating little transient ailments that go undiagnosed may in fact be allergenic conditions that are not yet understood as such.

REVIEW QUESTIONS

1. How and where are antibodies formed?
2. What is the immune reaction?
3. What is meant by natural active immunity?
4. How is a toxoid prepared?
5. Passive immunity is made possible by the use of immune serum. Explain.
6. Discuss the vaccines used in the treatment of rabies.
7. What is meant by an allergic reaction and what are its symptoms?
8. What is the difference between a local allergic reaction and a generalized allergic reaction?
9. Why is epinephrine effective in the treatment of a generalized allergic reaction?

REFERENCES

BOTTOMLEY, H. W., *Allergy, Its Treatment and Care.* New York: Funk & Wagnalls, 1968. (Paperback)

BRAMBELL, F. W. et al., *Antibodies and Embryos.* New York: Oxford University Press, 1951.

CRADDOCK, C. G. et al., "Lymphocytes and the immune response," *New England Journal of Medicine,* **285**: 324–331, 378–384 (1971).

CRIEP, LEO H., *Clinical Immunology and Allergy,* 2nd ed. New York: Grune and Stratton, 1969.

HILLEMAN, MAURICE R. et al., "Live, attenuated mumps-virus vaccine," *New England Journal of Medicine,* **278:** 227–232 (1968).

LANE, J. MICHAEL, and J. D. MILLAR, "Routine childhood vaccination against smallpox reconsidered," *New England Journal of Medicine*, **281**: 1220–1224 (1969).

MOVAT, HENRY Z., *Inflammation, Immunity, and Hypersensitivity*. New York: Harper and Row, 1971.

Additional Readings

HIRSCHFELD, HERMAN, *The Whole Truth About Allergy*. New York: ARC Books, 1963.

20
TISSUE AND ORGAN TRANSPLANTATION

The transplantation of organs has been the subject of fictional and scientific speculation for centuries. But only in recent years have organs been transplanted from one person to another with the reasonable expectation that they will become permanent "items of equipment." What are the problems which have been so difficult to solve and what future advances in organ transplantation might we expect?

In order to answer these questions we must consider again the nature of proteins—their complexity and their diversity. This diversity of proteins accounts for the differences exhibited by almost all living things. The differences among people, and the ways in which people differ from other mammals, are due to these protein differences. However, the more closely two people are related the greater the similarity of their proteins. The proteins of a brother and sister would show a greater similarity than those between cousins; those between cousins a greater similarity than those between unrelated individuals; and these protein differences can be demonstrated by immune reactions. For example, if a piece of skin is transplanted from a father to son, it will heal in place and grow well with all the appearances of vitality. But after a period of five or six weeks it will begin to appear infected, will degenerate, and after another week or so will become necrotic (dead) and will be sloughed off. This process is the result of *tissue transplantation immunity.*

What has happened? The reader is probably able to guess from what has been said in the previous two chapters. The donor's (father's) skin is composed of proteins which are, at least in part, foreign to the son's body. (One-half of the proteins of the son's body would be like those of the mother). This means that as soon as the skin transplant is in place the body responds by manufacturing antibodies against these "foreign" proteins. When the number of antibodies has been built up to a high level in the blood, the rejection process begins. If the graft is examined microscopically at this stage, it will be seen to be invaded by thousands of lymphocytes. (This is added evidence, incidentally, that lymphocytes either manufacture antibodies, or transport them to the blood, or both.) This lymphocytic army surrounds and eventually destroys the organized tissues of the grafted skin. The graft recipient also begins to run a fever, and shows an elevated white blood cell count, as if he were fighting an infection such as appendicitis, even though no infectious organisms are involved. A normal condition returns only after the transplant falls away or is reduced to a useless scar. If the transplanted skin were from an individual more distantly related, the rejection process would follow the same course, but, of course, would be more rapid.

On the other hand, if this same person were to have skin transplanted from one part of his body to another, no rejection reaction would occur. Obviously, no foreign proteins are involved. Actually, such transplants have been used for a number of years in the treatment of burns. Grafts between identical twins are similarly successful. Since identical twins are developed from a single fertilized egg, they are identical in their protein makeup. Also, a few tissues such as the cornea of the eye and cartilage can be transplanted without the development of tissue transplantation immunity. This is attributed to the fact that these tissues have no blood circulation; instead they derive their nourishment from surrounding tissues by diffusion. They are thus immunologically isolated, and if a cornea or piece of cartilage is carefully transplanted from one person to another it will usually survive.

The various types of transplants have been given specific terms: grafts within the same person are *autografts;* grafts between any two individuals, other than identical twins, are *homografts;* grafts between identical twins are, in essence, autografts, but are given a special name, *isografts.*

This, then, was essentially the state of affairs regarding our knowledge of tissue and organ transplantation shortly after World War II. Then in the 1950s it was discovered that newborn or fetal animals do *not* show immune responses. For example, if a skin homograft is implanted in a newborn rat, the graft is tolerated and survives for the life of the animal. However, if the transplant is made two weeks following birth, an immune reaction develops and rejection occurs.

It is as if the newborn animal takes a census of its proteins. All proteins present at the time of the "census" are regarded as "self-proteins." Any newcomers following this period are "foreign" and are regarded as antigens. This means that if a skin homograft were implanted from father to newborn son at birth, it might be that, later in life, other tissues or organs could be effectively transplanted. This does not appear to be a practicable solution to transplant problems. However, it does mean that if the body could somehow be returned to the immune-free state which exists at birth, it might be made to accept a transplant even as an adult, or become what has been termed *immunologically tolerant*. This was shown to be the case, and it was first accomplished by means of irradiation.

One of the effects of whole-body irradiation, by X-ray or by certain radioactive elements, is suppression of antibody formation and destruction of blood-forming cells. If the amount of irradiation is sufficient to bring about an inhibition of these antibody-forming cells (plasma cells), yet does not destroy them (or the other blood-forming cells), a transplant made during this period of suppression will "take." It is as if the plasma cells undergo a period of amnesia, and upon reawaking take a new census of body proteins and accept any new ones.

The problem is in measuring the exact level of irradiation—it must be sufficient to suppress antibody formation but not destroy the blood-cell forming capacity. Achievement of such a perfect balance proved extremely difficult. However, a few kidneys have been successfully transplanted following such whole-body irradiation.

More success has been achieved with the use of drugs which suppress antibody production. These drugs act by interfering with the mechanism by which proteins are synthesized, and since antibodies are proteins, the drugs interfere with antibody production. Obviously, such drugs are extremely poisonous and must be used in low dosage, and even then the balance of survival is still precarious.

The kidney transplantations now being made generally make use of these drugs. Better drugs are becoming available, and the transplantation of kidneys, although far from commonplace, no longer makes news headlines.

So far we have discussed only the transplantation of skin and of kidneys. Kidneys have attracted the attention of the surgeons because (1) we are almost all born with two kidneys, but can get along quite well with one, if the need arises; (2) each kidney is supplied with large blood vessels which can be readily united with the vessels of the recipient; and (3) the development of artificial kidneys has made it possible to keep patients with severe kidney disease alive until a donor kidney becomes available. The transplantation of other paired organs, such as lungs and adrenal glands, one of which a living, healthy donor can safely supply without gravely endangering his own well-being, is theoretically possible, and such transplantations can reasonably be expected.

Unpaired organs, such as the heart and liver, must come from cadavers, or possibly from other mammals. Progress here has been rapid in the past several years. However, the limitations necessarily imposed, such as determining irreversible dying processes and the time of clinical death in the donor individual, present major problems. The ultimate solution for the replacement of unpaired organs would seem more logically to lie in the development of synthetic mechanical organs. In the case of the heart we may expect such developments in the next decade. An artificial liver is much further away.

REVIEW QUESTIONS

1. Why are tissue or organ transplants from a close relative more likely to be successful than one from a person outside the family?
2. How is the normal production of antibodies against the transplanted tissue or organ prevented?

Additional Readings

BALNER, HANS et al., *Transplantation Today.* New York: Grune and Stratton, 1971.

BILLINGHAM, R. E., and W. K. SILVERS, *Transplantation of Tissues and Cells.* Philadelphia: Wistar Press, 1961.

LYONS, CATHERINE, *Organ Transplants: The Moral Issues.* Philadelphia: Westminster Press, 1970. (Paperback)

MOORE, FRANCIS D., *Give and Take: The Development of Tissue Transplantation.* Philadelphia: W. B. Saunders, 1964.

RUSSELL, PAUL S., and ANTHONY P. MONACO, *Biology of Tissue Transplantation.* Boston: Little Brown, 1965.

RUSSELL, PAUL S., and HENRY J. WINN, "Transplantation." *New England Journal of Medicine.* **282:** 786–793, 848–854, 896–906 (1970).

21
THE HEART AND THE CIRCULATION OF BLOOD

We have discussed many functions of blood, but none of these are possible without a system by which the blood can be transported. The series of tubes which serve this purpose are the blood vessels, and the blood is moved through these vessels by the pumping action of the heart. Collectively the heart and blood vessels constitute the circulatory system. If you are interested in your own health and those aspects of it that *are* relatively controllable, you should take particular note of your circulatory system. Here is a system upon which you *can* exert some influence. To be sure there can be inherited defects or weaknesses that predispose to defects, but there is much that we can do to ward off the degenerative diseases of heart and circulation.

The Two Great Circuits

The general arrangement of the major circulatory patterns is shown in Figure 21–1. As you undoubtedly already know, there are really two circuits of blood flow, and the heart is two pumps. The two circuits, called the pulmonary and systemic circulations,

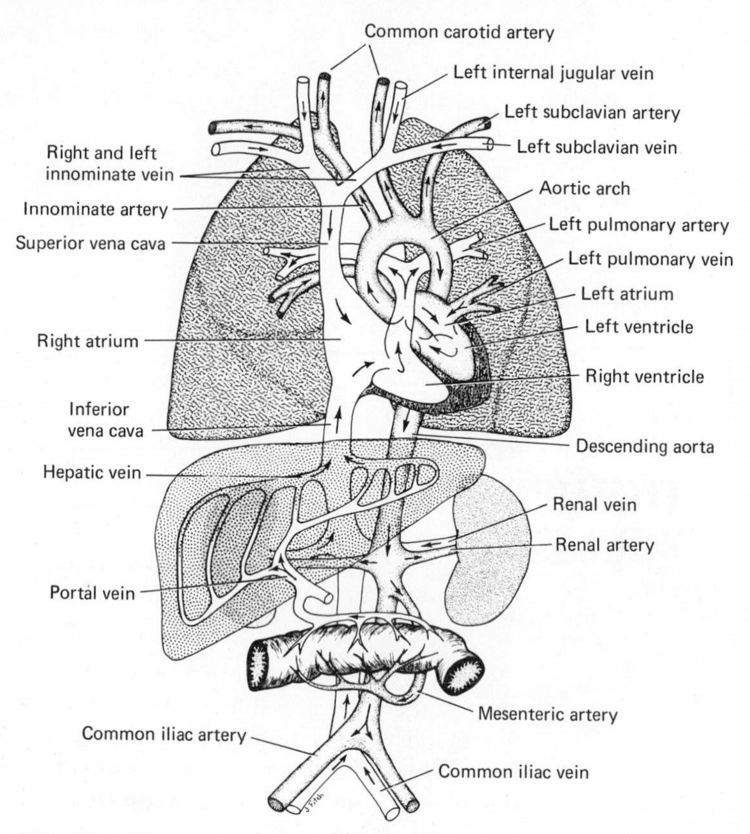

FIG. 21–1 Major circulatory patterns. The pulmonary and systemic circulations are shown together with the two subcircuits of the systemic circulation which serve the excretory and digestive organs.

each consist of arteries which carry blood *away* from the heart and veins which carry blood *toward* the heart. (This constitutes the only valid distinction between arteries and veins; definitions involving amounts of oxygenated and deoxygenated blood only confuse things.) The repeated branching of the arteries results in smaller and smaller subdivisions, eventually leading to the smallest vessels of the circulatory system which are classed as neither arteries nor veins, but are called *capillaries.* Capillaries feed into the smallest veins, which in turn form tributaries to progressively larger and larger veins, until eventually the blood is returned to the heart.

The pulmonary circulation utilizes the pumping action of the right side of the heart. Blood is pumped from the right atrium into the right ventricle. (We should pause here to note that the term *auricle* is frequently used as a synonym for atrium, but although firmly entrenched in the majority of physiology and health books is nevertheless incorrect. Auricle as an anatomical term properly refers only to the earlike appendages on the atria.) The contraction of the right ventricle then forces the blood into the pulmonary artery which branches into right and left pulmonary arteries leading to the lungs. Minor branches subdivide and eventually lead to the capillaries of the lungs. Oxygen in the air sacs of the lungs diffuses into the blood of the capillaries; carbon dioxide moves in the other direction—into the air sacs. Thus a gas exchange takes place, and the blood is said to be oxygenated. The capillaries then converge to form small veins which further converge until two main trunks, the pulmonary veins, leave each lung. These veins enter the left atrium, completing the pulmonary circuit.

The systemic circulation now takes over: During relaxation of the heart and finally with contraction of the left atrium, blood moves into the left ventricle, and contraction of the left ventricle forces blood into the large aorta which delivers blood to the entire body. Most of the systemic circulation serves to bring oxygen and nutrients to the tissues, and to remove from the tissues carbon dioxide and metabolic wastes. However, two subcircuits serve, in addition, rather special functions. One of these is the circulation through the kidneys which serves to filter out the metabolic wastes and excess fluid which results in the production of urine; and the other is the circulation through the digestive organs for the absorption of food and its transport to the liver.

The Heart

More specific comments concerning the heart are essential if we are to understand later the problems of heart diseases. Particularly we must direct attention to the mechanical features which are indispensable if the heart is to be an efficient pump.

Each "beat" of the heart actually consists of two contractions. In the first contraction the muscular walls of the atria compress the atrial chambers causing a final surge of blood into the two ventricles—blood from right atrium to right ventricle, and from left atrium to left ventricle. In the second contraction the ventricular walls compress the ventricles forcing blood from the right and left ventricles to the pulmonary artery and aorta, respectively.

Upon contraction of the ventricles, blood must be prevented from passing back up into the atria. This is accomplished by flaps or valves which are forced down against the ventricular walls as the blood passes from the atria to the ventricles (Fig. 21–2). However, as the atria relax and the ventricles begin their contraction, the blood pushes the valves back toward the atria. They are, however, restrained from being pushed up into the atria by strong tendons, the *chordae tendineae,* which anchor them just sufficiently so that they meet and close the atrio-ventricular channels. Blood is forced, as a consequence, to follow the only paths of exit, which are the pulmonary artery on the right and the aorta on the left.

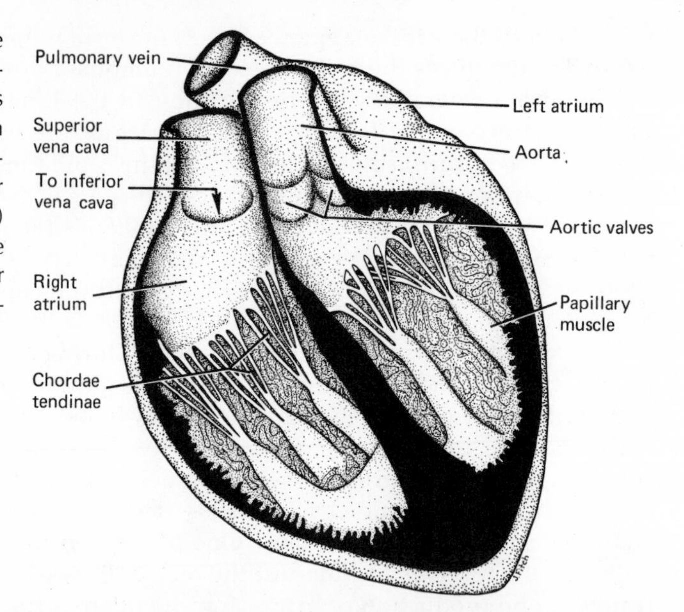

FIG. 21–2 Valves of the heart. The pulmonary artery and the anterior parts of the ventricles have been removed. The flaplike structure of the atrio-ventricular (tricuspid and bicuspid) valves, and the pocketlike structure of the semilunar (aortic) valves are emphasized.

Although the atrio-ventricular valves on the two sides of the heart are essentially similar, they are distinguished by the number of flaps forming the valve—three flaps or cusps on the right form the *tricuspid valve,* and two on the left form the *bicuspid valve* (or *mitral valve* since it has a fancied resemblance to a bishop's mitre).

Blood which is pumped into the pulmonary artery and aorta stretches the walls of these large vessels, and the recoil of their elastic walls serves to propel the blood toward the peripheral arteries. This recoil would tend to move the blood back into the ventricles during relaxation of the ventricular walls except for the presence of valves at the beginnings of these vessels. These particular valves are simpler in their structure than the atrio-ventricular valves. They each consist simply of three pockets whose mouths point in the direction of blood flow. When the ventricles relax, and the recoil of the vessel walls exert pressure on the blood, the pockets fill with blood, meet each other like a pie cut in three pieces, and shut off communication with the ventricles. These valves are called *semilunar valves.*

The closing of the valves of the heart, first the atrio-ventricular valves and then the semilunar valves, is the primary cause of the heart sounds that are heard as a "lub-dup" when listening to the heart by means of a stethoscope.

Heart Rate and Arterial Pulse

The *arterial pulse,* as felt in the artery passing over the wrist (and elsewhere), is the result of a wave of distention which passes over the arterial system with each ventricular contraction. The rate of the heartbeat can be determined, as well as irregularities in its rhythm, when "taking the pulse."

The heart rate is basically established by a small mass of specialized muscle tissue called the sino-atrial node or *pacemaker* which lies buried in the heart in the muscle of the right atrium (Fig. 21–3). The pacemaker is in turn regulated, though not fully controlled, by nerve impulses coming to it from the two divisions of the autonomic or involuntary nervous system. The sympathetic division (or sympathetic system as we have generally referred to it) accelerates the heart rate, as, for instance, in the "alarm reaction," whereas the parasympathetic system, specifically the vagus nerve, depresses the heart rate.

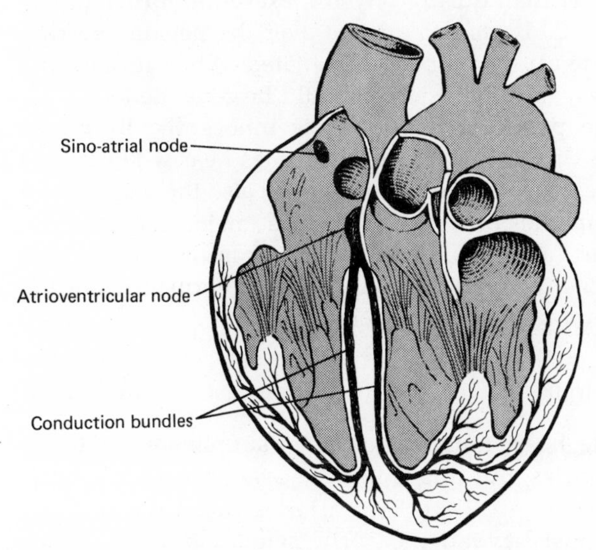

FIG. 21–3 The regulation of the heart rate. The heart beat is initiated in the sino-atrial (S-A) node or pacemaker located in the wall of the right atrium. Electrical activity spreads over the entire atrial muscle. A slight delay occurs before the electrical wave reaches the atrioventricular (A-V) node. When it does, the electrical activity is spread through the muscles of the ventricles by way of the large conduction bundles. The result is contraction of both atria followed briefly by contraction of both ventricles.

The average resting heart rate at rest is given as 72 beats per minute. This should not be taken as a normal rate. The normal resting rate may vary from the middle thirties to the upper eighties without cause for concern. In determining your normal pulse (heart) rate it is best to make the count in the morning upon waking, preferably while still in bed. It is frequently advantageous to know what your normal rate is, so that a relative *tachycardia* (increase) or *bradycardia* (decrease) can be identified.

Bradycardia, unless it results gradually from regular physical activity (Chapter 31), is so unusual that a physician should be consulted immediately. The causes can be of a serious nature, such as thyroid deficiency or pressure on the brain caused by intracranial bleeding or tumor growth. (Some persons have perfectly normal resting heart rates well below 72 (as low as 50-55) even though they are not very physically active.)

Tachycardia almost always has some apparent and generally harmless cause.

It is an obvious and necessary concomitant of exercise, and the normal heart of the youngster or the young adult can beat as many as 180 to 200 times per minute without harm or alarm. Beyond middle age this maximum heart rate tends to become lower, especially if vigorous physical activity has not been a part of one's life. Even then the maximum rate for the normal heart is likely to be between 150 and 170. Any experience which brings about an alarm reaction also increases the heart rate—emotional excitement, nicotine from cigarette smoking, environmental stresses such as high humidity, heat, or high altitude, and a host of drugs which either stimulate the sympathetic system, such as epinephrine and amphetamines, or depress the parasympathetic system, such as belladonna or atropine. Other than during physical exertion, normal tachycardia should not often exceed 20 beats per minute over the person's resting rate or persist for long after the cause has been eliminated. When it does, the condition may be more serious, and a physician should be consulted.

Minor irregularities of the pulse rate occur quite innocently. By minor irregularities are meant the "skipped" beats we sometimes hear when we are aware of our heartbeat, as occasionally when lying in bed; also the waxing and waning that sometimes accompanies breathing. However, major irregularities, such as periods when the heart rate is 150 to 300 beats per minute, or sudden episodes when the heart seems to "flutter," may mean abnormalities of the pacemaker, and warrant the attention of a physician.

The Blood Supply of the Heart

It might be expected that the heart muscle would be supplied with blood directly from the atria or ventricles, but this is not the case. Instead, the heart has its own circulatory system (Fig. 21–4). Two arteries called the *coronary arteries* arise from the aorta just beyond the aortic semilunar valve. These arteries branch over the surface of the heart and give rise to small arteries which pass into the heart muscle. The veins draining the blood from the heart muscle are the *coronary veins,* and these eventually deliver the blood to the right atrium.

The coronary arteries are the first branches of the aorta, and hence, the first systemic arteries to receive blood. Following ventricular contraction, blood rushes through the coronary arteries into the heart muscle. The next contraction of the heart moves the blood from the muscle into the larger coronary veins which lie on the surface of the heart. In other words, during contraction of the heart, the coronary vessels within the heart muscle are emptied, and during relaxation they are filled. So, after each contraction of the heart a new supply of oxygen and nutrients are available for use by the muscle cells of the heart. We shall have much more to say about coronary arteries in the next chapter when we discuss diseases of the heart.

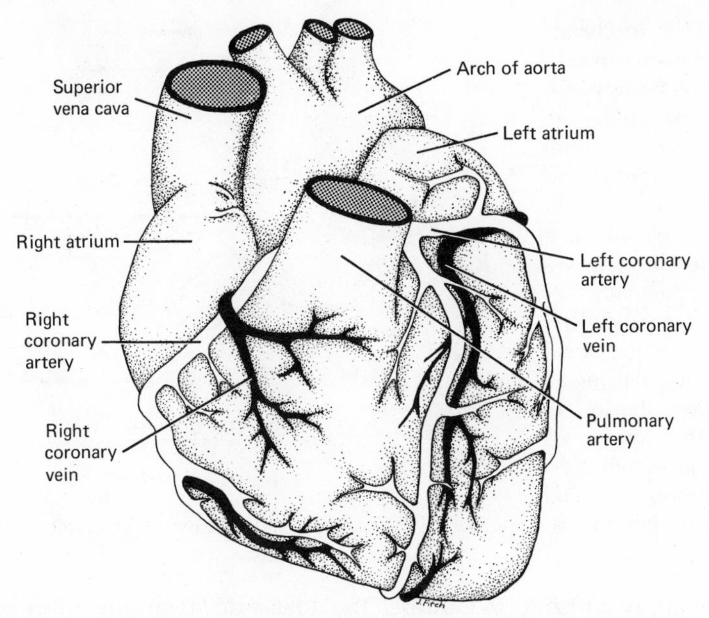

FIG. 21–4 The coronary circulation. The coronary arteries arise from the base of the aorta (hidden by the pulmonary artery).

The Arteries

To understand blood pressure and the movement of blood into the capillary beds we must know something of the nature of the arteries themselves. The largest of the arteries, the *aorta,* and its immediate branches which deliver blood to the major divisions of the body, are characterized by very elastic walls. But as the smaller branches are approached there is a gradual decrease of passive elastic tissue and an increase of actively contractile muscle in the wall. Ultimately in the smallest arteries, *arterioles* which lead directly into the capillaries, the walls are made up almost entirely of muscle.

The gradual shift from an elastic wall to a muscular one reflects a progressively changing function. The elastic walls of the large arteries are distended by the stream of blood forcibly pumped into them by contraction of the left ventricle of the heart. When the ventricles relax, the recoil of these elastic walls forces the blood on toward the smaller arteries and arterioles. But the muscular nature of the smaller arteries provides a further necessary feature to this movement of blood: it offers what is referred to as *peripheral resistance.* The importance of this peripheral resistance can be illustrated by analogy to a garden hose. The water leaving the open end of the hose will have a rather low pressure, and will fall in a rather abrupt arc. If a long, forceful stream of water is

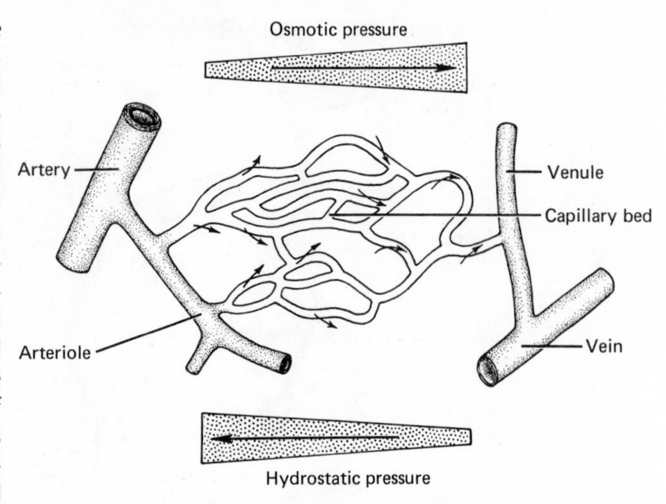

FIG. 21-5 The exchange of fluids and dissolved substances between the blood and the tissue fluid. The hydrostatic (blood) pressure is high at the arterial end of the capillary bed and drops rapidly by the time the blood reaches the venous end, principally because fluids and dissolved substances leave the blood and pass into the tissue fluid. However, the loss of fluids from the blood tends to increase the osmotic pressure, and at the venous end of the capillary bed this is sufficient to draw fluids and dissolved substances back into the blood.

desired we apply a nozzle. What does the nozzle do? It simply offers peripheral resistance to the flow of water within the hose, and the pressure of the water against the wall of the hose is increased. When the muscles of the arterioles contract, restricting movement of blood into the capillaries, the pressure is similarly increased. Just as the nozzle serves to spray water onto a garden, the arterioles serve to spray blood into the capillaries; and without this peripheral resistance offered by the arterioles, the heart would be unable to maintain pressure sufficient for circulation (Figs. 21-5 and 21-6).

The muscular nature of the arterioles serves further to regulate the distribution of blood. Blood must be preferentially directed to certain areas simply because there is not enough blood to supply fully all capillary beds simultane-

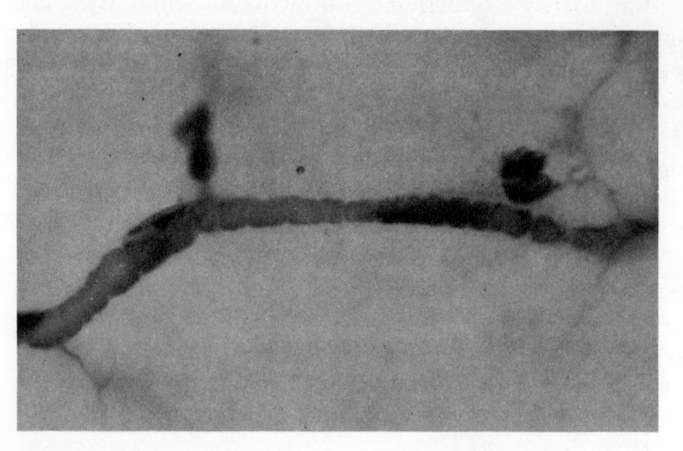

FIG. 21-6 Red blood cells within blood capillaries.

ously; in fact the volume of blood falls far short of any such dimension. This is reasonable in view of the dynamic nature of tissue needs. Obviously, the intestines require far more blood during the process of digestion than at other times; muscles require far more blood during exercise than at rest. This is the reason why extremely strenuous exercise after meals is unwise; it demands that two major beds be fully dilated at once, which is not practicable, and must result in strain on both systems. Probably the brain alone requires and receives a constant, undiminished blood supply.

Blood Pressure and Its Measurement

The contraction of the left ventricle culminates in a maximum pressure of blood in the large elastic arteries; this maximum pressure is referred to as *systolic blood pressure.* The distended elastic walls dissipate this pressure by forcing the blood on toward peripheral arteries. When the pressure reaches its lowest point, before onset of the next ventricular contraction, this minimum pressure is the *diastolic blood pressure.* The diastolic pressure represents the resistance against which your heart must work while the systolic pressure is, in a sense, a measure of the work accomplished by your heart. These concepts will take on greater meaning for you when we discuss hypertension (Chapter 23).

Anyone who has observed blood spurt from the cut end of an artery is impressed with just how high arterial blood pressure is. In 1711 Reverend Stephen Hales, an English clergyman and physiologist, showed that this pressure could be accurately measured (Fig. 21–7). After securing a white mare to a stable door he dissected free the carotid artery (the large artery carrying blood to the head), and inserted into it the windpipe of a goose to which he linked a vertical glass tube 12 feet 9 inches high. When he unclamped the artery, the blood spurted up the glass tube to a height of 9 feet 6 inches. The level of the blood rose and fell with each heart beat, and these high and low levels we could now designate as the levels of systolic and diastolic blood pressure, respectively. We measure blood pressure today in millimeters of mercury because it requires less space. If we convert Reverend Hale's 9½ foot column of blood to a column of mercury, we get a height of 224 millimeters of mercury, written as 224 mm Hg. (Blood is 1.05 times heavier than water; mercury is 13.6 times heavier than water—25.4 millimeters in an inch.)

The direct measurement of blood pressure is such a dramatic and instructive demonstration that it is frequently performed upon animals in physiology laboratories; but in spite of its colorful and instructive value the technique is hardly suitable for the measurement of human blood pressure, and therefore for this purpose we use the *sphygmomanometer.* This is the familiar inflatable cuff which is connected to a mercury gauge (manometer). The cuff is wrapped around the upper arm and air is pumped into it by means of a rubber bulb. The large brachial artery of the arm is collapsed when its pressure is

FIG. 21–7 In 1733, Stephen Hales, a minister turned physiologist, performed the first measurement of blood pressure. He used the windpipe of a goose to connect a long glass tube to the carotid artery of a horse. He noted that the column of blood rose and fell with each heart beat. (Bettmann Archive, © Otto L. Bettmann, 1956)

exceeded by the pressure exerted by the cuff. If the examiner applies a stethoscope to the front of the elbow (directly over the brachial artery below the cuff) at this time he will hear no pulse. Now the examiner loosens the valve of the bulb which allows air to escape from the cuff, and the pressure slowly falls. As the pressure falls a point will be reached when the examiner hears a "thump" as a jet of blood spurts through the artery. This point marks the highest blood pressure, or systolic blood pressure. However, when the blood pressure falls below the systolic level (following ventricular contraction) the artery collapses again. So the examiner hears a series of thumps, representing blood spurts, each time the blood pressure overcomes the cuff pressure. As the cuff pressure continues to fall, a point is reached when even the lowest pressure within the artery exceeds that of the cuff, and at this point the blood flows through continuously, and the thumping disappears. This point, when the blood flow is continuous, is the lowest level of blood pressure or diastolic blood pressure.

The blood pressure is represented, then, by two figures, a higher systolic pressure and a lower diastolic pressure, and these are written as a fraction. The systolic pressure in young people is usually between 110 and 140, and the diastolic pressure is usually between 70 and 90. So a typical blood pressure reading might read 120/75. Values may be higher for older people without being abnormal.

Capillaries, Veins, and the Return of Blood to the Heart

By the time the blood has passed through the capillary network and has entered the venules, the smallest of the veins, the blood pressure has dropped precipitously. This is due to two factors: one is the loss of the peripheral resistance offered by the arterioles; the other is the porous nature of the capillary wall which allows the fluid constituents of the blood to leave the capillary, reducing the volume of blood and consequently lowering its pressure. As might be expected, this escaping fluid contains the salts and nutrients required by the tissues, and indeed, is the source of the tissue fluid itself.

This reduction in blood volume lowers the blood pressure but also raises the *osmotic pressure* exerted by the dissolved proteins of the blood which normally do not escape through the capillary wall. So, as the pressure of the blood goes down the osmotic pressure goes up. Toward the end of the capillary network the osmotic pressure becomes sufficient to draw fluids back into the capillaries, and the exchange of tissue fluids has been accomplished. The blood pressure in the venules is about 8 mm Hg and decreases still further as the blood enters the small veins. By the time the blood enters the right atrium the venous blood pressure is 0 mm Hg!

How can these very low pressures account for the return of such large quantities of blood to the heart, especially from the legs where the return is opposed by gravity? The answer, of course, is that factors other than venous blood pressure account for much of the movement of blood toward the heart. Three of these factors are so important that they must be discussed in order to understand the disorders of the venous circulation.

Venous Valves and Muscular Contraction

Most of the veins of the body, but especially the veins of the legs and arms, possess a lining which is periodically thrown up and doubled upon itself as a cusp. Each cusp forms a semilunar pocket, and since the cusps occur in pairs, together they form valves (Fig. 21–8). Just as the gates of a canal lock, thus do they open—in one direction, that is, toward the heart.

The presence of these valves enables the veins to take advantage of the regular contraction of the surrounding muscles for the purposes of moving the blood toward the heart. When a muscle contracts, it shortens, and when it shortens it tends to crowd and compress the adjacent structures. When the veins are compressed their contained blood is squeezed into less compressed sections of the same vein, but there is a net gain in movement of blood toward the heart since the valves fill up and close in response to a backflow.

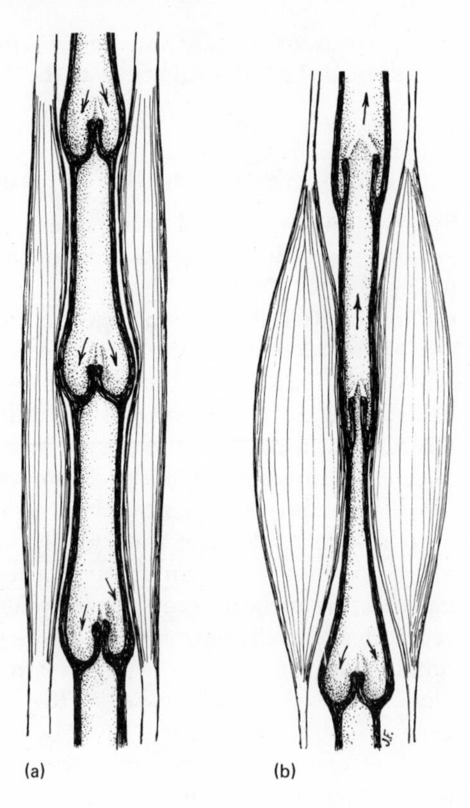

FIG. 21–8 (a) Venous valves form pockets that fill with blood and close when blood tends to back up. (b) Valves open during muscular contraction moving the blood toward the heart.

Intraabdominal Pressure

The low venous blood pressure together with the "pumping" action of the surrounding muscles serves to move the blood from the lower extremities into the large veins of the abdominal cavity. The problem of moving the blood on up through the diaphragm and into the thoracic cavity is also solved by muscular action. But here the important muscles are those of the abdominal wall whose contraction tends to compress the abdominal contents. The increased intraabdominal pressure is an essential factor in defecation since it forces the feces out of the abdominal cavity—that is, through the colon and into the rectum. The same increased intraabdominal pressure forces the blood within the large veins of the abdomen (principally the *portal vein* from the digestive organs and the *inferior vena cava* from everywhere else) in the only direction available to it, that is, through the diaphragm and into the thoracic cavity. In the case of the inferior vena cava the route is direct; the blood of the portal vein passes into and through the liver and joins the inferior vena cava just above the diaphragm. This is one reason, among others, why flabby abdominal muscles are unhealthy: they fail to play their part in venous blood flow.

Reduced Intrathoracic Pressure

At the same time as increased intraabdominal pressure is "pushing" the venous blood up into the thoracic cavity, the action of inspiration is "pulling" the blood in the same direction. During inspiration the thoracic cavity is enlarged by a lowering of the diaphragm and a raising of the rib cage. The increased size of the thoracic cavity provides space which has to be filled. The filling is provided mostly by the entering inspired air, but in part is provided by a filling of the inferior and superior venae cavae, and ultimately the right atrium, with blood.

The Lymphatic System and Tissue Fluid

The lymphatic system is a diverse and complex collection of organs and tissues related to circulation and blood formation. Its most conspicuous parts—the spleen, thymus, and tonsils—are poorly understood, and we shall have but little to say about them. They appear to be dispensible, for their removal little affects the health of most individuals. Of more consequence to us is the generally less conspicuous system of *lymphatic vessels* with their interspersed *lymph nodes.* A simple understanding of these structures is essential if a number of common and important disease processes are to be appreciated.

Anatomy of the Lymphatic Circulation

Lymph

The fluid contained within lymphatic vessels is termed lymph, and has much in common with blood plasma. It is almost colorless; contains large numbers of lymphocytes but few other white cells, and no red cells; and contains proteins (enzymes, globulins, and some clotting factors) but in less concentration than the plasma. The lymph vessels of the small intestine transport fat absorbed from the intestine; this fat-containing lymph has a milky appearance and is called *chyle.*

Lymphatic Vessels

The lymphatic vessels are extremely delicate and thin-walled, and extend their tiniest ramifications into all the tissues of the body. The smallest of the lymphatic vessels are the lymphatic capillaries which end (or really begin) in rounded or swollen bulbs which are perforated by spaces which bring their channels into communication with the tissue fluids. Lymphatic vessels drain the capillary beds and form vast interconnecting networks which eventually converge to form larger vessels in the inguinal, axillary, and cervical areas, and along the major blood vessels within the abdomen and thorax (Fig. 21–9).

All of these major lymphatic channels below the level of the diaphragm

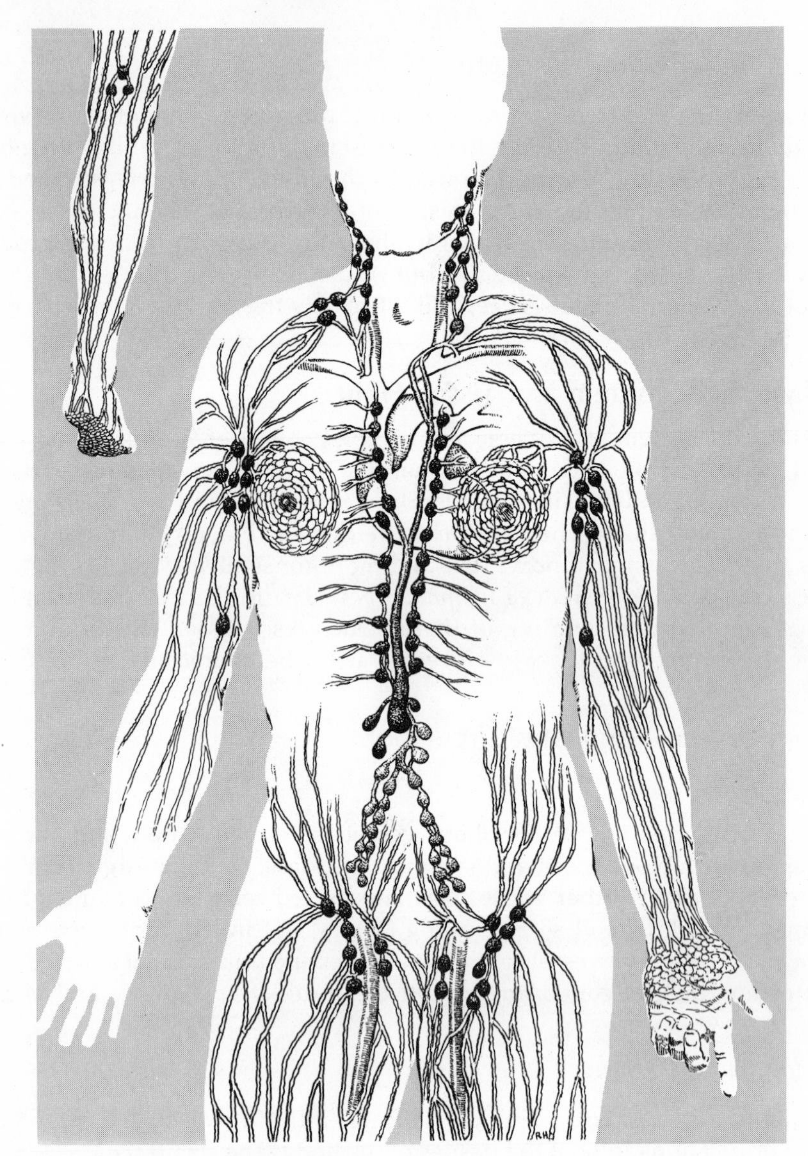

FIG. 21–9 The lymphatic system.

drain into a single large lymph trunk, the *thoracic duct.* The thoracic duct also drains the lymphatic vessels from the left half of the thorax, neck, and head before finally ending by emptying into the left subclavian vein (under the left clavicle or "collar bone"). The remaining upper-right quarter of the body drains into the right subclavian vein.

Lymph Nodes

Lymph nodes are found at many of the junctions of lymphatic vessels. They are particularly abundant in the areas where the large lymphatic channels are formed, that is, in the inguinal area, axillae, and neck. In the deeper parts of the body they occur principally along the large lymphatic channels that accompany major blood vessels. The structure of a lymph node is shown in Fig. 21–10. Two or more afferent lymphatic vessels enter the node at various places, and one or more efferent vessels emerge at a depression called the hilus.

Internally the lymph node is composed of a framework of fibrous tissue which divides the structure into a series of spaces called *lymph sinuses.* Lining the walls of the sinuses are phagocytic cells, and within the cavities of the sinuses are large numbers of lymphocytes.

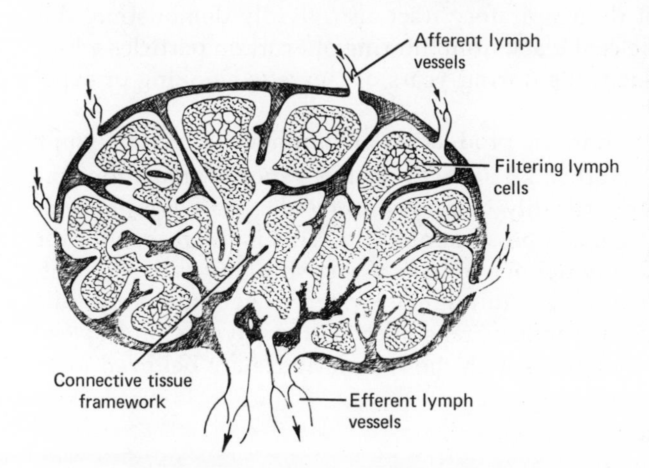

FIG. 21–10 A lymph node. Several afferent lymph vessels carry lymph into the lymph node. After filtration of foreign material or other debris picked up from the tissue fluid the lymph leaves the node by way of one or two efferent lymph vessels.

Physiology of Lymphatic Circulation

The lymphatic circulation functions as an integral part of a complex interaction between the blood vascular system and the tissue fluids. In our discussion on blood circulation we pointed out that the high blood pressure at the arterial ends of the capillary networks forces many constituents out of the plasma and into the tissue fluids, and nourishment to cells is effected. Much of this fluid, along with by-products of cellular metabolism, return to the blood at the venous end of the capillary network where the blood pressure is low and the osmotic pressure is high. It was noted that the differential in osmotic pressure was made possible by the fact that plasma proteins do not leave the blood, and so become more concentrated toward the venous end of the capillaries.

Now this is generally the case. However, some proteins do pass through the capillary walls and enter the tissue fluids. When they do they are unable to pass back into the blood—apparently they can go in one direction through the capillary wall but not the other. These proteins do not accumulate in the tissue fluids because they pass into the porous ends of the lymphatic capillaries and become a part of the lymph. Also removed from the tissue fluids are many types of particulate matter which sometimes occur in the tissue fluid—bacteria, remnants of broken down cells, white blood cells that have been attracted to sites of infection or trauma, red blood cells extravasated from blood vessels during tissue injury, and so forth.

What is the fate of the lymph laden with such a variety of debris? The lymph nodes immediately come to mind as a possible filtering mechanism, and this is exactly the case. The phagocytic cells lining the lymph sinuses engulf the bacteria and foreign material from the lymph as they filter through. All of us are aware that lymph nodes commonly become swollen during severe bacterial infection. Lymph nodes of the respiratory tract also vividly demonstrate this activity—they often become coal black from filtering out carbon particles which have passed through air sac walls during years of cigarette smoking or exposure to air pollution.

Lymph nodes also function in the production of lymphocytes. As the lymph passes from one node to another on its way to the venous circulation, the concentration of lymphocytes steadily increases. In fact far more lymphocytes are passed into the blood than can be accounted for. This is the reason for the supposition made earlier that lymphocytes have a very short life in the circulating blood. However, the matter is still a great mystery—we know very little about lymphocytes. Other lymphatic organs such as the thymus and spleen also produce lymphocytes and along with the lymph nodes are believed to be the site of antibody formation.

Circulation of Lymph

Two factors are largely responsible for the flow of lymph along the lymphatic vessels. These are the *pressure of the tissue fluid* and the *lymphatic pump*. The open communication of the lymphatic vessels with the tissue fluids usually allows an escape for excess tissue fluids. In fact, the lymphatics normally function to keep the volume of tissue fluid barely sufficient to fill the intercellular spaces, and whenever the pressure of the tissue fluid exceeds a few millimeters of mercury, the fluid is forced into the lymphatic capillaries, and eventually back into the circulating blood.

The lymphatic pump is similar to the venous pump discussed earlier. Valves are present in almost all lymphatic vessels, and serve to restrict movement of lymph to one direction, that is, toward the venous outlet. Each time a muscle contracts, the adjacent lymphatic vessels are squeezed against surrounding structures, and the lymph is forced to move in the direction allowed by the valves. Exercise of any sort increases the flow of lymph, and the quantity of fluid drained from the tissues is greatly increased.

Edema

An abnormal accumulation of fluid in the tissue spaces is termed *edema*. Any condition which causes an excessive flow of fluids into the tissue spaces from the blood capillaries or which interferes with the drainage of fluid by the lymphatic system can cause edema. For example, *lymphatic blockage* may occur as a result of destruction of lymphatic vessels from trauma or during surgery. Certain diseases, such as *filariasis,* also cause blockage of lymphatics. The sequence of events in this disease exemplify the development of edema when lymphatic vessels become blocked from any cause.

Filariasis occurs principally in the South Sea Islands, and is transmitted to humans through the bites of mosquitoes which introduce into the bloodstream the larval stage of a worm parasite. The young worms then migrate to the tissue spaces and grow to the adult stage within the lymph nodes. Here they plug the lymph circulation so that tissue fluid is no longer drained from the affected area. As a result proteins build up in the tissue fluid raising the osmotic pressure to the point that it equals the osmotic pressure of the blood in the venous ends of the capillary networks. Thus the principal mechanism for the return of fluid to the blood capillaries is lost. As proteins continue to build up in the tissue spaces, greater and greater quantities of fluid are retained. The extremities—which are most often affected by the disease—swell to enormous proportions, to the point that a leg may have a diameter two or three times normal. A common name for the disease, *elephantiasis,* derives from the grotesque appearance of the extremities of those afflicted.

Edema from other causes does not generally have such extreme results, but can still be quite serious. For example, *a low level of protein in the plasma* can lower the osmotic pressure of the blood to the point that fluid is not drawn back into the capillaries in sufficient amounts. Low plasma protein levels can result from starvation—one is reminded of pictures from underdeveloped nations of starving children with swollen abdomens, the site of fluid accumulation in this case. Also, loss of plasma protein in the urine in kidney disease, or through burned areas of the skin, can lead to edema. Among less serious causes of moderate edema are the swollen ankles brought on by tight garters which obstruct both venous and lymphatic flow. Swollen ankles also occur in some women late in the menstrual cycle (premenstrual edema). But any edema requires medical attention.

In this chapter we have discussed, for the most part, only the basic anatomy and physiology of the vascular system. Circulatory diseases are not only the most common cause of death in this country, but they also account for a great share of the disability and hardship associated with poor health. For this reason the next several chapters are devoted to the major circulatory diseases—their causes (to the extent that we understand them); the rationale of the treatments usually undertaken by the physician; and, most importantly, how they may be most effectively prevented or minimized.

REVIEW QUESTIONS

1. Itemize the sequence of events (the movement of blood; the opening and closing of valves) that occur during the contraction of the heart.
2. How is the rate of the heartbeat controlled?
3. How is the heart nourished?
4. How do the arterioles function in maintaining blood pressure and controlling the distribution of blood?
5. How is blood pressure measured?
6. What are the mechanisms that aid in the return of venous blood to the heart?
7. Briefly discuss several functions of the lymphatic system.

Additional Readings

ADOLPH, E. F., "The heart's pacemaker," *Scientific American*, **216**: 32–37 (1966).

FOLDI, MICHAEL, *Diseases of Lymphatics and Lymph Circulation*. Springfield, Ill.: Charles C Thomas, 1969.

GUYTON, ARTHUR C., *Textbook of Medical Physiology*, 4th ed. Philadelphia: W. B. Saunders, 1970.

WOOD, J. EDWIN, "The venous system," *Scientific American*, **218**: 86–96 (1968).

22
DISEASES OF THE HEART

About one half of the total deaths occurring in the United States today are a result of some form of heart disease. These diseases may arise in the heart itself, but more often the heart simply exhibits the most overt symptoms of more generalized disorders.

Coronary Artery Disease

If the coronary arteries are unable to supply an adequate amount of blood to the heart muscle, coronary heart disease results. This is most often due to a narrowing of the arteries through *atherosclerosis* (Fig. 22–1). The atherosclerotic arteries (Chapter 23) become lined with fatty deposits which gradually narrow the lumen and obstruct the movement of blood. If these deposits become hard, calcified plaques which destroy the normal elasticity of the artery, as is often the case in persons of advanced age, the condition is known as *arteriosclerosis.* Both conditions can attack all the arteries of the body, and can cause hypertension and coronary heart disease as well as brain damage in the form of strokes.

When the deposits narrow the branches of the

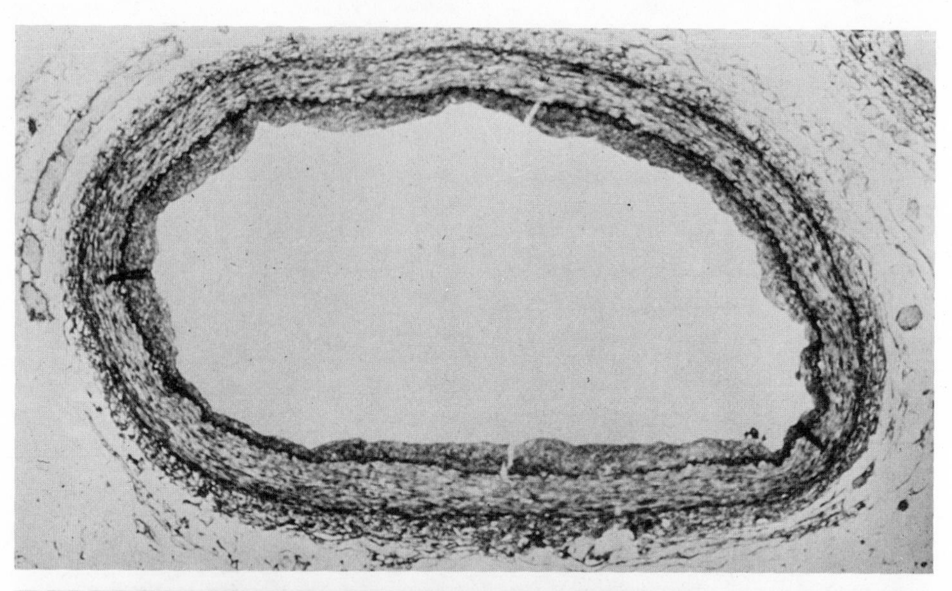

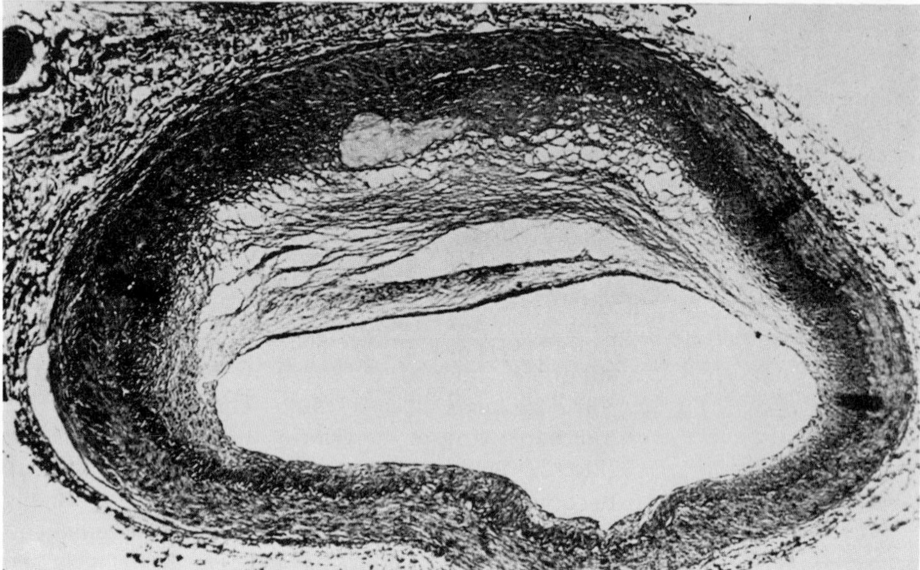

FIG. 22–1 Normal and atherosclerotic arteries. The fatty deposits form in the inner lining and cause a narrowing of the blood channel. (American Heart Association)

coronary arteries to such a point that the heart muscle is unable to meet the demands placed on it by increased physical activity or acute and sudden emotional stress, the patient suffers a sudden and severe pain in the chest. This pain, called *cardiac pain* or *angina pectoris,* is felt initially as a crushing pressure over the chest, and gradually radiates toward the left shoulder and upper arm. This cardiac pain is usually accompanied by a shortness of breath so that the patient complains of a sense of smothering.

Angina pectoris occurs when a portion of the heart muscle temporarily receives an inadequate blood supply but no actual tissue damage occurs. Although loosely referred to as a heart attack, such is not the case since there is no occlusion or actual heart damage, but angina pectoris certainly bodes no good for the patient and unless properly treated and dealt with can be the prelude to a future debilitating or fatal heart attack.

The treatment for angina pectoris is usually the administration of a drug which will dilate the coronary arteries and increase the supply of blood to the heart muscle. The compound glyceryl trinitrate (nitroglycerin) is a drug commonly used for this purpose; a tablet of the drug is allowed to dissolve under the tongue, and usually gives relief in two or three minutes.

Coronary Occlusion

A coronary occlusion is an obstruction of a branch of one of the coronary arteries. Such obstructions most frequently occur in arteries which have become narrowed by atherosclerosis, so it is not surprising that coronary occlusions often arise in persons with a history of cardiac pain, though they occur frequently in those who have never suffered from angina. The atherosclerotic plaques seem to promote the formation of blood clots, and such clots can easily wedge in the artery and complete the obstruction of an already much narrowed artery. Such a clot-induced coronary occlusion is called a *coronary thrombosis,* and is the most common cause of the familiar "heart attack."

A coronary occlusion generally leaves a part of the heart muscle permanently without a blood supply. Although circulation to the damaged area may be partially restored, there is almost always some heart muscle which dies and becomes replaced by scar tissue. Such scars can be identified by the pathologist in postmortem examinations. Most people who die of coronary artery disease show some evidence at postmortem of previous heart damage, and presumably many people have had "heart attacks" without realizing the nature of the experience.

The *electrocardiogram* (EKG) can sometimes give valuable evidence to the physician of the occurrence and extent of injury to the heart muscle. The recording apparatus, called an *electrocardiograph,* is able to pick up, through electrodes placed on the skin, a minute part of the electrical current which is generated by the heart muscle during its contraction (Fig. 22–2). The current is greatly amplified and recorded on paper as an electrocardiogram. The procedure is painless and exceedingly helpful in the diagnosis of the nature of the heart disorder. The difference between a normal electrocardiogram and one from a person with damaged heart muscle is shown in Figure 22–3. It is important to

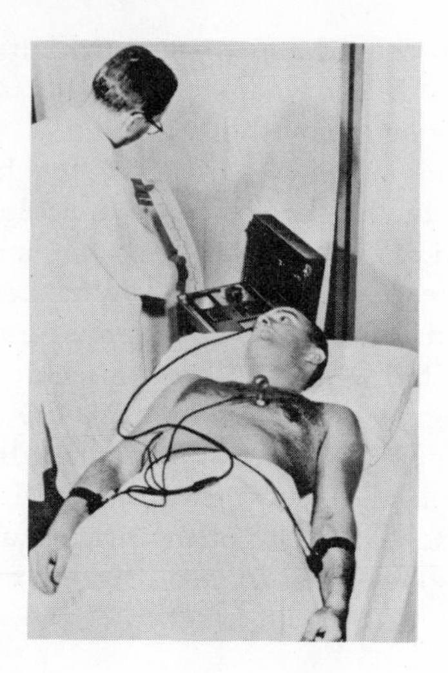

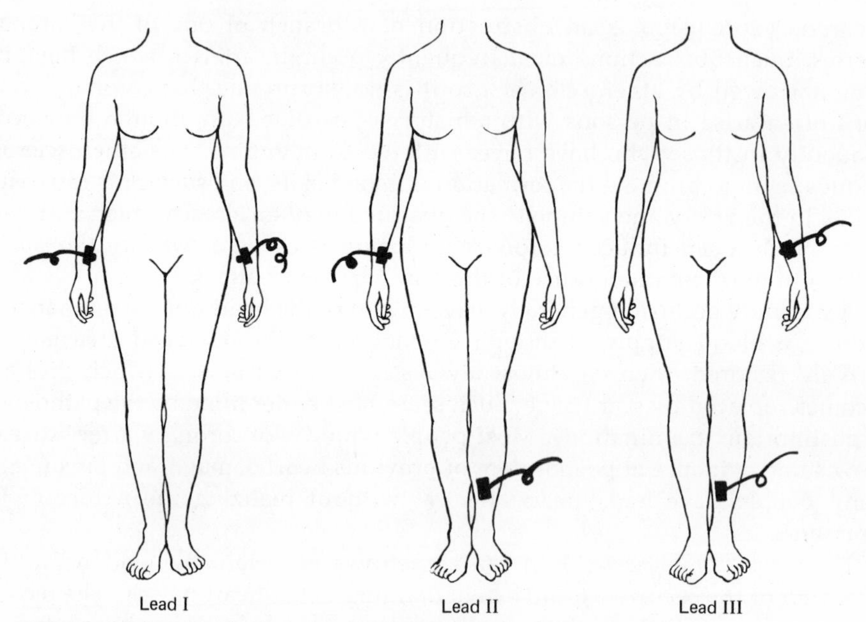

FIG. 22–2 The electrical changes generated in the heart muscle during its contraction are readily conducted to all parts of the body. In order to record the heart currents it is only necessary to connect two parts of the body to an *electrocardiograph* (*top*). The parts of the body usually employed for this purpose are shown (*bottom*). Each pair of connections is referred to as a *lead*. Lead I is right arm and left arm; lead II is right arm and left leg; and lead III is left arm and left leg. A fourth connection, between the chest overlying the heart and left leg, is usually recorded as well. (*Top*, American Heart Association)

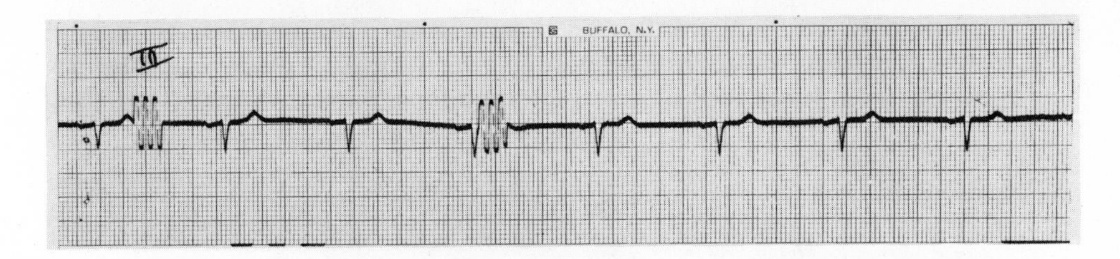

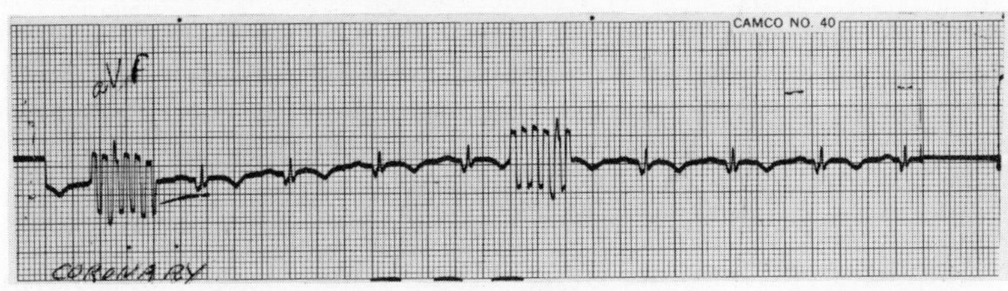

FIG. 22–3 Normal and abnormal electrocardiograms. *Below:* The changes following a coronary attack are apparent. Many times, the changes in the electrocardiogram are very subtle and require considerable experience to interpret. (American Heart Association)

realize that the resting EKG does not tell all about the health of the heart—a disorder may not be detected, but may show up only on an exercise EKG. Even then not all clinically silent disorders will be revealed. For example, the EKG tells the physician nothing about the valves, the size of the heart, or semi-occluded coronary arteries not yet sufficiently restricted to cause ischemia.

When a patient survives a coronary occlusion, complete rest is required for several weeks or months, time required for the heart to compensate for the loss of a functional part. Following this period the patient may have to adhere to a more limited regimen of activity, though he need not become an invalid. The fact is, of course, that almost all of us at middle age must begin to avoid sudden, strenuous physical exertion. Exceptions are possible in those people who maintain a routine of strenuous exercise; indeed, reasonable, regular exercise is an excellent, though not infallible, protection against developing heart conditions. Those who do not maintain body activity are the ones who collapse while shoveling snow or in hard running, and the collapse is usually the result of overworking a heart which has its blood supply made inadequate through atherosclerosis.

Valvular Heart Disease

Valvular heart disease generally results as a consequence of *rheumatic fever,* a disease which appears to follow infection by a bacterial organism designated by bacteriologists as *type-A hemolytic streptococcus.* Almost 98 percent of all heart disease in patients under 20 years of age is directly related to rheumatic fever.

Heart involvement apparently develops during the convalescent period following upper respiratory infections such as tonsillitis or pharyngitis. Injurious products (or toxins) of the infectious organism presumably cause the inflammation which develops in the joints and in the endocardium or heart lining. (The joint inflammations give rise to the arthritis and rheumatic pain which gives the disease its name.) However, some authorities contend that the endocarditis is the result of an allergic reaction between the bacterial toxin and antibodies attached to the endocardium. In any case the effect on the endocardium is the development of granular nodules on the cusps of the valves. When healing occurs, connective tissue forms around the injured valves, frequently causing a thickening or even a fusion of the cusps. Sometimes the chordae tendineae are shortened due to the accumulation of some scar tissue.

Scarred and stiffened valves obviously reduce the pumping efficiency of the heart. If the damage is minor, as it frequently is in children, it simply means that the child tires and becomes short of breath more quickly than normal, but by no means does it make him an invalid. However, particular attention is given to the prevention of further valvular damage by subsequent streptococcus infections. Fortunately, penicillin is effective against this particular strep organism, and is usually given at the first sign of an upper respiratory infection in such cases. All sore throats that persist for more than a few days should be examined by a physician so that he can take a throat culture to check out the possibility of a strep infection. If the test is positive, administration of an appropriate antibiotic such as penicillin will almost certainly prevent the rheumatic fever and secondary damage to the heart valves. If the damage is major, an artificial valve may be surgically implanted (Fig. 22–4).

In almost all valvular disorders, even in those people who are but slightly affected, there is a characteristic modification of the heart sounds called a *murmur.* The murmur is probably caused by the turbulence of the blood as it passes back through the defective valve. A murmur in many, perhaps most, cases is no cause for worry; only the judgment of a medical expert should decide whether, and how far, a heart murmur should curtail activity.

Congestive Heart Failure

Congestive heart failure refers to the general inability of the heart to maintain circulation, either through atherosclerosis, hypertension, serious valvular disease, or other causes. As an example, we can trace the development of congestive heart failure arising from a severely damaged bicuspid valve, the valve, incidentally, which is most often affected in rheumatic fever endocarditis. In this case the left side of the heart fails to pump the blood delivered to it from the lungs through the contraction of the right ventricle. The failure of the bicuspid valve means that blood regurgitates into the left atrium during ventricular contraction. This back-up of blood into the left atrium, in turn, results in a

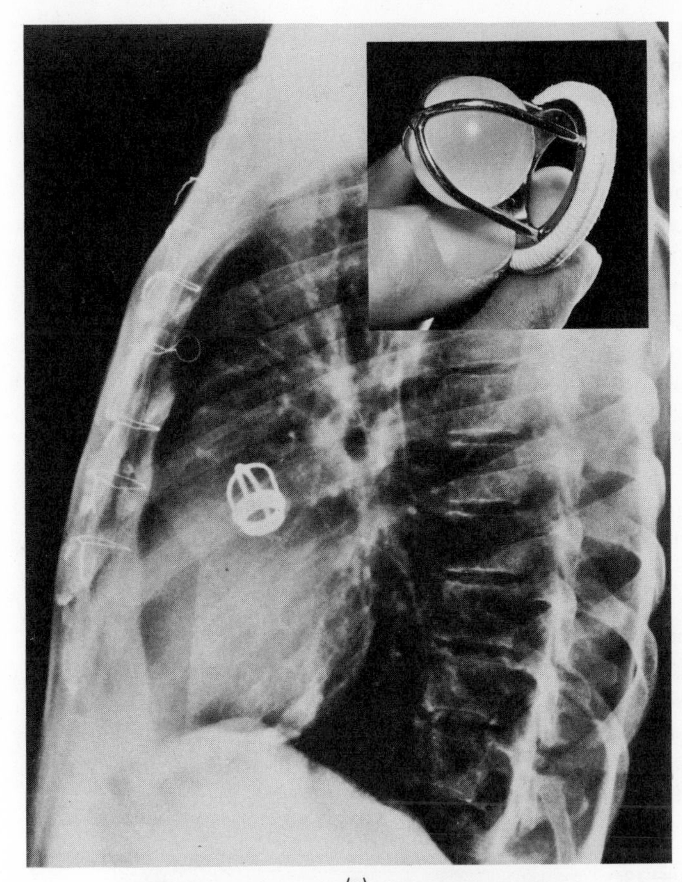

(a)

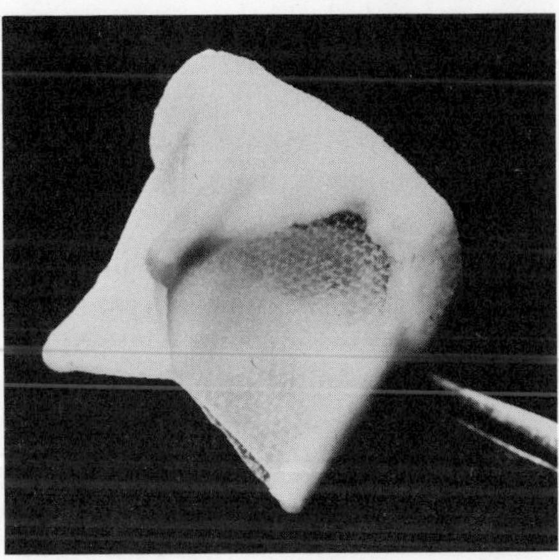

(b)

FIG. 22–4 Artificial heart valves. (a) The Starr-Edwards valve. The ball is not radio-opaque and therefore does not show in the X-ray. (American Heart Association) (b) The Hufnagel tri-leaflet aortic valve, developed in 1966 at the Georgetown University Medical Center, combines strength, long flex life, and excellent performance in a simple design that is anatomically very similar to living valves of the tricuspid configuration. (National Institutes of Health).

damming of the blood in the pulmonary circulation. The right ventricle *compensates* for this pulmonary congestion by enlargement (right ventricular hypertrophy) so that it does part of the work of the left ventricle, pushing blood not only through the lungs but through the left side of the heart as well. As long as the right ventricle compensates for the left ventricular inadequacy, circulation remains more or less adequate. Eventually, however, the hypertrophied right ventricle is unable to compensate. This may be due to the enlarged muscle outgrowing its blood supply, or to the encroachment of atherosclerosis on the coronary circulation, or perhaps to other causes. Whatever the cause, a state of decompensation sets in, that is, the right ventricle fails in its efforts, blood begins to accumulate in the pulmonary circulation, the pressure in the pulmonary capillaries increases, and fluid is forced into the lung tissues, resulting in pulmonary congestion. The patient may actually drown in his own fluids.

Disorders of the tricuspid valve or pulmonary semilunar valve result in the failure of the right side of the heart to pump all the blood returned to it—blood accumulates in the veins and the systemic venous pressure increases. Fluid tends to accumulate in the tissues, particularly in the lower extremities, causing the swollen ankles so characteristic of many people with congestive heart failure. The excessive fluid accumulation in the tissues (edema) is, of course, typical of congestive heart failure, regardless of the heart disorder involved. Accompanying and aggravating the edema is the abnormal retention of salt due to the sluggish flow of blood through the kidneys which normally eliminate the excess salt. The salt accumulation in the tissues draws fluid from the blood as a simple osmotic phenomenon. Salt-free diets are frequently given to these patients, along with *diuretic drugs* which increase urine production.

Congenital Heart Disease

The development of the heart is an event to behold: how it begins as a simple tube, how it folds itself into an S-shaped configuration, divides into four chambers, and forms the pulmonary artery and aorta from a single tube—all the time functioning as an efficient pump for the fetal circulation. That this feat of hydraulic engineering is achieved without mishap or error in the great majority of instances is truly remarkable indeed.

When an error does occur, it is usually in the nature of incomplete or arrested development of some structural characteristic essential for function in the fetus, which in the newborn acts as a derangement. Disorders of this sort are termed congenital heart defects. Some of these may be inherited, but considerable evidence suggests that they are most often caused by external influences such as a deficient oxygen supply to the fetus or a viral infection in the mother during the first three months of pregnancy (heart defects are just one kind of abnormality found to be associated with the viral disease rubella, German measles, as discussed in Chapter 9).

In order to discuss the nature of these congenital heart defects we must consider some of the special characteristics of fetal circulation just before birth. The fetus, suspended in amniotic fluid, has no functional pulmonary circulation since no oxygenation of the blood can take place through the lungs (the placenta is a combined lung, kidney, and gut, performing the function of each of these organs quite competently). Since efficient circulation would be inhibited by passage of blood through the nonfunctional lungs, bypasses develop which permit blood to pass (1) directly from right to left atrium and (2) from the pulmonary artery to the aorta. The first is an opening, or rather a valve, called the *foramen ovale;* the second is the *ductus arteriosus.* The foramen ovale may persist after birth, even into adulthood with but little consequence. A higher pressure gradually develops in the left atrium which causes a functional closure if not an anatomical one. Only rarely, when the opening is so enlarged as to form a gap in the interatrial wall, may a problem develop. Then, of course, a disability results, since blood from the left atrium tends to be pumped back into the right atrium. This, of course, weakens the systemic circulation and causes already oxygenated blood to be recirculated through the lungs, thus wasting valuable effort and overloading the right ventricle. The latter may enlarge in an effort to overcome the disorder.

The ductus arteriosus is of greater clinical importance. Following birth the fetus is dependent upon its lungs for oxygenation of the blood, and an efficient pulmonary circulation is essential. Normally, the ductus arteriosus constricts and closes during the first few days after birth, and is identifiable in the adult simply as a ligament connecting the two arteries. But occasionally the ductus persists, and when it does, it results in some rather complicated modifications of the circulation. Systemic, that is, aortic, blood pressure, is normally much greater than pulmonary pressure; thus when the ductus remains open, blood passes from the aorta into the pulmonary artery and through the pulmonary circulation a second time. Adequate systemic circulation can only be achieved by increasing the heart output. This means a heart enlargement. Also pulmonary congestion tends to develop because of the increased pulmonary pressure. These conditions usually lead to death between the ages of 20 and 40 unless the defect is corrected by surgery. The persisting ductus arteriosus is often accompanied by an opening in the wall separating the two ventricles. This condition brings about further complications in the pulmonary and systemic circulations.

Heart Surgery

The recent developments in open-heart surgery have largely been made possible by the heart-lung machine. This machine receives blood through tubes connected to the superior and inferior venae cavae. The blood is circulated through an oxygenator, and then pumped back into the body through the femoral artery, the principal artery of the thigh. The machine substitutes for both the heart and the lungs to keep these organs at rest during the complicated surgical procedures that are required permitting correction of the defects. The procedure also makes possible the replacement of scarred and defective heart valves by artificial valves.

Heart Block and the Artificial Pacemaker

Coronary occlusion and rheumatic fever, as well as several other conditions, can result in damage to the network of conductive fibers that transmit from the heart's pacemaker (page 299) the wave of contraction that spreads over the heart. Without control by the pacemaker, the ventricles settle down to a regular but abnormal rate as low as 20 beats per minute, while the atria continue the normal rhythm. This condition is known as *heart block.* Cardiac efficiency is greatly reduced and, if the block is permanent, death may follow.

The recent development of artificial pacemakers (Fig. 22–5) has made possible a normal life for many cardiac patients. This electronic device, implanted under the skin of the upper abdomen with wires leading into the heart,

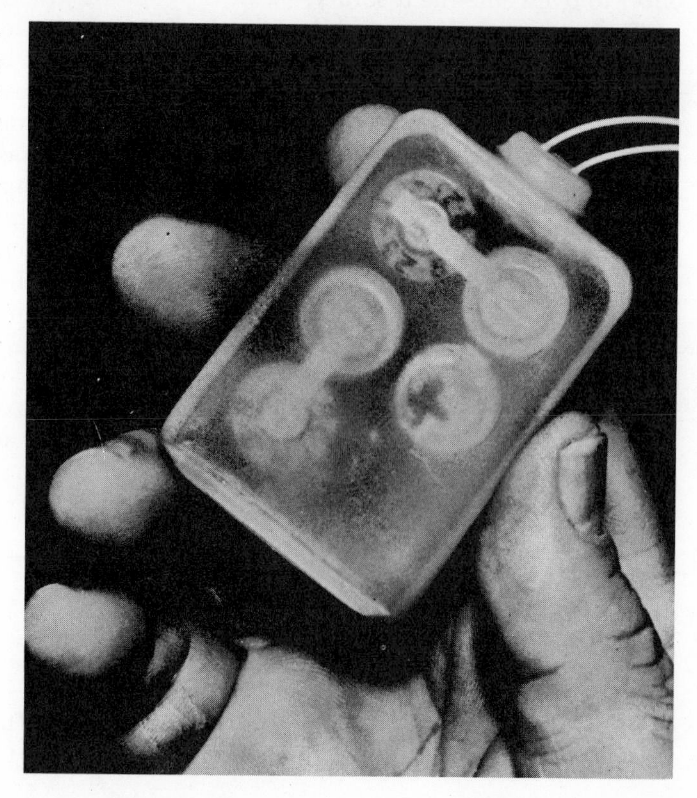

FIG. 22–5 The pacemaker shown here can be implanted under the skin, and the wires are inserted into the heart. Only minor surgery is necessary to replace the battery every five years. (Maimonides Hospital of Brooklyn)

provides a stimulus for a normal heart rate of 60 to 70 beats per minute and does not otherwise interfere with the normal beating of the heart.

Conclusion

Although there is some evidence that a tendency toward heart disease may be inherited, in this country we have a particularly high incidence of coronary artery disease. Here are steps anyone can take to *reduce the chances* that he or she will become a victim of cardiovascular disorders:

1. Do not smoke (especially cigarettes).
2. Limit fat intake to no more than 25 to 30 percent of total calories.
3. Limit saturated fats to no more than 50 percent of total fat intake.
4. Limit sucrose intake (sweets) to 100 grams per day.
5. Participate in reasonably vigorous exercise regularly.
6. Learn to cope effectively with anxiety and tension to relax!
7. Make an effort to reduce an excessively high blood cholesterol level if it exists.
8. Prevent or reduce excess body fat.
9. Control hypertension, diabetes, or gall bladder trouble in cooperation with competent medical guidance.

REVIEW QUESTIONS

1. What is angina pectoris, and what are the conditions that bring it about?
2. What is the most common cause of a heart attack?
3. What are the symptoms of valvular heart disease? What is its most likely cause?
4. What is the nature of some of the heart defects present at birth? What are some of the explanations for these defects?
5. What is heart block?

Additional Readings

BALDRY, P. E., *Battle Against Heart Disease*. Cambridge, Mass.: Cambridge University Press, 1971.

BRAMS, WILLIAM A., *Managing Your Coronary*, 3rd ed. Philadelphia: J. B. Lippincott, 1966.

JENKINS, C. DAVID, "Psychologic and social precursors of coronary disease," *New England Journal of Medicine*, **284:** 244–254. 307–316 (1971).

KOLFF, WILLIAM J., "An artificial heart inside the body," *Scientific American*, **213:** 38–46 (1965).

LIKOFF, WILLIAM, and BERNARD L. SEGAL, *Complete Family Heart Book*. Philadelphia: J. B. Lippincott, 1972.

MARSHALL, ROBERT J., and JOHN T. SHEPHERD, *Cardiac Function in Health and Disease*. Philadelphia: W. B. Saunders, 1968.

23
HYPERTENSION AND ATHEROSCLEROSIS

The degenerative changes occurring in the arteries called atherosclerosis and arteriosclerosis, and their common associate, hypertension or high blood pressure, are so prevalent in this country that more extensive remarks are in order. Although the public tends to associate these conditions with advanced age, we now know that young people are by no means immune. For example, autopsy studies of young American soldiers (average age 22.1) killed in action in Korea showed that over 77 percent had signs of coronary artery disease. The same authors [Enos et al.] observed that coronary artery disease was rare among young Koreans killed in action. High blood pressure had also been shown to be prevalent in young men in another study [Brouha and Heath]. Although hypertension is not always a sign of shortened life, when it is present with arteriosclerosis, it becomes serious indeed.

Hypertension

High blood pressure is usually referred to as hypertension. Since increased blood pressure can result from a variety of causes, hypertension must be considered a symptom rather than a disease. For this reason the condition is classified according to the cause. *Organic hypertension* simply means that the high blood pressure can be attributed to the malfunction of some system or organ: arteries (atherosclerosis and arteriosclerosis), kidneys (renal hypertension due to kidney inflammation or sclerosis of the renal arteries) or endocrine glands (excessive secretion by the pituitary, thyroid, or adrenal cortex). *Essential hypertension* is high blood pressure without detectable disease.

Essential Hypertension

Essential hypertension is by far the most common, accounting for approximately 90 percent of all hypertensive patients. About all we know is that it appears to be inherited as a Mendelian dominant characteristic. A possible explanation is that these people inherit excessively active arterioles which cause a generalized increase in peripheral resistance, and hence, increased blood pressure. Fortunately, if the disorder is controlled, it does not seem to progress or attack any second organ such as the kidneys, brain or retina, as organic hypertension so often does.

Generally, there are no specific drugs or diets for the treatment of essential hypertension. Certainly, a reduction in caloric intake is in order if the individual is overweight. This is most easily accomplished by the elimination of excessive fats from the diet (Chapter 17). The physician may prescribe a relatively salt-free diet, especially if there is a tendency toward edema, to control the blood volume and thus help reduce the pressure in the system. Appetite suppressant drugs are to be strictly avoided since they, almost all of them, cause an increase in the heart rate and a further increase in the blood pressure. Cigarette smoking will specifically aggravate the condition as well as cause more general debilitation. It is not sufficient to cut down; cigarettes must be given up entirely for any significant benefit.

Organic Hypertension

We might generalize by saying that organic hypertension is caused by abnormal function of blood pressure regulative mechanisms. In *arteriosclerosis* a thickening and hardening of the arteries may occur with or without deposition of fatty deposits in the arterial walls (atherosclerosis). When arteriosclerosis attacks the large arteries they become so hardened that they lose their elasticity. We noted earlier that during ventricular contraction the elastic aorta and other large arteries function to absorb much of the force developed by the ventricles and dissipate this force between heart contractions by pushing blood into the smaller arteries. When these large arteries become hardened (in advanced cases they may even become calcified), they cannot be distended, do not absorb any of the pressure developed by the ventricles, and systolic blood pressure rises, sometimes to 250 mm Hg or higher. The diastolic blood pressure may remain at normal or near normal levels, so that with each heartbeat there is a great rise and fall of blood pressure. This is *arteriosclerotic hypertension.* The resulting continual, long term hypertension may lead to a dangerous weakening and ballooning of an artery (aneurysm) or even the rupture of one of the thin-walled arteries supplying the brain (see discussion of stroke in Chapter 34).

In the case of renal hypertension, the kidney exercises a mechanism which increases its own blood supply—when blood flow through the kidney is inadequate, as may occur in atherosclerosis, it releases a hormone, *renin,* which has the effect of generalized constriction of the arterioles. This inadequate blood flow through the kidney might be due to atherosclerosis of the renal arteries; in such a case the release of renin is a secondary effect, and the atherosclerosis is the principal cause. Thus atherosclerotic hypertension leads to renal hypertension, and as the artherosclerosis progresses the renal involvement progresses. If the progression of the disease is rapid it becomes *malignant hypertension.* Damage to the kidneys through inflammation following infection (glomerulonephritis) can also lead to impaired renal function, which leads to increased blood pressure. Whether, in this case, the increase is due to the release of renin or to the accumulation of unexcreted salts is not clear, but the main fact is that hypertension can arise directly from kidney disease.

Disorders of the adrenal cortex or anterior pituitary can sometimes lead to the hypersecretion of adrenal cortical hormones which cause the kidney to retain excessive quantities of salts and water. The fluid volumes increase throughout the body, causing arterial blood pressure to rise to twice its normal value.

Though we certainly would not recommend the elimination of all salt from the diet, it is worth noting that animal experiments [Haight and Weller], as well as human population studies [Dahl], suggest the possibility that *excessive* salt intake may promote or contribute to hypertensive disorders.

In any type of organic hypertension intensive treatment is a must. This will almost always include weight reduction, no smoking, and the limitation of salt intake. But it will also generally involve the identification and evasion of the stresses which promote emotional tension and anxiety.

It is important to recognize a secondary effect of persistent hypertension, namely, enlargement of the left ventricle of the heart. This is not the normal heart muscle hypertrophy which results from endurance training (Chapter 31), but a pathological condition caused by the heart's working over a long period of time against a great resistance. In such an instance, the heart may double its weight. Unfortunately, the coronary blood vessels do not keep up with the muscle hypertrophy and such a heart becomes more susceptible to coronary heart disease.

The doctor may prescribe any of a number of "antitensive" drugs which reduce blood pressure. Some of these act by blocking the effects of the sympathetic nervous system whose nerves are responsible for the constriction of the arterioles; others act on the kidneys to increase fluid loss (diuretics); still others are tranquilizers, which act upon the vascular control center in the hypothalamus of the brain. All are powerful drugs, frequently imparting distressing side effects. Therefore it is understandable that physicians are cautious in their use and resist prescribing them until simpler and safer measures are attempted. The average person with essential (hereditary) hypertension should not expect his physician to put him on any such drugs and should understand the doctor's reluctance to do so.

Hypotension

Some people have a systolic blood pressure below the normal range (100 to 130 mm Hg), but which seems to be normal and compatible with vigorous physical activity and endurance. We can only conclude that the tissues are supplied with sufficient nutrition, that is, a sufficient supply of blood, with less demands on the heart. In fact, a naturally low blood pressure seems predisposed to increased longevity [Cowdry]. The term *hypotension* should probably be restricted to disease conditions in which the blood pressure falls below a previously higher level due to heart failure or to other pathological processes that bring about frequent fatigue and a lack of energy.

Atherosclerosis, Cholesterol, and Fat Intake

The controversy which rages over fat in the diet centers around the relationship of saturated fats to *cholesterol* levels in the blood and, in turn, the relation between such blood cholesterol and atherosclerotic disease. Atherosclerosis is a disease process in which deposits of fatty material, mainly cholesterol, appear on and in the inner lining of the arteries. Now cholesterol is not simply an injurious waste substance found in the blood. It is a fatlike substance (called a sterol by chemists) found as a normal and important constituent of all cells, comprising a part of the cell at its surface where the biochemical reactions associated with life processes occur. The steroid hormones, that is, those of the ovary, testis, and adrenal cortex are formed from cholesterol as well as the bile acids and vitamin D. Cholesterol is also an essential, in fact a predominant, constituent of nerves.

The lining of arteries, however, is one place where cholesterol does not belong; and when it does occur there, in atherosclerosis, a series of serious consequences is liable to develop. A progressive narrowing of the channel of the artery is certain, and this will offer increased resistance to blood flow and increased blood pressure. There will gradually develop a loss of elasticity or resiliency of the arterial wall (arteriosclerosis) so that the vessel no longer stretches and recoils in response to blood pumped into it by the heart's contractions.

These inelastic arteries tend to become tortuous, as is frequently evident on the temples of elderly people. Aside from these manifestations of hypertension, this fatty deposition causes a roughening of the vessel lining resulting in pits and crevices which promote the development of local blood clots or thrombi. Clots of this sort seem to be self-nurturing and grow larger, and can eventually plug the artery. This is a thrombosis, and if it occurs in a branch of one of the coronary arteries which supply the heart muscle, and it often seems to, a *coronary thrombosis* results. The part of the heart muscle supplied by the plugged artery is deprived of the oxygen and nutrients of the blood, and the person experiences a "coronary" which may or may not kill him, depending upon the size of the artery involved (the larger the artery, the greater the area of the heart muscle isolated), the degree of collateral circulation available, and the physical condition of the patient. These, then, are the general consequences of atherosclerotic heart disease.

The supposed causal relationship between saturated fatty acids and blood cholesterol, and between cholesterol and atherosclerosis has stirred up a controversy among various groups of nutritionists, heart specialists, and biochemists, which is far from being resolved. The basis of the controversy is as follows.

Atherosclerosis is an exclusively human disease; except under specific laboratory conditions all other living things are free of it. When this observation was first made, it was noted also that only human beings indulged in high cholesterol diets throughout life. Man alone continues to drink milk long after infancy, and includes eggs, butter, and cream in his daily diet. These foods happen to be the chief dietary sources not only of saturated fats but also of cholesterol itself.

Blood cholesterol is increased by high dietary intakes of fat, particularly of certain saturated fatty acids and by dietary cholesterol. However, blood cholesterol is also increased by high protein intakes, especially animal proteins. High blood cholesterol levels are lowered by high intakes of polyunsaturated fatty acids (fatty acids which are deficient in more than one hydrogen atom), and by strict vegetarian-type diets, as well as by stepped up energy metabolism such as results from regular exercise.

An indictment can be made, then, against saturated fats if we accept that high blood cholesterol is a direct cause of atherosclerosis. Here the evidence becomes uncertain. For instance, no one has been able to show that the cholesterol of atherosclerotic arteries comes principally from the blood. Cholesterol can be manufactured by all the cells of the body (including the cells forming the arterial walls) from nonfat substances. *In fact, nearly all the cholesterol in the blood is produced by the liver and very little comes from ingestion of cholesterol. However, increase in dietary saturated fats increases the rate of cholesterol synthesis by the liver.* But lowering the blood cholesterol level would not necessarily prevent cholesterol deposition in the arterial walls, and those who have indicated saturated fats have not yet proved that atherosclerosis can be prevented or even slowed down by thus lowering blood cholesterol levels. In fact the Federal Food and Drug Administration has ruled that "any claim direct or implied in the labeling of fats and oils and other fatty substances offered to the general public that they prevent, mitigate, or cure diseases of the heart or arteries is false and misleading and constitutes misbranding within the meaning of the Federal Food, Drug and Cosmetic Act."

At the present time, then, a causal connection between saturated fats and atherosclerosis has not been conclusively demonstrated, at least via the cholesterol route. And it could be that cholesterol, at least blood cholesterol, is innocent. Cholesterol happens to be a substance whose concentration in the blood can be fairly easily measured, and for this reason it might have had more than its share of attention. Study is now being directed more and more to the blood triglycerides, the term given to the true fats of the blood as opposed to the numerous fatlike substances such as cholesterol which are also in the blood. Some evidence suggests that it is the triglyceride level in the blood which determines whether or not the cholesterol circulates harmlessly or forms aggregates which are deposited in the arterial wall. And the triglyceride level of the blood is tied in closely with the total caloric intake rather than with the fat intake. This would mean that overeating, whether it means excessive intake of carbohydrates, proteins, or fats of any sort, is at the root of the problem. As a matter of fact, there is some evidence that atherosclerosis is associated with excessive intake of sucrose, that is, the carbohydrate in table sugar [McGandy].

In the meanwhile several respected research teams have given support to the adoption of a so-called prudent diet, which is principally a weight-controlling diet with an emphasis on unsaturated fats. This involves a shift from butter to vegetable oil margarine and from hydrogenated shortening or lard to a liquid oil. These researchers believe that unsaturated fats counteract the cholesterol effects derived from saturated fats. The objection has been raised that to accomplish this end would mean increasing the intake of unsaturated fats to a dangerous caloric level, which, of course, the prudent diet does not do. The critics claim that any successes achieved from this diet would be due simply to the reduction in the total caloric intake. Certainly no diet specifically designed to reduce blood cholesterol levels should be undertaken except under the direction of a physician.

There is little doubt that the average American diet could be benefited by a reduction in the total fat intake to where it constitutes about 25 percent of the total calories rather than 40 percent. This could be accomplished by consuming less butter and margarine, salad dressings, nuts, gravies, french fried foods, creamy desserts, and by eating a relatively greater amount of poultry and fish (Chapter 17). *Factors other than diet, such as stress from high tension living, general lack of exercise, and cigarette smoking, are certainly of considerable importance in atherosclerosis,* and have been discussed in Chapter 22.

REVIEW QUESTIONS

1. Distinguish between atherosclerosis and arteriosclerosis.
2. What is meant by essential hypertension?
3. What are some of the factors that lead to organic hypertension?
4. Why do some nutritionists believe that atherosclerosis is caused by dietary factors other than simply excess intake of saturated fat?

REFERENCES

BROUHA, L., and C. W. HEATH, "Resting-pulse and blood-pressure values in relation to physical fitness in young men," *New England Journal of Medicine*, **228:** 473–477 (1943).

COWDRY, E. V., *Problems of Aging*, Baltimore: Williams & Wilkins, 1942.

DAHL, L. K., "Salt, fat, and hypertension: The Japanese experience," *Nutritional Reviews*, **18:** 97–99 (1960).

ENOS, WILLIAM F., R. H. HOLMES, and JAMES BEYER, "Coronary disease among United States soldiers killed in action in Korea," *Journal of the American Medical Association*, **152:** 1090–1093 (1953).

HAIGHT, A. S., and J. M. WELLER, "Electrolytes and blood pressure of rats drinking sodium chloride solutions," *American Journal of Physiology*, **202:** 1144–1146 (1962).

MCGANDY, ROBERT B., "Dietary fats, carbohydrates and atherosclerotic vascular disease," *New England Journal of Medicine*, **277:** 186–192, 242–247 (1967).

Additional Readings

BLAKESLEE, ALTON, and JEREMIAH STAMLER, *Your Heart Has 9 Lives.* New York: Pocket Books, 1966. (Paperback)

HUTCHIN, KENNETH C., *Heart Disease and High Blood Pressure.* New York: Arco, 1964. (Paperback)

KATZ, LOUIS N. et al., *Nutrition and Atherosclerosis.* Philadelphia: Lea & Febiger, 1958.

MOSES, CAMPBELL, *Atherosclerosis: Mechanisms as a Guide to Prevention.* Philadelphia: Lea & Febiger, 1963.

24
DISEASES OF THE VEINS AND CIRCULATORY SHOCK

Phlebothrombosis

The presence of a thrombus or clot in a vein is termed *phlebothrombosis* (vein-blockage-condition). A related term, *thrombophlebitis* (blockage-vein-inflammation), refers to the inflammation of the vein wall which frequently accompanies thrombus formation, although it is difficult to determine when the thrombus is the cause and when it is the result of the inflammation. In any case, when either condition occurs it is the lower extremities that are most often involved, and the patient usually experiences soreness and swelling in the legs and may show a fever.

If a limb is immobilized for long periods (such as several weeks), thrombus formation may be promoted. The thrombi tend to break away and become *emboli* (plugs). In this case there is grave danger of *pulmonary embolism*, that is, the obstruction of a branch of the pulmonary artery by the floating clot, by which instant death can result. A few decades ago patients recovering from surgery were confined to bed for long periods of time and pulmonary embolism was a serious hazard. Upon arising from bed after such confinement the first few steps could break the clots loose and pulmonary embolism would quickly follow. This is one of the reasons that women are now urged to get out of bed and resume activity as soon as possible after childbirth. In fact this practice is believed to account for a significant drop in the rate of maternal mortality in recent years.

Varicose Veins

We earlier discussed the problems of venous return which has resulted from man's upright posture. Counteracting the force of gravity in the return of blood to the heart are the actions of the valves and the squeezing or pumping effects of the muscles. These factors work efficiently in the deeply placed veins, but the more superficial veins which lie in the fatty subcutaneous tissue between the muscles and the skin often become abnormally lengthened, tortuous, and swollen through lack of adequate support by the surrounding tissues. The sluggishly moving blood must be supported by the valves which, as a consequence, become swollen, and appear as nodules or beads along the bluish course of the vein. Such *varicose veins* almost always occur in the legs and are most frequently associated with, or at least aggravated by, pregnancy, excessive weight-bearing (usually from obesity), and as an occupational disorder resulting from standing for long periods of time without muscular activity, as is the case with salesclerks, barbers, and dentists. However, many cases seem to be due to congenital inadequacy of the valves, especially if the condition occurs early in life.

The discomforts arising from varicose veins are variable. Sometimes there is nothing more than swelling of the legs and ankles due to loss of fluid into the tissues. More often there are aching pains and itching over the affected areas. The circulation to the skin sometimes becomes so inadequate that open sores (ulcers) form.

Early cases are treated by bed rest, leg elevation, and elimination of prolonged standing or sitting in one position. It is also beneficial to provide external support to the distended veins, and this is best accomplished by the use of elastic stockings or bandages. These exert a squeezing effect on the vein similar to that of the muscles, and blood is forced into the deeper veins which are rarely affected. More severe cases are often treated by surgical removal of the affected veins. Blood then drains into the deeper veins and the problem is usually solved.

Hemorrhoids

Hemorrhoids or piles are varicose veins lying in the area of the lower rectum and anus. They result from poor venous return, just as the variocose veins of the legs, but in this case the cause is usually increased pressure within the pelvic cavity, sufficient to prevent normal venous return from the area but not arterial flow into it. Common causes are pregnancy and chronic constipation, and it seems that sedentary occupations are also conducive to the formation of hemorrhoids. It has been estimated that fully a third of adult population of this country have some degree of hemorrhoid formation.

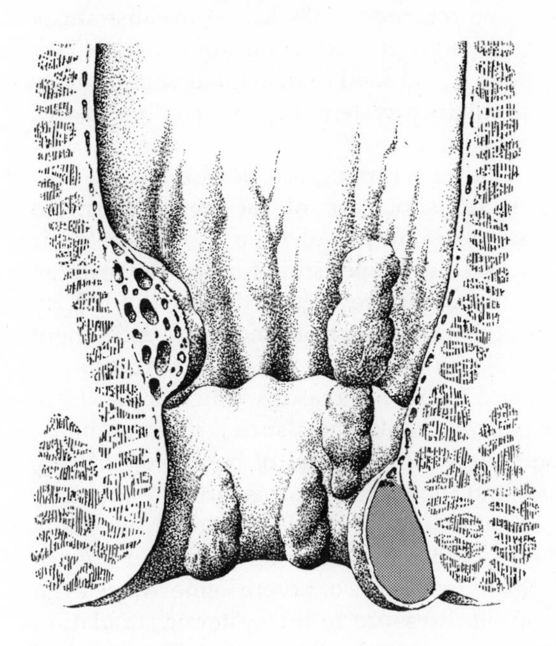

FIG. 24–1 Lower part of rectum and anus showing internal and external hemorrhoids.

Hemorrhoids may be external, where the distended vessels form soft protrusions of the pigmented anal skin, or internal where the vessels are covered by the thin moist membranous lining of the rectum. External hemorrhoids often appear suddenly and enlarge rapidly, and they may bleed profusely. Fortunately, they also seem to disappear spontaneously in many cases. Internal hemorrhoids may bleed during bowel movements, and may sometimes project from the anus. They also seem to promote excessive mucus secretion from within the rectum, and when this mucus leaks out onto the anus, it can cause severe itching. Both types can be painful to the degree that the individual avoids defecation.

Not every instance of itching or blood or mucus in the feces is indicative of hemorrhoids; there are other possible causes. But if any or all of these symptoms persist for more than several days or appear frequently, a physician should be consulted. Treatment is variable. The physician may recommend a change in diet as well as exercise for the avoidance of constipation, or he may even recommend their removal by surgery.

Circulatory Shock

If, for any reason, the heart suddenly fails to pump sufficient blood to maintain adequate circulation, a state of *shock* results. Of course, the most obvious cause of shock is heart failure due to coronary occlusion, but any condition that significantly reduces the amount of blood returned to the heart can cause shock – the heart cannot pump blood not returned to it. Such conditions include lack of blood due to severe hemorrhage, pooling of blood in distended veins, failure of the heart itself, leakage from the circulatory system due to infection, toxins, or burns, and many other causes.

The shock which we hear in association with surgery, severe injury, prolonged exposure to harsh weather, electric shock, or intense emotional strain is most often due to a sudden cessation of impulses from the sympathetic nervous system. These nervous impulses are responsible for the constriction of the arterioles supplying blood to the digestive organs during the "alarm reaction." Normally, in response to physical danger or emotional excitement these arterioles are constricted in order to insure maximum distribution of blood to the skeletal muscles, brain, and heart. For reasons which are far from clear, severe trauma, either physical or emotional, places such a stress upon the sympathetic system that it fails completely; and instead of being constricted, the arterioles become widely dilated, and the viscera, particularly the intestine and liver, become engorged with blood. Much of the blood which normally would pass to the skin, muscles, brain, and lungs becomes pooled in the abdomen, and the individual gives all the appearances of severe hemorrhage even though he may have lost no blood at all. Pressure in the systemic circulation falls so low that blood flow to the heart is greatly diminished. The shocked person is pale, his skin is cold and clammy, his muscles weak and limp, his pulse rapid but feeble, and his blood pressure low. He may be barely conscious but usually is too numbed to feel pain.

Some exceedingly helpful and possibly life-saving measures can be taken before a physician arrives. If there is no back injury and no serious head or chest injury, the shocked person should be placed flat on his back with his hips and legs elevated, his head and shoulders depressed so that whatever blood is available will go to the vital and highly vulnerable brain. If available, a blanket should be placed securely around his abdomen in an attempt to squeeze blood from the viscera, and he should be covered under and over his body to conserve body heat unless the environmental conditions are excessively hot and humid. Above all, the enthusiastic "first-aiders" should be prevented from getting him "on his feet," or from forcing whisky or brandy down his throat, or from throwing cold water in his face. Fluids given by mouth, carefully so as not to choke him, may help a little by increasing his blood volume.

The physician's treatment may be a blood transfusion, even though the patient may have lost no blood, since this would tend to increase blood pressure. He would also probably administer a drug which would constrict the arterioles of the viscera in an attempt to prevent further pooling of the blood in the abdomen. Such a drug would be epinephrin.

A number of other conditions can lead to shock. We have already discussed anaphylaxis (Chapter 19), which is a generalized reaction to an antigen. In addition, any situation leading to the reduction in total blood volume can cause shock. This would include direct loss through hemorrhage, loss of blood plasma in first degree burns, lack of water to drink, and excessive loss of water through the kidneys or gut. Each of these conditions diminish the blood volume to the extent that systemic blood pressure is dangerously low and inadequate quantities of blood are returned to the heart. The treatment in each varies, but the principal problem is the restoration of blood pressure.

It is a good precautionary measure to treat any seriously injured person for shock, according to the principles previously outlined, even if he does not show the symptoms. Shock may develop well after the incident causing injury and may often be prevented by precautionary treatment and some reassuring words.

REVIEW QUESTIONS

1. How is pulmonary embolism related to thrombophlebitis?
2. What are some of the factors that seem to lead to varicose veins?
3. What are some of the symptoms that might lead you to believe that you have hemorrhoids?
4. What is circulatory shock? What is appropriate first-aid treatment?

REFERENCES

Davis, Loyal (ed.), *Christopher's Textbook of Surgery*, 9th ed. Philadelphia: W. B. Saunders, 1968.

Freeman, James, *Shock*. Boston: Little, Brown, 1969.

Jacobson, Eugene D., "A physiologic approach to shock," *New England Journal of Medicine*, **278:** 834–838 (1968).

Liechty, Richard D., and Robert T. Soper, *Synopsis of Surgery*, St Louis, Mo.: C. V. Mosby, 1968.

Additional Readings

Rothenberg, Robert E., *Understanding Surgery*, rev. ed. New York: Pocket Books, 1965. (Paperback)

25
EXCRETION

Excretion is the removal of *metabolic* wastes from the body. Excretion should not be used to mean defecation which is the elimination of digestive wastes—these for the most part are not products of metabolism and were never a part of the body.

Excretory substances are the by-products of the metabolism of the body's cells. These products, largely carbon dioxide, water, a number of salts, and nitrogen-containing chemicals, are continually released from the cells and picked up by the blood. Carbon dioxide and water are lost through the lungs, and for this reason, the lungs can be considered to serve an excretory purpose. The same is true of the sweat glands which serve to eliminate a good deal of excess salt and water and even a small amount of carbon dioxide.

When we think of excretory organs, however, we usually think of the kidneys, the primary function of which is to rid the body of the nitrogen-containing substances, generally quite toxic if allowed to accumulate. In the chapter on digestion, we emphasized that amino acids are not stored, and those that are not used in the manufacture of new tissue proteins are burned for energy. Most of these excess amino acids can be fed into the pool of energy-producing substances, but only after they have been *deaminized.* In this process, which occurs in the liver, the amino group ($-NH_2$) is removed as ammonia (NH_3), and the ammonia is combined with carbon dioxide to form *urea.* The liver empties urea into the blood from which the kidneys filter it out in the production of urine.

The Kidneys and Associated Structures

The kidneys lie flattened against the back wall of the abdominal cavity near the spine and just below the last rib—a trifle higher than most people imagine. Leading from each kidney is a tube called the *ureter* which transports the urine to the urinary bladder. The bladder empties to the outside through a single tube, the urethra (Fig. 25–1).

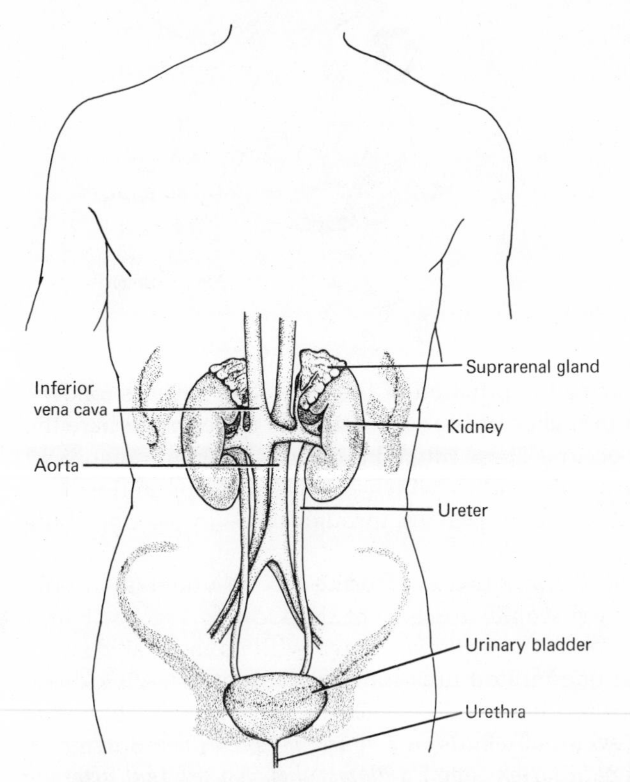

FIG. 25–1 The excretory system.

Each kidney is bean-shaped (conversely botanists describe beans as kidney-shaped!) with its concave surface facing the spine. Here the renal artery (from the Latin word *ren* meaning kidney) enters, and the renal vein and ureter leave.

The kidney in Figure 25–2 is sectioned lengthwise to show the basic structure. An outer, granular-appearing layer is called the cortex. Eight to 18 cone-shaped *pyramids* point toward the center and form the core or *medulla* of the kidney. A cavity, the *pelvis*, occupies the center, and cuplike extentions of the pelvis embrace the tip of each pyramid.

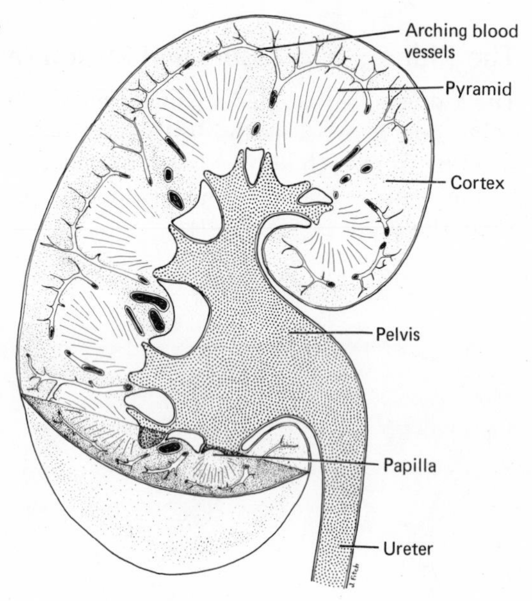

FIG. 25–2 A longitudinal section of the human kidney.

Basically, we can summarize the function of these various parts as follows. Blood is delivered through branches of the renal artery to the cortex where the primary filtering of wastes occurs. These filtered products are then transported through thousands of microscopic tubes which compose the pyramids, and finally exit into the pelvis as urine by passing through pores in the tips of the pyramids.

The filtering process and the production of urine can be understood only by a consideration of the microscopic structure of the kidney. The basic unit responsible for urine formation is the *nephron*, shown with its associated structures in Figure 25–3. About one million nephrons are present in each kidney. Each begins as a cup-shaped receptacle called *Bowman's capsule*, and continues as a tortuous *tubule* which eventually ends in a *collecting duct*. Each Bowman's capsule encloses a knot of capillaries called a *glomerulus*. An *afferent arteriole* carries blood into the glomerulus, and an *efferent arteriole*, exiting from the glomerulus, delivers the blood to a dense capillary network which envelops the tubules and collecting ducts. The blood passes from these capillaries into venules, then veins, and eventually into the renal vein.

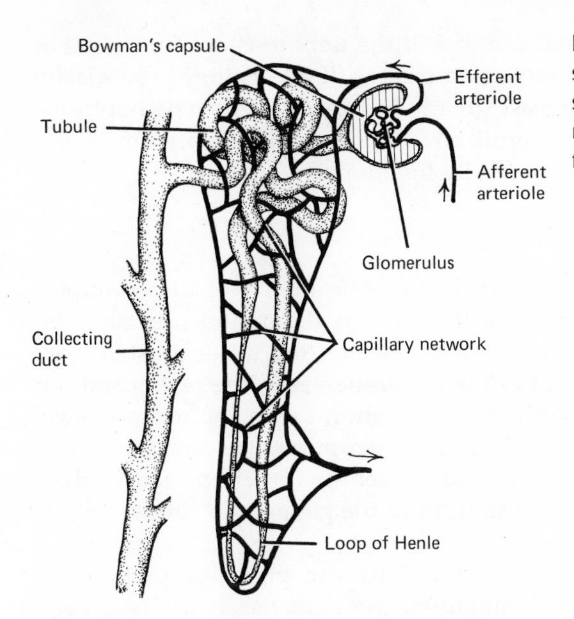

FIG. 25–3 A nephron of the kidney shown diagramatically with its blood supply and collecting duct. The arrows indicate the direction of blood flow.

Operation of the Filtering System

Much of the blood flowing through the glomerulus passes into Bowman's capsule. This is the *glomerular filtrate,* and includes all constituents of the blood except blood cells and proteins. The glomerular filtrate would include, then, not only waste products such as urea and certain salts, but also a considerable portion of the essential elements of the blood such as sugars and amino acids. Obviously, glomerular filtrate is not synonymous with urine; in fact, only a very small part of the glomerular filtrate actually becomes urine. Of the approximately 180 liters of glomerular filtrate formed each day, all but about one liter returns to the blood by passing back out of the tubules and collecting ducts and into surrounding capillaries. This return is called *reabsorption,* and by this process essential substances are brought back into the blood at a desirable concentration. This is a particularly effective mechanism in regard to salts. Since a person will eat variable quantities of salts, such as those of sodium, magnesium, phosphorus, and sulfur, reabsorption provides a mechanism for regulating the level of each in the blood according to the needs of the body. Certain other substances, such as amino acids and sugars, are reabsorbed entirely, unless their concentration reaches abnormally high levels.

Obviously, a great amount of water is reabsorbed, and it is largely water metabolism which accounts for most of the variation in our urine output. If we sweat a great deal or drink large quantities of water, the volume of urine is affected accordingly. It has been appropriately stated that the composition and volume of the blood are determined not so much by what we eat and drink but by what the kidneys keep.

It is probable that only a limited number of the nephrons are functional at any one time, and the remainder serve as a reserve. If one kidney is surgically removed the other kidney will increase greatly in size. Since no new nephrons are formed, we presume that glomeruli and tubules which were small and relatively inactive enlarge and become fully functional.

Regulation of Urine Output

Two hormones act upon the kidney to regulate the nature of urine output. *Aldosterone* is produced by the cortex of the adrenal gland, and regulates the rate of excretion of sodium and, secondarily, of water (the water must "carry" the sodium). If the aldosterone level in the blood increases, the rate of sodium reabsorption increases. Greater sodium reabsorption raises the osmotic level of the blood, which in turn increases water reabsorption. The end result is a decrease in urine volume. Aldosterone is so efficient in its action that if a deficiency of sodium exists almost all the sodium in the glomerular filtrate will be reabsorbed.

The *antidiuretic hormone* (ADH), produced by the posterior part of the pituitary gland, regulates directly the amount of water reabsorbed. The greater the level of ADH in the blood, the greater the amount of water reabsorbed. A rare disease, *diabetes insipidus,* occurring when ADH is absent or secreted in only small amounts, causes the patient to excrete two or three gallons of urine a day. To prevent dehydration the patient must drink an equivalent amount of water. Another problem at one time associated with diabetes insipidus was a constant complaint of chills. The consumption of cold water and the production of warm urine caused a severe drain on body heat making it customary now to advise these people to drink water slightly warmer than body temperature.

Diuresis

Increased urine production is termed *diuresis,* and drugs which cause diuresis are *diuretics.* Familiar examples of diuretics are caffeine and alcohol. Caffeine acts by suppressing reabsorption of water from the tubules. Alcohol has a more indirect effect by inhibiting the release of ADH from the posterior pituitary gland. This is the reason that considerable dehydration can follow a night of heavy drinking, and the great thirst experienced on the "morning after" is not surprising.

Certain very potent diuretics are sometimes prescribed by the physician for treatment of edema due to congestive heart failure or certain other causes. (They are generally *not* prescribed for edema due to kidney failure.) Chlorothiazide and its derivatives (Diuril, Hydrodiuril, Esidrix) and the mercury-containing diuretics (Mercuhydrin and Thiomerin) are the most important, and should be used only according to the recommendations of the physician. Particularly, they should be kept out of reach of children.

The Urinary Bladder and Control of Urination

Peristaltic waves pass the length of the ureter, each wave pushing a drop of urine ahead of it. Each drop requires about one minute to make the 12-inch journey to the bladder. The urinary bladder is a sac of involuntary muscle which serves as a storage reservoir for urine. This muscle can be gradually relaxed so that the bladder can be distended from a volume capacity of about 1 milliliter to a maximum of about 1 liter. When the volume of urine reaches about 300 milliliters (about 1¼ cups), the bladder wall becomes stretched sufficiently to cause volleys of nerve impulses to be sent to the spinal cord (Fig. 25–4). These impulses initiate a reflex in which the involuntary muscle of the bladder wall is stimulated to contract, expelling the urine into the urethra. This is the normal sequence of events in most mammals and the natural, uncontrolled reflex in infants.

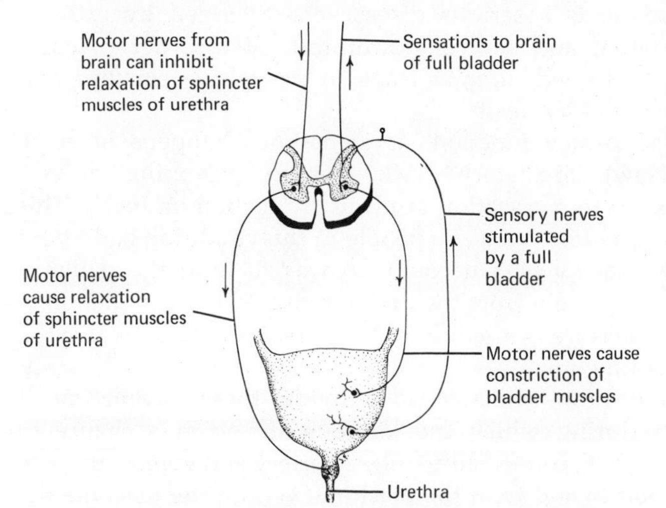

FIG. 25–4 The control of urination. Sensory nerves bring to the brain the sensation of a full bladder. Whether to urinate or not is under the voluntary control of the brain. In an infant the neural control is at the spinal cord level, and emptying of the bladder is a simple reflex.

During the first two years of life the child learns the use of a powerful voluntary muscle surrounding the urethra, the *external sphincter*. This muscle enables us to hold the urine in spite of bladder contractions. Extended periods of continence following control of the external sphincter are still dependent, however, on the size of the bladder. When the bladder reaches a maximum capacity of about one cup, the child can be expected to go through the night. But this volume is not reached until as late as the end of the fourth year in some children. The final achievement in bladder control is the ability to urinate at will without the sensation of a full bladder, and this requires some degree of control over the involuntary muscle of the bladder wall. Generally, this ability will develop between the ages of 3½ to 6 years.

Bladder control is one of the most difficult of our learning processes, and emotional stress of any sort can seriously retard its development. Individuals of abnormally low intelligence, at the so-called imbecilic level or below, only rarely are able to achieve continence, and remain life long at the reflex level of control.

Although bed-wetting, *enuresis,* can be due to a physical cause, it is far more commonly a psychiatric problem. Thus it should be regarded by parents as a symptom of some emotional disturbance rather than as a simple problem corrected by attention to it alone. If enuresis persists beyond kindergarten age, some professional assistance should be sought.

Kidney Diseases

Any disease which seriously interferes with kidney function results in the accumulation of waste products in the blood. This condition is called *uremic poisoning* or *uremia.* Uremia is a serious disease characterized by extreme fatigue, progressive anemia, and severe gastrointestinal disorder (nausea, vomiting, diarrhea). If continued, uremia leads to convulsive seizures and coma, followed in a few hours by death.

Sudden shutdown of kidney function, as sometimes happens in acute nephritis (discussed below), circulatory shock, mercury poisoning, or as a transfusion reaction, results in a cessation of urine production or *anuria.* This in turn will lead to uremia. A few years ago people in this condition had a poor prospect of survival. All the physician could do was to limit the patient's fluid intake, and eliminate protein from the diet to reduce the accumulation of urea in the blood. These measures served to prolong life with the hope that the kidney would resume its function.

Artificial kidneys are now available to provide these people many months of good health during which the kidneys are given a chance to resume function. Plastic tubes, connected to a large artery and vein in the arm or leg, serve to carry blood to and from the machine. Within the machine the blood passes through a fluid-filled chamber whose walls have the same properties as that of the glomerulus, that is, they allow passage of urea, salts, and other substances, but no cellular constituents or proteins. The fluid in the chamber has these same diffusible substances in solution, in the concentration desired in the blood. This means that urea, for example, could have a concentration of zero so as to permit passage from the blood into the fluid—from the greater to the lesser concentration. Calcium, on the other hand, might have a concentration set at 10 milligrams per 100 milliliter in order to reach an equilibrium at this concentration. The concentration of any diffusible substance of the blood can be controlled by adjusting the concentration of that substance in the bathing fluid of the chamber. Substances such as sugars and amino acids could even be fed into the patient by raising the concentration in the bath above that of the blood. The fluid in the machine is changed when its solute concentration begins to vary from the optimum range.

Kidney and Bladder Infections

With the exception of the urethra, the urinary systems are identical in the two sexes. In the female the urethra is very short (about one inch), and its opening is located close to the vaginal and anal orifices. For these reasons bladder infections (*cystitis*) are much more common in females.

Cystitis causes burning during urination, and a continual feeling of fullness of the bladder. The condition is sometimes associated with infections of the kidney itself, or in the male with infections in the genital ducts. Since neglect may lead to permanent kidney or bladder damage, or to possible sterility in the male, bladder infections should be seen promptly by a physician. Treatment with sulfa drugs or antibiotics, or both, usually is successful if started early.

Nephritis or Inflammation of the Kidney

Acute nephritis is sometimes a serious sequel of strep throat infections or scarlet fever. The toxins of the streptococcus bacteria cause inflammation of the glomeruli and tubules (*glomerulo-nephritis*). This results in decreased filtrate formation, and this in turn causes a decrease in the volume of urine. Fluids accumulate in the body tissues causing swelling (edema), especially noticeable in the face. Symptoms of uremia may develop if the infection is severe. Blood cells and blood proteins such as albumins may pass through the inflammed glomeruli, and their presence in the urine (*hematuria* and *albuminuria*) are important signs the physician looks for in the diagnosis of nephritis.

As in the case of valvular heart disease (the other serious consequence of streptococcus infections) acute nephritis is most commonly encountered in young people. The effectiveness of sulfa drugs and antibiotics in the treatment of streptococcus infections has reduced the incidence of the disease and the amount of permanent damage to kidney tissue.

Chronic nephritis is a more dangerous disease and leads to more extensive kidney damage. Although inflammation is less marked, it persists for long periods, and leads to the development of considerable scar tissue (renal fibrosis) and degeneration of glomeruli and tubules. No infectious organism has been identified with the condition although streptococcus infections may aggravate the disease. While many patients appear in good health for a number of years, progressive insufficiency of kidney function eventually leads to uremic poisoning and death. Many of the kidney transplants discussed elsewhere have taken place in such cases of chronic nephritis, where they offer some hope for those afflicted.

One particular type of chronic nephritis develops in association with hypertension and arteriosclerosis. Hardening of the renal arteries diminishes blood flow through the kidneys, and the kidneys respond by releasing a hormone called *renin* which acts to increase blood pressure by increasing blood volume (Chapter 23). This is obviously a protective mechanism to ensure proper blood flow through the kidney, but in this case the result merely aggravates the hypertension, and circulation through the kidneys is little improved. The consequence is progressive kidney and vascular failure.

Transfusion Reactions

One of the complications resulting from transfusion of mismatched blood is kidney failure which leads to death in many instances. The incompatible red blood cells transfused into the patient are agglutinated (stuck together) as they come into contact with the antibodies (agglutinins) naturally present in the patient's plasma. The clusters of red cells tend to become unstuck as they are carried through the capillary beds, and in the process their membranes are ruptured (hemolysis), spilling hemoglobin into the bloodstream. The hemoglobin passing through the glomeruli clogs the tubules, and filtrate movement through the tubules is prevented which results in kidney failure.

Kidney Stones

Certain calcium salts, for example, calcium oxalate, and a nitrogen-containing substance called uric acid (a product of the metabolism of purines, a constituent of cell nuclei) tend to form crystals or stones in the pelvis of the kidney. Suspensions of these crystals are present in the urine of everyone and generally cause no trouble, but, for unknown reasons, they seem to grow in some people into larger crystals with sharp edges, or even spinelike processes (Fig. 25–5). If such

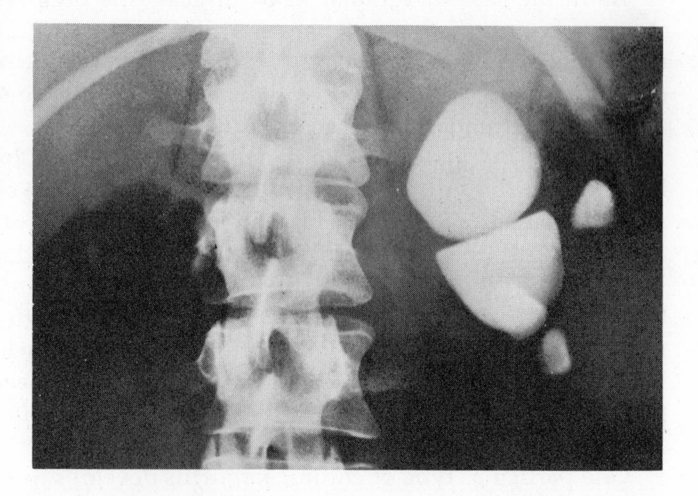

FIG. 25–5 An X-ray photograph showing kidney stones. Smaller stones may be passed in the urine, or in passing, may obstruct the ureter. Larger stones stay in the kidney. In the latter instances, they may have to be removed surgically. (Indiana University Medical Center)

large crystals pass through the ureter, they can cause extreme pain. This pain, called *renal colic,* comes in one or two minute spasms which probably represent the peristaltic waves pushing on the crystal or stone. Larger stones may take a day or more to pass through the ureter and may even cause complete obstruction requiring surgical removal. Once in the bladder the stones cause little difficulty. The urethra is much more distensible than the ureter, so the stones can pass during urination unnoticed.

Kidney stones are about three times more common in men than in women, and usually occur between the ages of 30 and 50. The tendency to form uric acid crystals in the kidney pelvis may be associated with their formation elsewhere, particularly in joints. In this case the person suffers from *gout,* a most painful affliction. Contrary to the common impression, gout is not exclusively a disease of overweight gluttons who also drink heavily. It is found in people of all weights and drinking habits, and the sufferer little deserves the derisive ridicule usually directed at him.

Physicians are able to prescribe drugs which reduce the uric acid level in the blood or which cause it to be more readily excreted in the urine. The patient may also have to go on a restricted diet which reduces the intake of high purine-containing foods, generally foods containing a high proportion of cells, and therefore cell nuclei.

Urinalysis

Much valuable information concerning a number of disease conditions can be gained by the examination of the patient's urine. A century ago urinalysis was one of the very few objective diagnostic tools available to the examining physician. Shakespeare made some comic references to what he thought was doctors' preoccupation with urine and urinalysis. Today urinalysis includes not only gross inspection, but microscopic examination and chemical analysis. Some findings and their significance are the following:

HEMATURIA: red blood cells in the urine; indicates hemorrhage somewhere in the urinary system. The condition can be caused by an ulcerated, cancerous growth, but also by a number of less serious conditions.

PYURIA: white blood cells in the urine; indicates an infection somewhere in the urinary system.

ALBUMINURIA: albumin is the commonest of proteins which may occur in the urine, thus is a specific type of *proteinuria.* Albuminuria is one of the characteristics of acute glomerulo-nephritis, and also of a disease of pregnancy called *eclampsia.* Regular urinalyses during pregnancy are highly desirable for this reason among others. Occasional albuminuria may have no significance, but when persistent or due to infection it requires investigation.

GLYCOSURIA: sugar in the urine. Glycosuria is one of the primary signs of diabetes mellitis, but may occur in healthy persons as an unimportant phenomenon following excitement, an athletic event, or in pregnant and nursing women when it is generally due to a sugar other than glucose such as lactose (milk sugar). Certain monosaccharide sugars such as fructose, levulose and pentose may also occur in the urine without particular significance. Glycosuria generally becomes of clinical interest when it is associated with a rise in blood sugar (hyperglycemia). This topic is pursued further in the discussion of diabetes mellitus (Chapter 32).

POLYURIA: the excessive excretion of urine. This is also one of the signs of diabetes mellitus, but, as mentioned earlier, may also be due to the less serious diabetes insipidus.[1]

In a urinalysis, cloudiness of the urine, in itself, has no significance. Albuminuria causes urine to be cloudy, but so do high, but normal, concentrations of certain salts such as phosphates. Neither is the odor of urine significant, since it varies with the diet and is generally not unpleasant until bacteria infest it to cause putrefaction. In fact, in a healthy person, urine is sterile, and is even useful as an antiseptic wash for wounds in the absence of soap and water. The conventional habit of washing the hands following contamination with urine has no medical justification, only a strong social one.

What can you do to protect the health of the excretory system? Perhaps the most valuable thing is to make a few observations which can be passed on to your physician. Certainly, any of the following conditions justify a doctor's examination: painful urination which may indicate infection of the urethra or bladder; difficulty in urination in the male, which may mean prostatic enlargement; any peculiar coloration of the urine; and a sore throat in a child which may mean streptococcus infection. Above all, do not attempt to treat yourself with "kidney pills" or tonics. Healthy kidneys need no stimulation or "purging."

REVIEW QUESTIONS

1. Why is it inappropriate to refer to feces as excretory wastes?
2. What organs other than the kidneys serve in excretion?
3. Where does urea come from?
4. Almost any chemical substance present in excessive amounts in the blood will be removed by the kidneys and become part of the urine. How is this accomplished?

[1] The origins of the two words diabetes mellitus and diabetes insipidus are interesting. Some astute diagnostitian long ago observed that in sugar diabetes the urine has a sweet taste, while in diabetes insipidus there is only a flat, salty taste (mellitus, sweet; insipidus, tasteless). One wonders if modern doctors are made of as strong a stuff as their forebears!

5. What are some of the factors that regulate the volume of urine you produce?
6. What are stages we go through in learning bladder control?
7. How does an artificial kidney work?
8. What are some disease conditions that can be identified through examination of the urine?

REFERENCES

GUYTON, ARTHUR C., *Texbook of Medical Physiology*, 4th ed. Philadelphia: W. B. Saunders, 1970.

GOLDEN, ABNER, and JOHN F. MAHER, *The Kidney: Structure and Function in Disease*. Baltimore: Williams & Wilkins, 1971.

Additional Readings

SMITH W., "The kidney," *Scientific American*, **188**:40–48 (1953).

WOLF, A. V., "The artificial kidney," *Science*, **115**:193–199 (1952).

PART SIX
THE AIR WE BREATHE

26
RESPIRATION

At the beginning of this century, respiratory diseases were overwhelmingly the most common cause of death. Tuberculosis, pneumonia, bronchitis, influenza, and their kin were the classic killers, as can be seen even in the literature of the era when fictional characters typically died of "consumption," that is, pulmonary tuberculosis. Today, cancer, heart disease, and stroke have displaced the respiratory groups from the head of the killer list, but several types of respiratory diseases still rank among the top twenty-five. They have been abated but far from conquered—as, say, smallpox in this country.

The Nature of Respiration

The combustion of foods by the trillion or so cells of the body requires oxygen, and the end-products of the combustion, besides energy, are carbon dioxide and water. It is obvious that for this process to proceed efficiently, oxygen must be transported to the cells and carbon dioxide from them. This transport, plus the combustion process, is *respiration,* and may be conveniently divided into four phases:

(1) ventilation (breathing); (2) external respiration, which is the two-way movement of these gases between the lungs and the blood; (3) internal respiration, which is the transport of the gases by the blood to and from the tissue fluids and the cells; and (4) true respiration (or biological oxidation), which is the actual utilization of oxygen by the cells and their liberation of energy, carbon dioxide, and water. We shall be largely concerned here with ventilation and external respiration since internal respiration and oxidation are adequately discussed in the chapters on digestion and circulation.

The Respiratory System

External respiration requires the movement of gases in and out of the lungs; and the series of tubes and passageways which serve this purpose is collectively called the respiratory system. Generally included in this system are a number of air-filled cavities, the *paranasal sinuses,* which communicate with the air passageways (nasal cavity) but are not themselves involved in the air transport. From the outside in, the air passages are the *nasal cavities, nasopharynx, oropharynx, larynx, trachea, bronchi, bronchioles,* and the terminal *air sacs* or *alveoli.*

Nasal Cavities

The paired nasal cavities are separated by a thin sheet of cartilage and bone called the *nasal septum.* The nasal cavities become confluent posteriorly as the *nasopharynx* which lies above the soft palate (Fig. 26–1). The nasal cavities are not nearly as large as the size of the nose would suggest. Much of the space is occupied by three curled and shelflike bony projections from the outer wall called the *nasal conchae* which serve to divide each nasal cavity into three narrow passageways. The conchae, as well as the other walls of the nasal cavities, are covered by a mucous membrane containing a very rich and elaborate blood supply—the source of the typically copious "nose bleeds." The air moving along the restricted nasal passages is warmed by proximity to the circulating blood, and a sticky secretion produced by the glands of the mucous membrane serves to filter out dust particles and to moisten the air. The mucous membrane above the upper nasal conchae, high in the nasal cavities behind the bridge of the nose, contains the nerve endings for smell.

The Sinuses

Some rather large, membrane-lined, air-filled spaces in some of the bones of the face communicate with the nasal cavities, and are called *paranasal sinuses.* These sinuses serve the primary purpose of reducing the weight of the skull

FIG. 26–1 The relationships of the nasal cavity and mouth to the divisions of the pharynx.

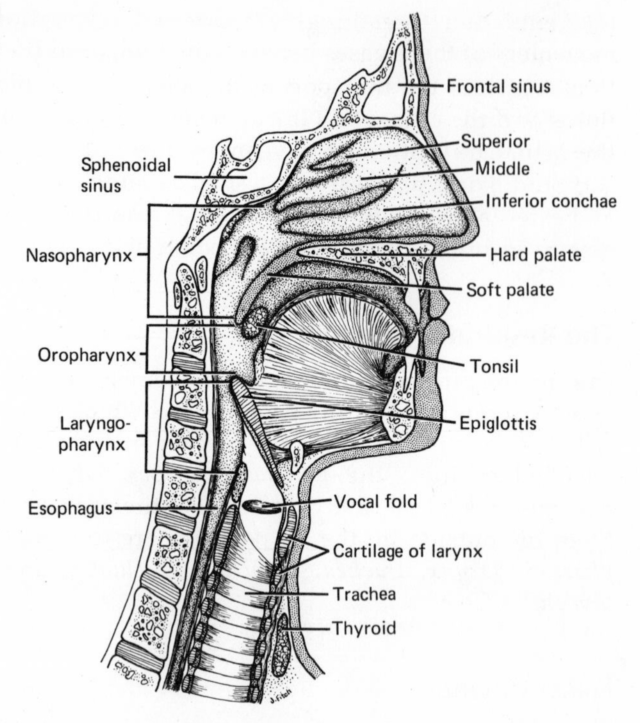

and facilitate the growth of the face. The fact that they contain air necessitates communication with the outside so that the air pressure remains equalized. The sinuses probably contribute to the warming and moistening of the air during breathing, and also act as resonating chambers affecting the quality of the voice. Unfortunately, they may become infected, and when they do the fluid exuded by the inflamed tissues is not efficiently drained, particularly from the large maxillary sinuses which lie in the cheek bones. Sinusitis is the consequence.

The Pharynx

The pharynx is the cavity lying behind the nose (*nasopharynx*), mouth (*oropharynx*), and larynx (*laryngopharynx*). On the sidewalls of the nasopharynx are the openings of the Eustachian tubes which lead from the middle ear cavities. The roof of the nasopharynx contains masses of tonsil-like tissue commonly called the *adenoids.* The adenoids are of little importance unless they are enlarged to such an extent that they block the Eustachian tube openings and prevent equalization of pressure between the middle ear and the pharynx. This will cause a partial deafness such as is experienced in upper respiratory infections which plug the Eustachian tubes. Sometimes enlarged adenoids may even block air movement sufficiently to force the individual to breathe through his mouth. The continually gaping mouth characteristic of the mouth-breather not only leads to a rather dull, stupid expression, but may even cause deformity of the upper jaw. Adenoid removal is a simple operation, and should be performed when the physician suggests it.

What most of us call our "throat" is the oropharynx, that is, the pharynx posterior to the mouth. Both air and food must pass through the oropharynx, and their routes cross as shown in Figure 26–1—air entering the *larynx* or voice box; food or liquids passing posteriorly to enter the laryngopharynx and below into the esophagus.

The Larynx

The larynx caps the *trachea* or windpipe. It functions in speech, but also as an adjustable entryway tending to prevent anything but air from entering the trachea. In addition, the larynx serves as a pressure valve for the air within the bronchial system below, making possible the mechanism of coughing. Finally, the larynx is exquisitely sensitive, as anyone knows who gets even a droplet of water into it; any such stimulus sets off violent coughing, effectively ejecting the irritant and thus protecting the lower passages.

The larynx is composed of a number of cartilages held together by muscles and ligaments. One of the cartilages, the leaflike *epiglottis,* stands over the opening (*glottis*) of the larynx. In the process of swallowing, the larynx moves upward against the base of the tongue bringing its opening under the shelter of the epiglottis, thus preventing the aspiration of food or liquid. Hold your "Adam's apple" (thyroid cartilage) between your fingers, swallow, and note the rise and fall of the larynx and the fact that breathing stops during the process.

Most of the muscles and cartilages of the larynx function to control the tension on the *vocal cords.* The latter are really folds of the mucous membrane covering a band of tough connective tissue. They project from the sidewalls of the larynx, like shelves, forming between them a slitlike opening. When the vocal cords are tensed and brought close together and air is forced between them, they vibrate and produce sound as shown in Figure 26–2. The degree of tension and the length of the folds that participate in the vibration determine the pitch; the larger and longer vocal cords of the adult male produce a lower voice range than the smaller and shorter cords of the female or child.

The Trachea and Bronchial System

The trachea is a tube about an inch in diameter and 4½ inches long (Fig. 26–3). It ends behind the upper end of the sternum by dividing into two tubes, the *primary bronchi,* which lead into the lungs. The trachea is kept open by rings of

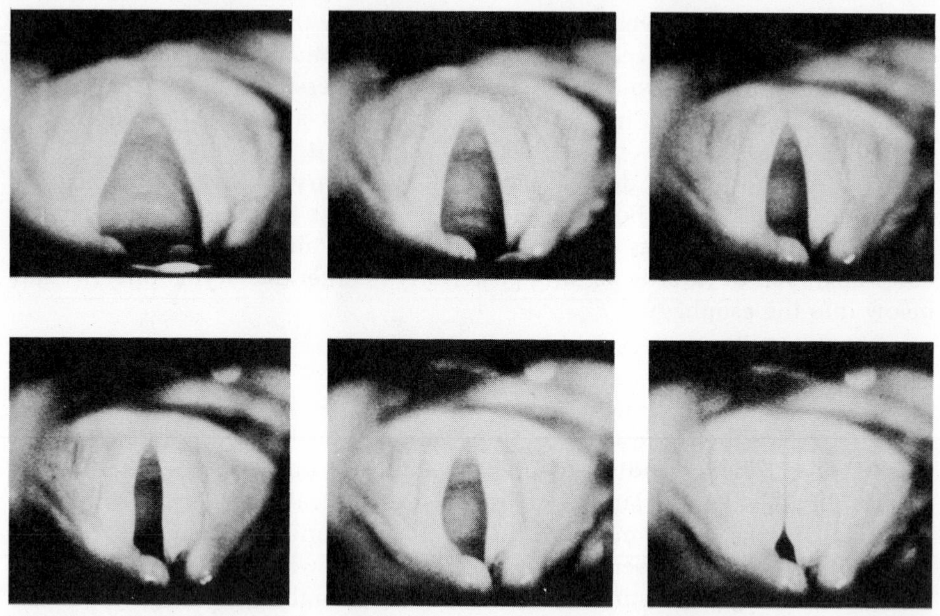

FIG. 26–2. Photographs of the vocal cords. *Top* (*left to right*): whispering; loud, low tone; and soft, low tone. *Bottom* (*left to right*): loud, high tone; soft, high tone; and coughing. (Bell Telephone Laboratories)

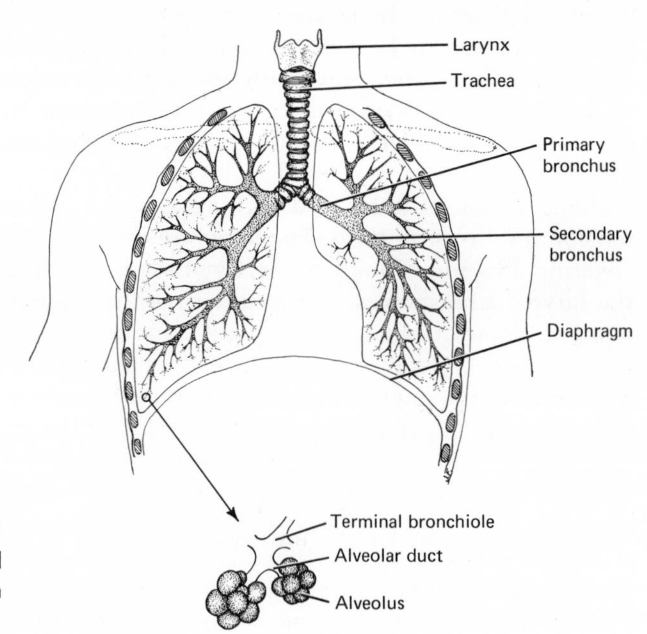

FIG. 26–3 The bronchial system. The pleura and pleural cavity have been omitted.

cartilage which can be felt through the skin. These rings are incomplete, more like horseshoes, with the open ends facing the esophagus. This allows some degree of encroachment of the esophagus into the trachea during passage of food. Solid food does not, then, trip along the rings as though sliding down the rungs of a ladder. The structure of the trachea is rather like that of a goose-neck lamp, and serves the same purpose—combined firmness and flexibility. A solid tube would obstruct movement; a soft one would collapse on inspiration like a soggy drinking-straw.

Upon entering the lungs each bronchus immediately divides and redivides much like the branches of a tree, and at each branching the total area of the two subdivisions exceeds that of the parent branch. From a mechanical standpoint this means that air moves more slowly through each successive subdivision, and in the smallest subdivision the air movement is sluggish indeed. The bronchi also have walls containing cartilage, but as the branches become smaller, the cartilage decreases in amount, and muscle replaces it so that the diameters can be regulated to some extent.

Throughout the bronchial and tracheal system the mucous membrane lining contains mucus-secreting glands and hairlike *cilia.* The sticky covering catches dust particles from the air and the cilia, in continual motion, moves the dust-laden mucus upward in a steady wave until it reaches the upper opening of the larynx where it is either swallowed or, if present in excessive amount, is coughed up.

The final subdivisions of the bronchial tree are termed *bronchioles,* distinctive in that the cartilage of the walls has been wholly replaced by muscle so that they can close almost entirely. Each bronchiole ends in a chamber whose walls are bulged out to form a cluster of thin-walled air sacs or *alveoli.* The subdividing bronchial tree is accompanied along its course by subdivisions of the pulmonary artery and vein. The pulmonary capillaries form a dense network around the alveoli and gases are easily exchanged between the blood and the alveolar air. It is estimated that the lungs contain as many as 600 million alveoli which provide a total surface area of 600 square feet. If you can imagine the floor area of a 20-foot by 30-foot room covered by a thin fluid film through which gases may diffuse, you arrive at some idea of the extent of gas transport in the lungs.

Lungs and the Chest Cavity

The thoracic spine, the ribs, sternum, and diaphragm enclose the chest or *thoracic cavity,* and most of this cavity is occupied by the lungs and heart. A part of the heart (principally the left ventricle) occupies a notch in the left lung so that the latter is somewhat smaller. Separating the lungs from the inner thoracic wall is the double-layered *pleura.* The inner layer is firmly attached to the lungs but folds out around the root of the lung, where the bronchi and vessels enter, to continue as the outer layer which lines the thoracic wall. Movement occurs between the pleural layers (in the *pleural cavity*) during breathing, and is facilitated by a thin fluid film. Inflammation of the pleura causes a roughening and irritation which makes breathing difficult and painful, a condition called *pleurisy* or *pleuritis*. Occasionally, the inflammation is so severe or long-standing that the pleural layers fuse together during the healing process, and *pleural adhesions* result.

Mechanics of Breathing

The lungs are, for the most part, passive in breathing. The filling and emptying of the lungs are accomplished by changes in the size of the thoracic cavity. A typical laboratory apparatus which effectively illustrates the breathing process is shown in Figure 26–4. A bell jar is equipped with a pliable rubber diaphragm across the opening and with a rubber stopper in the top through which is passed a short length of glass tubing ending in a balloon. When the diaphragm is depressed, the volume of the bell jar is increased, creating a partial vacuum that forces air to pass through the glass tube to inflate the balloon. When the diaphragm is pushed up into the bell jar, the air within the jar becomes compressed, and as the balloon collapses the air is forced out. In breathing, the volume of the thoracic cavity is increased by the lowering of the diaphragm and the elevation of the rib cage and pressure is equalized only by air passing through the respiratory passages into the lungs. As long as the pleural cavity is a *closed* cavity, the expansion of the thoracic volume tends to develop a vacuum and the lungs are forcibly expanded. To exhale, the diaphragm relaxes, the rib cage is lowered, thoracic volume decreases, and air is forced out.

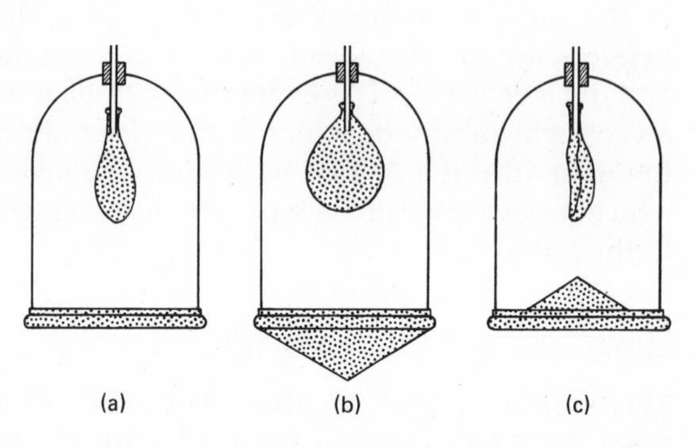

FIG. 26–4 Laboratory apparatus used to demonstrate the mechanism of breathing: (a) is a bell-jar covered below with a plastic diaphragm; a cork in the top of the bell-jar contains a glass tube which leads to a balloon. In (a) the air in the bell-jar is in equilibrium with atmospheric pressure and the balloon hangs limply. In (b), the plastic diaphragm has been lowered, increasing the volume of the bell-jar and thereby tending to lower the pressure below that outside; atmospheric pressure forces air through the glass tube and balloon inflates. In (c) the plastic diaphragm has been raised decreasing the volume of the bell-jar and thereby tending to raise the pressure above that outside; the increased pressure forces air out of the balloon and it becomes inflated.

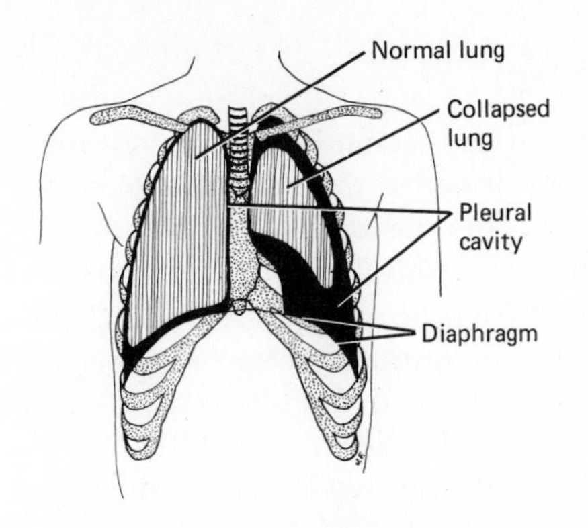

FIG. 26–5 If the chest wall is punctured so as to admit air into the pleural cavity, the lung will collapse. This condition is called *pneumothorax*. In tuberculosis, pneumothorax is sometimes deliberately induced in order to put the infected lung to rest. The pleura will eventually absorb the air and the lung be inflated to its normal size.

If the pleural cavity is pierced, as might occur in a stab wound in the chest, air rushes through the opening and the elasticity of the lung causes it to collapse. It is as if the rubber diaphragm of the bell jar in Figure 26–4 were pierced—the outside air would fill the bell jar and the balloon would collapse. The completely deflated lung and the air-filled pleural cavity constitute, in medical terms, the state of *pneumothorax* (Fig. 26–5). Pneumothorax is occasionally induced in cases of tuberculosis. The collapsed lung is, essentially, put to rest in order to promote the healing process. Pneumothorax may also occur spontaneously if a defect in the wall of a bronchus (usually a bronchial cyst) breaks through to bring the bronchial spaces into communication with the pleural cavity. In either instance, the lung will reinflate itself as the air in the pleural cavity is absorbed by the pleura.

Hiccups

Hiccups are repeated, involuntary contractions of the diaphragm followed by a sudden inrush of air into the larynx. The closure of the glottis then blocks this air movement and produces the characteristic sound. Hiccups may be infrequent and simply a nuisance or, to the other extreme, recurring and debilitating. Although a wide range of mostly useless remedies are familiar to all of us, recently one received attention in a leading medical journal (see Engleman), and is apparently gaining wide acceptance. The remedy was described as "exceedingly simple, benign and successful" and consisted of swallowing "dry" one teaspoonful of ordinary white granulated sugar. The investigating physicians claimed the remedy to be immediately successful in 19 of 20 patients who had showed up in the emergency room of a hospital for treatment of their hiccups. One of us (KF) has found the technique effective for himself and his family and recommends it.

Volumes of Air Breathed and Lung Reserve

At rest the average adult takes a breath about 14 times each minute. The volume of air which rushes in and out with each breath is called the *tidal volume*, and amounts to about half a liter (about one pint). This means a total of about 7 liters of air breathed every minute. If, however, the individual makes a maximum inspiration—breathing in all the air of which he is capable—and then follows this with a maximum expiration, it will be found that he can breathe out about 4.5 liters. This total breathing effort is termed the *vital capacity*. It may normally range as high as 6.5 liters in a trained athlete, or as little as 3.0 liters in a frail woman. A very small vital capacity (under 2.5 liters) may be indicative of pulmonary disorder.

The difference between tidal volume and vital capacity is our lung reserve, or in other words, the additional air which we can breathe in during increased physical activity. By increasing our breathing rate and utilizing more of our vital capacity, the volume of air breathed every minute by an average adult can exceed 100 liters, or over 14 times our resting rate. Loss of this lung reserve, that is, a decreased vital capacity, means a greater difficulty moving air in and out of the lungs during strenuous exercise and may mean a longer recovery period before breathing returns to normal following strenuous exercise.

A residual volume of air, about 1.5 liters, is present in the bronchioles and air sacs, even after maximum expiration, and this is turned over rather slowly, about once very minute, by mixture with the inflowing air. This residual air is involved in gas exchange: A hard blow on the abdomen will thrust the diaphragm upward, compress the lungs, and forcibly expel part of this residual air, seriously impairing gas exchange. The agonizing efforts required to restore this residual air when our "wind is knocked out" is familiar to most of us.

Gas Exchange in the Lungs

The changes which air undergoes in passage through the lungs may be studied by comparing the content of inspired and expired air. The air we breathe in is approximately 21 percent oxygen, 0.04 percent carbon dioxide, and 78 percent nitrogen. The exhaled air in the resting state is about 16 percent oxygen, 3.5 percent carbon dioxide, and 78 percent nitrogen. Thus at rest our blood picks up only about one-fourth of the oxygen taken into the lungs. During exercise, oxygen removal can approach 40 percent of that taken in with carbon dioxide removal increasing proportionately.

The oxygen used is partially replaced by carbon dioxide in the expired air which increases by over 4 percent. The remainder of the absorbed oxygen is replaced by water vapor which amounts to almost a pint of water lost through the lungs per day; the evaporation of this water involves a corresponding loss of heat. Some nitrogen is dissolved in the blood, but since it is not used in any metabolic reactions it leaves the blood at the same rate it enters, making the nitrogen content the same in both inspired and expired air.

The Air We Breathe

Room Ventilation

The discomforts we all have experienced while sitting in a crowded, poorly ventilated room are sometimes attributed to the lack of oxygen and/or high levels of carbon dioxide. Experiments carried out on human volunteers suggest that the discomfort is caused, not by the rebreathing of stale air but by the lack of cooling power normally present in circulating air. The oxygen level of the air can fall to as little as 12 percent, and the carbon dioxide level rise to as high as 3.5 percent without causing discomfort. (Neither homes nor schoolrooms could be expected to reach these levels, even with the windows closed for long periods.) What is required for maximum health and comfort is an efficient system which circulates and filters the air. The only physical justification for keeping windows open in cold weather is the possibility that bacteria might be more efficiently dispersed. Modern office buildings, stores, and schools are now usually equipped with systems which maintain a clear atmosphere of filtered and circulated air and require no further purification by opening windows. Indeed, modern air pollution may render outdoor air inferior to the filtered product, at least in urban areas. However, the psychological effect of freshness and freedom which comes with an open window may be difficult to duplicate.

Adjustment to High Altitude

The total mass of air held to the surface of the earth by gravitational force exerts upon us a pressure we call the *barometric pressure.* As we travel upwards this pressure is reduced as the air becomes rarer. At 50,000 feet barometric pressure is only about one-tenth of what it is at sea level as the pressure of oxygen is proportionately reduced.

Of importance to breathing is not only the net amount of oxygen present, but also the pressure it exerts. At sea level oxygen composes about one-sixth of the air, therefore exerting about one-sixth of the barometric pressure. If the total barometric pressure is reduced, so is the oxygen pressure. Oxygen pressure is what causes the diffusion of oxygen from its higher concentration in the alveoli of the lungs into the blood—an area of lower concentration. At sea level this pressure is sufficient to cause a 97 percent saturation of the blood with oxygen. (A normal individual breathing pure oxygen at sea level would bring about a mere 3 percent increase in the oxygen level of his blood.) If oxygen pressure is reduced, the amount of oxygen absorbed will be reduced, so that at 10,000 feet our blood absorbs about 93 percent of the oxygen it absorbs at sea level. This will be hardly noticeable to a healthy person unless he undertakes some sort of endurance exercise. At 15,000 feet oxygen absorption is about 82 percent of normal. Many people remaining at such an altitude for extended periods will experience one or more of the symptoms of mountain or altitude sickness: dizziness, nausea, increased heart and respiratory rates, blurred vision, and incoordination [Lenfant and Sullivan]. Some people will experience a euphoric excitation while others will become sleepy. At 30,000 feet (approximately the altitude of Mt. Everest) oxygen saturation is only 20 percent of normal, and without a source of pure oxygen, an individual has only about a minute of consciousness, and will live but a few minutes longer.

The development of pressurized interiors in modern aircraft has made possible today's high altitude flights. The artificial environment of passenger planes is generally pressurized to maintain approximately the same pressure as that at 7000 feet.

Adjustment to Increased Pressures

As stated earlier, a person on land is exposed to the pressure of all the air above the earth, or, more concisely, to one atmosphere of pressure. The weight of 33 feet of water is equal to the whole of the atmosphere, so that a person at a depth of 33 feet below the water's surface is under a pressure of two atmospheres. At 100 feet below the surface this rises to about four atmospheres; and at 300 feet, to about ten atmospheres.

This means that the pressures exerted upon atmospheric gases at these depths are proportionately increased. Since solids and liquids cannot be compressed only the air-filled parts of the body are affected. For example, air in the mouth or respiratory passages will be compressed, and exert pressures in proportion to the depth. At a depth of 300 feet below water level, a person breathing atmospheric air will be exposed to a pressure of oxygen and nitrogen ten times that at sea level, and these gases will enter the blood at ten times the normal rate. We can compare the process to the manufacture of carbonated beverages. Water normally dissolves only a small amount of carbon dioxide, but the amount of the gas dissolved increases with the pressure, and under this increased pressure the beverage is bottled.

These extreme oxygen levels in the blood supply the tissues with far more oxygen than is normal—this causes cellular metabolism to increase so greatly that the cells literally burn up and cause death. The most obvious effect is on the brain—the individual may go into a euphoric delirium ("rapture of the deeps") under which rational behavior is impossible. He may even go deeper or perform senseless acts which lead to his death. This oxygen poisoning can be prevented by supplying proportionally less oxygen to the diver's air mixture as he goes to greater depths.

Divers can generally descend to a depth of about 200 feet before the effects of oxygen poisoning become serious. At about this same depth there develops an effect, due to the increased nitrogen dissolved in the blood, that will act as an anesthetic and may put the diver to sleep. Nitrogen poisoning is frequently prevented by substituting helium for nitrogen in the breathing mixture.

Nitrogen is also the cause of "the bends" or decompression sickness. At sea level there is about 1 liter of nitrogen dissolved in the blood; at 200 feet below water level this increases to 7 liters. When the diver ascends, the decreased pressure allows the nitrogen to leave the blood and enter the expired air; but if the diver ascends too rapidly, the nitrogen comes out of solution as bubbles, much as carbon dioxide comes out of solution in a carbonated beverage when the cap is removed. The bubbles may form in any tissue of the body, and wherever they form they cause cellular damage and severe pain. The most rapidly damaged tissue seems to be the central nervous system. The mechanical rupture of the brain substance can lead to serious mental disturbances and permanent paralysis or both.

Decompression sickness is easily prevented by allowing the diver to come to the surface very slowly or by placing him in a decompression chamber. This gives time for the dissolved nitrogen to be expired. The use of helium as a substitute for nitrogen in the breathing mixture lessens the severity of decompression sickness since it does not dissolve in the blood as readily as does nitrogen and diffuses out of the blood much more rapidly

Scuba (self-contained underwater breathing apparatus) diving has become an extremely popular sport and recreational activity in this country. There is much to be learned before one actually attempts scuba diving, too much to be covered here. Suffice it to say that, because both oxygen toxicity and decompression sickness *can* occur with scuba diving, you should *never attempt it without very thorough instruction and practice beforehand.* This instruction should come from a well-trained and reliable instructor in whom you have complete confidence.

Anoxia

The failure of oxygen to reach the tissues results in oxygen starvation or anoxia. The trouble may be in the respiratory passages, in the blood, or in the tissues themselves. Of the many classifications of different types of anoxia, there are two that are of particular interest.

ANOXIC ANOXIA. This condition (or the more common term, asphyxiation) refers to the failure of oxygen to reach the blood. This could occur because of lack of oxygen in the air, obstruction of the air passages, or congestion within the lungs through heart failure, pneumonia, or drowning.

A piece of food or any foreign object can block air movement if it becomes wedged in the larynx or trachea. This is a hazard to anyone who puts things in his mouth, either out of curiosity (children) or for lack of a "third hand" (adults). Coughing, with the head down, may dislodge such objects, but if this fails, the object should be removed with fingers or an instrument within a few minutes or death or permanent brain damage will result. Slapping the victim with a "sharp blow to the back, between the shoulder blades," is a questionable practice. This may cause a reflex gasp by the victim and serve to suck the object farther down into the bronchi.

If the object does continue downward into one of the bronchi, the immediate danger is not so great since at least one whole lung remains functional. In this case the individual should be taken immediately to a physician for treatment. Emergency rooms of hospitals frequently display the wide variety of objects which have been removed from lungs: peanuts, popcorn, nails, whistles, bobby pins, safety pins, buttons, erasers, bottle caps, or almost anything of a size which may pass through the larynx.

ANEMIC ANOXIA. This condition is simply a failure of the blood to transport sufficient oxygen, either because of too few red blood cells or too little hemoglobin or in carbon monoxide poisoning.

Carbon monoxide is peculiar in that it has an affinity for hemoglobin 250 times that of oxygen. This means that a mixture of air with a concentration of 0.07 percent carbon monoxide (1/250th the concentration of oxygen in the air) will cause one-half of the hemoglobin to combine with carbon monoxide and one-half with oxygen. The harmful effects of carbon monoxide poisoning are related to time of exposure as well as to the concentration, that is, a low concentration over a long period of time can be as dangerous as a short exposure to a high concentration. For example, an exposure to a 0.07 percent concentration of carbon monoxide in air for three hours can be fatal as well as an exposure to a concentration of 0.2 percent for one hour. These are both approximately minimum exposures which may cause death. Less serious exposures frequently lead to severe headache, nausea and vomiting, or loss of consciousness.

Carbon monoxide is a combustion product whenever carbon-containing substances (coal, wood, gasoline, oil) are burned with insufficient oxygen. Since it has about the same characteristics as unpolluted air—colorless, odorless, and nonirritative—exposure is usually unsuspected, and there are probably many who suffer from its chronic effects (headache, dizziness, and weakness) without an appreciation of the cause. Although carbon monoxide is present in the fumes emanating from almost all industrial plants, the biggest source is automobile exhaust. Industrial toxicologists contend that almost all automobile mechanics and parking garage attendants suffer some harmful effects; and that not all the victims are indoors, since the carbon monoxide content of busy metropolitan streets, where the traffic moves slowly, may reach dangerously high levels. Traffic policemen who stand long hours at busy intersections often show effects. This is just one aspect of the air pollution problem which is becoming a threat to most large cities, and which, as yet, is unsolved.

In short, carbon monoxide is nothing to fool with! Most people are aware of the danger in running a car's engine in a closed garage, which can run the content to high levels in a very short time. Fewer, however, are aware of the amount of carbon monoxide that can build up within a moving automobile which has a defective muffler or tailpipe. Such conditions probably cause more deaths and accidents than is generally supposed. Every winter defective flues in various sorts of fuel-burning space heaters account for deaths of entire families. One should be suspicious, even hypersuspicious, of all headaches, nausea, or dizziness in the presence of any combustion, and should take prompt measures to get fresh air.

Pulmonary Resuscitation

Your knowledge of a method for restoring breathing to a person who has for any reason ceased to breathe could mean saving a life. Actual "biological death" does not occur until some four to six minutes after breathing and circulation have ceased. If resuscitative measures can be initiated within this six-minute period, brain damage is proportionately minimized, and there is a chance that "life" may be restored. Medical experts recommend against such procedures if the victim *is known* to have been without respiration and/or heartbeat for more than six minutes. With this limitation in mind, anytime breathing has stopped, from any cause, or when breathing is so shallow or difficult that stoppage of breathing or even unconsciousness is imminent, begin mouth-to-mouth resuscitation *immediately* as shown in Figure 26–6. If possible, no more than ten seconds should be spent in preparation. This means you must be prepared and well-rehearsed in the procedures.

1. Make sure victim is not a "neck breather" from a surgical hole in the windpipe. This requires a different procedure.
2. Remove foreign material from throat and mouth of victim (including false teeth) and keep tongue pulled forward (see Fig. 26–6a).
3. Keep air passage open by tilting victim's head back and pulling up on his jaws; place yourself to the side (see Fig. 26–6b).
4. Prevent air leakage from victim's nose and around his mouth by pinching nostrils shut with your free hand and after a deep breath blow forcefully into his mouth with your mouth pressed tightly over the victim's. (See Fig. 26–6c.) (*Note: Blow gently into child's mouth, covering both nose and mouth with your mouth.*) (See Fig. 26–6d.)
5. When you see the victim's chest rise, take your mouth away and allow natural exhalation.
6. When exhalation is completed, repeat the blowing procedure; repeat about every two to five seconds.
7. Be sure air exchange with the lungs is actually taking place.
8. Continue until the victim recovers breathing or until he is pronounced dead by a physician.

9. Be prepared for the victim to vomit. If it occurs, remove debris from throat and mouth immediately and begin again.

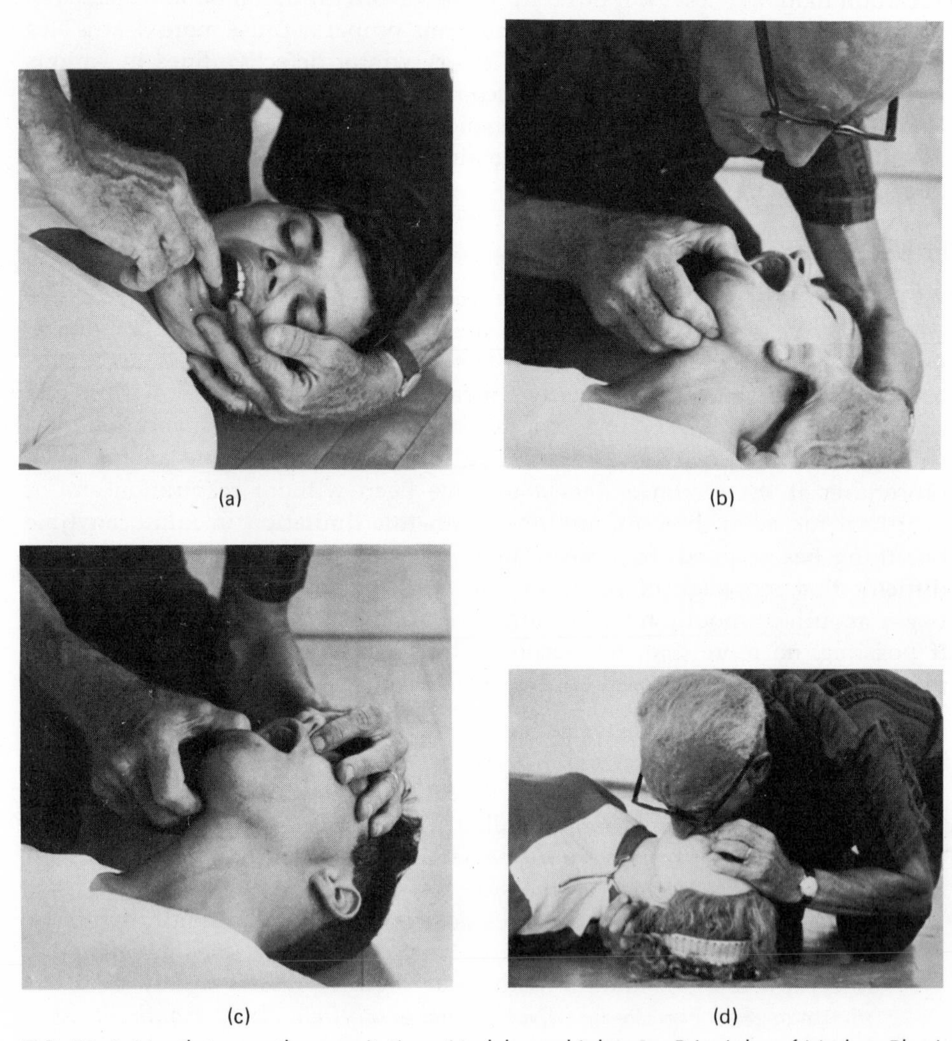

(a) (b) (c) (d)

FIG. 26–6 Mouth-to-mouth resuscitation. (Updyke and Johnson, *Principles of Modern Physical Education, Health and Recreation*. New York: Holt, Rinehart and Winston, 1970)

Quite often you will need to initiate external cardiac massage in an effort to restore the heart to action. Henderson states that "between 7500 and 9000 people die in the United States each year from cardiac arrest, which need only have been temporary had the proper resuscitative measures been applied quickly." It is known that in many instances the heart can be "restored" by properly squeezing it rhythmically between the sternum and spinal column. In order to apply external cardiac massage, the following procedures should be carried out (see Fig. 26–7).

1. Place victim on his back.
2. Place yourself at right angles to his chest, kneeling.
3. Prepare him for mouth-to-mouth resuscitation.
4. Blow air into his lungs three times.
5. *For adults,* use heel of hand only (with heel of other hand on top of it) on lower one-third of breastbone and press firmly down so that breastbone is depressed about 2 inches. *Do not press with fingers on ribs.* Repeat once every second. *For babies and young children,* use only the fingertips, applied to the center of the breastbone, and do not press too hard. For children nine to ten and up, use heel of one hand only.
6. Have someone give mouth-to-mouth resuscitation. *Cardiac massage is useless without respiratory resuscitation measures.* If no one else is available, stop massage every one-half minute and give mouth-to-mouth resuscitation for four deep breaths.
7. Continue until relieved or until the victim has recovered or is pronounced dead by a physician.

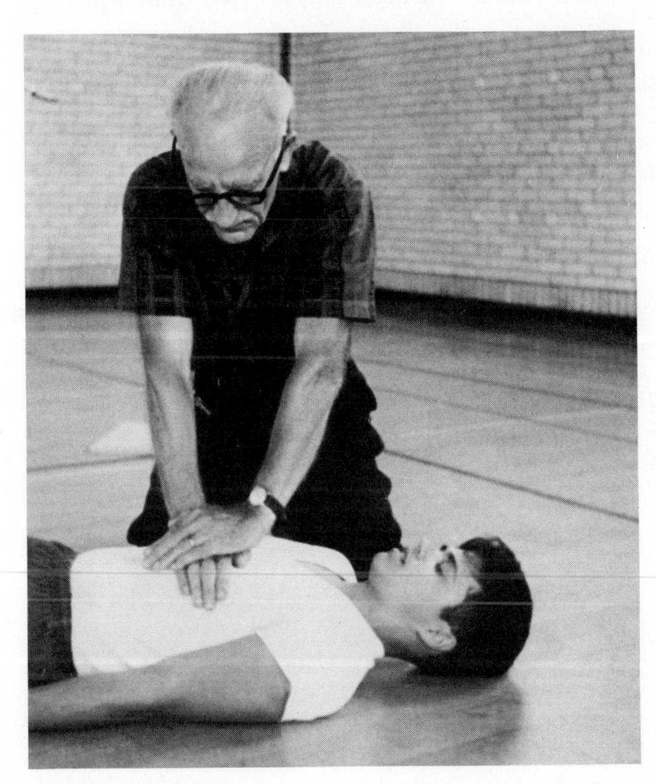

FIG. 26–7 External cardiac massage. (Updyke and Johnson, *Principles of Modern Physical Education, Health and Recreation.* New York: Holt, Rinehart and Winston, 1970)

REVIEW QUESTIONS

1. Why do you suppose that the air sinuses are included as part of the respiratory system?
2. Trace the movement of air from the nasal cavity to the air sacs of the lungs.
3. Why is cartilage present in the walls of the larynx, trachea, and bronchi?
4. What purpose is served by mucus-secreting glands in the lining of the respiratory passages? of cilia?
5. What is the cause of our discomfort when we are in a poorly ventilated room?
6. What is barometric pressure and how does it affect our breathing? Illustrate your answer with examples of the effect of high altitudes and of deep sea diving?
7. How does carbon monoxide cause death?
8. What resuscitative measures could you provide to someone who has stopped breathing?

REFERENCES

BOKONJIC, N., *Stagnant Anoxia and Carbon Monoxide Poisoning.* New York: American Elsevier, 1969.

COMROE, J. H., *Physiology of Respiration,* Yearbook Medical Publishers, 1965.

COMROE, J. H., and J. A. NADEL, "Screening tests of pulmonary function," *New England Journal of Medicine,* **282:** 1249–1253 (1970).

GOLDSMITH, JOHN R., and STEPHEN A. LANDAW, "Carbon monoxide and human health," *Science,* **162:** 1352–1359 (1968).

GUYTON, A. C., *Textbook of Medical Physiology,* 4th ed. Philadelphia: W. B. Saunders, 1970.

HENDERSON, J., *Emergency Medical Guide.* New York: McGraw-Hill, 1963.

LENFANT, G., and K. SULLIVAN, "Adaptation to high altitude," *New England Journal of Medicine,* **284:** 1298–1308 (1971).

NUNN, J. F., *Applied Respiratory Physiology.* New York: Appleton-Century Crofts, 1967.

SAFAR, P., and M. C. MCMAHON, *Resuscitation of the Unconscious Victim: A Manual for Rescue Breathing,* 2nd ed., Springfield, Ill.: Charles C Thomas, 1961. (Paperback)

ENGLEMAN, E. G. et al, "Granulated sugar as treatment for hiccups in conscious patients," *New England Journal of Medicine,* **285:** 1489 (1972).

Additional Readings

COMROE, J. H., "The lung," *Scientific American,* **214:** 56–68, 1966.

STEINCROHN, P. J., *You Live As You Breathe.* New York: Daniel McKay, 1967.

27
RESPIRATORY DISEASES

Upper Respiratory Infections

The numerous infections involving the nasal cavities, pharynx, larynx, trachea, or bronchi are collectively referred to as upper respiratory infections, particularly when uncertainty exists concerning the causative agent, or when more than one of these areas is involved. A few of the diseases conveniently treated under this heading are colds, influenzas, throat infections, laryngitis, whooping cough, and bronchitis. Sinusitis, previously discussed, might reasonably be included in this classification.

Common Cold

The cold, this commonest of upper respiratory infections, is a virus disease, and a number of distinct viruses, perhaps as many as 15, are known to cause it. In most instances the incubation period (time between exposure and appearance of symptoms) is short, and subsequent immunity is brief, lasting perhaps only a few months. In the meanwhile, the person is susceptible to the viruses other than the one which has caused his most recent infection.

Probably only a few of the different viruses are extant at one time in any small stable human population, and presumably we are relatively immune to these few which we are constantly exposed to. This is the reason for the tendency to come down with colds when we make an abrupt change in our personal contacts—when moving, when returning to school in September, or when entering military service, for example. It has been found that small groups of servicemen living together on the South Pole for many months, quite remote from other human contact, will develop a group immunity to the viruses they bring with them, and become remarkably free from upper respiratory infections. When returning home, however, they frequently suffer severe colds, supposedly from contact with viruses they left behind, and to which all immunity has been lost.

The symptoms and course of the disease are familiar. The virus attacks the mucous membranes of the nasal cavities and pharynx, and the first indication of the cold is usually a scratchy sensation in the throat. The mucous membranes become inflamed and secrete large quantities of watery mucus which is discharged from the nose. These symptoms normally persist four to ten days; then the membranes shrink, the nose dries up, and recovery is rapid. In many instances, however, the infection may pass down into the larynx, trachea, or bronchi (laryngitis, tracheitis, bronchitis) and produce a cough and other symptoms which persist longer. The infection may also spread to the sinuses, or through the Eustachian tube into the middle ear.

The only effective preventative measure is to avoid close contact with people who show cold symptoms. Certainly, anyone in the symptomatic stage of the disease should exercise great care in his handling of contaminated objects such as paper handkerchiefs. Coughs and sneezes are probably responsible for the transmission of most cold viruses (Fig. 27–1), and it is obviously good hygiene, as well as good manners, to cover a cough or sneeze. In passing, it is interesting that in our culture a blow or other physical assault is bitterly resented and even cause for lawsuit; but transmission of a cold, which may be far more damaging and cause greater misery and loss, is accepted as "just one of those things." In a truly civilized community, a person would feel, and be considered, guilty of at least ignorance or indifference (if not a crime) if he infected another party. Civilized individuals might well act accordingly themselves, and should feel no hesitation in warning off a carrier of infection.

In general, supplementary vitamins will not prevent or cure a cold though a run-down condition from poor diet or any other cause seems to favor infection. Linus Pauling offers convincing evidence for the efficacy of ascorbic acid (vitamin C) in the prevention and treatment of colds. However, the matter is far from resolved and needs considerably more investigation. And despite their popularity, antihistamines have never been shown to be of any value, either in abating a cold or in relieving its symptoms. Antibiotics are completely ineffective against viruses, and unless there is accompanying bacterial

FIG. 27–1 High speed photograph of a violent unstifled sneeze (American Society for Microbiology)

infection, they should not be taken. Aspirin gives the best relief, and other medication should be taken only on advice of your physician.

Influenza

Influenza, or the "flu," is also a virus infection of the upper-respiratory passages, but generally gives symptoms a good deal more severe and persistent than those of a cold. Fatalities so common in influenza epidemics (an estimated 20 million persons died in the worldwide epidemic of 1918) are usually the result of bacterial pneumonia superimposed upon a respiratory system already weakened by the virus.

In addition to the symptoms of a cold, the patient suffers chills and fever (uncommon in a cold except in children), aches in the neck, arms, and legs, and general prostration. Obviously, fever and aching extremities accompanying a "cold" warrant a visit to a physician.

One frequently observed characteristic of influenza is that many deaths are due to relapse: The patient feels better, gets up, or exerts himself too soon, and succumbs to a return of the flu itself or to a secondary infection. Many of the 1918 victims perished from trying to help others while still sick themselves. The lesson is obvious: flu requires a long convalescence.

The development of effective vaccines is hampered by the same problems encountered with cold viruses. Some investigators claim there are as many as ten different viruses responsible for influenza classified into two principal groups designated A and B. Immunity, again, is short-lived. However, there are effective vaccines against a few types, and "flu shots" can give some protection.

STREP THROAT. Strep throat is a severe infection caused by a member of a bacterial group called streptococci. The symptoms of a strep infection, though painful, are often not severe enough for some persons to feel justified in consulting a physician, and herein lies the danger of the disease. If untreated, the toxins produced by the organism may be spread to other parts of the body, notably the bones, joints, and heart valves (Chapter 22) and the condition is then referred to as rheumatic fever. Inflammation of the kidneys (glomerulonephritis) is another serious complication of strep throat. Certainly, any serious throat infection deserves the attention of a physician. If the cause is a strep infection, effective antibiotics are available.

Whooping Cough (Pertussis)

Whooping cough is an acute bacterial infection of the trachea and bronchi which produces a severe irritation of the mucous membrane. The cough becomes progressively worse over a period of one or two weeks as a sticky mucus accumulates on the tracheal and bronchial walls causing breathing to become more difficult. The violent though ineffectual coughing episodes can end in strangling gasps for air and vomiting. The characteristic whooping sound is caused by the sudden inhalation of air following each cough. The cough may continue for as long as two months.

The whooping cough vaccine consists of a suspension of the dead bacteria which have a high antigenic ability to achieve a relatively long-term immunity. The vaccine is customarily administered during the first year of life, usually along with the vaccines against diphtheria, tetanus, and poliomyelitis. Thus the disease, though once common, is now becoming rare in our country.

Bronchitis

Bronchitis is inflammation of the mucous membrane lining the bronchial tree. It may be acute or chronic. *Acute bronchitis* is usually a bacterial infection accompanying an upper respiratory virus infection. The latter apparently lowers natural resistance to bacteria normally present in the nose and throat, and allows them to invade the trachea and bronchial tree. A mild fever is generally present for a few days, followed by a cough which may persist for several weeks.

Chronic bronchitis is a more serious, long-term disease of the trachea and bronchial tree, usually with inflammation and degenerative changes developing in the mucous membrane. No specific organism is involved although recurrent infections may be a factor in the development of the disease.

The degenerative changes in the bronchial lining makes drainage inadequate, leading to the retention of pus and mucus. This, in turn, is responsible for the dependence upon coughing for bronchial clearance. The condition may show itself simply as a persistent cough such as a "cigarette cough" each morning. Obviously, the condition makes upper respiratory infections much more severe, and bronchopneumonia becomes a particularly dreaded complication. Chronic bronchitis accompanies emphysema and lung cancer as a frequent consequence of cigarette smoking, and will be discussed again in that connection.

Lung Infections

Our lungs are constantly exposed to many kinds of bacteria and viruses which are able to cause serious disease. They generally do not do so because of a number of protective mechanisms which operate efficiently in the healthy body—our so-called resistance to disease. But when an individual becomes weakened by another infection, physically exhausted, undernourished, or chilled, this resistance lowers, and a lung infection can result. The most common lung infections are pneumonia and tuberculosis.

Pneumonia

Pneumonia refers to any inflammation of the lungs, usually as a result of infection, and a variety of organisms is responsible. *Bronchopneumonia* is the result of invasion of the bronchioles and alveoli by bacteria which have come from infections in the upper respiratory passages, such as a sore throat or bronchitis. The symptoms are fever, painful and difficult breathing, and general weakness.

Lobar pneumonia is the formerly dreaded disease which only a few decades ago was the leading cause of death. The disease organism is the pneumococcus bacteria commonly present in the respiratory passages. The infection develops when the organism becomes localized in the bronchioles and alveoli of one or two of the five lobes of the lungs. The involved lobes become almost solid with fluids and pus. The symptoms of lobar pneumonia come on swiftly. Typically, the disease begins with a high fever following a chill; the patient's breathing becomes shallow, rapid, and painful. A brown, sticky mucus is often coughed up from the lungs. Fortunately, the pneumococci are very susceptible to a number of antibiotics, and lobar pneumonia is no longer of such serious consequence, although it is still a terminal cause of death in many elderly persons already weakened by cancer or heart disease.

Virus pneumonia, sometimes called primary atypical pneumonia, often follows a severe cold. Rather than being restricted to one or two lobes the infection is scattered throughout both lungs. The disease is often diagnosed by X ray and may be confused with tuberculosis. Symptoms include fever, headache, and weakness; however, breathing is often more or less normal in the early stages which makes diagnosis difficult. Although not often fatal in adults, this disorder can involve a long convalescence.

Tuberculosis of the Lungs

The tuberculosis organism, *Mycobacterium tuberculosis,* or tubercle bacillus, may attack almost any organ of the body, but the usual site of infection is the lungs. Pulmonary tuberculosis, once called "consumption," was the leading cause of death up to the beginning of this century. Infection usually begins in a group of neighboring alveoli. The damaged area or lesion is called a *tubercle*

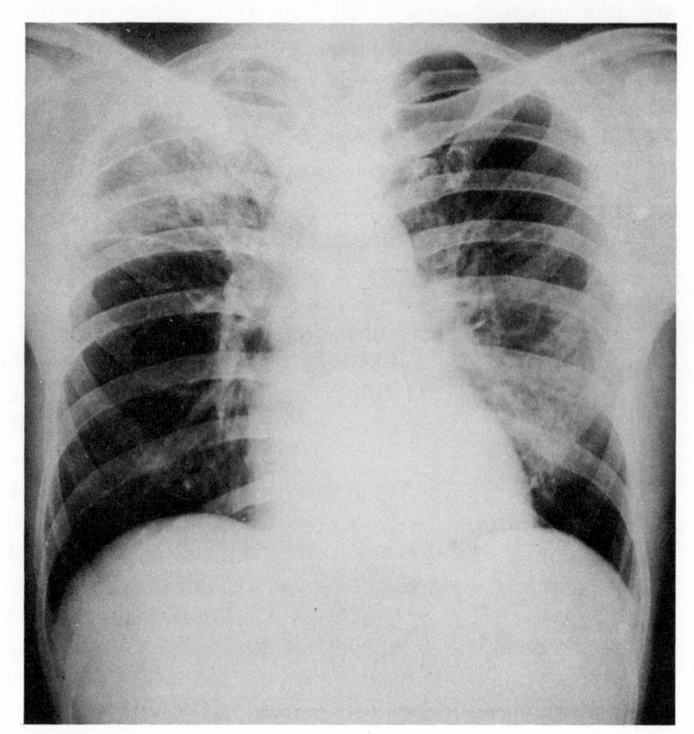

FIG. 27–2 Chest X-ray of a patient with moderately advanced tuberculosis. The diseased area in the upper part of the right lung appears cloudy. (National Tuberculosis and Respiratory Disease Association)

(Fig. 27–2). The surrounding tissues respond by walling off the tubercle, which may arrest the disease, sometimes permanently. The walling off can, however, cause death of the tissues, which then empty their contained bacterial toxins into the blood stream. This causes more generalized body symptoms: fever, night sweats, and loss of appetite. From this stage the disease may progress very rapidly, or quite slowly, sometimes destroying a whole lung or much of both lungs. Coughing is unusual in the early stages, but as the disease progresses and excavates lung tissues, considerable liquefied material may be coughed up, sometimes with hemorrhaged blood.

Occasionally, early cases of tuberculosis can be treated at home, but it is usually advisable for patients to enter tuberculosis hospitals where rest and treatment is most effectively carried out. Treatment may involve surgical removal of the affected area, pneumothorax (Chapter 26), or antibiotic drugs such as isoniazid and streptomycin. If the disease is in its early stages, the chances of recovery are good. However, when extensive excavation of lung tissue has occurred the prognosis is poor. Alcoholism is a common concomittant among older tubercular patients in ghetto areas, and these patients are not only difficult to enlist in effective treatment programs, but also serve as a continual reservoir of contagion.

The highly contagious nature of tuberculosis makes its control difficult. Persons with an active infection may believe they have a "chronic bronchitis," and transmit the disease to those about them by their constant coughing. The use of routine chest X-rays and tuberculin tests are an effort to identify persons with the disease, not only for their own well-being, but as a public health measure.

The tuberculin or Mantoux test is a method of determining if an individual has been exposed to the disease (Fig. 27–3). The invasion of the organism causes the production of antibodies, a response which normally follows the introduction of any foreign protein. When a small amount of the tuberculin toxin is injected under the skin of an infected person, a typical allergic reaction will result—the skin will become reddened and swollen at the site of the injection. This is a "positive reaction" and indicates that the person has been exposed to the disease. It does not mean that he has the disease, or even had it previously, but it does warrant a chest X-ray and other examinations which are necessary to determine if an active disease is present. A "negative test" means no reaction or only a slight reaction at the site of the injection, and indicates the individual is definitely free of the disease.

The close and sometimes overcrowded living conditions found in student housing in many college and university campuses is ideal for the spread of tuberculosis, and for this reason most institutions require a tuberculin and/or chest X-ray as part of the routine entrance physical examinations.

A vaccine which develops active immunity against tuberculosis is available —the Bacillus Calmette-Guerin or BCG vaccine. It consists of a mutant form of

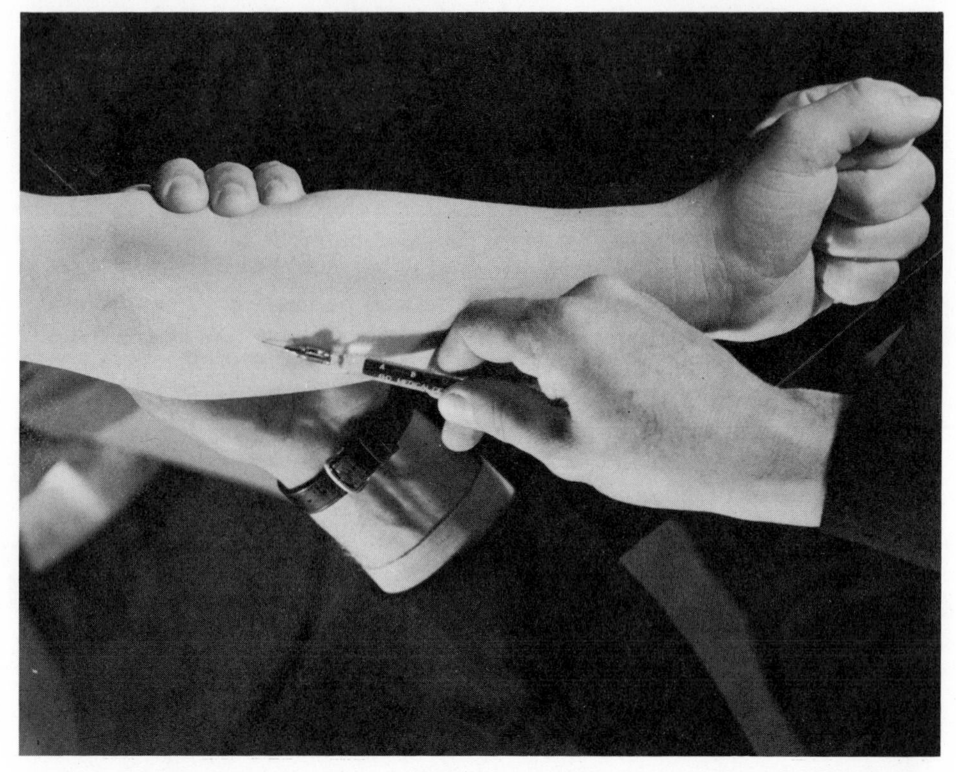

(a)

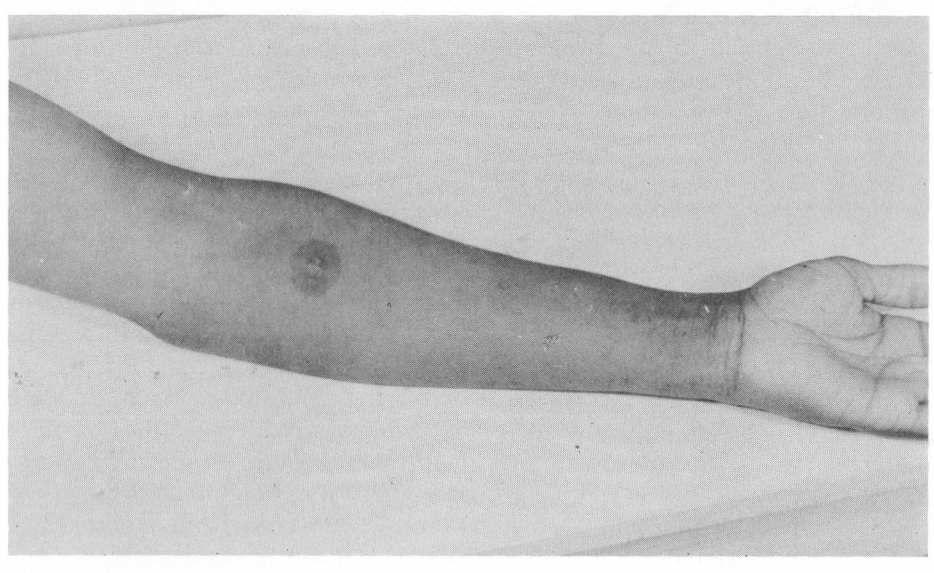

(b)

the tubercle bacillus which is unable to cause disease, but which effectively induces antibody formation. Its usage is not recommended except under conditions where exposure to the disease is likely to occur, as for example, in children of a household where tuberculosis is known to be present.

Bronchial Asthma

The word *asthma* comes from the Greek word meaning "to pant" and emphasizes the distress that the asthmatic suffers. Asthma can also refer to other conditions that result in panting, but in common usage it means bronchial asthma. In its most common form it is an allergic condition and is described as such in Chapter 19. However, other etiological factors can be involved including heredity and emotional factors.

Whatever the cause the mucous membrane of the bronchial passages becomes swollen and the muscular walls of the bronchioles undergo spasms. Breathing becomes difficult and the characteristic wheezing results. Inspiration is an active muscular activity involving the strong intercostal and diaphragmatic muscles while expiration is a consequence of relaxation of these muscles. This means that the asthmatic is usually able to breathe in more easily than he can breathe out. This accounts for the old practice of giving artificial respiration to the sufferer by pressing on the rib cage.

Acute attacks of asthma are now effectively treated by inhalation of epinephrine-like drugs that cause dilation of the bronchioles. Aminophylline and ephedrine are examples of such valuable drugs. The almost instanteous relief—usually in a matter of seconds—is remarkable, particularly when viewed in relation to the suffering which was the lot of the asthmatic just a few decades ago.

Emphysema

Emphysema is a chronic and progressive disease of the lungs, and is believed to develop as a consequence of chronic bronchitis, asthma, or other conditions which affect the bronchial tree. In such diseases muscles of inspiration are capable of forcing air into the alveoli through partially blocked bronchi or bronchioles, but the expiratory mechanism is not able to empty the alveoli to the same degree. As a result the alveoli become overinflated and rupture. The

---

FIG. 27–3 The Mantoux test. (a) Tuberculin, a sterile fluid containing antigenic material from the tuberculosis bacterium, is injected under the skin. (b) Inflammation around the site 48 hours later indicates a positive reaction. This means the individual has been at some time exposed to tuberculosis and may or may not have the disease. (National Tuberculosis and Respiratory Disease Association)

breakdown of neighboring alveoli combines many air spaces into large cavities which trap "dead" air, raising the volume of residual air and making expiration still more difficult (see Fig. 27–4). Adequate ventilation is possible only by ever more forcible inspiration which results in the rupture of more alveoli, which explains the progressive nature of the disease. The lungs increase in size as the residual volume increases (actually an increase in nonfunctional dead-air space), the diaphragm becomes flattened and less efficient, and the rib cage is expanded—the typical "barrel-chested" appearance of the emphysematous individual. Yet ventilation becomes steadily poorer, and the shortness of breath (dyspnea) becomes more acute.

The condition is difficult to detect in the early stages, and usually by the time it is diagnosed considerable lung damage has already occurred. Since alveoli cannot be regenerated, no cure of the disease is possible. However, inhalation of oxygen several times a day can bring some relief; the use of a constricting belt around the upper abdomen which forces the diaphragm upwards can aid breathing movements.

Although formerly associated with old age, emphysema is now encountered with growing frequency in younger people. In fact, the incidence of emphysema has taken a sudden spurt, much as lung cancer did a generation ago. For example, in California, the death rate has increased from 1.5 per 100,000 in 1950 to 11.6 in 1964, nearly an eight-fold increase, and this increase has occurred primarily among men. Although emphysema is not a common cause of death, it has been estimated that more than 10 million Americans are afflicted, and many are disabled. The Social Security Administration pays more disability allowances to workers aged 50 through 64 for emphysema than for any other cause except heart disease. The probable relationship of emphysema to cigarette smoking is considered in the next chapter.

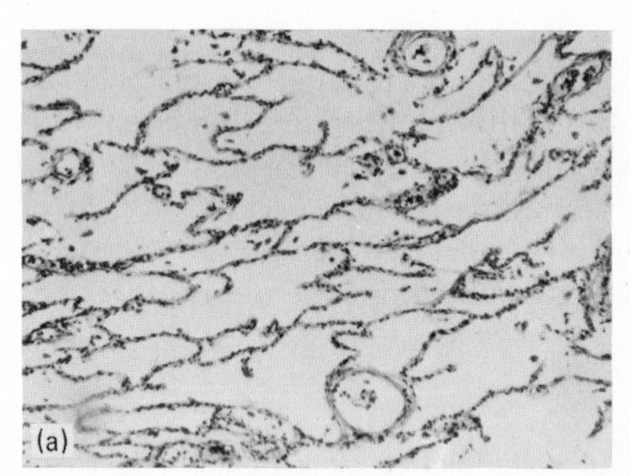

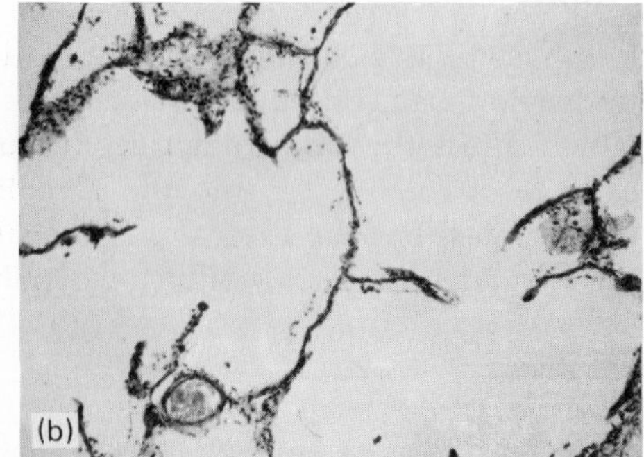

FIG. 27–4 (a) Appearance of normal air sacs of lung when prepared in very thin slices and examined under a microscope. (b) In emphysema, the air sac walls thicken and rupture resulting in large dead air spaces. (Dr. Oscar Auerbach)

Bronchiectasis

Bronchiectasis may be a congenital disease, but more often is the result of infection in the mucous membranes of the bronchial passages—bronchitis, tuberculosis, whooping cough, or any disease which leads to retention of infected material. In any case the bronchial wall thickens in response to the inflammation, and the circular muscles of the bronchial walls are much weakened and are unable to expel the mucus and pus during coughing. As the material collects the bronchial channels widen and coughing becomes increasingly ineffective. The obstruction of the respiratory passages leads to emphysema—a condition often accompanying bronchiectasis.

Conclusion

It is evident that modern medical treatment and public health practice have come a long way toward eliminating many of the respiratory diseases which in former times were the cause of such high mortality and general misery. In recent times, however, we have seen a tragic increase in respiratory diseases which were formerly of little consequence—lung cancer and emphysema. These diseases, and a number of others, have a causal relationship to cigarette smoking and this relationship is examined in the next chapter.

REVIEW QUESTIONS

1. Under what conditions are we most likely to come down with a cold?
2. What are some of the causes of chronic bronchitis?
3. There are three principal forms of pneumonia. What are the nature and cause of each?
4. Discuss some of the problems encountered by public health efforts to control tuberculosis.
5. Why is the "barrel-chested" appearance typical of long-term sufferers of asthma?

REFERENCES

BATES, D. V., "Chronic bronchitis and emphysema," *New England Journal of Medicine*, **278:** 546–551, 600–604 (1968).

BELINKOV, STANTON, *Emphysema and Chronic Bronchitis*, Boston: Little, Brown, 1971.

BOUHUY, AREND, and JOHN M. PETERS, "Control of environmental lung disease," *New England Journal of Medicine*, **283:** 573–582 (1970).

DUBOS, RENÉ, and JEAN DUBOS, *The White Plague: Tuberculosis, Man, and Society*, Boston: Little, Brown, 1952.

PAULING, LINUS, *Vitamin C and the Common Cold.* San Francisco: W. H. Freeman, 1970. (Paperback)

PETTY, THOMAS L., and LOUISE M. NETT, *For Those Who Live and Breathe with Emphysema and Chronic Bronchitis*, Springfield, Ill.: Charles C Thomas, 1969.

STUART-HARRIS, C. H., *Influenza and Other Virus Infections of the Respiratory Tract.* Baltimore: Williams & Wilkins, 1965.

TYRRELL, DAVID A., *Common Colds and Related Diseases.* Baltimore: Williams & Wilkins, 1965.

Additional Readings

ADAMS, JOHN M., *Viruses and Colds: The Modern Plague.* New York: American Elsevier, 1967.

MITCHELL, ROGER S., "Control of tuberculosis," *New England Journal of Medicine,* **276:** 842–848 (1967).

THOMPSON, PAUL D., *The Virus Realm.* Philadelphia: J. B. Lippincott, 1968.

WAKSMAN, SELMAN A., *Conquest of Tuberculosis.* Berkeley, Calif.: University of California Press, 1965.

28
SMOKING AND HEALTH

The subject of smoking has received much public attention over the past decade. However, it is frequently discussed, for and against, with more heat than light, and more self-interest than self-examination. Certainly, any scientific evaluation is inevitably on the negative side; but the arguments given are often unconvincing and are not strengthened by the approval, even if unsolicited, of the self-righteous who do not argue but only prohibit. The problem, as will be seen, is of such unexpected magnitude that it merits more careful and judicious treatment. The following is a selection of the most conclusive findings.

History of the Problem

Although the use of tobacco dates from the early part of the 16th century, when explorers introduced it into Europe from the New World, it was not until World War I, when cigarettes gained widespread popularity, that inhalation of tobacco smoke became customary. Smoke from cigars and pipes is dense and slightly alkaline, and few people can inhale it without coughing or becoming ill. Cigarette smoke, on the other hand, is easily inhaled; and the pleasurable effects of the nicotine are more immediate and intense due to the absorption of the drug directly into the blood through the alveoli of the lungs. Although some nicotine is probably absorbed by those who do not inhale, most of the pleasurable effects of cigar and pipe smoking are related to taste and smell rather than to the metabolic effects of the nicotine. For those, then, who crave the pharmacological effects of tobacco, cigarettes are the obvious preference.

In any case cigarette smoking has obviously become the tobacco habit of choice. In the period from the early twenties to 1970 the consumption of cigarettes rose from 750 per adult (15 years or older) to 3970 (198½ packs) per adult per year. The Department of Agriculture estimates that about 70 million persons regularly smoke cigarettes in this country. If we divide this into the domestic consumption of 534 billion cigarettes, we find that the average smoker consumes a little over one pack per day (Fig. 28–1). This is big business. Over $8.5 billion were spent on cigarettes in 1968. This is more than the total *combined* general expenditures of the state governments of New York, California, and Illinois for the same year!

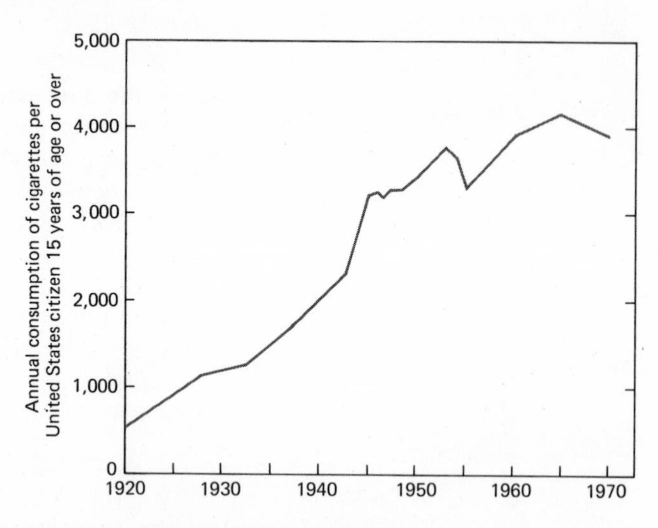

FIG. 28-1 The increase in cigarette consumption, 1920–1970. (United States Department of Commerce data)

This public embracement of the cigarette habit was followed by the sudden appearance of lung cancer as a regularly occurring disease instead of a relatively rare one (Table 28–1). During the period from 1900 to the present, when death rates from most infections and some degenerative diseases were showing a decline, lung cancer climbed to the status of an epidemic. In 1914 only 371 deaths were attributed to lung cancer. The results of cigarette consumption and lung cancer for selected years 1930 through 1970 are shown in Table 28–1. In order to get an idea of the magnitude involved, we can point out that motor vehicle accidents resulted in 52,924 deaths in 1967 (estimated at 54,000 in 1968).

TABLE 28–1 Lung Cancer Deaths and Consumption of Cigarettes in the United States

Year	*Deaths* [a]	*Domestic Consumption of Cigarettes (in billions)*
1930	2,357	136
1940	7,121	182
1950	18,313	332
1960	36,420	388
1962	41,376	395
1964	45,838	511
1966	54,101	542
1967	58,086	551
1968	62,134	546
1969	66,100	529
1970	69,180	534

[a] Classifications 161–163 of General Mortality Tables. U. S. Department of Health, Education, and Welfare: *Vital Statistics of the United States*.

The rise in lung cancer deaths run counter to the 70-year generally downward trend in total death rates. In 1970 lung cancer accounted for an estimated 69,180 deaths. A nearly steady death rate for heart and circulatory diseases conceals a significant rise in coronary artery disease (also connected to cigarette smoking) which is offset by a long-term decline in other forms of heart disease.

It is of interest to note, also, that lung cancer deaths and motor vehicle deaths were almost the same in 1964, 45,838 and 45,825, respectively. Lung cancer is now first among fatal cancers, having overtaken cancer of the colon and rectum, the longtime leader, in 1966. Lung cancer deaths far exceed those from cancer of the breast or uterus: the American Cancer Society estimates that lung cancer will have killed over 59,000 males and 13,000 females in 1972; 13,300 females will have died of cancer of the uterus, and 32,000 of breast cancer in the same year.

Statistical Studies

Cigarette smoking is just one of several agents associated with the rise of lung cancer. Certain industrial dusts and vapors breathed in large amounts in occupational exposures can cause lung cancer, but the association with cigarette smoking is of far greater significance in terms of the numbers of people affected. One of the first observations in this country linking cigarette smoking and lung cancer was published in 1939 by Alton Ochsner and Michael DeBakey (now famous as a heart surgeon). They noted that nearly all their lung cancer patients were cigarette smokers, and suggested a causal connection between the two—interestingly, a report dated as early as 1761 associated cancer of the nasal cavity with snuff (see Redmond).

However impressive such "retrospective" studies may appear, they lack the scientific requirements of nonbias and control. To this end several studies were begun in the late 1950s in which larger numbers of presumably healthy persons were studied over a number of years. Deaths among these individuals were recorded and efforts were then made to associate cause of death with habits reported *before* becoming ill. The most comprehensive of such studies was the one reported on in 1966 and under the supervision of Dr. E. Cuyler Hammond, Director of the Statistical Section of the American Cancer Society.

Between October 1959 and February 1960, 68,000 volunteer workers of the American Cancer Society enrolled 1,078,894 men and women for the prospective study. Enrollment was by families, with the specification that there be at least one person over the age of 45 in each family enrolled. All members over the age of 30 filled out detailed, confidential questionnaires containing questions on family history, past diseases, present physical complaints, occupation, occupational exposures, various habits, and other factors. Only illiterates, persons too ill to answer a questionnaire, and persons who could not be traced were excluded. The subjects were traced annually, and once every two years were requested to fill out a follow-up questionnaire. Whenever a death was reported, the investigator obtained a copy of the death certificate from the state or local health department.

Of the 1,078,894 subjects originally enrolled, a number were lost because volunteer workers failed in the follow-up or because the questionnaires were unusable. This left a total of 1,045,087 subjects effectively enrolled. Of the 1,045,087 subjects 99.6 percent were successfully traced through September 30, 1962, and 97.4 percent were successfully traced through September 30, 1963. At the end of this last follow-up, 971,362 were reported to be alive, 46,212 dead (26,448 men and 16,773 women) and 27,513 "lost."

On the basis of this information Hammond made the following observations. Men who never smoked regularly and men with a history of only pipe smoking had the lowest death rates. Men with a history of only cigar smoking had slightly higher death rates. Men with a history of cigarette and other smoking had far higher death rates; and those with a history of only cigarette smoking had the highest rates. In the age group 45 to 54, the death rate of men with a history of only cigarette smoking was more than double that of men who had never smoked regularly. (If 1.0 is the death rate of men who never smoked regularly, the *mortality ratio,* in this case, is 2.2:1.)

Death rates were also found to be higher among women who smoked. In the age group 45 to 54, the death rate was about a third higher (mortality ratio 1.31:1) than in women who had never smoked. (The reason that cigarette smoking appears to be less harmful to women than to men is open to a number of explanations. Women generally smoke fewer cigarettes per day, inhale less deeply, and have continued the habit over a shorter period of time.)

Death rates for both men and women were also found to increase with the number of cigarettes smoked per day, with the number of years smoked, and with the the depth of inhalation of the smoke. Furthermore, the data suggest that cigarette smoking increases the likelihood of dying from almost *any* cause. Table 28–2 shows the excessive death rate or mortality ratio for various causes of death among cigarette smokers aged 45 through 64 as compared to those of the same age who have never smoked.

TABLE 28–2 Mortality Ratios for Selected Causes of Death for Cigarette Smokers in Comparison to Nonsmokers

Men *Cause of Death*	*Mortality Ratio to Nonsmokers*
Cancer (total)	2.14
Lung	7.84
Mouth and pharynx	9.90
Larynx	6.09
Pancreas	2.69
Heart and circulatory disease (total)	1.90
Coronary heart disease	2.03
Hypertensive heart disease	1.40
Cerebral vascular lesions (stroke)	1.38
Emphysema	6.55
Gastric ulcer	2.95
Duodenal ulcer	2.86
Cirrhosis of liver	2.06

Women (Heavier Smoking[a]) *Cause of Death*	*Mortality Ratio*
Cancer (total)	1.12
Lung	3.63
Mouth, pharynx, larynx, esophagus	3.17
Pancreas	2.58
Heart and circulatory disease (total)	1.80
Coronary heart disease	2.10
Cerebral vascular lesion (stroke)	2.09
Cirrhosis of liver	3.25
Emphysema	7.38

[a] "Heavier" women smokers are those who (1) smoked 20 or more a day at time of interview regardless of age they began smoking or (2) smoked 10 or more and began smoking before age 25.

The mortality ratio of cigarette smokers to nonsmokers refers to the number of smokers who die of a given cause relative to each nonsmoker. Thus 7.84 smokers died of lung cancer to each nonsmoker. (E. Cuyler Hammond, Scientific American, **207:** *39–51)*

Matched Pair Analysis

An earlier report by Hammond [1958] covered findings after 44 months of follow-up of 187,783 men and included a comparison of death rates from various causes of cigarette smokers and nonsmokers who were alike in many characteristics other than their smoking habits.

Men who had never smoked regularly were matched by a computer against men currently smoking 20 or more cigarettes a day at time of enrollment. The two men in a pair had to be alike in the following characteristics: race; height; nativity (native or foreign born), residence (rural or urban); occupational exposure to dusts, fumes, vapors, chemicals, radioactivity, and so on (yes or no); religion (Protestant, Catholic, Jewish, or none); education; marital status; drinking habits; sleep per night; amount of exercise (none or some); nervous tension (yes or no); history of cancer other than skin cancer (yes or no); history of heart disease, stroke, or high blood pressure (yes or no).

TABLE 28–3 Causes of Deaths among Equal Numbers of Men Matched into Pairs in All Respects except Smoking Habits

Cause of Death	*Number of Deaths Never Smoked Regularly*	*Cigarettes 20 or More per Day*
Cancer (total	96	261
Lung	12	110
Mouth; pharynx	1	3
Larynx	0	3
Esophagus	0	6
Bladder	1	2
Pancreas	6	16
Liver	1	7
Stomach	9	10
Colon and Rectum	20	25
Other specified sites	43	64
Site unknown	3	15
Heart and Circulatory Disease (total)	401	854
Coronary heart disease	304	654
Other heart disease	30	64
Aortic aneurysm	8	30
Cerebral vascular lesion (stroke)	44	84
Other circulatory disease	15	22
Other Diseases (total)	73	127
Emphysema	1	15
Gastric ulcer	3	5
Cirrhosis of liver	9	17
Other specified disease	59	86
Ill-defined diseases	1	4
Accidents, violence, suicide	58	66
No Death Certificates Received	34	77
Grand Total	662	1,385

The computer was able to match 73,950 men (36,975 smokers matched to 36,975 nonsmokers). Table 28–3 shows the number of deaths from various causes among the matched pairs.

Hammond summarizes:

> Since the two men in each matched pair were selected for their similarity in all characteristics mentioned, they were almost certainly similar in many other respects. However, they differed greatly in smoking habits: One never smoked regularly and the other smoked 20 or more cigarettes a day. During the same period, far more deaths occurred among cigarette smokers than among the nonsmokers. It is hard to escape the conclusion that this difference in number of deaths was due to the difference in smoking habits.

The most startling result of the matched pair analysis was the difference in deaths from coronary heart disease—it claimed over twice as many cigarette smokers as nonsmokers. Coronary heart disease killed approximately 345,000 men in 1967. If the matched pair data is applied to this figure, and assuming that half the male population smokes cigarettes (60 percent is actually closer to the official estimate), we can conclude that roughly 230,000 of these were cigarette smokers and 115,000 nonsmokers. This means that coronary heart disease killed 115,000 more men than would be expected on the basis of the death rate among nonsmokers—in other words, it is probable that 115,000 cigarette smokers died of coronary heart disease who would not have died except for the cigarette habit, and this still does not include a formidable and growing number of women. In terms of gross numbers, this is a far more serious consequence of smoking than is lung cancer.

If we were to make similar calculations for the other diseases which claim excessive numbers of lives of cigarette smokers, we could conclude that at least 200,000 men are prematurely dying each year because of cigarette smoking. Certainly, the number of deaths from motor vehicle accidents (approximately 54,000 in 1968) pales in comparison. *In the United States no every day hazard to health and well-being approaches the seriousness of cigarette smoking—this includes overeating, drinking, narcotics, accidents of all sorts, or contagious disease.*

Hammond's classic and exhaustive studies do not stand alone. The studies are numerous and the results are consistent. Of the more than forty studies reported to date, *all* have shown that the lung cancer death rate for cigarette smokers is greater than for nonsmokers.

The Scientific Basis

Statistics can not stand alone to prove a medical cause and effect relationship. Reasonable and convincing mechanisms must be demonstrated. In the present case, these are abundant and conclusive.

As it enters the mouth, cigarette smoke is an aerosol with billions of suspended particles. This particulate matter can be collected when cooled in a vacuum to very low temperatures and the resulting material is known as "tobacco tar" (Fig. 28–2). The tar contains, besides the particulate matter, a number of gases which are condensed at these low temperatures.

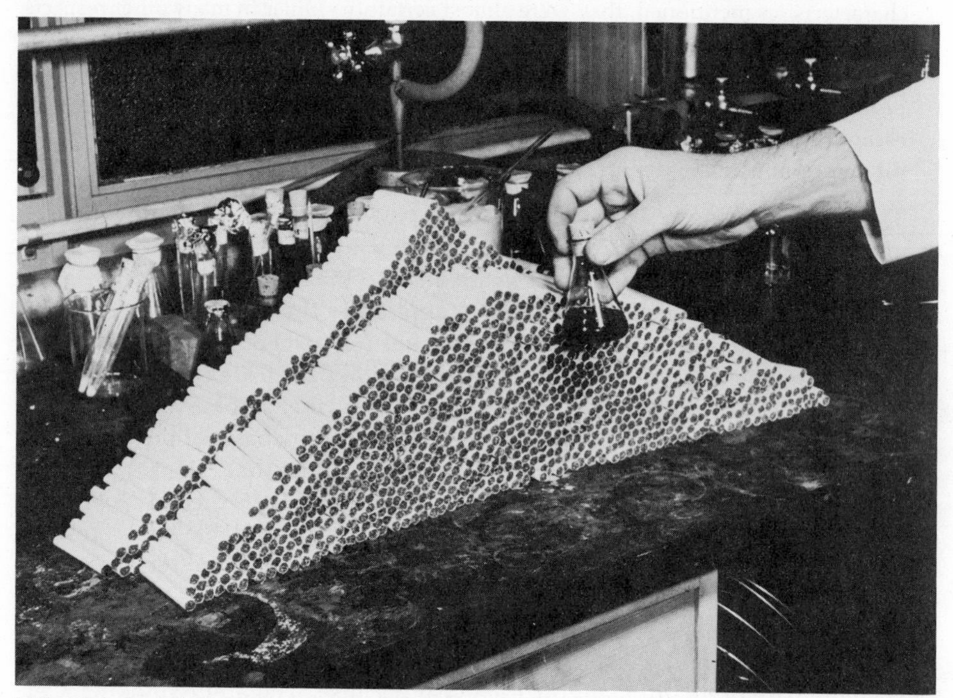

FIG. 28–2 Two thousand cigarettes, heaped on the table produce the amount of tobacco tar shown in the flask. (American Cancer Society)

The analysis of this tar has yielded several hundred different compounds [Surgeon General's Report] and includes numerous poisons such as nicotine and carbon monoxide; a number of chemicals which are exceedingly irritating to human tissues (including phenol, a component of embalming fluid; acrolein and formaldehyde, also used to preserve dead tissue; and many others); and a number of carcinogenic (cancer-producing) compounds. True, these various agents are present only in minute quantities, and similar minute quantities of many other poisons commonly swallowed, breathed, or otherwise absorbed, are harmless, or even beneficial in low concentrations. The difference here is the frequent and persistent administration of these toxins. The cumulative amounts, taken in weeks or mere days, is not minute, but considerable (Fig. 28–3).

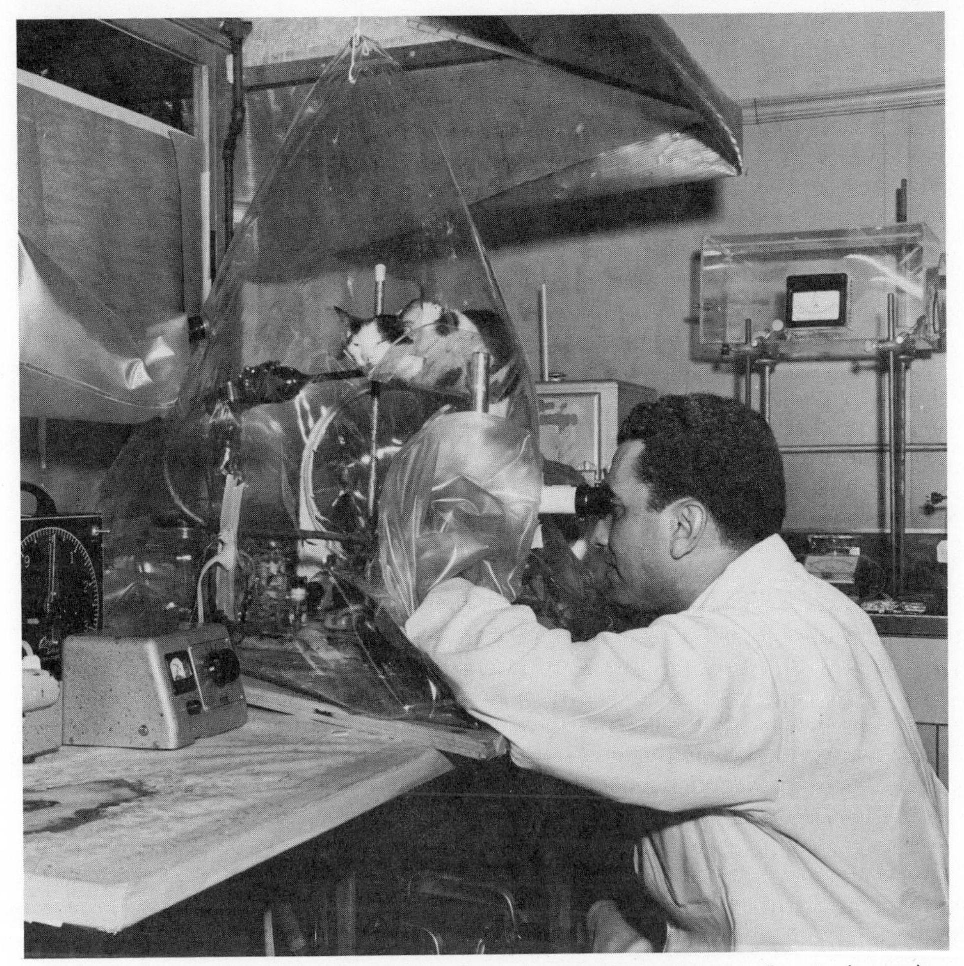

FIG. 28–3 An anesthetized cat is used in studies of the cilia and mucus flow in the trachea to learn how tobacco smoke changes action of mucus. (*Today's Health*, published by the American Medical Association)

A number of investigators have been able to produce skin cancer in mice and rabbits by applying the tobacco tar to the skin repeatedly over a long period of time. However, no laboratory has, as yet, been able to induce lung cancer in a mouse or rabbit by inhalation of cigarette smoke. Much has been made of this fact by spokesmen of the tobacco companies. The difficulty is presumably due to the fact that mice and other small rodents breathe entirely through their noses, and the nose in these animals is an exceedingly efficient filter. Hammond [1962] exposed mice to cigarette smoke under conditions in which they were forced to breathe smoke of approximately the same concentration as that taken in by human smokers. Many of the animals immediately went into convulsions and died, and the remainder lived only a short time. By reducing the concentration of smoke the animals could be kept alive, but Hammond doubted that their lungs were any more heavily exposed to the particulate matter of cigarette smoke than are the lungs of a nonsmoker sitting in a small room with several heavy smokers.

More recently, Dr. Oscar Auerbach [1967; see also Hammond and Auerbach], a Veterans Administration pathologist, induced cigarette smoking in ten beagles by means of tubes connected directly to the trachea. The dogs were broken in gently on filter cigarettes, and over a period of seven months worked up to as many as twelve regular-length nonfiltered cigarettes per day. The twelve-cigarette dosage was considered the equivalent of heavy smoking in humans.

The experiment was terminated after 14 months. Five dogs died during the experiment, and the surviving five were sacrificed for autopsies. In two dogs which died spontaneously, the lungs were damaged so severely that the physicians were unable to assess the nature of the disease process. In the others, the effects were very similar to severe emphysema in man. It was not expected that cancer would develop in the short duration of the experiment.

Studies of Human Tissues

The first tissues exposed to tobacco smoke are the lips, mouth, and tongue. And a number of studies have demonstrated a strong association between smoking and epidermoid cancer of these areas even though their covering is many-layered and relatively impermeable. Significantly, this is the same type of cancer produced when tobacco tar is applied to the skin of mice.

However, most cancers associated with cigarette smoking develop in the lining or epithelium of the respiratory passages—the larynx, trachea, and bronchial tree. This lining is composed of a single layer of cells which rest upon a thin mat of fibers called the basement membrane (Fig. 28–4). Most of these cells resemble tall columns (columnar cells) but squeezed in among them are small "basal" cells and "goblet" cells (resembling wine goblets) which secrete mucus. Projecting from the free surfaces of the columnar cells are the hairlike cilia which move the dustladen mucus up and out of the respiratory passages. Tobacco smoke seems to have its most damaging effects on these lining cells.

Dr. Auerbach [1963] and a group of co-workers have conducted a remarkable study in which they have examined tissue from the bronchial tubes of more than 1000 individuals. They were able to correlate three major changes in the lining cells with cigarette smoking: hyperplasia (an increase in the number of layers of cells); loss of ciliated columnar cells; and changes in the nuclei of cells. Hyperplasia usually occurs in any surface epithelium which is exposed

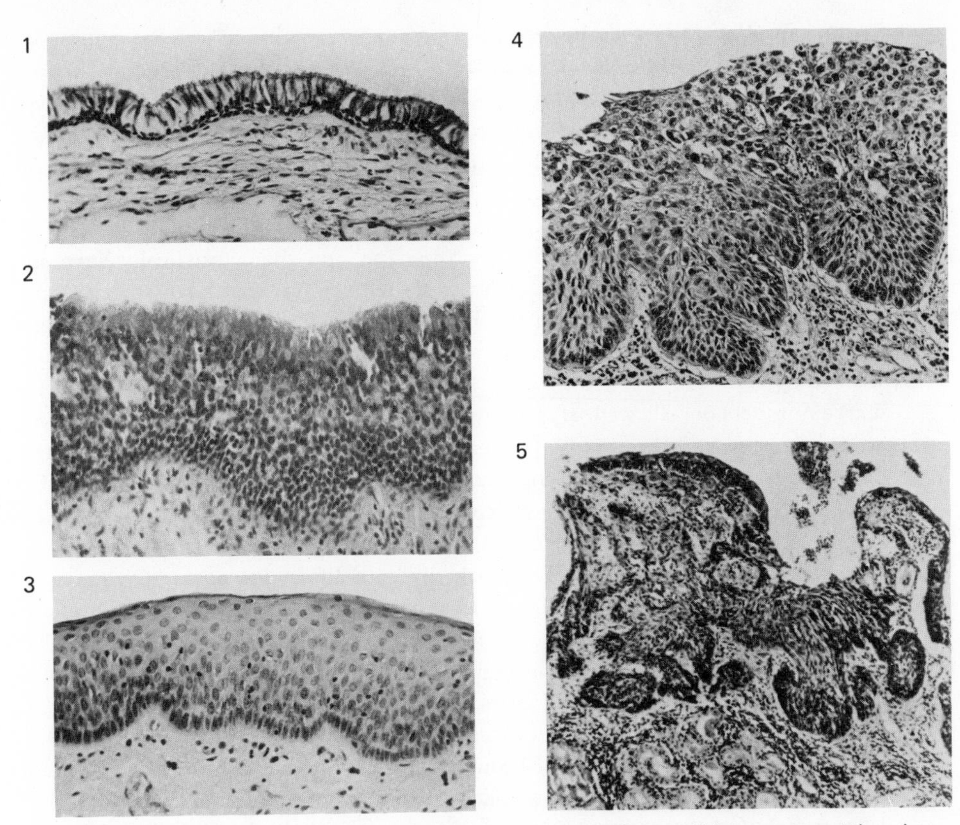

FIG. 28–4 Bronchial epithelium is the developing site of almost all lung cancer. The photographs (1 through 5) show the changes in the epithelium which are believed to lead to cancer. One of the first effects of smoking on normal epithelium (1) is hyperplasia (2), an increase in the number of cell layers. Later the typical epithelium is replaced by numerous layers of flattened or squamous cells (3) so that the bronchial lining takes on the appearance of esophageal epithelium. When the cells develop atypical nuclei and become disorganized (4), the result is called carcinoma. When the cancerous cells break through the basement membrane and penetrate into the underlying tissues (5), the cancer may spread through the lung and to the rest of the body. (Dr. Oscar Auerbach)

to chronic irritation, and cigarette smoke is undoubtedly an irritant. The picture cannot be better expressed than in the words of one of the investigators.[1]

> We found some degree of hyperplasia in 10 to 18 percent of slides from nonsmokers, in more than 80 percent of slides from light cigarette smokers, and in more than 95 percent of slides from heavy cigarette smokers. Extensive hyperplasia (defined

[1] From E. Cuyler Hammond, *The Effects of Smoking*. Copyright July 1962 by *Scientific American*, Inc. All rights reserved.

as five or more layers of cells between the basement membrane and the columnar cells) was frequently found in heavy cigarette smokers but rarely in other subjects.

Loss of ciliated columnar cells was observed in non-smokers but far more frequently in cigarette smokers, and the frequency of this observation increased with the amount of cigarette smoking. The implication is that foreign material tends not to be removed, and thus can accumulate where the cilia have been destroyed.

An important finding was the occurrence of cells with atypical nuclei. The nuclei of cancer cells are usually large, irregular in shape, and characteristically have an abnormal number of chromosomes. A few cells with nuclei that have such an appearance are occasionally found in the bronchial epithelium of men and women who have never smoked. In non-smokers the frequency of such cells does not increase with age.

Large numbers of cells with atypical nuclei of this kind were found in slides from cigarette smokers, and the number increased greatly with the amount of smoking. In heavy cigarette smokers we found many lesions composed entirely of cells with atypical nuclei and lacking cilia. Fewer such lesions were found in light smokers and none were found in non-smokers. Among heavy cigarette smokers the number of cells with atypical nuclei increased markedly with advancing age.

Somewhat more changes were found in slides from ex-cigarette smokers than in slides from men who never smoked. . . . The study indicated that the number of cells with atypical nuclei declines when a cigarette smoker gives up the habit. This probably occurs slowly over a period of years.

In order to determine the significance of these changes we studied the bronchial epithelium of men who had died of bronchogenic carcinoma. Carcinoma is defined as a tumor, composed of cells with atypical nuclei, that originated in the epithelium and has penetrated the basement membrane and "invaded" the underlying tissue. Once such an invasion has occurred, the tumor grows—often to considerable size—and spreads to many parts of the body. In men who had died of lung cancer we found large numbers of cells with atypical nuclei, as well as many lesions composed entirely of such cells, scattered throughout the epithelium of the bronchial tubes of both lungs. In a few instances we found tiny independent carcinomas in which the tumor cells had broken through the basement membrane at just one small spot. These carcinomas looked exactly like many of the other lesions composed entirely of cells with atypical nuclei, except that in the other lesions we did not find any cells that had broken through the basement membrane. We are of the opinion that many, if not all, of the lesions composed entirely of atypical cells represent an early, preinvasive stage of carcinoma. This is a well-known occurrence in the cervix of the uteri of women and is called carcinoma *in situ*.

Judging from experimental evidence as well as from our findings in human beings, we are of the opinion that carcinoma of bronchial epithelium originates with a change in the nuclei of a few cells; that by cell division the number of such cells gradually increases; that finally lesions composed entirely of atypical cells are formed and that occasionally cells in such a lesion penetrate the basement membrane, producing the disease known as carcinoma. Apparently the process is reversible up to the time the cells with atypical nuclei break through the basement membrane.

Hammond is of the opinion that tobacco smoke produces a change in the local environment of the bronchial epithelium that favors the survival and reproduction of certain mutant cells with atypical nuclei which in turn become an invasive cancer. The data seem to suggest that far more people would be expected to die of lung cancer than really do, and the most probable explanation is that coronary heart disease and other circulatory ailments simply kill these people before they develop lung cancer.

Curiously, a high rate of cancer of the bladder has been associated with cigarette smoking [Cole et al.]. This association has not been explained, but the authors expect the incidence of bladder cancer to increase, particularly in women who smoke.

Other Lung and Bronchial Diseases Related to Smoking

Hyperplasia and the loss of cilia increases the tendency of the small bronchioles to become plugged with mucus. Since cigarette smoking also seems to increase the activity of the mucus-secreting glands (goblet cells), lung congestion would tend to become a problem, and the growth of infective organisms would be favored. The higher death rates among cigarette smokers from infectious disease can certainly be explained on this basis.

Emphysema seems to be related to still other pathological changes. In the studies by Auerbach [1963], lungs were obtained at autopsy from 1340 patients. Nonsmokers were matched with various categories of smokers by age, race, and occupation, and then placed in random order for microscopic examination. (This insured that the pathologists studying the slides were unaware as to identity or smoking habits of the patient.) The incidence of certain pathological changes—fibrotic thickening of the alveolar walls; rupture of alveoli; thickening of the walls of arteries and arterioles; and the presence of padlike attachments to alveolar walls—increased according to smoking habits, and was significantly higher than in nonsmokers even in those who smoked less than one pack per day (Fig. 28–5).

The rupture of the alveoli is probably related both to the thickening (fibrosis) of the alveolar wall and to the hyperplasia which tends to trap mucus. Air is trapped in the alveoli, and when the person attempts to cough the pressure of the trapped air is increased to such a degree that the thickened and consequently less distensible walls rupture. At the same time the small arterioles and venules which transport the blood to and from the air sacs are destroyed.

The rupture of the alveoli leads to various degrees of emphysema. If the condition is severe, the loss of blood channels can put increased pressure on the right ventricle, which must force the same amount of blood through fewer

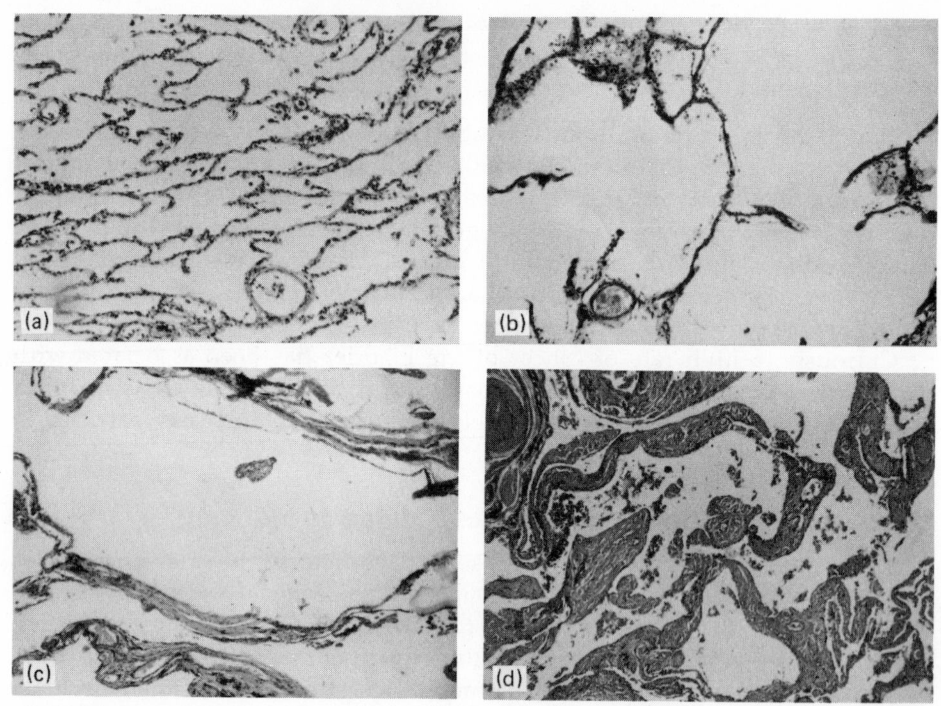

FIG. 28–5 Changes in the air sac walls associated with cigarette smoking. (a) Normal lung tissue. (b) Rupture of air sacs (emphysema). (c) Large dead air spaces (emphysema) and the disappearance of normal air sacs. (d) Formation of padlike thickenings in the walls of the air sacs. (Dr. Oscar Auerbach)

vessels. The thickening of the alveolar walls also decreases the rate at which gases can diffuse between the blood and inspired air—all of which leads to a decrease in respiratory reserve.

Of greatest interest, perhaps, is the finding that many of these pathological changes tend to disappear when a person gives up smoking. Althought ruptured alveoli can never be replaced, the thickening process can apparently be reversed and the progressive tendency of emphysema possibly averted.

Chronic Bronchitis

The American Thoracic Society defines chronic bronchitis as "a clinical disorder characterized by excessive mucus secretion in the bronchial tree. It is manifested by chronic or recurrent productive cough." The causal relationship with cigarette smoking is almost beyond contention [Anderson]. The hyperplastic changes in the bronchial lining and the destruction of cilia, accompanied by excessive secretion of mucus, would undoubtedly promote a cough—about the only remaining mechanism available to the person to clear his lungs. The symptoms of chronic bronchitis tend to come and go, but each recurrence usually persists a little longer. The infections which often accompany bronchitis can generally be treated with antibiotics, and this is the only treatment usually required. However, a smoker with bronchitis, particularly a cigarette smoker, can expect little benefit from antibiotics beyond a short respite (Fig. 28–6).

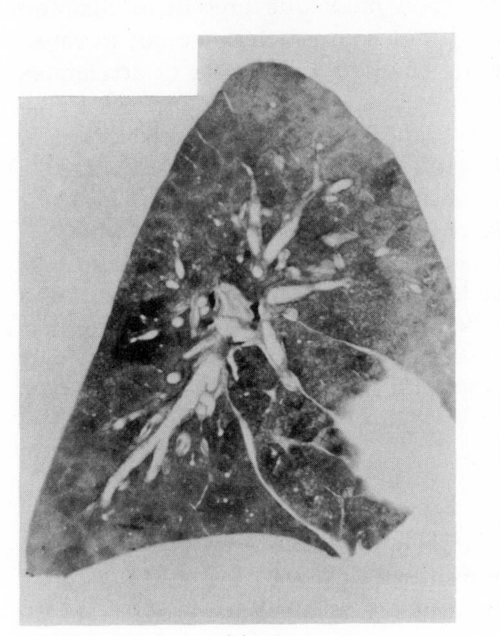

(a)

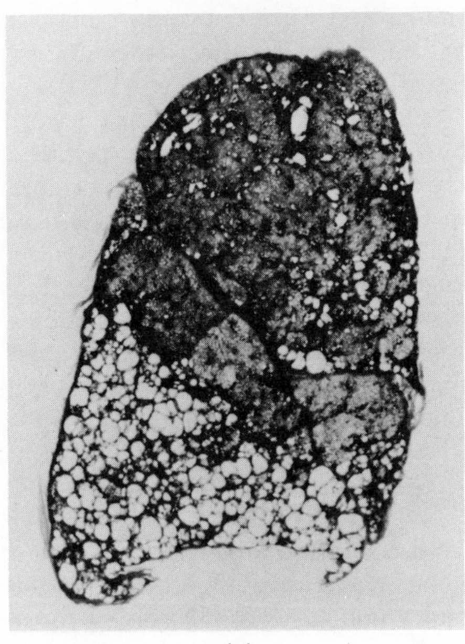

(b)

FIG. 28–6 Slices through the whole lung of (a) a nonsmoker and (b) after many years of heavy cigarette smoking. The loss of functional tissue is evident. The large cavities in the lower part of the smoker's lung is due to emphysema and bronchiectasis. The black areas are accumulated carbon particles. (Original tissues provided by Dr. Oscar Auerbach)

Effects on Circulation

The cause-and-effect relationship between inhalation of cigarette smoke and respiratory damage would seem to be a reasonable one. However, the high incidence of circulatory disorders and the high arteriosclerotic heart disease death rate in males and females among cigarette smokers is not so easily explained [Blackburn; Seltzer]. We can assume that gas exchange is not so efficient, and this would mean that the heart is required to pump a greater amount of blood through the lungs for a given saturation level of hemoglobin with oxygen. It would also mean a faster breathing rate, and this is easily enough observed among heavy cigarette smokers.

All this would cause additional strain on the heart and blood vessels, but would not explain the high rate of atherosclerosis, particularly atherosclerotic heart disease, among cigarette smokers. Auerbach demonstrated that atherosclerotic thickening of arteries of the lungs was much more common among cigarette smokers than among nonsmokers. That atherosclerosis is promoted by cigarette smoke seems indicated, but why can only be speculative. Auerbach [1962 studies] suggests that the carbon monoxide present in cigarette smoke and carried by the blood, though not concentrated enough to cause anoxia, might be the element in cigarette smoking that causes or accelerates the sclerotic process.

One particular circulatory disorder worthy of special discussion here is Buerger's disease or *thromboangiitis obliterans* [McKusick et al.]. In this disease the relationship with cigarette smoking is certain. Nicotine apparently causes, in these people, excessive constriction of the arterioles of the extremities so that circulation is seriously impaired. Gangrene, amputation, and ultimately death result. Repeatedly, it has been demonstrated that progression of the disease is halted when the patient stops smoking. It is indeed amazing that some patients who have already lost toes or feet, and who have been warned that continued smoking may result in loss of their legs, nevertheless persist in smoking.

Conclusions

Most cigarette smokers do not need medical evidence to tell them that the habit is harmful. However, they do tend to minimize the effects, and hold the rather fatalistic belief that the gravest consequences will befall someone else.

The most reasonable and intelligent view based on presently available evidence is that cigarette smoking is harmful to virtually *all* persons who indulge in it; that the effects are cumulative—the longer and more intense the habit, the graver the effect; that many of the pathological changes are reversible when the habit is broken. The evidence also suggest that mere cutting down is of little benefit. Abnormalities such as hyperplasia and loss of cilia seem to persist if there is any exposure to cigarette smoke at all.

This should be sufficient reason to persuade anyone to quit smoking; and probably most people who do smoke have conceded, to themselves at least, the desirability of quitting. In fact, the great majority of cigarette smokers have tried at one time or another to quit, with varying degrees of success. Although there is no true withdrawal sickness the person suffers acute "nicotine hunger" and excessive nervousness; but these symptoms rarely persist more than a few days. Still, success requires considerable determination and self-restraint. Probably most effective is the "one-day-at-a-time" method. Resolve to have no cigarette today. Let tomorrow take care of itself, but none today. This has been the approach to drinking used so effectively by Alcoholics Anonymous, and works in dieting and smoking as well (no desserts today—maybe tomorrow, but not today).

One of the difficulties commonly encountered in persons who have given up smoking is the tendency to put on excess weight. Food tastes better (the taste buds are no longer scorched), the olfactory sense (smell) is more acute, and there is some evidence that digestion is more efficient. So a resolve to control diet is sometimes necessary also—perhaps later, if the double effort is too much. But some excess weight, temporary or even permanent, is certainly less harmful to health than cigarette smoking, and this may be some consolation. Above all, anyone who does not smoke should weigh carefully the consequences of smoking against its scant pleasure. It is quite easy to decide never to take up the habit, but difficult to quit once the habit is begun.

REVIEW QUESTIONS

1. Although tobacco has been used in various forms for centuries, why has lung cancer become such a major problem only since about 1930?
2. Although lung cancer is a major cause of death, heart disease is probably the cause of a greater number of deaths among cigarette smokers. Explain.
3. Could you give an explanation for the fact that practically all causes of death show a higher rate among cigarette smokers?
4. What are some of the pathological changes that occur in the bronchi and air sacs in response to cigarette smoke?

REFERENCES

ANDERSON, D. O. et al., "Role of tobacco smoking in the causation of chronic respiratory disease," *New England Journal of Medicine,* **267:** 787–794 (1962).

AUERBACH, OSCAR, "The pathology of carcinoma of the bronchus," *New York State Journal of Medicine,* **49:** 900–907 (1949).

AUERBACH, OSCAR et al., "Changes in the bronchial epithelium in relation to smoking and cancer of the lung," *New England Journal of Medicine,* **256:** 97–104 (1957).

AUERBACH, OSCAR et al., "Bronchial epithelium in former smokers," *New England Journal of Medicine,* **267:** 119–125 (1962).

AUERBACH, OSCAR et al., "Changes in bronchial epithelium in relation to sex, age, residence, smoking and pneumonia," *New England Journal of Medicine,* **267:** 111–119 (1962).

AUERBACH, OSCAR et al., "Smoking habits and age in relation to pulmonary changes: Rupture of the alveolar septums, fibrosis and thickening of walls of small arteries and arterioles," *New England Journal of Medicine,* **269:** 1045–1053 (1963).

AUERBACH, OSCAR et al., "Emphysema produced in dogs by cigarette smoking," *Journal of the American Medical Association,* **199:** 241–246 (1967).

BLACKBURN, H. et al., "Common circulation measurements in smokers and nonsmokers," *Circulation,* **22:** 1112–1124 (1960).

COLE, PHILIP et al., "Smoking and cancer of the lower urinary tract," *New England Journal of Medicine,* **284:** 129–133 (1971).

DOLL, RICHARD, and A. BRADFORD HILL, "Lung cancer and other causes of death in relation to smoking," *British Medical Journal,* **2:** 1071–1081 (1956).

HAMMOND, E. CUYLER, and DANIEL HORN, "Smoking and death rates: Report on 44 months follow-up on 187,783 men," *Journal of the American Medical Association,* **166:** 1159–1172, 1294–1308 (1958).

HAMMOND, E. CUYLER, "The effects of smoking," *Scientific American,* **207:** 39–51 (July 1962).

HAMMOND, E. CUYLER, "Smoking in relation to the death rates of one million men and women," *National Cancer Institute,* Monograph 19, 1966.

HAMMOND, E. CUYLER, and OSCAR AUERBACH, "Effects of cigarette smoking on dogs," *Ca—A Cancer Journal for Clinicians,* **21:** 78–94 (March-April 1971).

MCKUSICK, V. A. et al., "Buerger's disease: A distinct clinical and pathological entity," *Journal of the American Medical Association,* **181:** 5–12 (1962).

OCHSNER, ALTON, and MICHAEL DEBAKEY, "Primary pulmonary malignancy. Treatment by total pneumonectomy; analyses of 79 collected cases and presentation of seven personal cases," *Surgery, Gynecology, and Obstetrics,* **68:** 435–451 (1939).

REDMOND, DONALD E., JR., "Tobacco and cancer: The first clinical report, 1761," *New England Journal of Medicine,* **282:** 18–23 (1970).

SELTZER, C. C., "An evaluation of the effect of smoking on coronary heart disease," *Journal of the American Medical Association,* **203:** 193–200 (1968).

U. S. Department of Health, Education, and Welfare, *Smoking and Health, Report of the Advisory Committee to the Surgeon General of the Public Health Service,* 1964.

Additional Readings

BLAKESLEE, ALTON L., *It's Not Too Late to Stop Smoking,* Public Affairs Pamphlet, 1966.

BREAN, HERBERT, *How to Stop Smoking.* New York: Pocket Books, 1963.

Consumer's Union Report on Smoking, Consumer's Union, 1963.

CORTI, E., *A History of Smoking.* London: George G. Harrap, 1931.

DIEHL, HAROLD S., *Tobacco and Your Health: The Smoking Controversy.* New York: McGraw-Hill, 1969.

HORN, DANIEL et al., "Cigarette smoking among high school students," *American Journal of Public Health,* **49:** 1497–1511 (1959).

NEUBERGER, MAURINE B., *Smoke Screen—Tobacco and the Public Welfare.* Englewood Cliffs, N. J.: Prentice-Hall, 1963.

SALBER, E. J. et al., "Smoking habits of high school students in Newton, Mass.," *New England Journal of Medicine,* **265:** 969–974 (1961).

TERRY, LUTHER L., and DANIEL HORN, *To Smoke or Not to Smoke.* New York: Lothrop, Lee and Shepard, 1969.

WYNDER, E. L., and F. R. LEMON, "Cancer, coronary artery disease and smoking. A preliminary report on differences in incidence between Seventh-Day Adventists and others," *California Medicine,* **89:** 267–272 (1958).

PART SEVEN
MUSCLES, BONES, and MOVEMENT

29
THE LOCOMOTOR SYSTEM

Have you ever been plagued with sore muscles and wished ardently that you could exercise painlessly or that you could find some quick-acting analgesic balm to put an end to your misery? Are you numbered among the masses who seemingly worship physical prowess and strength? Or, less dramatically, is it true that muscle turns to fat as you age or become less active? Is strength associated with better health? Are the little occasional uncontrollable muscle spasms of the eyelid (or almost anyplace else) cause for concern? What is the best treatment for a sprain or strained muscle? These and many other questions have more than likely crossed your mind at some time, or if not have already seriously concerned you. You know muscles allow us to move, but there are also less obvious functions of muscles: communication (words, facial expression, gestures, writing); breathing; and even eating and defecation. In short, though in our culture we seem really to know little about our muscles and what we do know is often vague and misinformed, we cannot do without them for it is in fact the working tissue of the body. It is the *only means by which we can express our thoughts and wishes, and satisfy our needs, in the world around us.* As such, muscle is vitally important, and its health should be a major concern.

Muscles and Their Accessories

The Structure of Muscles

Human muscle is of three kinds. What is called *smooth,* involuntary, or unstriated muscle occurs in many internal organs, in bloodvessels, in the iris of the eye, and in other places. *Cardiac* or heart muscle is found only in the heart, but these two types are not usually thought of as muscles, and they are discussed elsewhere. *Voluntary* (striated or skeletal) muscle tissue forms the muscles as we generally think of them. It alone will be discussed in this chapter.

Voluntary is the best name for this type of muscle. Most voluntary muscles are attached to the skeleton, however some are not, so that the word skeletal is inaccurate. The fibers of voluntary muscle are striated or striped crosswise, but those of cardiac muscle are also striated, hence making the term striated muscle ambiguous. Since voluntary (Latin *volens*: willing) muscle *is* normally under control of our will—whereas both the other types are not—this name is both descriptive and accurate.

A muscle is made up chiefly of cells very unlike those of any other tissue. These muscle cells are long threads, tapering at the ends called muscle fibers. Those of heavy muscles that do hard work, as in the hip and shoulder, are comparatively coarse, up to several inches long and a hundredth of an inch thick, and could be seen by the naked eye, if separated from their fellows like strands of silk; whereas those of small muscles that make precise, skillful movements, as in the hand or around the eyeball, would be visible only by microscope. Thousands of muscle fibers are needed to form most muscles, and hundreds for even very small ones.

Each muscle fiber, as noted above, is striped crosswise with alternate dark and light bands, the striations. These striations are rather complex when studied closely, and are related to the subtle mechanism by which a fiber contracts (Fig. 29–1). Each fiber can contract by about a third of its resting length. Of course, as a fiber contracts it also bulges. The contractile pull of myriads of delicate fibers makes up the powerful pull of each muscle, and their bulging adds up to cause the visible bulging of the muscle.

Muscle fibers are held together, and attached to bones and other structures, by a harness of *connective-tissue threads* or *fibers.* These are made of a substance called collagen produced by living cells called fibroblasts; but, unlike the muscle fibers, the collagen fibers are not themselves living. Collagen, though flexible, is extremely strong, stronger than steel wire for its size. Very fine collagen threads form a web around each muscle fiber, continue beyond its ends, and join the harness of following muscle fibers. (Muscle fibers in most muscles

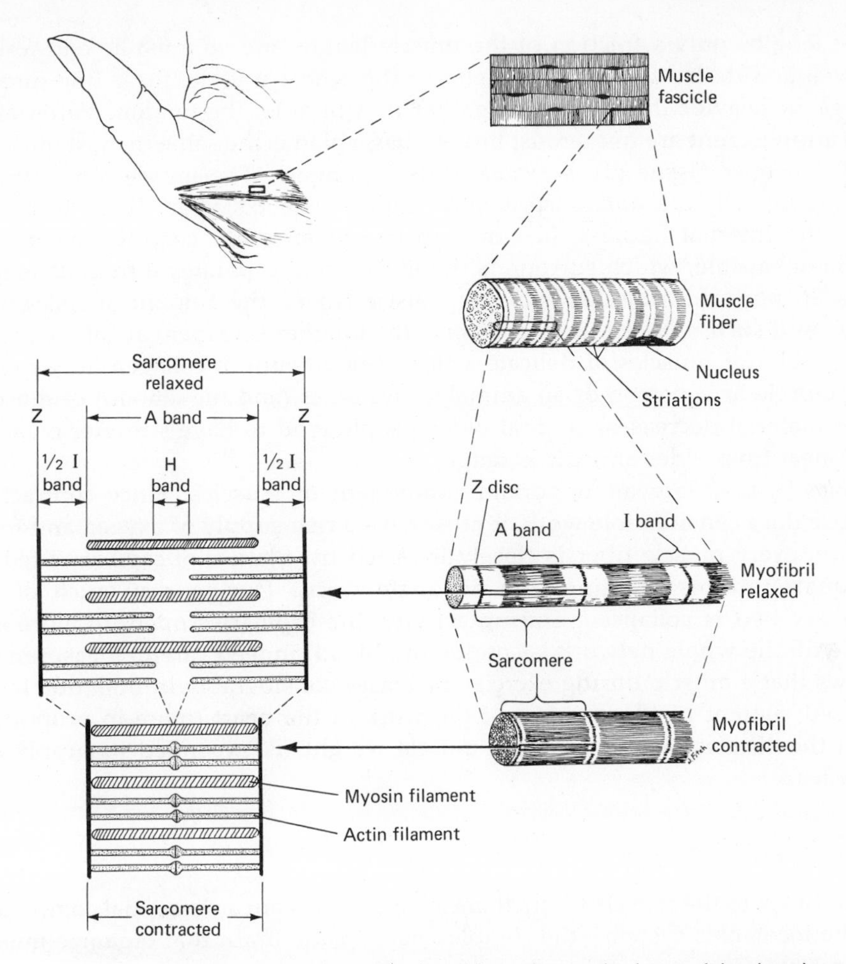

FIG. 29–1 Voluntary muscle. A muscle such as that at the base of the thumb can be dissected into *fascicles*. When a fascicle is examined under a microscope it is seen to be composed of specialized cells called *muscle fibers*. A muscle fiber shows cross stripes or *striations,* and with further magnification (under the electron microscope) it can be seen that the striations are formed by the exact alignment of hundreds of *myofibrils*. The functional unit of a myofibril is called a *sarcomere,* and is bounded by adjacent Z discs. The striations are present in each myofibril and reflect the molecular arrangement of *myofilaments*. These myofilaments are of two kinds: the thicker *myosin filaments* and the more delicate *actin filaments*. The myosin and actin filaments interdigitate to form the two broad dark parts of the A band. The myosin filaments are absent in the I band, and the actin filaments are absent in the H band. It has been found that muscle contraction involves a sliding and further interdigitation of the myosin and actin filaments so that in contracted muscle the I bands become much narrower and the sarcomere shortens.

have lengths only a fraction of the muscle length and so must lie end-to-end as well as side-by-side.) At each end of the whole muscle, these fine threads merge in heavier threads which gather in a bundle, the tendon. Varieties of this arrangement are numerous; but the basic plan is the same in most muscles.

Connective tissue gives a muscle its toughness. The muscle fibers themselves are jellylike and surprisingly fragile. The collagen fibers form not only the internal harness, just described, but an outer coating, the muscle fascia or capsule, which surrounds the muscle and separates it from its neighbors. In powerful muscles with big muscle fibers, the amount of collagen is large, and such muscles in animals give the tougher, cheaper cuts of meat; and conversely for muscles of delicate action. The amount of collagen increases in any muscle as a person or an animal grows older (and the amount of muscle-fiber material decreases), so that older people tend to have stringier muscles, and meat from older animals is tough.

Blood vessels are an important component of muscles. Since contracting muscle does considerable work, it must have a rich supply of oxygen and food. Hence, every muscle fiber is closely invested by a lattice of capillaries fed by adequate arteries draining into adequate veins. During rest much of the capillary bed is collapsed. During activity, the capillaries open up more and more till the whole network is conducting blood; indeed careful measurement shows that a muscle during exercise increases considerably in bulk due to increased content of blood. Most of the work of the heart (more in proportion than the 45 percent due to gross muscle weight) is expended to supply our muscles.

Muscle and Nerve

In addition to the muscle fibers themselves, nerves are an essential component of the locomotor system—the complex parts of the brain that organize muscle actions are discussed in Chapter 33. Indeed, without nerves, the muscles are as useless as a car without an ignition. The messages planned in and dispatched from the brain reach the muscles only through nerves. For most of the body, these messages run down the spinal cord. Here they are picked up by special large nerve cells, the motor neurons, to be relayed along the outgoing (motor) nerve fibers of those neurons, which run in the spinal nerves. The latter emerge from the cord as a series of 31 on each side, and spread to all parts of the body. Messages to head muscles and to a few others run through certain of the cranial nerves emerging from the brain itself.

When a motor nerve fiber reaches the muscle to which it is bound, it branches. In the large, coarse muscles, a nerve fiber may have 200 or more branches; in a small, delicate muscle, it may have only half a dozen. Each branch runs to one muscle fiber and ends on it in a flattened, oval expansion, the *motor end-plate* (Fig. 29–2). The latter transmits commands from the brain to the muscle fiber.

The vitally important point here is that the motor nerve cells, and their fibers in the nerves, are absolutely the only channel through which the brain can control the muscles. For that reason these nerve cells in the spinal cord and their fibers are sometimes called collectively the *final common path.* If this path to any muscle is cut off, the brain, however normal and healthy, cannot reach the muscle and control it.

For example, in infantile paralysis (poliomyelitis), the virus kills many of the motor cells. The muscle fibers supplied by the nerve fibers from those cells

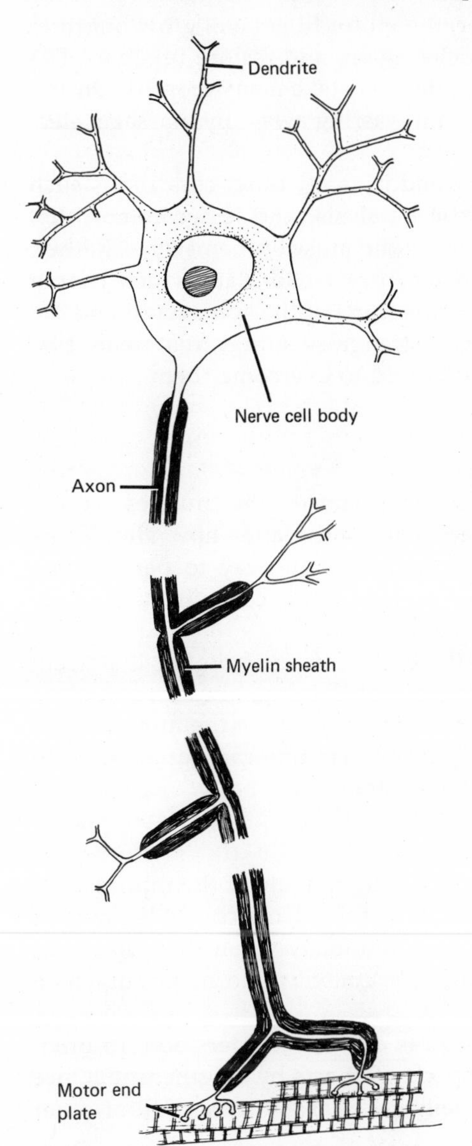

FIG. 29–2 Motor nerves end in a very intimate contact with the muscle fiber, the *motor end plate*. The arrival of a nerve impulse results in the release of a chemical at the nerve ending that triggers the contraction of the muscle.

are left without control. They may, at first, go into aimless twitchings and spasms, but soon become limp and eventually waste away leaving only their collagen harness. Since no power can replace a nerve cell once it is destroyed, this condition is permanent; but nerve cells only stunned by the virus may recover, and their muscle fibers recover with them. Hence proper treatment of polio cases may produce considerable improvement.

Similarly, if a nerve is severed or crushed, all its motor fibers are interrupted and the muscles to which they run are paralyzed. In this case, however, if the cut ends of the nerve are sutured together, the motor fibers will grow out from their cells along the nerve, reach the muscles again, and restore function. The final common path is reestablished, though this may take many months. During that time, the muscle must be prevented from wasting away, by massage, electrical stimulation, or other means.

The nervous control of muscle is disturbed in many other conditions such as multiple sclerosis (MS), forms of spastic paralysis, and less common diseases. Indeed, few "muscular" disorders are in the muscles themselves. Rather, some, such as myasthenia gravis, are in the motor end-plate which cannot transmit messages properly; and some in other parts of the spinal cord and the brain. This explains the many, sometimes strange-seeming, treatments that are given such disorders and the frequent failure to overcome them.

The Muscular System

Though each muscle is enclosed by its collagen sheath, the muscles are not isolated units. Rarely does any muscle act alone, no matter how simple the task. Rather, the muscles work together in an elaborate way to perform the countless actions of which our bodies are capable. Thus collectively they are very properly spoken of as a muscular system.

The muscular system is the largest in the body. It forms roughly 45 percent of our body mass, depending on state of muscular training, obesity, and other factors. The muscles fill out almost all the familiar contours of the human form; in this sense they almost seem to *be* the person. Yet one can get along with greater loss of muscle than almost any other system in the body.

How a Muscle Acts

A knowledge of muscle action is valuable in avoiding many of the injuries discussed in the next chapter—not to mention how trained and controlled muscle action enables us to treat such injuries more rationally when they do occur; how it guides us to more effective exercise and to athletic performance in which a small advantage is often decisive.

Even the simple basis of muscular action is often misunderstood. In brief, a muscle always acts by shortening; no muscle ever acts by lengthening. Once shortened, a muscle is lengthened again only by relaxation, the pull of other muscles, by the pull of gravity, or by other forces outside it. In short, muscles always pull; they never push. Even though a muscle functions only by shortening, we can accomplish certain tasks by the reverse process. For example, when you let yourself down slowly from a "chin-up" position on a horizontal bar, you do so by gradually lengthening or *releasing* the tension of the elbow flexor muscle group. This is sometimes called eccentric contraction or negative muscle work.

Most muscles act by pulling on bones. A few, such as those around the mouth, are circular and act by contracting like a draw-string, as in whistling. However, the most typical muscle has a tendon at each end, attached to a bone. The bone attachment at the end that is most solidly fixed and that serves as the base from which the muscle pulls, is the *origin* of the muscle; the attachment to the bone that moves most freely is the *insertion*. In limbs, where most muscles lie, the origin is almost always closer to the trunk. Thus in the biceps (Fig. 29–3) the origins on the scapula (shoulder blade) are usually less mobile and are certainly closer to the body than the mobile insertion in the forearm. Most muscles can act in reverse of their origin and insertion: If you grip a horizontal bar above your head and draw yourself up, your forearm is relatively motionless and your shoulder is pulled up to meet it. However, these unusual actions need not influence generalizations on origins and insertions.

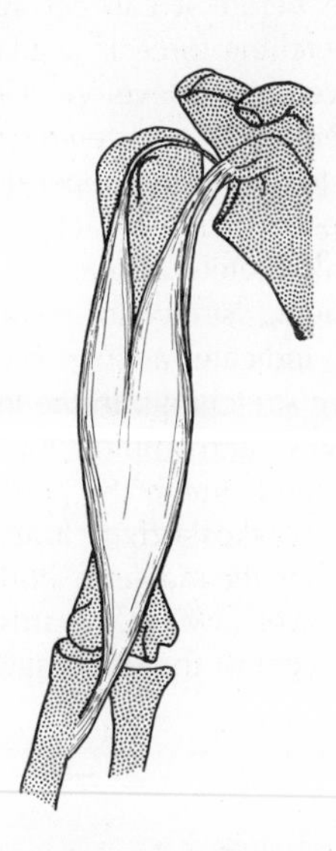

FIG. 29–3 The *biceps brachii* muscle drawn to show the tendinous attachments to bone. In the biceps the origins on the scapula (shoulder blade) are usually less mobile and are closer to the center of the body than the mobile insertion in the forearm.

A layman tends to think in terms of one-muscle-one-action. In this he is often wrong. Straighten (extend) your elbow, palm down, put your other hand on your biceps, and flex your elbow keeping the palm down; you will feel the biceps bulge somewhat, showing that it does assist in the flexion (though in fact, other muscles perform most of this action). Now, keeping the elbow flexed, turn your palm up and you will feel the bulge of the biceps double itself; evidently it is even more active in this movement (supination). In fact, the biceps act weakly on the shoulder joint too.

In general, if a muscle crosses any joint, it can move that joint more or less. Thus the biceps crosses the shoulder joint, the elbow, and the less familiar joint whereby the radius (or the thumb side of the forearm) flips over the ulna (little finger side) to turn the hand palm-down, and back, palm-up; it acts on all of these joints. If a muscle is acting strongly on one joint, it cannot act effectively on others. An example of the rule, well-known to small boys, is that if you flex a person's wrist, you weaken the grip of his fingers: The muscle tendons that flex the fingers also cross and help to flex the wrist, and if they are involved in wrist-flexion, they have less reserve pull for their proper work of finger-flexion. This rule can be applied to many parts of the body by anyone interested in muscle actions (Fig. 29–4).

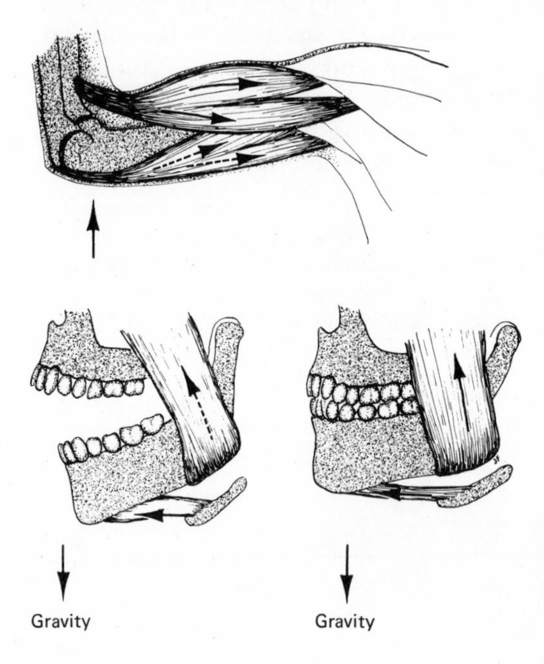

FIG. 29–4 How muscles act. A muscle always acts by shortening and pulling; a "push" muscle is impossible. When a muscle has shortened, it must be stretched by opposing muscles, gravity, or other force before it can act again. If no such stretching force is available, the muscle eventually develops a *contracture*, a permanent shortening with loss of ability to act. In the upper drawing, the biceps and triceps act against each other, alternately shortening actively, and being stretched passively (solid arrows indicate action; broken arrows, passive stretching). In the lower drawing, gravity and minor muscles lower the jaw and stretch the powerful masseter which takes origin from the cheekbone. When the masseter shortens, it draws the lower jaw up towards the cheekone and against the upper jaw.

How Muscles Work Together

The best way to explain muscle team-work is by an example: Say that you simply extend your index finger to point at something (Fig. 29–5). The muscle chiefly responsible for this action has the imposing name *extensor indicis proprius* (particular extender of the index). It can be felt stirring on the back of the forearm as the finger moves. This muscle is the *agonist* (actor) in the movement.

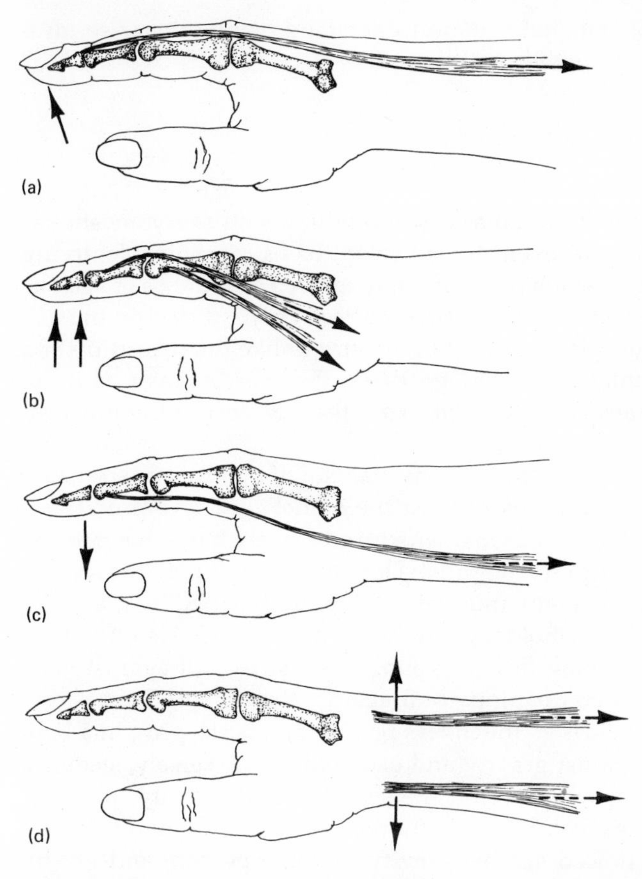

FIG. 29–5 The muscles involved in extending your index finger. (a) The *agonist* or *prime mover* (extensor indicis proprius). (b) The *synergists,* muscles that assist in the action of the agonist. (c) The *antagonist,* the flexor muscle that directly opposes the agonist, (d) The *stabilizers,* muscles that prevent undesired, accompanying movements.

Further, however, small muscles in the hand itself also send tendons to the back of the finger and help to extend it. These can vary the action skillfully so that only the first joint, only the second and third, or all three joints of the finger may be extended. They are *synergists* (with-workers) in the movement.

Next, suppose you are carefully flicking a bit of dust with your fingernail. You now want a very precise action for which you use certain muscles on the front of your forearm. These usually flex the fingers; but now they act rather as brakes, slowing and steadying the extension. Such muscles are *antagonists* (counter-actors) in the movement.

Finally, undersirable, accompanying movements must be prevented. Since the *extensor indicis proprius* crosses the wrist-joint, it tends to bend the wrist back, which would spoil the precision of the finger-flick. To prevent this, other muscles, on both the front and the back of the forearm, act to steady the wrist. These muscles are *stabilizers* in the movement.

Thus the apparently very simple action of extending a finger may, in fact, involve a dozen muscles. And of course, even more muscles are required for more complex actions. This fact helps us to understand many otherwise mysterious things about muscle exercise and disorders.

Special Forms of Muscle Contraction, Normal and Abnormal

Tonus, or tone, is a basic quality of muscle action with which everyone should be familiar. Any muscle at rest, even during sleep, has a recognizable, living firmness, as you can readily test for yourself. Only in deep anesthesia or serious forms of unconsciousness do the muscles become really slack; a person in such conditions show a limp, inert looseness of body very unlike the relaxation of a sleeping person. In motionless but alert postures, the muscles may be quite firm and tense. Such firmness and tension, from the slightest to the greatest, is called tonus.

Tonus is due to a simple but important mechanism. If a muscle is stretched even very slightly, it sends messages back to the spinal cord or brain. These messages pass directly to motor neurons, which in turn discharge back to the same muscle and cause it to tighten slightly. This reaction is called the stretch reflex. Now, in most postures, many muscles tend to be stretched by gravity—those that straighten the knee, hold up the head, and so on. Also, in most postures, even when one is lying down, groups of muscles pull against each other, as the biceps and its companions pull against the triceps. This is one reason why sleeping in a chair is so much less restful than is sleeping in a bed: more muscles are pulling against gravity and each other. Conversely, sleeping seated on a couch or bench, with feet on the seat, knees drawn up, head on knees, and arms around legs, can be perfectly restful if correctly done. In this posture, all the joints are braced against gravity by mere posture and not by muscle pull, and muscles are balanced so as to excite few stretch reflexes and little tonus. But the stretch reflex is active, even if only slightly, in most muscles most of the time. This is why some degree of tonus is normally always present—to maintain the body as a more or less firm, stable object, rather than as a flabby, tottering one, without our having to attend to the matter. Since most action in a human or animal body is really due to a disturbed balance of tonus, then tonus is the basis of action.

The stretch reflex is used by the doctor in certain tests. Everyone is familiar with the knee-jerk test, where a doctor taps the leg with a rubber hammer just below the knee (Fig. 29–6). The tap gently but suddenly stretches the tendon below the knee cap (patella), which in turn stretches the big knee-extending muscles attached to the knee cap, exciting the stretch reflex and causing the leg to kick out. The tendons of other muscles are often tapped in the same way with similar results. The degree and nature of the jerks give important information about conditions in the spinal cord through which the reflex must pass. Exaggerated or depressed reflexes can reveal a great deal to an experienced observer.

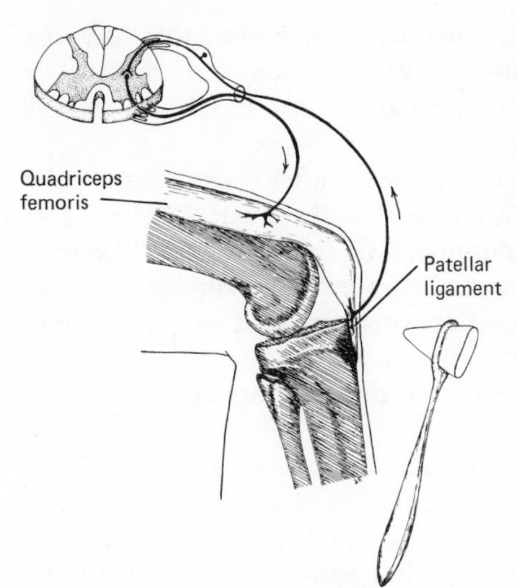

FIG. 29–6 The stretch reflex. A gentle tap below the knee cap stretches the tendon of the large muscle (the quadriceps femoris) that occupies the front of the thigh. Sense organs in the tendon send impulses through sensory nerves to the spinal cord where they make contact with motor neurons. The stimulated motor nerve sends impulses to the quadriceps muscle causing it to contract and the leg to kick out. The pathways mediating such stretch reflexes involve just two neurons.

Clonus is an abnormal disturbance of tonus. Many people have experienced a rhythmic jerking of some joint, such as the knee when the lower leg is hanging free. This is because the antagonistic pulls of the knee flexors and extensors, for example, are unbalanced for some reason, such as excitement or fatigue acting on their nervous controls. Thus, one group produces not an invisible twitch but a visible motion at the joint before the antagonistic group reacts and, in turn, overpulls the joint the other way. The resulting gross, rhythmic movement is clonus. In the great majority of cases, clonus is harmless and soon stops. But if it persists, or occurs repeatedly, it should be reported to your doctor.

Shivering is rather like a fine clonus, but occurs chiefly when one is cold. It too is an exaggeration of the back-and-forth pull of muscle groups in the stretch reflex, but it is triggered by higher controls in the brain—those dealing with heat regulation. Muscle action burns food and in the process liberates heat to counteract chilling; and though the shivering movements may seem trivial, they sometimes involve almost half the body weight, are rapid and continuous, and thus substantially contribute to maintaining body temperature. Shivering can also be triggered by other conditions in the brain, such as anxiety or fear.

Contracture is a permanent uncontrolled shortening of a muscle and is a more serious disorder of muscle action. Some forms are due to chemical changes in muscle that is exhausted, poisoned, or diseased, or has lost its nerve supply; but these are beyond the scope of this book. Contractures may also occur in a normal muscle of which the antagonists are paralyzed. In such a case, though the muscle is no longer stretched by the pull of the antagonists, it may be stretched by gravity, and so on, and will then contract in obedience to the stretch reflex. Now, however, it has no balancing counterpull from the antagonists and so may cramp and shorten permanently. Therefore, not only the paralyzed antagonist but the normal muscle must be massaged, relaxed by heat, and so on. Contractured muscles can be painful and disabling as well as the cause of deformity.

Small muscular *aches* and *pains*, "muscle stiffness," after exercise, are quite different from contractures and will be discussed in the next chapter. Most ordinary muscle *cramps*, such as those experienced in swimming and "charley horses," are probably due to other causes. Cold water, overexercise, reduced blood supply, or prolonged unnatural positions may make a muscle unduly sensitive to messages reaching it along the nerves, or even make it prone to cramp "on its own." Some abnormal cramps are due to more complex, even unknown, causes.

Tetany is a state of maximum contraction in a muscle. It can be perfectly normal in extreme exertion, though it can never be maintained for long naturally. Even weight-lifters, for example, perform their feats by skillfully shifting stress from one group of muscles to another, so that none is in tetany for long. Uncontrolled tetany, due to abnormal causes and persisting or frequently repeating, is very painful, exhausting, and damaging, as severe overstress inevitably would be. Tetanus, for example, is overwhelming tetany caused by a bacterial toxin, and is most often fatal due to sheer exhaustion of the victim. Fatally exhausting tetany can also be caused by other diseases and by poisons such as strychnine. Deadly and agonizing agents, like these, are nothing to gamble with; tetanus should be averted by routine preventive shots and by prompt treatment of contaminated wounds. Strychnine, of course, should simply be avoided.

The Skeletal System and Joints

A brief discussion of bones and joints is appropriate here. They are closely associated with the muscles in body activity. Indeed, muscles, bones, and joints are sometimes grouped as the locomotor system. *Bones*, though familiar to everyone, are usually not well understood. People tend to think of them as inert objects that simply hold the body up mechanically. Actually, a bone is largely living tissue and collagen fibers. If soaked in acid to remove its hard, mineral components, a bone loses only about half its weight and changes very little in appearance. Even the minerals in a bone are continually being withdrawn for use by other parts of the body, and are replaced when convenient. The minerals provide hardness, the fibers provide toughness, and the living elements attend to upkeep and change.

A typical bone has a shaft and two ends. In growing children each end is separated from the shaft by a thin plate of cartilage which is simpler in structure than bone and can grow more easily, thus providing for growth in length; as this plate expands, its surfaces are continually replaced by bone, so that it never becomes very thick. Nevertheless, it does provide an elastic cushion which makes children's bones resistant to falls and blows that would fracture those of an adult. At maturity, the plate stops growing and disappears, and the bone tissues on each side of it meet and fuse. After this, growth in length is no longer possible. In some disorders the cartilage plates either disappear too early or grow too fast and too long. These abnormalities result in certain types of dwarfism or giantism, which are further discussed in Chapter 33.

Some bones differ from this plan more or less. Those of the upper skull are flat or very irregular in shape, and grow in a complicated way without cartilage. Hence injuries to the dome of the skull do not heal as do fractures of the limbs and, if extensive, must be repaired by artificial plates. Bones of the face, the shoulder and pelvic girdles, the vertebrae of the spinal column, the ribs, and others have their own peculiarities; but they heal in the usual way.

Most bones are hollow, enclosing either large cavities or a spongy mass of delicate bone and a network of cavities. This arrangement makes the bones lighter while reducing their strength very little. (A tube with moderately thick walls is almost as strong as a solid rod of the same diameter.) The cavities, however, are not waste space. In most adult bones and in the shafts of children's long bones, they are a storehouse of fat, the yellow marrow; in adult and children's ribs and sternum (breastbone), and in the ends of children's long bones, red marrow manufactures blood cells. Certain blood disorders can be detected by puncturing the sternum and withdrawing a drop of red marrow for study. This use of hollow bones for fat storage and blood manufacture is a curious but economical arrangement.

Joints between bones are of many sorts. In the skull, jigsaw borders unite the bones immovably. In the spinal column and pelvis, connections partly fibrous and partly of rubbery cartilage permit a degree of twisting and bending, but are not freely movable. Movable joints controlled by muscles, in the limbs and elsewhere, are our chief concern here. They are indeed what we usually think of as joints.

In such a joint, typically, the ends of two bones come together. Each end is covered with cartilage like that of the growth-plate, but with a smooth and slippery free surface. The two cartilage surfaces play on each other more easily than would bone. Around the edges of the cartilage surfaces, a sleeve of connective tissue, the *joint capsule,* connects the two bones and encloses the joint. This sleeve is formed of thin, loose tissue where free motion is necessary, as on the front and back of elbow or knee; but it is thickened into heavy bands, the *ligaments,* where strength is needed, as at the sides of elbow and knee. The interior of the capsule is lined with a smooth, delicate *synovial membrane* which secretes a small amount of slippery *synovial fluid* to lubricate the joint surfaces. The whole structure is a beautifully sturdy yet efficient mechanism (Fig. 29–7).

Muscles and ligaments are attached to bones very strongly. The collagen fibers of the tendons and ligaments actually penetrate the bone and continue

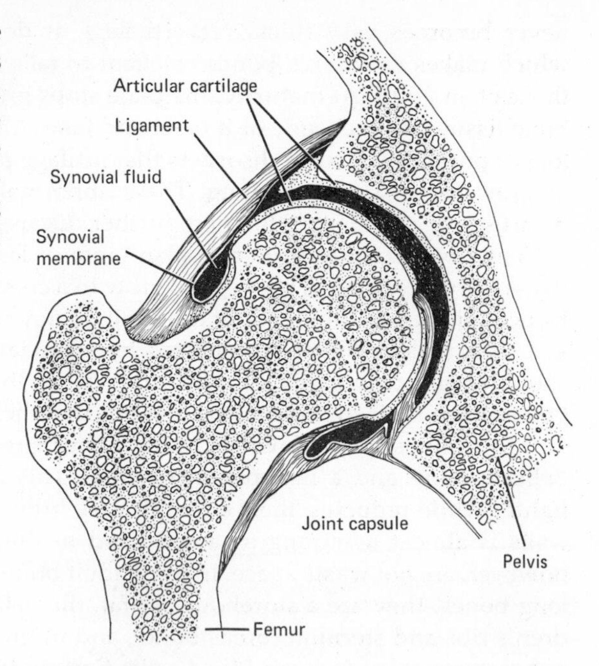

FIG. 29–7 A section through the hip joint. The ends of long bones are filled with a spongy framework of interlacing bone plates. The spaces are filled with red bone marrow, the tissue that manufactures blood cells. The shafts of long bones are hollow, and filled with yellow marrow which is largely fat.

through it as part of its fibrous component. Since these fibers are extremely tough, an injury rarely tears the tendon or ligament from the bone, but more often tears it from the muscle, or even fractures the bone by pull on the tendon. These injuries are forms of *sprain* (discussed in the next chapter). They are sometimes caused by an extreme and sudden muscle action, especially in a person whose muscles are unaccustomed to such action.

Bursae are small pockets of connective tissue with slippery linings (*bursa* is Latin for pocket or purse). They are usually found around joints and lie between structures that should slip easily over each other. In hands and feet, bursae are drawn out into long double tubes surrounding muscle tendons. Bursae are of great value in promoting easy motion, but can give great trouble if they are unduly irritated or if infection gets into them (bursitis).

This much information about the locomotor system serves to explain many conditions that one may have to deal with and should understand. Other conditions involving the muscles, but originating in the central nervous system, will be discussed along with that system in Chapter 33.

REVIEW QUESTIONS

1. The disease poliomyelitis was at one time regarded as a muscle disease rather than a nervous disease. Explain why this confusion might have occurred.

2. Show how several muscles are involved even in simple movements.
3. What is muscle tonus, and why is it important?
4. Explain the terms clonus, contracture, and tetany.
5. Why are most bones hollow?

Additional Readings

BASMAJIAN, J. V., *Muscles Alive: Their Function Revealed by Electromyography,* 2nd ed. Baltimore: Williams & Wilkins, 1967.

GALAMBOS, ROBERT, *Nerves and Muscles.* New York: Doubleday, 1962. (Paperback)

KARPOVICH, PETER V., and WAYNE E. SENNING, *Physiology of Muscular Activity,* 7th ed. Philadelphia: W. B. Saunders, 1971.

30
INJURIES TO MUSCLES, BONES, AND JOINTS

Injuries of the locomotor system are probably among the most common of major human ills. Few people indeed go through life without suffering some form of sprain, fracture, muscle bruise, dislocation, or other such insult to the system; and most of us endure them many times. These disorders are also among the easiest to deal with by common sense; any sensible parent, out-of-doors-man, soldier, or general first-aider, can cope with the ordinary, minor forms. But even with these, and in more serious cases, expert advice should be sought and as soon as available. A review of what these injuries really mean will show why this is so, as well as help to direct intelligent first aid.

Sprains

Sprains will be discussed at length because they are so common and because much of the discussion applies to other injuries. A sprain is an injury to the fibrous tissue around a joint, ligaments, capsule, and even tendons. Often it is associated with a small fracture when a ligament or tendon tears off the bit of bone to which it is attached. And almost always it is accompanied by internal bleeding. These complications explain much of the treatment for sprains; the treatment, of course, depending largely on the severity of the sprain. A simple stretched or slightly torn ligament—a "turned ankle" or such—may require no more than a bit of support and rest for a few days; rupture of a massive ligament with tearing-off of bone and severe hemorrhage into and around the joint may require surgery. The great problem is to know right away whether a sprain is slight or severe; for the treatment in the first ten minutes or so may make a big difference in the rate of recovery. All sprains hurt, and a serious one may hurt less at first than a moderate one. Therefore, one should not take needless chances with any sprain; carelessness may greatly aggravate the damage, and aggravated sprains are often harder to manage than are fractured bones.

Several things can be done for a sprain by anyone; indeed they usually should be done by the patient and his friends before a doctor comes, since their effectiveness depends on promptness.

For one the area should be *chilled.* This cuts down on bleeding and on swelling of the tissues. Fluid oozing from even uninjured blood vessels and from the irritated joint capsule, as well as hemorrhage, can distort, swell, and damage the tissues around them and take a long time to be absorbed. Chilling also cuts down pain. Any competent person can apply packs of crushed ice, or if this is not available, water as cold as possible. There appears to be some advantage to wrapping the joint or area during cooling with an elastic bandage, preferably with a flat piece of spongy material (directly over the affected area and under the wrap) to further inhibit swelling. Care should be taken that the wrap is not so tight as to interfere with circulation.

Immobilization is more tricky and should be carried out by someone with at least basic training. Strapping with adhesive tape by amateurs may look picturesque but gives trouble when it has to be stripped from the injured area; and it interferes with the really useful chilling treatment. Wooden or metal splints require skill to apply. Probably the best amateur splint is a pillow bound to the limb firmly enough to prevent loose movement but not so tightly as to cramp the part; also, this is easily combined with the cold packs. In no case should the splinting be done carelessly, which may cause the very aggravation of bleeding and further injury that the splint is supposed to prevent.

Analgesics (pain-killers) are a problem. Aspirin, for instance, will not kill severe pain, but will take the edge off it. Stronger analgesics, if available, may be valuable if a person in great pain must wait or be transported some distance, but if professional help is near, such drugs may make accurate diagnosis more difficult by dulling pain symptoms. In any case, drugs should be kept to a minimum.

After-care of sprains is a matter for judgment. Cold applications should be continued for 24 hours after which you can begin using heat and gentle massage. Some trainers actually prefer the cold treatment for at least one week in severe sprains. Too much movement too soon delays healing; unlike fractures, discussed below, the soft tissues in a sprain do not splint themselves by anything like a callus, but remain vulnerable so that the injury can be reopened by injudicious activity. This is one reason why sprains are so hard to treat. On the other hand, unnecessary prolonged lack of motion can cause stiffness of the joint, weaknesss of muscles, and other bad effects. The best guide is the advice of an experienced doctor; lacking that, careful trial is necessary.

No sprain of any severity should be neglected. X rays are usually advisable to make sure no fracture is present, since a sprain-fracture may fail to heal due to continual disturbance by joint movement, and prolong the disability indefinitely. What seems to be a sprain may be an outright dislocation or even major fracture. Doctors are constantly amazed by the severity of injuries that people will put up with for a long time before reporting them. Such delay always hampers healing, and may even be very dangerous.

One special warning is needed for athletes. The athlete is understandably eager to get back to his beloved sport, to fill his place on the team, and to show his disdain for petty ills. But he must remember that sprains are treacherous, that a joint from which swelling and pain have vanished is not necessarily healed. The ligaments have been stretched and, being non-living collagen fibers, require a long time to return to normal; a joint with stretched ligaments is very prone to respraining; and repeated sprains often weaken a joint permanently, putting an end to all athletics. Rely on what the doctor and/or an experienced trainer advise and not on how the joint feels to you yourself.

Other Joint Disorders

Dislocations are, in effect, sprains so severe that the bones of a joint are displaced grossly, even out of the joint capsule. Some people, especially slightly built, nimble ones, dislocate easily, but for the same reason can slip some bones back into position with little ill effect. Heavily built people, on the contrary, suffer more from any dislocation. In all cases where a dislocation is accompanied by pain from torn ligaments and other tissues, first-aid treatment should be the same as for a severe sprain: chilling, immobilization, and very cautious use of analgesics. If severe pain is present it is wise to treat for shock (Chapter 24). *In no case should an untrained person attempt to put the dislocated bone back into place (to "reduce" the dislocation) however ugly it looks.* This will not usually relieve the pain very much. But inexpert meddling may move the dislocated bone so as to tear blood vessels, with great aggravation of the injury, or nerves, with possibly severe paralysis.

Arthritis is really a symptom, not a disorder. The name means "joint inflammation," and this can come from many causes. Whatever the origin, the pain and incapacity it causes make arthritis a major medical problem.

The very common and tragic *rheumatoid arthritis* is a chronic inflammation of the joints of the hands and feet, although larger joints may be later affected. An estimated 4,500,000 people in this country, three-fourths of them women, suffer from this disease, and, of these, 200,000 are totally disabled [Bland]. Little is known of the cause of rheumatoid arthritis. Our present state of knowledge suggests a chronic bacterial infection with a development of tissue sensitivity. That a hypersensitive (immune) reaction may be involved has been shown by the presence of a *rheumatoid factor* in the blood which has many of the characteristics of an antibody.

The joint inflammation involves the synovial membrane which becomes swollen and pulpy causing an enlargement of the whole joint. The articular cartilages are eventually destroyed and the joint surfaces become fused. Although remissions often occur, the injury to the joint is permanent and relapses are generally the rule. Second and third attacks are usually more stubborn and have a tendency to become progressive.

Since the cause of rheumatoid arthritis is unknown, no cure is available. Rest, relief of pain with aspirin (highly effective in arthritis of all types) or other analgesics, and maintenance of joint function by physical therapy are the usual methods of treatment. Cortisone and ACTH (Chapter 32) and their synthetic relatives sometimes give dramatic relief. Unfortunately, these drugs have side-effects which indicate against long-term use except under the careful supervision of a physician.

Osteoarthritis is the other major form of arthritis. It usually occurs in the elderly, and is probably due to degeneration of the joint tissues through "wear and tear." A lifetime of severe overwork seems to wear out the cartilage or ligaments of a joint, or tendons crossing it, with resulting pain and breakdown especially in later life. Happily, few people have to carry on the prolonged, brutal labor that once was the lot of many, and such arthritis is now less common. Conversely, many aging people become very inactive so that they may never move some joints through the full normal range of motion. This too can result in degeneration of the joints with loss of function and growing disability. Here is a major reason for continued, judicious, well-organized exercise from youth on through life. Treatment of osteoarthritis is similar in many ways to that for rheumatoid arthritis. However, since osteoarthritis is not usually progressive it is important that the patient understands the distinction between the two diseases. Such reassurance can give considerable relief from anxiety.

The prevalence of arthritis and the absence of a cure make arthritis of all types a fruitful field for quacks. Since remissions are common, it is not surprising that all the nostrums have their "cure" testimonials. Useless treatments include special diets (diets of any sort are utterly ineffective except in cases of obesity); electrical gadgetry; medicated bath salts; and a wide variety of salves. The practice of wearing copper bracelets as a preventive measure (by supposedly "drawing out the poison") is seen even among college-trained athletes. This superstition is probably harmless unless it gives the wearer a false sense of security. Since trauma is the most important predisposing factor to osteoarthritis, the bracelets should prove to be more effective among professional golfers than among professional football players.

In a disease where pain plays such an important part and where the doctor is able to give at best only a cautious hope for control, much less a cure, it is to be expected that the afflicted would seek out those who promise relief. The conventional treatments involving rest and aspirin, or even cortisone, seem mundane in comparison to the alluring panaceas of the quack. Nevertheless, only the physician who thoroughly understands the patient can apply the most reliable and effective treatments.

Bursitis is often confused with arthritis since bursae are usually associated with joints, especially the shoulder, elbow, and knee. This can be very painful and disabling but is often easy to treat. Rest, warmth, and massage may clear up a bursitis; or a simple operation may be needed. But a self-diagnosis of "bursitis" and home treatment may postpone recognition of something much more serious. Let the physician confirm whether or not it is bursitis and prescribe the appropriate treatment.

Fractures

Fractures may seem much more important but are hardly amenable to amateur or even trained first aid beyond what was said for sprains. But the special features of fractures do require detailed discussion.

First, fractures are of several sorts (Fig. 30–1). Especially in young people, a fracture may be a cracking of the bone without actual separation of the parts, expressively called a *green-stick* fracture; this is the kind of which one hears when a player finishes a game without realizing he has a broken bone—he becomes aware of it as soon as his attention is free! *Evulsion fracture* is the type mentioned earlier where a ligament or tendon tears off a bit of bone. *Simple fracture* is the usual breaking of a bone into two parts, although it may be held close together by the soft tissues. *Comminuted fracture* is one in which the bone is broken into several parts. *Compound fracture* occurs when any fracture, simple or not, breaks through the skin.

Compound fracture is the most serious. Once a bone penetrates the skin, or is uncovered by a break in skin and soft tissues, it is exposed to infection. This is worse than infection of most tissues because once established it is very hard to overcome, even with antibiotics, and may smolder, spread, and do great damage. This means two things. *Any uncompounded fracture should never be moved except with the greatest care by trained personnel; it might be transformed from a minor disaster to a major one.* Second, *a compound fracture is a greater emergency than most other fractures and should stimulate every effort to get the victim to skilled help as soon as possible.* Splinting of fractures, simple or already

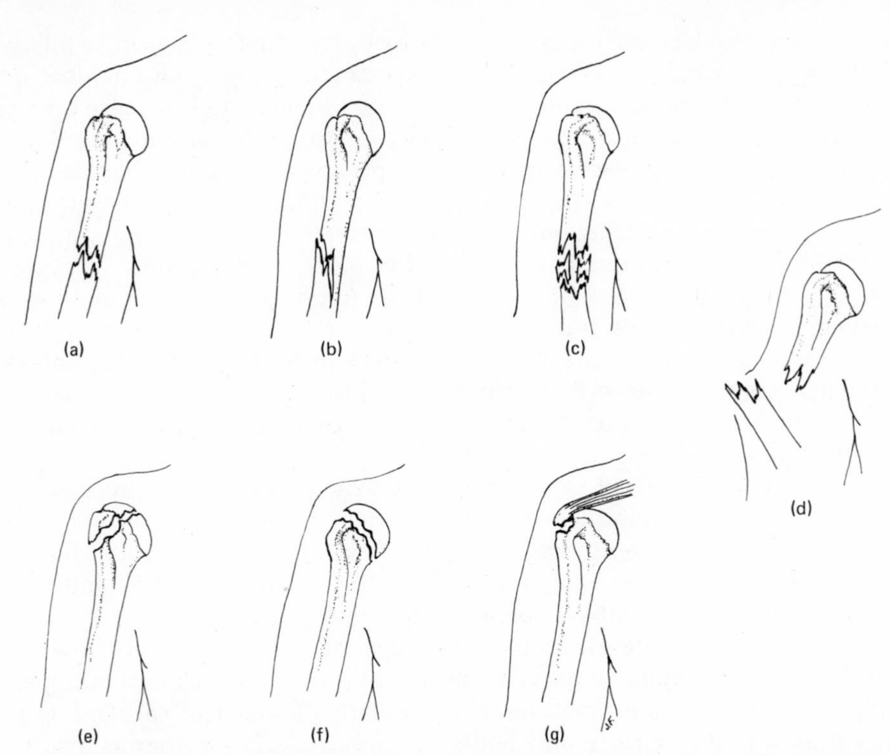

FIG. 30–1 The forms of bone fracture: A simple fracture (a) is what one usually pictures, and what, in fact, usually occurs. In most cases, it can be satisfactorily treated by competent setting and immobilization for a few weeks. A layman, however, cannot be sure that a fracture is not of a more serious sort. A *green-stick fracture* (b) most commonly occurs in youthful bones. It may not cause complete disability and may be mistaken for a sprain or contusion. If it is even suspected, an X-ray examination may save much trouble. A *comminuted fracture* (c) involves multiple breaks. It is harder to treat and may require surgical intervention, especially if any of the fragments lack blood supply and are apt to degenerate. A *compound fracture* (d) must be carefully distinguished from (c), which its name might suggest. It entails a break in the skin due to protrusion of the bone itself or to other lacerations, exposing the fracture site. It is extremely serious since it presents a high probability of bone infection. Bone infections are hard to control and tend to prevent union of the bones. Any compound fracture demands all possible care to avoid infection, including restraint of amateur "cleaning" (see text). And all fractures not compounded should be handled so as to avoid their becoming so. An *articular fracture* (e) is serious because the articular cartilage heals slowly and often badly. Expert care is needed. An *epiphyseal fracture* (f), through a growth-plate of an immature bone, may affect later growth of the bone. Here too, expert care is needed. An *evulsion fracture* (g) is more common in regions such as the heel bone, but is shown here for comparison. Severe stress by muscles or wrenching of ligaments may tear off a fragment of bone rather than rupture the soft but tough tendon. The result is a combination of sprain and fracture that may heal under routine treatment or may require surgery to fix the bone fragment in place.

compounded, can be carried out cautiously by trained first-aiders, and a pillow-splint may be used in an emergency by any careful person; but one does not want to increase exposure of an already exposed bone and so increase the risk of infection. Unqualified people should not attempt to clean a compound fracture or to put it back in place, which may simply spread infection; but they may carefully and gently cover it to prevent further contamination. Of course, all rules may have to be intelligently modified in cases of accident far from help.

Not only compound but all other fractures must be moved only with great care. In addition to the risk of compounding, injudicious moving may result in severe internal injuries. If the rounded end of a dislocated bone can damage blood vessels and nerves, one can imagine the effects of jagged and splintered fragments, especially since there are many vessels and nerves closely wound around bones. Bear this picture clearly in mind when dealing with even a possible fracture.

In all such cases (except very special emergencies) the correct procedure is: make the victim as comfortable as possible where he lies, even if this is a very inconvenient place; prevent unskilled busybodies from meddling; do what you can for other injuries; do what you can for shock or to prevent it; and make sure that everything possible is being done to bring expert aid.

Another point of interest and value in dealing with fractures is how they heal. Most bones originally grow by means of cartilage plates expanding and being replaced by bone; fractures heal similarly. A collar of cartilage forms quickly around the fracture and holds the broken ends together as a sort of natural splint. A fully developed callus is formed by 30 days, but will not bear much weight, although it usually resists minor movements of the body. Thus it must be reinforced by an artificial splint or cast. This collar is called a *callus,* to be distinguished from *callosities* of the skin, though both mean "touch." It too, as in the growing bone, is slowly replaced by bone tissue which then is modeled back to normal shape and size. A callus can sometimes be felt on a healing bone and it is a good sign.

Fractures sometimes grow together without having been properly set. The two fragments may overlap or join at an angle producing a deformity in spite of "healing." Unless the victim is resigned to keeping this deformity, it must be treated by separating the badly jointed fragments and letting them heal properly; this is a serious and tiresome business. If a bit of muscle or other tissue gets between the broken ends of a bone, healing may not take place at all till the obstruction is removed. Adequate treatment by an expert, in the first place, is surely better than risking such trouble.

Some fractures give special difficulty. The neck of the femur (thigh bone) where it joins the pelvis, is such an example, especially in older people whose bones heal slowly. A fractured femoral neck is almost impossible to splint and is disturbed by any body motion before callus forms. Thus, in former days, the unhappy patient had to lie motionless, banked with sandbags, for a long time;

and very often he would develop bed sores, pneumonia, or extreme depression, and even die. Recently, methods have been devised for fastening the bone fragments together with large, specially designed, metal nails and bars (Fig. 30–2). This may sound horribly uncomfortable, but in fact the operation is relatively simple, the nail is not even felt, and can be left in place for life. Other fractures are sometimes treated in the same way.

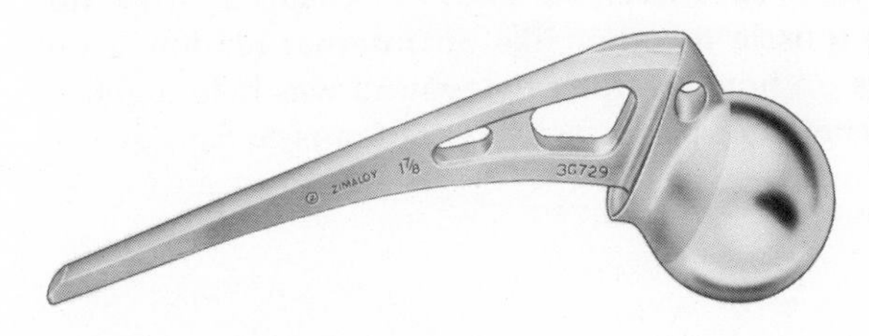

FIG. 30–2 Example of a metal hip prothesis where a new neck and head of the femur is required. (Zimmer Company, Warsaw, Indiana)

Injuries to Muscles

Muscle strains and pulls or direct blows to the muscle require immediate cold application for a period of an hour or two if serious. They usually need no splinting or bandaging but will, if severe, compel the patient to keep them still by a sling or other means.

After using cold for 24 hours, heat may be used although some trainers recommend that the cold treatment be continued for several days. Muscle aches after unaccustomed exercise are probably due to numerous tiny tears in the collagen harness or even the muscle fibers themselves, and possibly to minute hemorrhages. Other evidence suggests that excessive potassium concentration in the tissues may be responsible.

Muscle soreness seems to follow unaccustomed endurance activity much more so than a few maximum-type efforts (for example, from all-out situps or pullups, running for some distance, swimming, and so on). It usually sets in some 18 to 24 hours after the activity. The best practice would seem to be to participate in *moderate* activity involving all of the body as much as possible and gradually increase intensity over a period of days.

A muscle cramp is a sudden spasm or unrelenting, hard, contraction of a muscle. The calf muscles and those on the front of the thigh seem to cause difficulty most often, and they can be extremely painful. The cause is not clear but seems to be associated with conditions that cause the muscle to become especially irritable. When caused by exercise, it may be related to excessive potassium in the muscle but this, even if true, cannot account for the cramps you get when lying in bed or sitting perfectly still.

Regardless of the cause, treatment calls for immediate *stretching* of the muscle. For example, if the calf muscle cramps, you would grasp your toes and pull toward the shin. Unless a cramp is unusually severe, once the spasm is relieved there is no further problem.

A "charley horse" is a larger injury due to extreme effort or to violence from outside; it is marked often by severe cramps, but in normal people will heal spontaneously. If due to degeneration from disease or age, it may become chronic. In the latter cases heat, skillful massage, and gradual resumption of movement form the usual treatment.

Cuts and other wounds of muscle also usually heal spontaneously. The muscel fibers themselves do not grow back over the gap, but collagen fibers do, forming a scar that holds the muscle together like an internal tendon. How good this "tendon" is depends on how carefully the wound was held together during healing. Proper care permits surprising recovery of muscle from severe injury.

REVIEW QUESTIONS

1. What is a sprain? How should it be treated?
2. What are the two principal forms of arthritis, and how do they differ?
3. Discuss some of the hazards that are involved in the first-aid treatment of fracture.

REFERENCES

AEGERTER, ERNEST, and JOHN A. KIRKPATRICK, *Orthopedic Diseases,* 3rd ed. Philadelphia: W. B. Saunders, 1968.

BLAND, JOHN H., *Arthritis: Medical Treatment and Home Care.* New York: Collier, 1958. (Paperback)

LAYZER, ROBERT B., and L. P. ROWLAND, "Cramps," *New England Journal of Medicine,* **285:** 31–39 (1971).

O'DONOGHUE, DON H., *Treatment of Injuries to Athletes,* 2nd ed. Philadelphia: W. B. Saunders, 1970.

31
PHYSICAL FITNESS AND EXERCISE

Everybody knows something about exercise. We do it, talk about it, watch it. Some would say that we unfortunately do it less than we talk about it and watch it. Others would say exercise is for children only. In short, we all have *some* concept of exercise, but probably nothing in our culture that is so commonly conceptualized by so many people is at the same time so *variously* conceived and improperly understood. We form concepts about exercise not only from what we experience but from what we hear and read and see. What we experience affects how we feel about exercise and especially as it relates to what we hear about it, what we see others do, and so on. For example, if a child sees other children enjoying a sport and upon trying it himself is unsuccessful, he will not enjoy the sport and will have an entirely different feeling about exercise.

Whatever your concept of exercise, this chapter may help to separate truth from fiction and fact from fancy so that you might have a more nearly *correct* concept of exercise—what it *is*, what it *can* do, and what it *cannot* do.

We know that certain problems exist concerning exercise. Among the more important are these:

1. For many of us, increased automation and technology have decreased the amount of exercise necessary to exist.
2. Exercise is *not* a habit, in fact there may be a natural tendency to conserve energy by avoiding activity as much as possible; yet there is a great deal of evidence that physical exercise is second in importance only to eating and sleeping.
3. Even those who are educated and supposedly "exposed" to exercise opportunities (therefore educated about exercise) often have serious and limiting misconceptions about exercise.
4. Even those whose concept of exercise and its importance is more nearly correct often find exercising too inconvenient.
5. The concept of "exercise for fun," though certainly not in itself harmful, unfortunately dominates the concept that exercise is important even if it is *not* fun.

A fuller and more informed concept about exercise should result from a careful study of this chapter.

Essential Terms and Definitions

An understanding of the following terms, often misused and misunderstood, is essential if you are to attain a reasonable level of understanding about exercise and physical fitness.

EXERCISE. As used in this context, exercise means gross physical activity of the muscles causing gross external body movement. Thus, for example, typing in *not* exercise; running, walking, lifting objects, and so on are examples of exercise.

REGULAR EXERCISE. Generally taken to mean planned, daily exercise.

TRAINING. A program, regular and planned, intended to improve some skill or ability (that is, training to improve jumping ability or to improve time in the mile run).

CONDITIONING. Usually means a program directed at more general kinds of improvement, not so much in skill or ability but in general physical fitness.

PHYSICAL FITNESS. A general term that includes certain specific components, all of which in some way relate to health and none of which are limited by one's motor skill or athletic ability. The specific components are (see Fig. 31–1):

Circulatory-Respiratory Capacity is the capactiy to persist in large muscle activity of an endurance nature. The limits to this endurance are set by the respiratory and/or circulatory system rather than by local impairment of any muscle or muscle group. Example; running.

Strength is the maximal force that can be exerted by a muscle or muscle group. Example: biceps curl (elbow flexion with a barbell held by the hands at arms' length with palms facing away from the body); if the maximum weight that can be moved by bending the arms until the barbell is at the shoulders is 90 pounds then biceps strength is 90 pounds.

Muscular Endurance refers to the ability to persist in a given muscular action when the resistance is less than maximal. It is specific to the percent of maximum effort required and depends upon the rate of contraction. Example: chinups; pushups; situps; if the man in the example above is curling 45 pounds (half of his maximum of 90), he can do more than one. The number of repetitions he can complete provides a measure of his biceps muscular endurance at 50 percent of maximum.

Flexibility is the capacity of a joint or series of joints to move through a normal range of motion. Example: hamstring flexibility is measured by the well-known "toe-touch" test.

MOTOR ABILITY. A general term connoting high motor skill level; not necessarily related to physical fitness. It is composed of specific, often interrelated components as follows (see Fig. 31–1):

Agility is the ability to change direction quickly while moving rapidly.

Balance is the ability to maintain body equilibrium while stationary and while moving in various ways.

Reaction Time is the time between stimulus and initiation of response; *movement time* is the time from initiation to completion of a specific movement; *response time* is the total time from stimulus to completion of the movement.

Speed generally refers to the ability to move the entire body from one place to another quickly as in running.

Power is the ability to exert force very quickly thus resulting in an explosive movement as in the shot put or broad jump.

Coordination is involved in all of the other motor ability components and is essentially the efficient blending of several movements to produce the desired result.

MOTOR CAPACITY. The *potential* one has for performing motor skills.

MOTOR EDUCABILITY. The ability to learn motor skills.

SPORT. This word actually means play or diversion but usually refers to physical activity and not to quiet play or diversion (tennis and hunting are sports, bridge and movies are not sports).

ATHLETICS. Though often used interchangeably with sports, it is correctly used only to connote *organized sports competition.* Thus hunting, playing tennis, and playing touch football are all examples of sports participation but not necessarily athletic participation.

Exercise, Physical Fitness, and You

Behaviorally, the terms exercise and physical fitness take on importance *only* as they relate to *you.* Unfortunately, most of our exercise and sports experiences drive us away from this very important concept of the importance of the *individual*—his capacity, his needs, and his interest—in determining the meaning and importance of exercise and physical fitness. What you can do in sports com-

FIG. 31–1 Examples of the specific qualities of physical fitness and motor ability.

petition, in comparison to your neighbor or anybody else, may be of interest or of importance to your ego or emotional state but hardly has anything to do directly with your physical health. No one quarrels with the value of sports competition as a means of relaxation or enjoyment for some people. We do suggest that everybody, including the successful sports participant and the unsuccessful or uninterested, accept the fact that *everybody* is *not* a games player. There are other ways of relaxing, utilizing leisure time, and even competing, that are equally enjoyable and satisfying to some. In other words, if one is capable of combining exercise, recreation, and competition in some form of sport and if this makes him a better person, this is fine. If another person who is not capable of this efficient combination meets his recreation needs another way, that is also acceptable; but he may still need exercise in his life, and there are nonsports and noncompetitive ways of getting it. The decision should rest with the individual without his fearing ridicule from people whose decision is different from his. The key concept here is that regular exercise is important whether or not it is for fun, just as eating is essential to life whether or not the food *tastes* good!

MOTIVATION. For those who *enjoy* sports competition and have the time and space available, motivation is no problem so long as enjoyment, time, and space remain. There are other forms of motivation to exercise—concern about excess body weight and concern about appearance are examples—and something must motivate us, otherwise an apparently universal drive to conserve energy will win out in the long run. If you are motivated to exercise regularly, the next question concerns the appropriateness of your choice. Does it contribute to anything other than enjoyment? Does your selection meet all of your exercise needs? To answer this, you need to know two things: What are the body's general needs? What is my status now? Then you can decide whether you want to attempt to meet those needs.

General Needs

Every reasonably healthy person needs at least minimal levels of circulatory and respiratory capacity, strength, muscular endurance, and flexibility. The minimal levels provide a definite measure of protection against a number of health disorders related to respiratory and heart disease as well as low-back pain. You may wish to add to this minimal list of needs those based more particularly upon your own personal needs and interests. For example, you may wish to develop more than minimal levels of these particular qualities so that

---

All fitness qualities (parameters) can be tremendously improved in the reasonably healthy person with nothing more than adequate effort; fitness qualities are also more directly related to health and preventive medicine. Motor ability cannot be improved as much and, for the average person, is not as directly related to health.

you might be better prepared to meet possible emergencies which may involve better than average strength or endurance or both.

Your Current Status

Certain simple tests may be used to determine your current status with regard to your fitness needs. If you are in reasonably good health, you can attempt the Cooper Twelve-Minute Field Test or the 1½ mile modification of this field test. The former is accomplished by determining how much distance you can cover by running and/or walking in 12 minutes. The modified test involves determination of the time required to run and/or walk 1½ miles. Appropriate classifications are listed in Table 31–1.

In the event you cannot or choose not to use the Field Test, a mild bench stepping test such as the Tecumseh Submaximal Exercise Test, may be useful. In order to take this test you need a watch with a second hand, a stable stepping surface 8 inches above the floor, and a metronome for establishing the stepping rhythm. It is best to take this test in a reasonably well-rested condition, not having exercised vigorously within the previous three hours, and a reasonably comfortable environment, since the index involved is the recovery heart rate and heart rate is affected by factors such as activity, temperature, and humidity. The test is taken by alternately stepping up and down on the 8-inch bench to the rhythm established by the metronome (24 complete steps per minute, a metronome rate of 96) for three minutes. Immediately upon completing the test, sit down and locate the pulse at either the wrist or carotid artery (just to either side of the Adam's apple) in preparation for taking a ten-second pulse count exactly from 55 to 65 seconds after exercise. The scoring table may be found in Table 31–2.

Muscular endurance cannot be appropriately tested without also testing minimal strength. We find it less conjectural to take a position regarding minimal strength levels than minimal muscular endurance levels; the reason should be obvious. The minimal strength tests, most easily self-administered, involve your capacity to contract certain key muscle groups using your own

TABLE 31–1 Rating Scale for 12-Minute Field Test

If You Cover:	*Fitness Category*
Less than 1.0 mile	I Very poor
1.0–1.24 miles	II Poor
1.25–1.49 miles	III Fair
1.50–1.74 miles	IV Good
1.75 or more	V Excellent

TABLE 31–2 Tecumseh Exercise Test

Test:	Tecumseh Submaximal Exercise Test[a]		
Bench:	8 in.		
Rate:	24 steps per minute		
Time:	Three minutes		
Pulse count:	Method A: Time 10 heart beats beginning at 55 seconds after exercise Method B: Count heart beats for 10 seconds beginning at 55 seconds after exercise		
Scoring	Method A: $\frac{600}{x}$ where x = time (to nearest tenth of a second) for 10 heart beats Method B: Multiply count by 6		
Norms[a]	Percentile	Male	Female
	95th (exc.)	67	75
	75th (ab.avg.)	79	85
	50th (avg.)	90	95
	25th (poor)	100	110
Age group	10–69 unless in poor health		

[a] Approximation based on average of age groups 10–11, 12–13, 14–15, 16–17, 18–19, 20–29, 30–39, 40–49, 50–59, 60–69. Actual figures were no more different than 0–4 beats per minute except one case (Female 25th percentile) where difference was 8 beats per minute between 10–11 and 60–69 age groups.

SOURCE: H. J. Montoye, P. W. Willis III, and D. A. Cunningham, "Heart Rate Response to Submaximal Exercise: Relation to Age and Sex," *J. of Gerontology*, **23:**127 (1968).

body weight as the resistance. This is also the most practical since it is your own particular body weight that you must handle under ordinary circumstances. Men and women should be able to execute at least one bent knee sit-up while lying on the back, with the lower back flattened against the floor and beginning the situp with the head coming up first, then a few vertebrae at a time, with the fingers laced behind the head. As shown in Figure 31–2, knees should be bent so that the heels are within at least 6 inches of the buttocks; it will usually be necessary to have somebody or something hold your feet down. If you can do one of these, you have minimal abdominal strength. Abdominal muscle endurance is a very individual matter, but we suggest that ten situps of this type is a reasonable number for a *minimal* level.

For the male, minimal arm strength requires that he be able to execute at least one pullup or chinup from a *full* hanging position, with the reverse grip (palms turned toward himself) and execute one full dip on the parallel or dip-

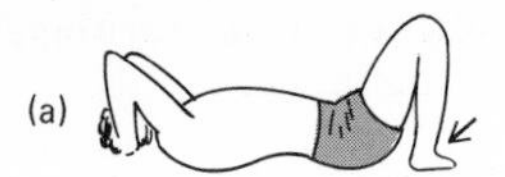

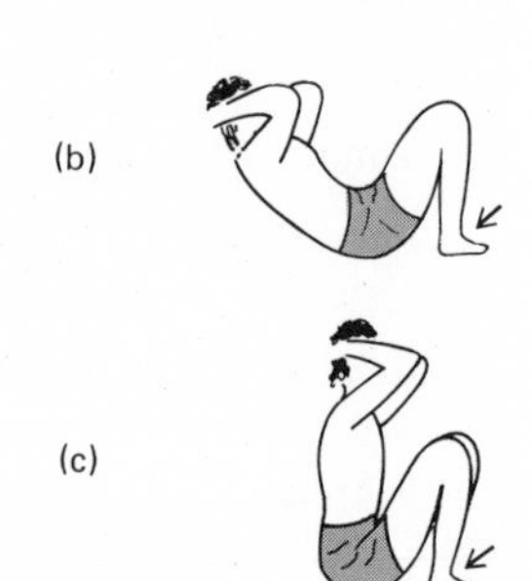

FIG. 31–2 Proper method of testing minimal abdominal strength. It may be necessary to have somebody hold your feet down (see arrow). Keep lower back flat and "curl" up.

ping bars. The test is taken by supporting your weight with arms locked and extended and hands a little wider apart than shoulder width. You then lower yourself, keeping the head up and the center of gravity not too far forward or backward, until the body weight is as low as it can be supported (elbows at least 90 degrees flexed) and then extend the arms again to the upright position. Again, endurance is an individual matter, but we suggest that three complete pull-ups and six dips might be considered minimal.

For minimal leg strength a man should be able to lift his body weight from a complete squatting position to the standing position using only one leg. Each leg should be tested; this can best be accomplished on some sturdy surface such as a first step or a sturdy chair so that the opposite foot can be maintained more easily without contacting the floor. It may be necessary for some people to secure the assistance of another person to help them balance during the test, but care should be taken not to use the arms in attempting to rise. Obviously, this test should not be attempted by a person with any sort of knee disorder. Endurance of the leg extensors can be measured by doing full leg squats using both legs and 25 would seem to be reasonable.

Ideally we feel minimal strength for women should be the same as for men with the possible exception of leg strength, but experience has taught us that some modifications are necessary. Ideally, we would like to see a woman be able to execute the pullup as described for men, the pushup from the floor rather than the dip on the parallel bars, and the leg extension test exactly as administered for men with one exception; the leg straightening is executed from a semisquat position rather than from a full deep squat (beginning from knee flexion at about 90°). For women who cannot execute one pullup or one floor pushup, the following might be acceptable as minimal strength standards although they are obviously unreliable as strength tests because they really include a measurement of muscular endurance: (1) at least 20 modified pullups and (2) at least 15 modified knee pushups as shown in Figure 31–3.

FIG. 31–3 Modified knee pushups (a) and leaning pullups (b) for women.

The most important flexibility measure is that of the hamstring muscle group; minimal trunk flexion and hamstring flexibility is measured by the simple toe touching test. Care should be taken not to try this test without due precaution because it is possible that painful, if not serious, damage may result. You take the test by keeping the knees straight and attempting to touch the toes with the fingertips of both hands by gently bouncing down several times, gradually building up the ability to stretch the backs of the legs until you actually attempt to see how far you can reach toward the toes without undue discomfort to the back of the legs. Minimal standard for men is to be able to touch the floor with the fingertips, whereas women should be able to do considerably better than this as a minimum standard. If proper testing equipment is available (see Fig.31–4), the male should score at least zero (nothing less) and a woman should be able to score at least + 3 inches.

Of these simple tests just described, the most important in terms of health are circulatory-respiratory capacity, abdominal strength, and flexibility. The reasons for the importance of these qualities will be discussed at greater length later in this chapter. Now that you know what the general fitness needs are and you have determined what your current fitness status is, your individual needs and preferences will help you to determine the nature of your exercise and fitness program.

Selection of Individual Program

If you should decide that you wish to improve upon your less than minimal or just barely minimal fitness qualities (or that you wish to maintain optimal levels), there are certain factors that will help you make a decision about your own personal program. The first decision you will need to make is whether you

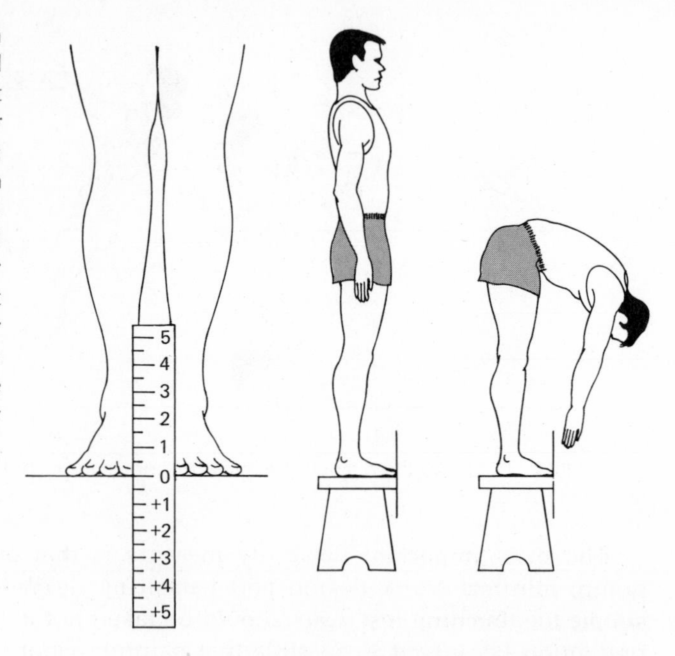

FIG. 31–4 Trunk flexion test. (1) Heels together and toes pointed to side at about 45° angle; (2) toes near or even with edge of bench; (3) bounce gently down twice (be careful not to pull too hard); (4) hold the third bounce (with hands together and fingers on ruler) for at least 2 seconds so that your partner can read your score; (5) record as minus above bench and plus below bench to nearest half inch.

also wish to improve the fitness qualities which relate more to possible emergency situations than to health per se. We refer to strength and muscular endurance of the arms and legs. Greater than minimal strength and muscular endurance of the legs and arms would seem to be necessary only in emergency situations in which one might be required over extended periods of time to use these muscles. Thus the decision as to whether you wish to be prepared for such emergencies is certainly a very personal one. But in terms of health, regardless of potential emergency situations, circulatory-respiratory fitness, abdominal strength, and flexibility of the hamstring muscle groups are of primary importance. If your purpose is to develop a high level of physical fitness for any reason (for a feeling of well-being, personal satisfaction and achievement, for sports success, and the like,) the program will obviously involve greater intensity and duration than if you are interested simply in maintaining a fitness level compatible with good health.

You need also to take into account certain prerequisites for different kinds of programs. For example: what kind of equipment is required? are facilities of any special kind required? is another person or a group of persons necessary? You will want to consider also your psychological, emotional, and sociological needs. There must be some feeling of accomplishment involved, whether it is simply the feeling that you are achieving greater fitness or whether you are having fun. Without some kind of motivation it is unlikely that you will continue for very long.

Different Approaches to a Minimal Program for Fitness

There are different ways of working at individual fitness improvement, but we suggest that they all should include (as a minimum) methods to improve your circulatory-respiratory fitness, abdominal strength, and hamstring flexibility. The minimum program should also include attention to weight control (see Chapter 17) and, either as part of the exercise program or as a separate entity, something that might be considered as recreational or relaxing in nature. In addition, if you have chosen to improve abdominal endurance, strength, and muscular endurance of the legs and arms, these things can be worked into your exercise program.

Physical Fitness Improvement

The best way we know to improve circulatory-respiratory fitness is to participate regularly in some form of rhythmic, endurance-type activity. This means activities such as walking, running, swimming, cycling, rope skipping, and certain continuous sports activities such as basketball, handball, badminton, tennis, squash, and paddleball.

You may apply some rule of thumb based on exercise heart rate or follow some prescribed program involving these activities. Generally speaking, in order to achieve a significant circulatory-respiratory fitness gain, one needs to increase heart rate to about 140 beats per minute for a minimum of 7 to 8 minutes at least three times per week. If you choose to follow a prescribed program, we recommend Cooper's "Aerobics" program, in which most of the activities mentioned above are prescribed on a week-by-week basis, depending upon your initial fitness level as determined by the Twelve-Minute Field Test. Cooper's program is a tidy and neat package that provides a good and easily understood guideline for the achievement of optimal circulatory-respiratory fitness levels. Be sure that you do not confuse exercise and activities that are not of the endurance, rhythmic type with the appropriate activities listed above (see Fig. 31–5); weight training, bowling, lying at pool side and achieving a suntan, table tennis, and the like are not activities which develop fitness.

Hamstring flexibility can be increased or maintained by simply doing the gentle toe touching exercise at least ten times per day. Abdominal strength and muscular endurance may be achieved or maintained by a regular program of bent knee situps. Remember that it is essential that the knees be bent, that the lower back be flat against the floor, and that the situp begin with the head and be gradual, one vertebra at a time. If there is no way to hold your feet down during this exercise, do not be overly concerned since moving past the half-way mark no longer involves the abdominal muscles anyway. It is a good idea to alternately sit up straight, then twist to the right, then twist to the left, when

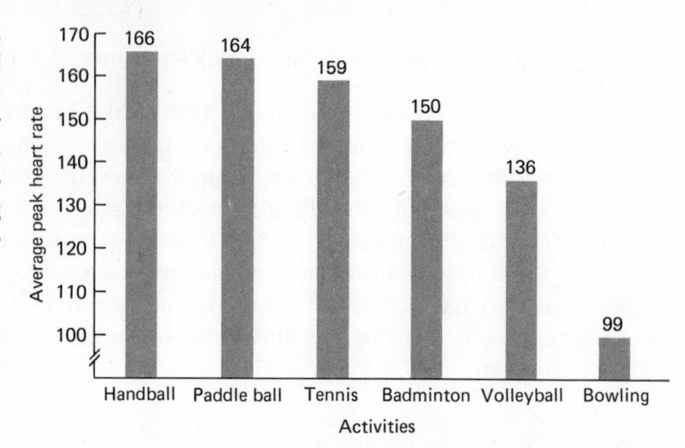

FIG. 31–5 Peak heart rates in college males attained during various intramural contests (data from Kozar and Hunsicker). Obviously, handball is more vigorous than bowling and volleyball, as are paddleball, tennis, and badminton; these activities are, therefore, better activities to promote C-R fitness than bowling and volleyball.

you do situps in order to involve the oblique abdominal muscles. Abdominal strength will not be improved beyond a certain point by continuing to do more situps each day, but can best be increased by progressively adding small amounts of weight behind the head, beginning with about 5 pounds when you can execute 25 situps with ease. Be sure to use an object (a sandbag is recommended) that will not cause you any harm while exercising.

It is obvious that one cannot do situps for abdominal muscular endurance until one can do at least *one*. If you do need to work up to *minimal* strength before you can even begin to be concerned with muscular endurance, you may need to utilize some sort of weight training in order to achieve at least minimal strength. If weight training equipment is not available and you do not wish to purchase it, you might try using isometric or eccentric training principles to build up the minimal strength level of any particular muscle group. The isometric principle involves the attainment of the desired end point of the movement in any way you can get there, then holding that position for as long as you can, taking a rest and repeating this as often as you can until fatigue sets in. For example, you might use your arms to assist you in achieving the halfway position on a bent knee situp and then hold that position as long as you can (Fig. 31–6); you can lift yourself in some way to a full chinup position on the bar and then hold yourself in that position as long as possible. The eccentric principle involves the attainment of the *end* position, then gradually returning to the *beginning* position, doing this as many times as possible, for example, by easing *down* from a bent knee sit-up position, or letting yourself slowly *down* from the chinup position on the high bar. A detailed discussion of weight training is beyond the scope of this text, but if you choose to use weight training to improve strength or muscular endurance, there are many books available that will aid you in setting up appropriate exercises.

You can accomplish your circulatory-respiratory work and the other aspects of your program, whatever they may be, separately or you can combine them

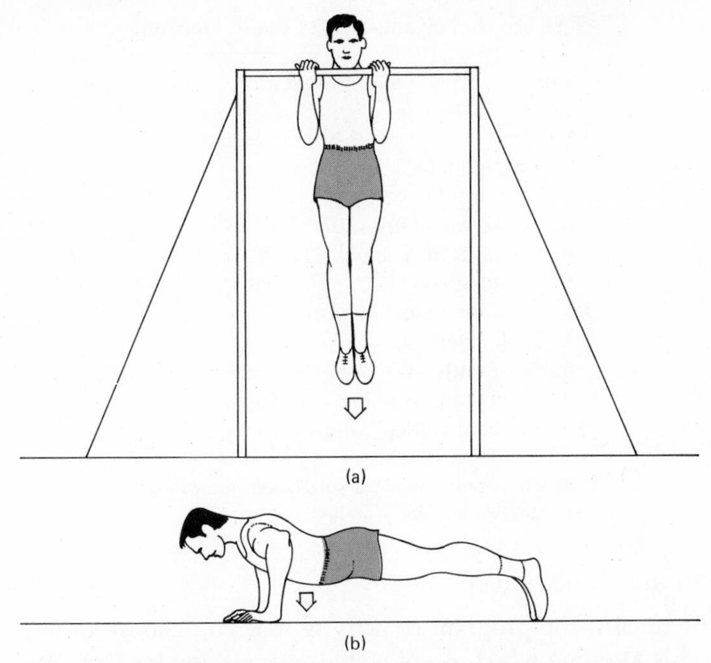

FIG. 31–6 "Holding" exercises for those unable to execute one pullup (a) or one pushup (b). Get to the desired position any way you can and then hold as long as you can. (Think ahead to prevent injury when you reach the "end point"; a broken nose could result from falling to the floor from the pushup position if the head is not pulled back or turned to the side.) "Eccentric" exercises, denoted by the arrows, involve a slow release after reaching the "up" position. The same principles of "holding" and "eccentrics" can be applied to other ME exercises.

into a program known as *circuit training* (see the sample in Table 31–3). Circuit training is a form of combining walking or running with other exercises interspersed (such as situps, pushups, pullups). *Interval training* (see the example in Table 31–4) is another method of applying the principles of aerobics and involves repeated laps of varying intensities (running, swimming, and such) with interspersed rest periods, gradually decreasing the length of the rest periods and/or increasing the intensity of the exercise spurts.

Careful records should be kept so that you can periodically evaluate your progress. A useful index of the circulatory-respiratory fitness level is the resting heart rate. If you take this several mornings each week before becoming active (let us say immediately upon arising), you will usually find that as circulatory-respiratory fitness improves, the resting heart rate is reduced. This is indicative of a stronger or more efficient heart or both.

TABLE 31–3 Example of Circuit Training

Sample of a Beginning Circuit for College Women
1. Jumping jacks—12
2. Bent-arm hang on high bar—20 seconds
3. Bent-knee sit-ups—10
4. Push-ups from knees—12
5. Toe-touches—10
6. Rope jumping—30 turns
7. Back extendors—12
8. Wrist curls—6
9. Jogging and walking—.5 mile.[a]
10. Maximal vertical jumps—6

[a] Bench stepping may be substituted in an indoor program when space is limited.

Recreation and Relaxation

It would be ideal if the program or activity that you choose to improve your fitness level is also one which is enjoyable and relaxing for you. Thus it would serve not only to improve your fitness level but would serve at least as a part of your program of recreation. If it does not, however, you should not ignore the importance of recreation and relaxation, and you should select another activity or activities which can provide the opportunity for you to relax. This is a highly individual matter because the same activity might be relaxing for one person and almost a nightmare for the other. Some people benefit very nicely from what is known as progressive muscular relaxation. This system of relaxation

TABLE 31–4 Example of Interval Training

Element	*Running*
Rate	1 mile/5 minutes (440 yards/75 seconds)
Distance or time	440 yards
Rest interval	5 minutes
Bouts	4
Variations	When 4th bout does not produce excessive discomfort, shorten interval to 3 minutes

was developed during World War II to assist fighter pilots in relaxing between missions. It involves a very simple principle: We can be tense without knowing it and the best way to recognize this is to go from the one extreme of excessive muscle tension to the other extreme of near absence of muscle tension. You begin by lying on your back with plenty of space and in a comfortable way (preferably where it is quiet). All the muscles of the right leg are contracted as tightly as possible, but without movement, for two seconds, then completely relaxed. Immediately they are contracted about half as tightly for two seconds, then released, then about one-fourth as tightly, then relaxed, and finally you only imagine the contraction and relaxation. This procedure is repeated with the left leg, buttocks, lower back, arms, neck, shoulder and finally the small muscles of the face (wrinkling the forehead, shutting the eyes, tightening the lips), in each case starting out with complete contraction, then one-half, then one-fourth, then imagining contraction. You will find that with very little practice this can be an extremely relaxing experience and has often been used by persons with insomnia, as well as by people in high-tension jobs who need to be able to relax when the tension mounts.

It might be well to note here that the "tensions" we complain of are almost certainly functions of our central nervous system. When we are "keyed up" it is probable that our central nervous system is in a state of readiness—for picking up sensory stimuli or for acting upon the slightest need. Long periods of such alert can be exhausting, especially without physical activity, and it is often difficult to go "off" the alert. When we ease off through progressive muscular relaxation, we are probably discharging or dissipating the nervous system circuits that have been "on the ready."

At any rate, it is wise to consider the importance or recreation and relaxation of some sort in your minimum program of physical fitness.

The Benefits of Regular Exercise

Some of the physiological effects of exercise are increased heart rate (Fig. 31–5) and blood pressure, increased rate and depth of breathing, increase in the cardiac output, changes in distribution of blood to the various parts of the body, and increased body temperature if exercise is intensive and of sufficient length. All of these are, of course, adaptive mechanisms essential for increased metabolic activity. There are also well established physiological effects of training or regular exercise (assuming the activity is of sufficient intensity and duration to result in significant and lasting physiological adaptations). For example (see Fig. 31–7), the resting heart rate may be reduced, the increase in heart rate with a standard exercise is significantly reduced, the maximum cardiac output can be increased, the maximum ability to use oxygen is increased, hemoglobin and blood volume can be increased, the ability to build up an oxygen debt can be increased, energy required for a given task can be reduced,

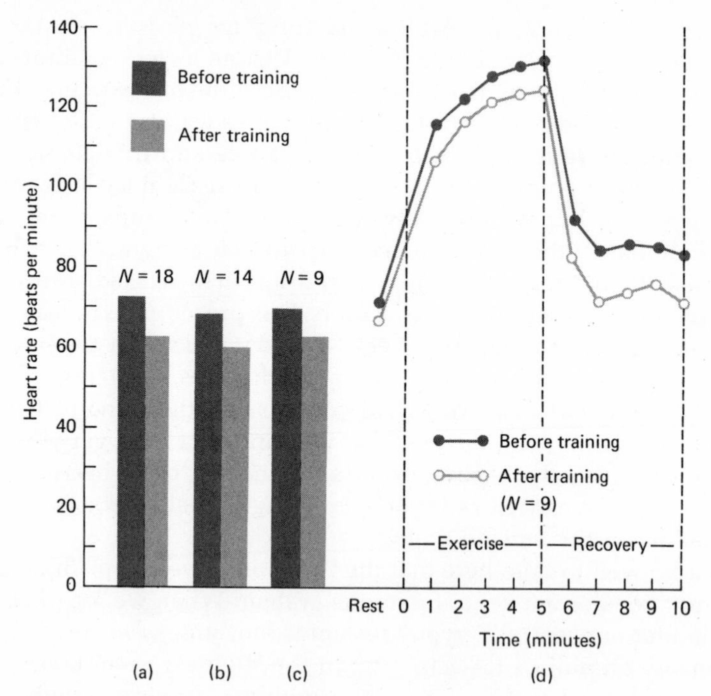

FIG. 31–7 Effect of training on resting heart rate (a, b, c) and on exercise and recovery heart rates (d). [(a) data from F. Henry; (b) data from Knehr; (c) and (d) from Johnson, Updyke, and W. Henry]

and, of course, specific adaptations to work can be expected, depending upon the nature of the task or tasks involved (for example, pullups, situps). These facts are interesting and have a practical value in that they provide tangible ways of evaluating progress in fitness programs, and they also do demonstrate that positive physiological changes do result from regular exercise; but of even greater interest and concern are the practical and healthful benefits of regular physical activity. We must caution you at the outset not to think in terms of glib generalities when considering the possible benefits of your own personal exercise program. Exercise programs and their benefits are highly specific. The results that you obtain depend to a significant extent upon exactly what you are doing. Thus, we cannot assume that every benefit that we are about to discuss accrues as the result of any and every kind of exercise program. Benefits depend upon the type of program, the intensity of the activity involved, the duration of the activity, and the frequency of exercise periods. Keep in mind that the benefits of exercise, in every case, depend entirely upon the factors just mentioned. For example, you would not expect pushups and pullups to contribute significantly to a program to prevent heart disease nor would you expect jogging to contribute to a low back disorder prevention program.

Physiological Benefits

Cardiovascular Health

There is an abundance of evidence that regular physical activity may prevent or postpone the onset of coronary artery disease. Evidence suggests that persons who are physically active are statistically less likely to have a heart attack, will have less severe heart attacks when they *do* occur, will recover more often, and will have fewer recurrent attacks than persons who are physically inactive. It may be that regular physical activity helps to promote the growth of collateral channels for coronary circulation when the gradual process of occlusion is taking place, but has not yet reached heart attack proportion. Thus the attack never occurs or, when it does, it is less severe because of this collateral circulation. Exercise may also decrease the rate of deposition of fatty substances into the coronary and other arteries. At any rate evidence strongly suggests that regular physical activity of a sufficient degree and nature can be one factor which helps to prevent or delay the atherogenic process (see Fig. 31–8). (Other coronary preventive measures are discussed in Chapters 22 and 23.)

Low-Back Disorders

The minimum exercise program recommended for the prevention of muscular disorders of the lower back includes toe touching and bent knee sit-ups—traditional calisthenic-type activities. Low-back pain afflicts at least half of the population of this country. Approximately 80 percent of these disorders are related to muscular problems rather than to those of the spine itself. These muscular problems often appear to be related to the degenerative process in which the abdominal muscles become too weak or in which the hamstring muscle group and associated tendons become too tight and inflexible or both (Figs. 31–9 and 31–10). The net result of either abdominal weakness or hamstring inflexibility (or both) is an incorrect tilt of the pelvis (forward, or downward in the front) so that the lower-back curvature is extremely exaggerated inward. One needs only to exaggerate purposely this lower spinal curvature to feel the extreme tension and discomfort. When this becomes the fixed position of the pelvis, chronic pain or discomfort can be the result. Our minimal program therefore includes maintenance of abdominal strength and hamstring flexibility as activities of the first order of importance along with the circulatory-respiratory exercise.

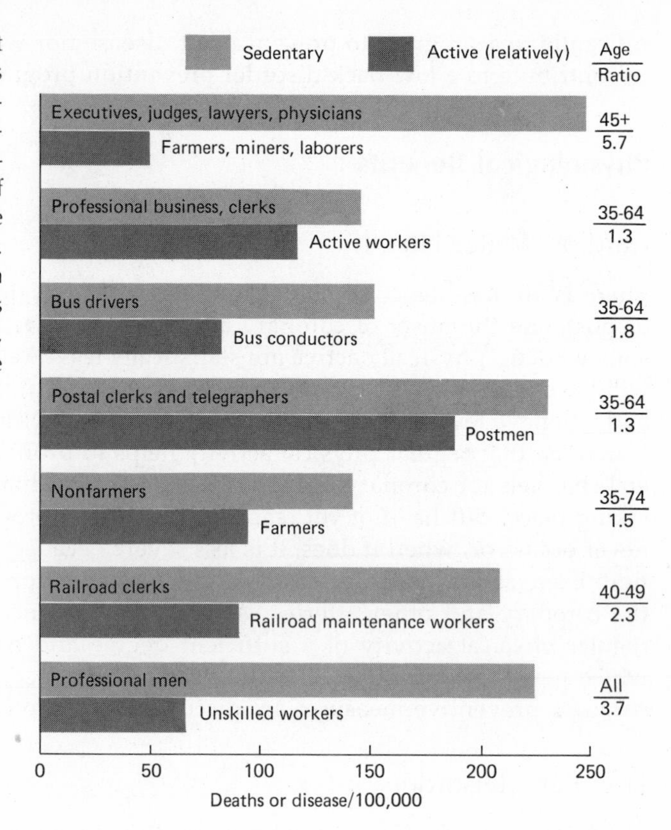

FIG. 31–8 Coronary heart disease and CHD deaths related to occupation, sedentary versus active. A summary of studies. "Age" refers to the age range of subjects; "ratio" is the ratio of sedentary to active. (Data from (reading from top): Pedley, Hedly, Morris and Crawford, Morris, et al., Zukel, et al., Taylor, Ryle and Russell).

Weight Control

You may have heard it said that "since you would need to chop wood for seven hours in order to lose a pound of body fat, exercise is ineffective as a contributing force in weight control programs." This is an excellent example of misapplication of a correct calculation. Think of this is two other ways: You did not put that extra pound *on* in one hour and, second, if one were to choose to chop wood, by chopping for a half hour per day, one could lose one pound in two weeks which is 26 pounds in a year!

There are three basic ways in which regular exercise can contribute to a program of weight reduction or weight maintenance. We have already mentioned that exercise does increase the caloric expenditure and thus can contribute to a negative caloric balance which ultimately will mean reduced fat stores. One pound of weight gain or weight loss is approximately equivalent to 3500 calories; thus any kind of a caloric deficit, whether it is composed entirely of added activity, reduced caloric intake, or a combination of both, will ultimately result in weight loss. You need only to look at Figure 31–11 to see that caloric expendi-

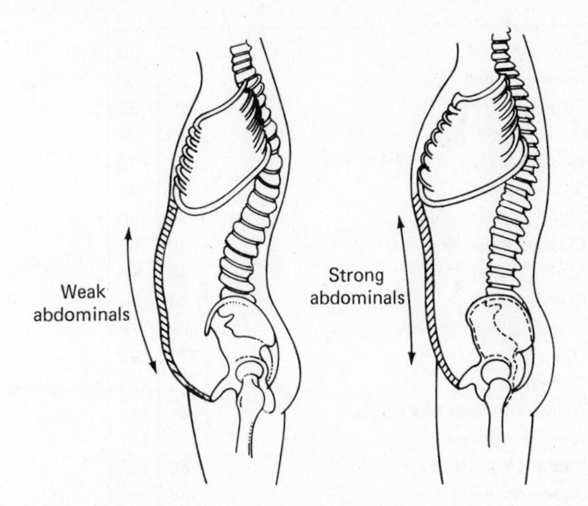

FIG. 31–9 Attachments of the abdominal muscles. It should be pointed out that the abdominal muscles do not attach to the legs, but to the pelvis and the ribs. Situps and leg lifts are only "indirectly" effective in strengthening abdominal muscles. Note exaggerated curve in lower back of figure at left.

ture can be greatly increased. It is quite obvious that exercise can contribute to either a negative caloric balance (or to caloric balance so that body weight will be maintained and not increased).

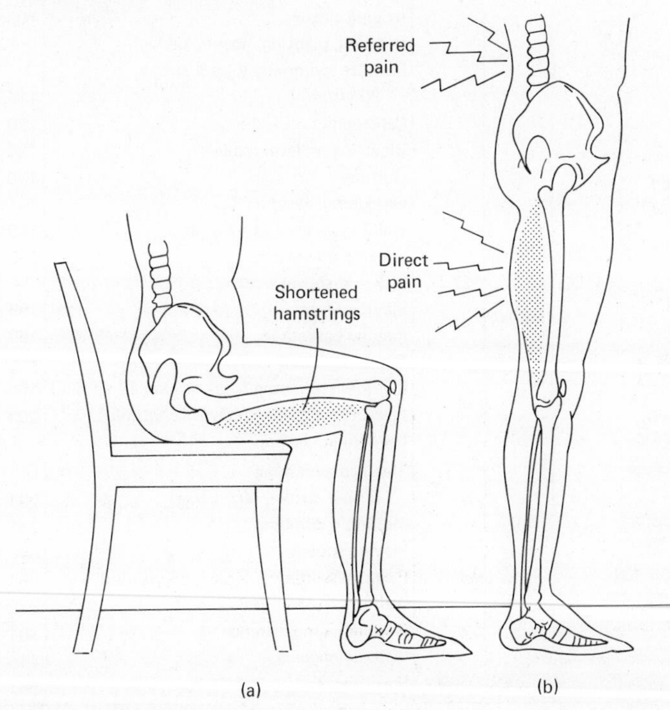

FIG. 31–10 Contracture ("permanent" shortening) of the hamstring muscles as a result of sedentary occupation, especially with prolonged sitting.

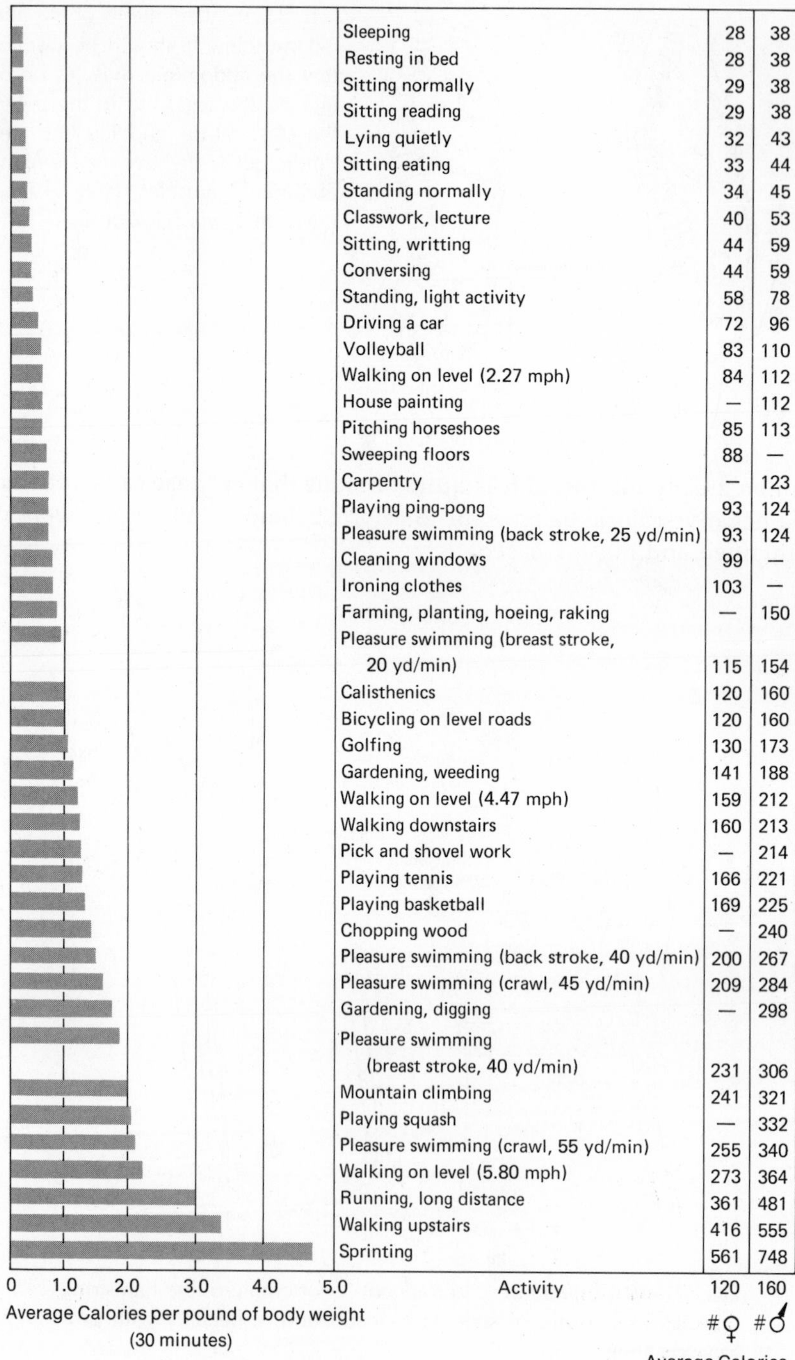

Activity	120 #♀	160 #♂
Sleeping	28	38
Resting in bed	28	38
Sitting normally	29	38
Sitting reading	29	38
Lying quietly	32	43
Sitting eating	33	44
Standing normally	34	45
Classwork, lecture	40	53
Sitting, writting	44	59
Conversing	44	59
Standing, light activity	58	78
Driving a car	72	96
Volleyball	83	110
Walking on level (2.27 mph)	84	112
House painting	—	112
Pitching horseshoes	85	113
Sweeping floors	88	—
Carpentry	—	123
Playing ping-pong	93	124
Pleasure swimming (back stroke, 25 yd/min)	93	124
Cleaning windows	99	—
Ironing clothes	103	—
Farming, planting, hoeing, raking	—	150
Pleasure swimming (breast stroke, 20 yd/min)	115	154
Calisthenics	120	160
Bicycling on level roads	120	160
Golfing	130	173
Gardening, weeding	141	188
Walking on level (4.47 mph)	159	212
Walking downstairs	160	213
Pick and shovel work	—	214
Playing tennis	166	221
Playing basketball	169	225
Chopping wood	—	240
Pleasure swimming (back stroke, 40 yd/min)	200	267
Pleasure swimming (crawl, 45 yd/min)	209	284
Gardening, digging	—	298
Pleasure swimming (breast stroke, 40 yd/min)	231	306
Mountain climbing	241	321
Playing squash	—	332
Pleasure swimming (crawl, 55 yd/min)	255	340
Walking on level (5.80 mph)	273	364
Running, long distance	361	481
Walking upstairs	416	555
Sprinting	561	748

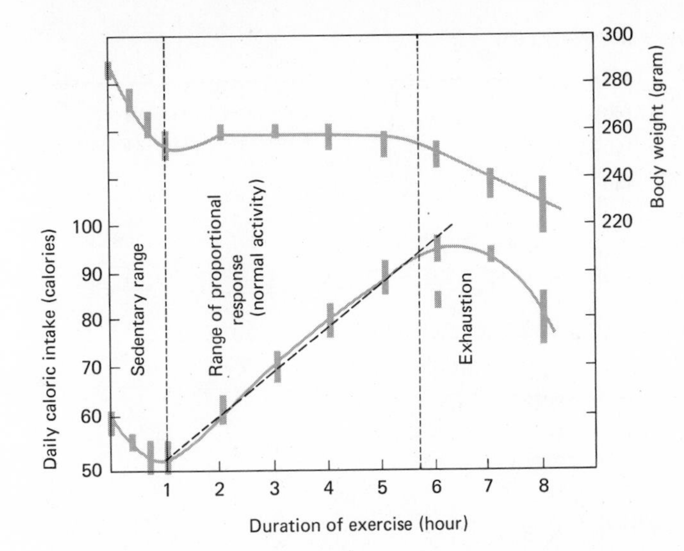

FIG. 31–12 Voluntary caloric intake and body weight as functions of amount of daily exercise in normal rats (Mayer). Note that in the sedentary range, the more sedentary animals ate more and were heavier, while in the normal activity range, more daily exercise led only to the increased caloric intake actually needed to *maintain* body weight. (Reproduced by permission of The Athletic Institute)

The other two contributions that exercise makes to weight control are not so well known. There is evidence that regular physical activity may very well promote more efficient operation of what is known as the body's "appestat." This tiny area in the hypothalamus is responsible for controlling the amount of food taken in with reference to the amount of energy expended, much as a thermostat controls the amount of heat needed in a home. Apparently when animals or people are too sedentary in their living habits, there is a tendency for the "appestat" to be set too high, thus causing a greater intake than is necessary and a resultant excess body fat deposition (Figs. 31–12 and 31–13).

Finally, exercise is apparently a more effective means of establishing a caloric deficit to lose weight (or maintaining a caloric balance so that weight will not be gained) than is a simple caloric restriction. In other words, when exercise contributes to the maintenance of a caloric balance or to the creation of a negative balance, even though the body weight may be the same, the body composition is more optimal. Animal studies have rather convincingly demonstrated this (Fig. 31–14), and there is some evidence that the same is true for humans (see Fig. 31–15).

FIG. 31–11 Approximate energy expenditure in various physical activities. Calories per pound of body weight per minute. Average expenditure is also listed for 120 and 160 pound person for 30 minutes. (Data taken from summary by C. F. Consolazio, R. E. Johnson, and L. J. Pecora, *Physiological Measurements of Metabolic Functions in Man,* New York: McGraw-Hill, 1963, pp. 330–332)

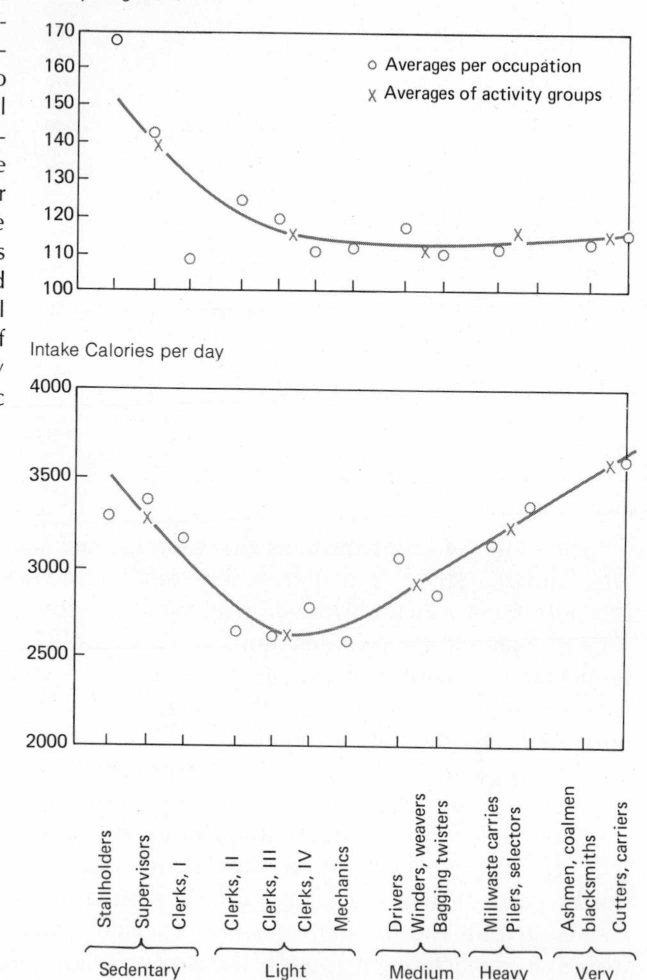

FIG. 31–13 A phenomenon almost identical to that depicted in animals (Fig. 31–12) was demonstrated to occur in humans as well (Mayer). Those in very sedentary occupations ate more than needed and were heavier (to the left of Clerks III), while those more active persons ate only what was required by the increased activity level and no more (to the right of Clerks III). (Reproduced by permission of The Athletic Institute)

General Health and Longevity

According to an interesting kind of folklore, a physically fit person is rarely expected to succumb to the ordinary infectious or contagious diseases, but there is no evidence to support this belief. Most of the research relating to this subject is of the 1920–1930 vintage and raises as many questions as it answers. There is some evidence that the extremely fatigued organism is somewhat more susceptible to at least certain kinds of infectious agents. If this is true, then

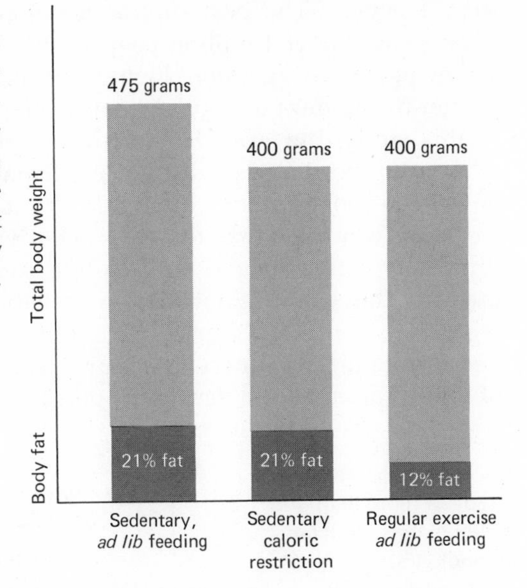

FIG. 31–14 Effect of 15 weeks of daily exercise on body composition of adult male rats, contrasted with caloric restriction per se, and with no exercise, *ad-libitum* feeding. Note that per cent body fat was far less in the exercise group even though average body weight was the same in the caloric restriction group. Percent body fat in the caloric restriction group was actually equal to the heavier, *ad-libitum* fed group. (Jones et al).

the physically fit person might in fact be better able to resist these agents if his fitness is of the kind that postpones fatigue. At any rate we cannot say with certainty that the trained or physically fit person is necessarily less susceptible to infectious diseases such as colds, virus infections, and bacterial infections.

As we have already pointed out, persons who are regularly active tend to be more resistant to at least one form of degenerative disease, atherosclerosis and its related disorders. But with respect to longevity, there is no conclusive evidence that a physically fit person lives any longer or dies any earlier than the

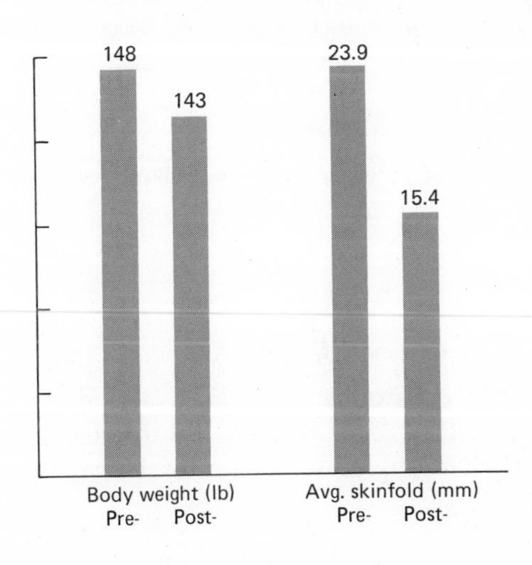

FIG. 31–15 Effects of six weeks' moderate exercise program (about 500 Calories per day, treadmill walking and jogging, six days per week) on body weight and average skin fold fat measurement; no diet control required, $N = 11$ overweight college women. (Data from Moody, Kollias, and Buskirk)

less fit person. The best support for regular exercise resides in Hammond's study of well over 1 million people which indicated that the longevity of more active persons is greater. Figure 31–16 shows rather dramatically that the greater the amount of regular exercise, the longer the lifespan of the individuals in this study; but even Hammond admits the possibility that those persons who indicated that they did not participate in physical exercise regularly may have been those persons in poor health to start with. In such a large sampling, we must conclude that there is some weight thrown behind the claim that physically active people tend to live somewhat longer. Though research to quantify this claim is not readily available, one needs to call upon very little logic to reason that a person with a high degree of circulatory-respiratory fitness, strength, and muscular endurance is better prepared to meet certain kinds of emergencies than a very unfit person.

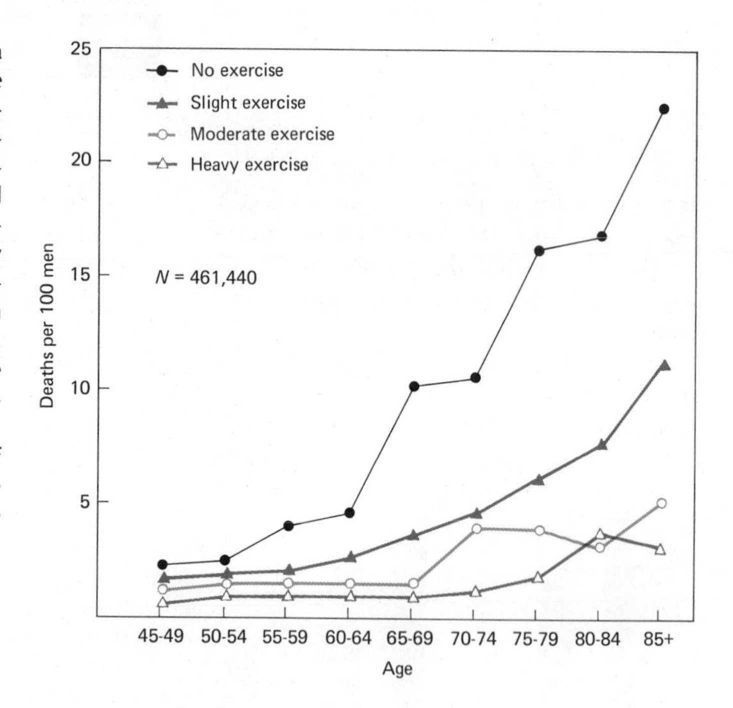

FIG. 31–16 From a study of more than one million persons (Hammond), this graph depicts death rate by degree of exercise and age group during a period of approximately one year after first survey. For example, in the age group 60 to 64, 1 percent (1/100) of the men in the heavy exercise category died, whereas 5 percent of those in the no-exercise group died during the follow-up year.

Readiness for Emergencies

In cases of emergencies, the need to run a mile or two, or to carry out some particular act involving strength and muscle endurance for a long period of time, can become a reality. To stretch the point (but still keep within the realm of possibility) if you were to find yourself hanging from a window ledge, it would be nice to have minimal elbow flexion strength in order to pull yourself up at least one time in order to climb into the window. The odds are against such emergencies occuring in the lifespan of the ordinary citizen, and each person needs individually to decide whether he wishes to be prepared for such potential emergency situations. But there is little question that the physically fit person's life and the lives of others with whom he may be associated may some day depend upon appropriate physical training.

Psychological and Sociological Benefits

To be sure, the psychological and sociological benefits of regular physical activity are less measurable than the physiological benefits; but they are none the less real and perhaps of equal importance. Though it has not been documented (and may never be because of the complexity of the phenomena involved), anyone who has ever improved from a state of poor physical fitness has invariably reported what is referred to as some form of "feeling better." That physiological benefits occur is certainly well-established (see page 433), but to say with certainty that these account for one's "feeling better" is not possible at this time. It is interesting that the report we commonly get is not just a casual observation but one which is fervently expressed and quite obviously a significant, important feeling to those who have experienced it.

The other psychological and sociological benefits of physical activity depend a great deal upon the individual and the particular form of exercise employed. With a little thought you can see the many possibilities for improved self-image, social efficiency, mental outlook and state of relaxation, and so on. However, there is a great danger in assuming that *any* kind of exercise program in *any* person automatically achieves improved social efficiency or an improved and delightful personality. Though these benefits certainly can result, they are certainly not consistently observed and are far from automatic benefits of regular physical activity.

Myths and Common Problems

We will discuss very briefly some of the more common myths and misunderstandings associated with regular exercise and physical fitness and also comment upon some of the more common problems.

The Masculine Woman

There is absolutely no evidence, either direct or indirect, that supports the myth that girls and women who are physically fit run the danger of becoming masculine in appearance.

Athletics and Physical Fitness

There seems to be a rather prevalent misconception that athletics and sports participation are synonomous with physical fitness. As you should know, this is far from the truth. An athlete may be very physically fit, but this depends primarily upon the sport and the intensity with which he approaches conditioning for it. Very unskilled, unathletic, nonsports participants can be much more physically fit than the finest athletes.

Strength Equals Physical Fitness

That physically strong or large-muscled persons are prime specimens of physical fitness may be true, but only if such a person also spends the time and effort developing circulatory-respiratory fitness as well as strength and muscular endurance.

Effortless Exercisers

A little common sense should provide one with the conclusion that the two words "effortless" and "exercise" are diametric opposites! We cannot exercise without effort. There is no shortcut to developing any of the physical fitness qualities described in this chapter.

Athlete's Heart

The term "athlete's heart" can be used correctly only in connection with a stronger, healthier heart. The pathologically enlarged heart (usually resulting from persistent and untreated high blood pressure) is not to be confused with the mythological, so-called athlete's heart.

Sweat Index for Effort

It is easy to confuse sweating with exercising. We often have the feeling that if we are exerting some effort and we are perspiring profusely that this means that we are exercising "hard." Such may be the case, but it is not necessarily so and the amount of sweat produced is much more likely to be related to temperature and relative humidity than to the intensity of the exercise.

Exercise-Induced Heart Attacks

Exercise does not *cause* heart attacks. A healthy heart cannot be damaged by exercise. It is obvious that the diseased heart may be embarrassed by excessive demands and a heart attack may be the result, but this is only because the heart is already in trouble. Exercise may be the precipitating factor but is not the cause. The logic behind the belief that exercise causes heart attacks rapidly loses credence when you recognize that more people suffer coronaries in their sleep than while exercising. Do we then conclude that sleep causes heart attacks?

You Can Be A Champion

Various highly successful athletes and athletic coaches have attempted to motivate youngsters and oldsters alike with the cliche "you can be a champion if you really want to." The general idea is that all it takes is "the will and the hard work to support the will." While such a philosophy may often have spurred youngsters on to great accomplishments in life, the truth of the matter is that everybody cannot be a champion—the philosophy dismisses heredity as a factor of significance; but it is obvious that some people can never develop the physical coordination and quick reaction time that are required in some sports. To imply that such an individual fails to excel because he has not "tried hard enough" is unjust.

Exercise Increases Appetite

Exercise won't do me any good because it will simply increase my appetite. As noted earlier (page 439), sedentary persons tend to overeat. Mayer (1968) has shown that moderately active people not only eat less than very active people but also less than sedentary ones up to the point where they actually *need* more, in which case excess weight would not be added.

The Annual Picnic Sports Hero

The problem of sudden and unaccustomed exercise is a very real and present one. Literally thousands of people each year end up at the hospital or in a physician's office (or in some other way laid up for at least a few days) because valor took over where wisdom should have been in command. The unfit individual who decides to participate in a vigorous game of softball, touch football, basketball, and the like, is simply asking for trouble and, as often as not, his request is granted! As one ages, the risk becomes even more serious because the very real possibility of heart attack exists and this is, of course, of greater consequence than the usual assortment of sprains, strains, and aching muscles.

Physical Fitness Declines with Age

Though physical fitness inevitably declines with age, there are many examples of very highly fit older persons, enough so that we know that a significant reason for deterioration of fitness is lack of physical activity and not the mere fact of increasing age. In fact there is some evidence that regular physical activity delays both the physical and mental aging process [Mateef].

The Manual Laborer is Physically Fit

Another common misconception is that any person who makes his living with his hands is necessarily physically fit. The plumber, the carpenter, the electrician, and so on, are often thought of as very highly fit individuals *by nature of their work*. People in such occupations may very well be physically fit, but not because of their job. Working with the hands obviously does not render a person more physically fit than anybody else.

REVIEW QUESTIONS

1. List some activities that would be of significant benefit in promoting circulatory-respiratory fitness?
2. What is the best way to monitor your progress in a physical fitness program?
3. What are the most dramatic and consistent effects of regular exercise?
4. Why do you suppose exercise aids in mental health?

REFERENCES

BORTZ, E. L., "Exercise, fitness, and aging," *in Exercise and Fitness*, The Athletic Institute, 1960.

COOPER, K. H., *Aerobics*. New York: Bantam Books, 1968. (Paperback)

HAMMOND, E. C., "Some preliminary findings on physical complaints from a prospective study of 1,064,004 men and women," *American Journal of Public Health*, **54:** 11–23 (1964).

HEDLEY, O. F., "Analysis of 5116 deaths reported as due to acute coronary occlusion in Philadelphia 1933–37," United States Weekly *Public Health Reports*, **54:** 972–1013 (1939).

HENRY, F. M., "Influence of athletic training on the resting cardiovascular system," *Research Quarterly*, **25:** 28–41 (1954).

JACOBSON, E., *You Must Relax*. New York: McGraw-Hill, 1948.

JOHNSON, P. B., W. F. UPDYKE, and W. HENRY, "Effect of regular exercise on diurnal variation in submaximal metabolism," *Abstracts of Research Papers*, AAHPER Convention, 1965.

JONES, E. M. et al., "Effects of exercise and food restriction on serum cholesterol and liver lipids," *American Journal of Physiology*, **207:** 460–466 (1964).

KNEHR, C. A., D. B. DILL, and W. NEWFELD, "Training and its effects on man at rest and at work," *American Journal of Physiology*, **136:** 148–156 (1942).

KOZAR, J. J., and P. HUNSICKER, "A study of telemetered heart rate during sports participation of young adult men," *Journal of Sports Medicine and Physical Fitness*, **3:** 1–5 (1963).

MATEEF, D., "Morphological and physiological factors of aging and longevity," in *Health and Fitness in the Modern World,* The Athletic Institute, 1963.

MAYER, JEAN, "Exercise and weight control," in *Exercise and Fitness,* The Athletic Institute, 1960.

MAYER, JEAN, *Overweight: Causes, Cost, and Control,* Englewood Cliffs, N.J.: Prentice-Hall, 1968. (Paperback)

MOODY, D. L., J. KOLLIAS, and E. R. BUSKIRK, "The effect of a moderate exercise program on body weight and fatness in overweight college women." Paper presented at American College of Sports Medicine meeting, University Park, Pennsylvania, 1968.

MORRIS, J. N., and M. D. CRAWFORD, "Coronary heart disease and physical activity of work; evidence of National Necropsy Study," *British Medical Journal,* **2:** 1485–1493 (1958).

MORRIS, J. N., et al., "Coronary heart disease and physical activity of work," *Lancet,* **2:** 1053–1057 (1953).

PEDLEY, F. G., "Coronary disease and occupation," *Canadian Medical Association Journal,* **40:** 147–155 (1942).

RYLE, J. A., and W. T. RUSSELL, "The natural history of coronary disease," *British Heart Journal,* **11:** 370–380 (1949).

SHARKEY, B., and J. P. HOLLEMAN, "Cardio-respiratory adaptations to training at specific intensities," *Research Quarterly,* **38:** 698–704 (1967).

TAYLOR, H. L., "The mortality and morbidity of coronary heart disease of men in sedentary and physically active occupations," in *Exercise and Fitness,* The Athletic Institute, 1960.

UPDYKE, W. F., and P. B. JOHNSON, *Principles of Modern Physical Education, Health, and Recreation,* New York: Holt, Rinehart and Winston, 1970.

ZUKEL, W. J. et al., "A short-term community study of the epidemiology of coronary heart disease," *Journal of Public Health,* **49:** 1630–1639 (1959).

Additional Readings

ANDERSON, K. N., "How your own strength can hurt you," *Today's Health,* April 1965.

JOHNSON, P. B., and D. C. STOLBERG, *Conditioning,* Englewood Cliffs, N. J.: Prentice-Hall, 1971.

PART EIGHT
CONTROL AND REGULATION

32 THE ENDOCRINE SYSTEM

The human body has two controlling systems, the nervous and the endocrine. These do not overlap, much less rival, each other though they do interlock, as will be seen. The nervous system can be compared to a telephone exchange as far as speed goes; the endocrine system can then be compared to a somewhat slower messenger service.

Hormones and Endocrine Glands

The endocrine system exerts its controlling effects by action of its various hormones, chemicals secreted by a body tissue and acting on other tissues. Hormones can be classified in two groups. *Local* hormones are produced by many nonendocrine tissues to act on their neighbors. Thus neurons transmit nerve messages at synapses by secreting acetylcholine or norepinephrine to act on the next neuron; and the small intestine secretes secretin and pancreozymin to act on the pancreas. Local hormones are discussed with the relevant systems, especially nervous and digestive, and will not be mentioned further here. *General* hormones are produced to act via the blood on distant organs or even on the whole body. Thus epinephrine is secreted by the adrenal medulla to act on all sympathetic nerve-muscle and some synaptic contacts; thyroxin, from the thyroid gland, affects metabolism in most if not all tissues. A true hormone acts *only* to stimulate or facilitate action. Thus glucose, though secreted into the blood by the liver and certainly affecting distant tissues, playing a major role in cell metabolism, is not a hormone.

The glands which produce general hormones are the *endocrine* or *ductless glands* (Fig. 32-1). They include anterior and posterior pituitary (hypophysis), thyroid, parathyroids, islets of Langerhans in the pancreas, adrenal cortex, adrenal medulla, ovarian follicles, ovarian corpora lutea, and testicular interstitial cells. The placenta secretes hormones incidentally; the hypothalamus secretes hormonelike substances. The thymus gland, once supposed to be endocrine, is now known to have quite different functions. The pineal body has often been assigned endocrine functions of various kinds, but these have never been confirmed to general satisfaction. The adrenal medulla is discussed with the autonomic nervous system, and the ovarian, placental, and testicular endocrine elements with the reproductive system; these glands and their hormones will therefore be referred to only incidentally here.

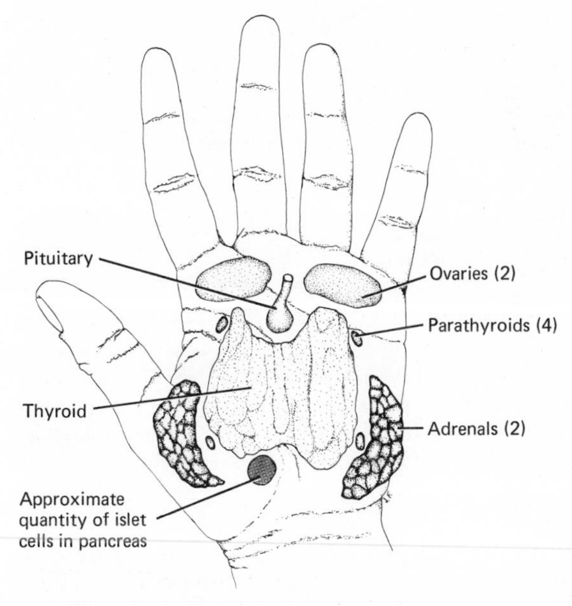

FIG. 32-1 The relative size and amounts of the body's endocrine tissue.

The endocrine glands, as is generally known, secrete their hormones directly into the blood—whence their name, meaning internal-secreting and the alternative term, ductless. For a long time they were a mystery since all glands supposedly required a duct, a tubular channel, to carry off their secretion, as in the case of the salivary and sweat glands, the pancreas, prostate, and others. This latter group is now distinguished as *exocrine,* external-secreting, though, of course, "external" may refer to any surface such as the lining of the stomach (Fig. 32-2).

The endocrine glands have vital importance in spite of their small size. They control growth, distribution and utilization of foodstuffs, metabolic rate in routine and emergency situations, sexual maturation, and a number of lesser functions. Their disorders therefore can lead to serious and even fatal consequences.

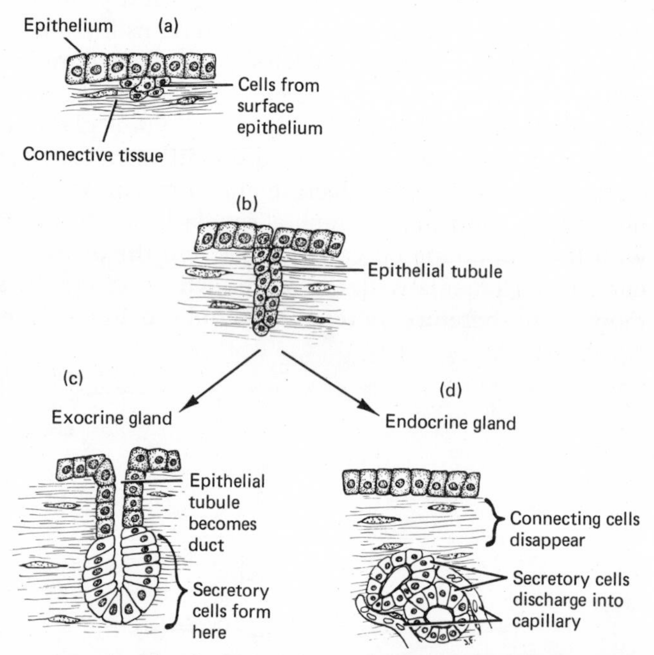

FIG. 32–2 Endocrine and exocrine glands compared. In an exocrine gland, an *acinus,* the saccular or tubular unit, discharges through a duct as in the case of a sweat gland (an embryonic stage shown here in c), or into a system of converging ducts, as in the case of a salivary gland. The ducts open on some surface, exterior (mammary, tear, sebaceous glands) or interior (pancreas, prostate, uterine glands). In contrast, in an endocrine gland, the secreting cells are arranged in a spongework or in cords or plates, every cell facing on at least one capillary. The secretions (hormones) pass through the cell membranes into the blood which distributes them throughout the body to act on their "target" cells. The example (d) represents the thyroid gland, which is peculiar in having its cells arranged in hollow spheres filled with a jellylike colloid. The hormone, thyroxin, can pass directly outward into surrounding capillaries or inward to be stored in the colloid from which it can be rapidly mobilized by passing through the cells into the capillaries.

The Pituitary Gland

The alternative name for the pituitary gland, hypophysis, is widely used, but it has the disadvantages of being too easily confused with the closely related hypothalamus and of producing such clumsy terms as hypothalamohypophyseal tract.

The pituitary gland is a structure about the size and shape of a lima bean —and like lima beans, pituitaries differ somewhat in size. It weighs only half a gram (one-fiftieth of an ounce) more or less. It lies on the floor of the cranial cavity in a bony pocket, the pituitary fossa or, more picturesquely, the *sella turcica* (turkish saddle), and underlies the mass of the brain, to which it is connected by a slender stalk. It is furthermore surrounded by a complex mass of arteries, veins, and nerves. Thus surgical procedures on the gland are extremely difficult and have evoked great ingenuity.

Small as it is, the pituitary is divided into two quite distinct parts: the anterior or glandular lobe, and the posterior or nervous lobe. Other subdivisions are perhaps of importance in some lower animals, but seem to be mere relics without special functions in man.

Anterior Lobe of Pituitary

The anterior lobe of the pituitary gland is the key endocrine gland. It is sometimes poetically called "the conductor of the endocrine orchestra." More accurately, it controls a group of hormone feedback systems involving several, though not all, other endocrine glands, as well as producing at least two hormones that have no known feedback. Its various actions can be best explained by individual examples. Most, if not all, of these actions are further influenced by the hypothalamus (Fig. 32–3).

The hormones of the anterior lobe comprise six major and an uncertain number of minor members. The major ones are the following:

1. A *growth* or *somatotrophic* (body-promoting) *hormone* (SH for short).
2. A *thyrotrophic* or *thyroid stimulating hormone* (TH or TSH) which we shall discuss with the thyroid gland.
3. A hormone stimulating the adrenal cortex, *adrenocorticotrophic hormone* (ACTH) which we shall discuss with the adrenal cortex.
4. An ovarian *follicle-stimulating hormone* (FSH).
5. A corpus luteum-stimulating or *luteinizing hormone* (LH).
6. A hormone, *prolactin,* having to do with milk production.

The latter three hormones also have effects in the male, but are today called by their "female" names even in males. In addition, there is a minor hormone that causes pigment cells to expand in the skins of some lower animals, and is said to cause darkening in the skin of some humans by proliferation of the pigment melanin—hence the name *melanizing hormone,* MH; there are possibly others not yet widely accepted. The hormones SH and MH are those without known feedback. The hormones FSH, LH, and prolactin are discussed with reproduction.

Growth or Somatotrophic Hormone

The effects of the growth hormone excite great popular curiosity which does not, however, justify disproportionate treatment here. For our purposes, one

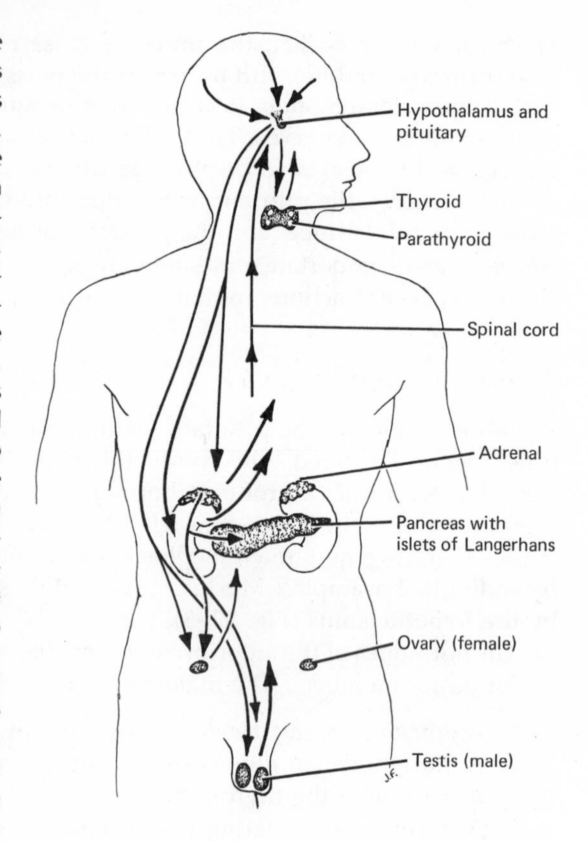

FIG. 32–3 The hormone web. The hormones of the endocrine glands form a series of feedback systems, as well as semiindependent agents. They must also be geared to the needs of the whole organism in emergency, stress, increased physical activity, reproduction, and so on. The key to all this is the hypothalamus-pituitary complex. The hypothalamus is a brain center with some characteristics of an endocrine gland. As a brain center it receives reports indirectly from our sense organs and other brain centers; it responds to these automatically, having little capacity for discrimination. As an endocrine gland, it acts on the pituitary by chemical "reactor substances", themselves, in effect, hormones. The pituitary, in turn, acts by hormones on other endocrines—the thyroid, adrenal cortex, and gonads—as well as on nonendocrine tissues—kidney, breast, and others. As secretions from the stimulated endocrine glands increase in the blood, they exert a negative feedback suppressing the corresponding pituitary secretion (recurrent arrows). Some also act on other endocrines, as for example the adrenal gland acts on the pancreatic islets and on the ovary-testis. The great complexity and subtlety of endocrine interactions explain the many pitfalls of hormone therapy and the great skill required by such medical specialists. The parathyroids and, to a great extent, the pancreatic islets, have simple functions and are controlled by concentrations of calcium and sugar in the blood, respectively. The adrenal medulla is under separate, direct control by the sympathetic system.

should know that the hormone exerts a widespread effect on metabolism causing a diversion of metabolites from storage or energy output to growth and generating an extra demand for nourishment; that this growth affects all tissues of the body more or less equally, but is not concerned with the growth-plates of the bones, which remain active or disappear independently; and that prolonged illness, starvation, and other stresses seem to restrict growth by inhibition of the hormone. We do not know what, besides these stresses and heredity, controls output of the hormone nor why growth stops at maturity though SH continues to be secreted, even if at a slower rate.

Popular curiosity centers on the results of over and underproduction of growth hormone. The first produces, at ages when growth is normal, an overgrowth resulting in *giantism* (or gigantism). True giants, ranging up to 8 feet in height, are more or less normal in form. They are not usually strong in proportion to their size, however, since body mass increases in proportion to the cube of the height, whereas strength is dependent on cross-sectional area of muscles and is therefore proportional only to the square of height; thus a giant may actually be handicapped by having to carry much more bulk with only somewhat more strength—maybe Goliath could not have been such a threat to the smaller and nimbler David after all. Giants usually stop growing at maturity like everyone else for the same unknown reason.

Abnormal growth (benign tumor) of the cells which produce growth hormone occasionally occurs in an adult and gives rise to the disease *acromegaly*. There is an increase in the size of all the organs of the body: the heart, the lungs, the kidneys, the spleen, and other parts. There is a marked increase in the size of the peripheral parts of the skeleton: the chin, the brow, and the feet and hands.

Too little growth hormone during childhood produces the other extreme of human size, the midget. Unlike the hypothyroid cretin, the midget is more or less well-proportioned, though the head may be disproportionately large as in a child. One might think that this condition, or simply moderate underdevelopment, could be treated by administration of SH, as diabetes is treated with insulin. Unfortunately, SH from animals other than monkeys is ineffectual in man; and, of course, available monkey (much less human) pituitaries are too scarce to provide adequate supplies of the hormone as animal pancreases do of insulin. Such treatment must therefore await synthesis of human-type SH in the laboratory.

Posterior Lobe of Pituitary Lobe

This curious structure is associated with two hormones, *antidiuretic hormone* (ADH or *vasopressin*) and *oxytocin*. It does not itself secrete these hormones but acts as a depot for them. Actually, the hormones are secreted by nerve cells in the overlying hypothalamus and are transmitted to the posterior lobe by the axons of these cells which run down the slender stalk of the pituitary. If this seems a peculiar arrangement, it should be compared to the discharge of *transmitter substances* at the terminals of all axons; the hypothalamic discharge is simply an enlargement and a specialization of this universal neuron activity. The only real difference is that the secretions do not act on other neurons or on muscle fibers, but are stored in the spongy tissue of the posterior lobe. There they are picked up as needed by a web of capillary blood vessels.

The antidiuretic hormone is the more important of the two. Its function, as its name implies, is to control excessive excretion of urine (diuresis). This it does by a feedback mechanism. If blood plasma becomes even slightly concentrated by loss of water, the resulting slight excess of salts acts on the relevant cells in the hypothalamus, causing more of the antidiuretic hormone to be produced; when the hormone is transmitted to the posterior pituitary, picked up by the blood, and finally reaches the kidneys, it acts on the excreting tubules to stimulate greater retention (actually reabsorption) of water. Thus, in increasing thirst, the urine becomes increasingly scanty, concentrated, and dark in color due to the increased production of ADH. Conversely, a slight thinning of blood plasma due to too much water intake (or other cause) inhibits secretion of the hormone, which permits the kidney to excrete more water, giving a greater quantity of dilute, pale-colored urine. The reason for this complicated arrangement is unknown. But the feedback system is very delicate and efficient.

Results of damage to the system, usually by severance of the pituitary stalk but also for rarer reasons, can seriously disturb kidney function. When antidiuretic hormone does not reach the bloodstream, much of the control over water output is lost. The kidney then excretes vast amounts of highly dilute, almost colorless urine. This condition is known as *diabetes insipidus* and has been discussed with excretion (Chapter 25).

Oxytocin is probably less important. It stimulates contraction of the uterus during parturition, and it is sometimes used medically for this purpose. Recent experiments suggest that oxytocin causes "ejection" of milk from the glandular parts of the breast into the ducts when a baby is nursing. The baby's sucking apparently stimulates the mother's nipple, causing nervous impulses to be sent to the spinal cord, and then through the brain stem to the hypothalamus. The hypothalamus releases the hormone to the posterior lobe where it is absorbed into the blood and milk is "let down." (Sometimes this reflex is so strong that simply the baby's crying will result in the release of oxytocin and milk spurts from one or both breasts before nursing begins.) Nursing, then, stimulates release of oxytocin, which in turn affects not only the breasts but also the uterus. This is the reason that nursing is said to hasten return of the uterus to normal size following birth.

The Thyroid Gland

The thyroid lies at the base of the throat where its narrow middle portion, the *isthmus,* crosses the trachea; its broad *lateral lobes* extend up on the sides of the larynx (Adam's apple) and down almost to the thorax. When enlarged, as in goiter, the gland bulges under the skin and down over the collarbone, but not upward because of muscles binding it down. In such cases, it does not usually obstruct breathing because it is soft and the trachea is firm; however, it may obstruct swallowing by pressing on the soft esophagus behind, especially in the narrow opening of the thorax.

The thyroid gland is by far the largest of the endocrines; it normally weighs about 25 grams (nearly an ounce), compared to roughly half a gram for the pituitary and only a gram for all the pancreatic islets. Its structure is unique, being composed of thousands of tiny spheres, called follicles, their shells built of cells and their interiors filled with a yellowish jelly, the thyroid *colloid*. This peculiar structure has one curious result: The colloid is rather well insulated from the blood and its antibody-forming elements; hence, in later life, a person may produce antibodies to his own thyroid proteins if these are exposed to the blood stream, with destructive effects on the gland. This is fortunately rare.

Thyroxin, the product of the gland, is well-known as a regulator of body metabolism. Too much of it, *hyperthyroidism*, overexcites the metabolism and produces a restless, excitable, tense personality, and usually a body in constant activity, tense and restless even at rest, and lean because it burns up even a larger than normal food intake. Great excess, due to pathological overgrowth of the thyroid, causes *thyrotoxicosis* characterized by protruding eyes, tremor, mental and emotional instability, and fatigue. The heart is strained in many cases. Too little thyroxin fails to maintain normal activity and produces opposite effects which, however, differ according to age. *Hypothyroidism* in infancy produces *cretinism*. (The origin of the term cretin is curious. It is not from the island of Crete as is sometimes incorrectly stated. Rather, it came from Switzerland where severe iodine deficiency made thyroid disorders, including cretinism, common. It appears to be a corruption of the French word for Christian—*chretian;* an idiot was considered incapable of sin and therefore a natural "child of God.") The infant fails to grow well or to develop mentally, and presents a "pot belly," bloated tongue, thick, waxy skin, and other features; he often succumbs early to infectious disease. Hypothyroidism in later life produces a condition called *myxedema*. The symptoms are increasing mental dullness, sometimes with a tendency to misconduct, a puffy face and body due to accumulation of a gelatinous fluid under the skin, a curious light-brown tinge to the skin, and often hoarseness due to puffiness of the vocal cords. Often a mild degree of myxedema in women will cause a simple listlessness and an amenorrhea (absence of menstruation). It is the latter difficulty which usually brings these patients to a physician and makes possible the diagnosis of their condition.

These various disorders are associated with changes in the size of the thyroid gland. Hyperthyroidism of any degree is usually associated with an enlargement, visible as a swelling or *goiter*, at the base of the throat (see Fig. 32–4). Cretins, on the other hand, lack thyroid tissue, sometimes entirely, although it may be present but not functional. Myxedematous persons, paradoxically, also usually have goiters.

The key to all of these problems is iodine, as is well known but not often understood in application. Thyroxin is a rather simple compound, characterized by four iodine atoms; a major activity of the thyroid gland is the capture of iodine from the blood, which it does so effectively that it normally contains

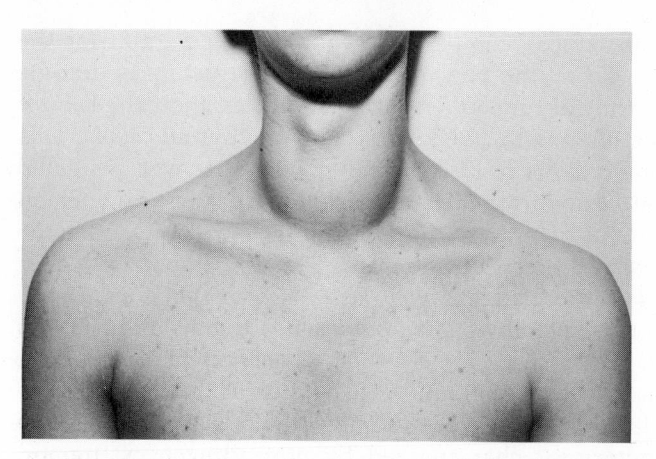

FIG. 32–4 Simple goiter is an enlargement of the thyroid gland. This condition may develop from lack of iodine in the diet. (Armed Forces Institute of Pathology)

25 to 600 times the concentration found in the rest of the body. This action is very efficient and provides a wide margin of safety, since not all thyroxin is secreted immediately and any surplus is stored in the interior jelly of the follicles. Disregarding cases where the thyroid is absent, atrophied, or nonfunctional, the mechanism can go wrong in several ways.

First, and formerly most common, is iodine deficiency. When this occurs, thyroxin is low in the blood and the thyroid-stimulating hormone of the pituitary increases to excite the thyroid to greater production. The gland responds by enlarging, and the enlargement of the thyroid is the nontoxic or benign goiter, and it may reach an enormous size. Iodine is associated with the sea and seafoods and is scarce in regions that have long been separated from the sea geologically, such as mountains and the interior of continents. Hence, nontoxic goiters were once endemic in such regions (though not unknown elsewhere); they were most common in women, who have a special need for thyroxin during sexual maturation and pregnancy. Rare forms of nontoxic goiter occur when ample iodine is present, but the thyroid chemistry is faulty and cannot concentrate it.

The cure for nontoxic goiter is simple addition of iodine to the diet. Not much, only about one milligram weekly, a twentieth of a gram a year, is required, as the body retains and recirculates iodine to a great extent. The introduction of a mere trace of iodine salts to ordinary table salt has been so effective that cases of ordinary nontoxic goiter are seldom available for medical study in most countries. (A curious side light on this achievement was the furor, now almost forgotten, that was aroused by iodized salt; almost exactly the same emotional objections and fictitious arguments were used as have been raised against introduction of that other halogen, fluorine, into drinking water; one may hope that the present obstructionism will pass into the same oblivion.) If thyroid chemistry itself is at fault, simple iodine therapy is ineffective; feeding of dried thyroid, or thyroxin, supplies the hormone directly.

Feeding of these substances not only halts but reverses growth of the goiter. They react on the pituitary and suppress the abnormally high secretion of TSH, which in turn ceases to stimulate thyroid growth. The thyroid then shrinks more or less slowly. Surgical removal of excess thyroid, in this form of goiter, is seldom used today except for urgent cosmetic reasons or to relieve pressure on other structures such as the esophagus.

Another form of goiter is due to tumors of the thyroid. The reason for such tumors, as for all others, is unknown, but they develop without pituitary control. They are usually benign but can become malignant. These tumors may be formed of primitive, embryonic thyroid-precursor cells, which do not produce thyroxin and are therefore nontoxic. However, they can develop into malignant or toxic forms which are usually removed. The toxic forms are chemically active, producing excessive thyroxin and can cause thyrotoxicosis; thus they *must* be removed. Removal may be surgical or by administration of radioactive iodine (Fig. 32–5). The latter concentrates in the thyroid sufficiently to cause controlled destruction while remaining too dilute elsewhere to have any effect. However, since side effects do tend to follow this treatment after many years, it is usually reserved for elderly people.

Finally, a goiter may appear spontaneously without known cause, though heredity is suspected. This is known as Graves' disease. The goiter is not tumorous, as shown by its structure, but is chemically overactive, producing excessive amounts of thyroxin with resultant toxicosis. This disease occurs chiefly in youth, rarely as late as 50 years of age, and is most common in females. Partial destruction of the thyroid, by surgery or radioactive iodine, is the usual treatment. You should be alert for any of the typical symptoms described for hypo- or hyperthyroid conditions and report these to a physician if they persist.

Parathyroid Glands

The parathyroid glands are normally four in number, two on each side (see Fig. 32–1). They are yellow-brown bodies about 5 millimeters in diameter and lie at the posterior edges of the thyroid lobes, just outside the thin capsule of the thyroid. Occasionally displaced parathyroids occur at any level of the throat or in the upper thorax.

The parathyroids secrete *parathormone* which raises the concentration of calcium in the blood. This function has wide repercussions throughout the body. It entails withdrawal and redeposition of bone-calcium, affects indirectly the concentrations of phosphorus which depend on those of calcium, and thus determines the balance of these elements in all body tissues — a balance vital to the performance of all tissues, especially muscle.

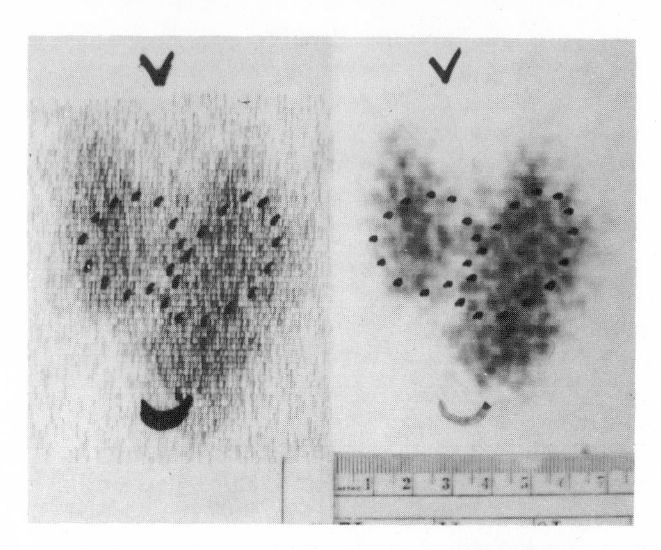

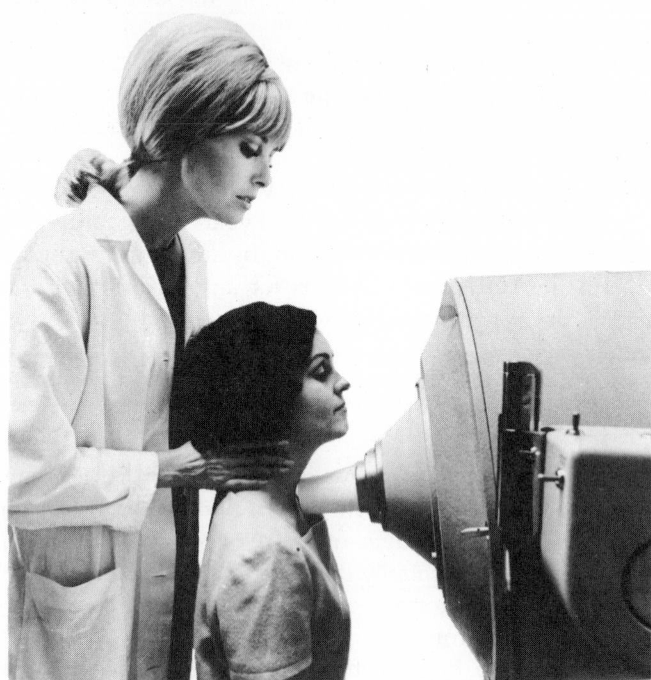

FIG. 32–5 Measurement of the uptake of radioactive iodine in the thyroid gland of a patient. The counter records the areas of concentration of iodine in the gland as shown on the right. In this case the uptake of radioactive iodine is being used to measure the activity of the thyroid gland. (Brookhaven National Laboratory; Nuclear-Chicago, Corp.)

A lack of parathormone shows itself first, and most drastically, by muscular cramps which develop into a form of tetany. Without treatment, this condition can be rapidly fatal; but the tetany can be relieved by injection of parathormone or, more simply and directly, of calcium. In many cases, the treatment can be discontinued after a time. Whether this is because the body adapts to the lack of hormone or because the parathyroid tissue regenerates is not known. When the parathyroids are barely adequate for routine life, a reduction in body acid (as when severe vomiting washes out hydrochloric acid from the stomach or when heavy breathing expels much carbon dioxide from the lungs) can bring on an attack of tetany. Attacks, from whatever cause, are sometimes mistaken for epilepsy.

The activity of parathormone is closely associated with that of vitamin D. Without vitamin D calcium cannot be properly absorbed from the intestine, so that the hormone must draw all its supplies from bone. This, with further complications, can cause the disease rickets (see Chapter 14).

Too much parathormone causes excessive withdrawal of calcium from bone. In extreme cases, as when an active parathyroid tumor is present, this causes cavitation and fragility of the bones, a racing heart, headache, nausea, and loss of appetite. The calcium in the blood reaches such a high level that the kidneys are overloaded in their efforts to excrete it and it is precipitated as a form of kidney stone. The bones themselves develop a peculiar "moth-eaten" appearance in X-rays. The condition can easily be treated by removal of the tumorous gland since the three others are more than adequate to sustain normal function.

Parathyroid disorders are uncommon, especially since surgeons have learned to recognize the glands and to carefully preserve them in operations on the thyroid.

Pancreatic Islets and Insulin

The islets (often called islets of Langerhans after their discoverer) in the pancreas are widely known. This is because their failure produces the all too common disease, *diabetes mellitus,* and because the dramatic story of insulin treatment, and its success, is such common knowledge.

The islets vary greatly in number in quite normal people. An average count would be a million. They are, however, so small—about a tenth of a millimeter in diameter—that their summed mass is only about a gram, a twenty-eighth of an ounce, compared to the full ounce of the thyroid. Even this small mass of tissue is not devoted exclusively to secreting insulin.

Insulin, the chief secretion of the islets, supervises utilization of glucose (blood sugar), the basic energy fuel of all tissues. It does so, apparently, by facilitating passage of the glucose through the membranes of certain cells. Without it, concentration of glucose may be abnormally high in the blood but hardly any passes into the cells. Complete failure of glucose absorption does not occur, since a certain amount passes through without the aid of insulin, and a certain amount of insulin is still present in most diabetics; hence the results of insulin lack are not acute but are slowly progressive. Special mechanisms, such as transformation of proteins into sugar through the action of the adrenal glucocorticoid hormones, may strive to supply the starving cells, but in vain; without insulin, the most lavish supplies cannot get into the cells adequately. This is especially true of muscle cells, both voluntary and cardiac, of fibroblasts and of fat cells; it is hardly, if at all, true of nerve cells, absorptive cells in the intestine, and excretory cells in the kidney. These various effects explain the nature of diabetes mellitus.

The most familiar symptom, of course, is excessive urination. When glucose accumulates in the blood to above 180 milligrams per 100 milliliters of blood (180 milligram percent), the surplus is excreted by the kidney. When blood sugar reaches high levels, approaching 500 milligrams percent, excretion also reaches high levels; and this requires extra quantities of water to carry if off, unless the urine is to become a viscous syrup. Even so, the sugar content justifies the name diabetes mellitus "honeyed through-flowing," in contrast to diabetes insipidus (tasteless). The loss of water generates a constant thirst which may be raging in severe cases.

The loss of sugar and water, however, is less significant than accompanying internal disorders. The muscles, lacking adequate amounts of fuel, can act only weakly, and waste away. The wasting may be accelerated by the adrenal effect described earlier. The digestive tract continues to absorb sugar and to convert starch into absorbable sugar, but to no avail since the supplies only go to increase the amount excreted along with the loss of water. Fatty tissue also cannot maintain itself and disappears, eventually leading to emaciation. Appetite may be ravenous, but, of course, no amount of intake halts the steady wasting and weakening. Indeed, sugary and starchy foods only worsen the condition.

Disorganization of sugar metabolism throws general metabolism into disorder also. In particular, abnormal amounts of chemicals called ketones, acetoacetic acid, and others are formed and accumulate faster than they can be broken down or excreted, until they reach toxic concentrations. Further, the body's efforts to get rid of the toxins use up sodium and results in acidosis of the blood. As noted, nerve cells are little affected by insulin lack, but they may be severely affected by the toxic by-products and acidosis, so that mental disorder and coma result. Also, loss of water and salts, together with weakness of muscle in blood-vessel walls, leads towards a condition of shock. Coma and shock are terminal results of diabetes, leading to death. Prolonged diabetes, even when carefully controlled, can cause an excess of fat metabolism and this seems to cause an accelerated rate of atherosclerosis. This is the reason that the most common disabling consequences of the disease are related to the vascular system. The poorer circulation may lead to coronary artery disease, stroke, or kidney disease. One of the most common results of prolonged diabetes is diabetic blindness caused by the degeneration of the delicate arteries which supply the retina.

The administration of insulin (Fig. 32–6), however, and a carefully controlled diet can do much to make the disease tolerable. Insulin cannot be taken orally because the digestive juices would break it down; hence it must be injected daily. However, once the diabetes has been controlled, certain drugs, which stimulate the residual islet tissue, can be taken orally and these, together with the carefully prescribed diet, can often hold the diabetes under control without insulin.

Also, we should make a distinction between the juvenile and the adult

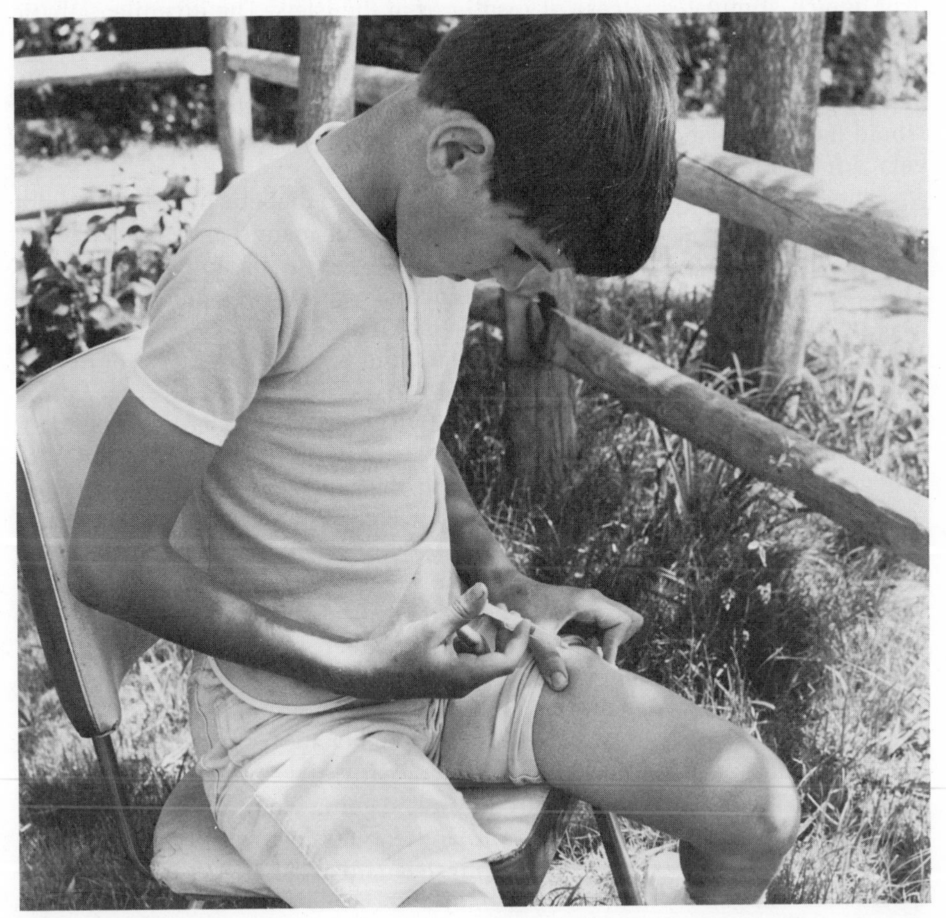

FIG. 32–6 This diabetic boy has learned to give himself hypodermic injections of insulin at the required times. (Michal Heron)

forms ot the disease. The child diabetic often has very poorly developed islets, and sometimes they are completely absent. The adult diabetic, on the other hand, is often found to have completely normal islets. Both suffer from a deficiency of insulin, but the deficiency found in the adult diabetic seems to be related to an inhibition of its release or a suppression of its production—perhaps, it has been suggested, because of a "diabetagenic" hormone produced by the anterior lobe of the pituitary or by the adrenal cortex. In general, adult diabetes results from a lifetime of overeating and a consequent stress on the insulin-producing mechanism. The juvenile diabetic is born with a deficiency or absence of islet cells.

This accounts for the difference in the severity of the two forms of the disease, and also the reason that the adult diabetic can often take the oral diabetic drug which stimulates release of insulin from the islets. The juvenile diabetic cannot benefit from the drug because he has little or no insulin to be released. If the vascular defects resulting from the disease develop only after 20 or more years, as seems to be the case, it would be expected that the juvenile diabetic would more often show these disabilities, since the adult diabetes usually develops in late middle age.

Diabetes is probably hereditary. That is to say, a diabetic heritage does not make development of the disorder inevitable; but many persons in diabetic families do inherit a *tendency* which requires a certain amount of precaution. These so-called prediabetics can often completely avert development of the disease by medication and mild restrictions on diet.

Insulin shock is the converse of diabetes and is due to an excess of insulin in the blood. It is usually caused by one of two conditions: an actively secreting tumor of islet cells, which can be treated by surgery if detected soon enough; or by insulin therapy itself when not conducted with proper care. Insulin, of course, helps the body to use up sugar in the blood; and if this occurs too rapidly without immediate replacement, the concentration of blood sugar can be driven too low to support cellular activity especially in the neurons which are highly vulnerable to sugar lack. The results, in degree of severity, are tremor, mental confusion, convulsions, and unconsciousness. Since this condition develops rapidly, it is a major hazard of diabetes; it, and not the diabetes itself, is the reason why all persons taking insulin should carry an easily found identification card. A person in insulin shock may be mistaken for a drunk and treated accordingly by authorities, which may have fatal results. Treatment is simply administration of sugar, orally in early stages or intravenously in later stages. The diabetic should carefully avoid going without food particularly about the time of an insulin injection, and should always carry an adequate supply of sugar, as prescribed by his physician, to take when exertion or other emergency brings on premonitory tremor and confusion.

Although insulin is a specific treatment for diabetes mellitus, one should here, as in all endocrine matters, avoid the fallacy of "one gland, one hormone." Other hormones act on the blood glucose level; certain islet cells themselves produce another hormone *glucagon*—glucose activator—that raises the level; glucocorticoids, as already noted, also tend to raise the level, and pituitary growth-hormone tends to depress it. All these hormones and others must act in balance with insulin. Sometimes their abnormalities cause less common disorders of sugar metabolism.

The Adrenal Cortex

The *adrenal glands* are also known as the *suprarenal glands*. These two glands, right and left, lie on the upper poles of the kidneys, whence their name—*ad*:at, (*supra-* or *epi-*:above) *ren*:kidney. They are inevitably compared to cocked hats tilted rakishly toward the inner borders of the kidneys. The combined weight of the glands varies with age, being high at birth, falling off during childhood, increasing again up to maturity and then remaining fairly constant throughout life. The mature weight is around 5 grams for each gland, roughly a third of an ounce for the two, which makes these glands the largest endocrines next to the thyroid.

The adrenal medulla (Chapter 33) is much smaller than the cortex. The latter forms about 85 percent of the total mass, which makes it alone still the second largest endocrine. Hence one would expect the adrenal cortex to play an important role in the body; in fact, destruction of the adrenal cortex, by experimental removal or by disease, leads to death within a day or so unless therapy is promptly initiated. This is not true of other endocrines, even the pituitary. The functions of the adrenal cortex have been aptly, if roughly, summed up as "salt, sugar, stress, and sex."

The adrenocortical hormones belong to a family of chemicals called *steroids*. The basic steroid molecule, cholesterol , is not very complex as organic molecules go; but it is capable of great variation both in its basic framework and in the accessory atoms attached to it, without losing its essential steroid character. A great number of steroids (30 to 40 according to various standards) can be isolated from the adrenal cortex, but most of these are evidently stages in development of the final products from cholesterol. Only a very few are established as major hormones.

Mineralocorticoids

Aldosterone and perhaps one or two other closely related hormones are collectively referred to as the mineralocorticoids, and they are together probably the most important of the adrenal steroids, their loss causing the sudden death in adrenal failure. They regulate the metabolism of the metallic ion, sodium, which in turn controls the concentrations of chloride and potassium ions. The crucial importance of these substances, especially sodium, is known even to most laymen. Sodium must be maintained at a very precise concentration in all body fluids and cells if normal function is to continue, so that sudden excessive or long-continued loss of the salt is certain to produce serious repercussions throughout the body. Excess sodium causes intense thirst and shriveling of tissues as water is withdrawn from them by osmosis (see discussion of sodium in Chapter 14).

Aldosterone acts through the kidney by stimulating the retention of sodium. By a feedback mechanism, an excess of the salt in the blood depresses secretion of the hormone so that more salt is excreted, and vice versa. The balance of salt retained and excreted determines the amount of water needed to maintain a precise salt concentration in the body fluids and, so, indirectly regulates the amount of water excreted; this control counterbalances the control exerted by the antidiuretic hormone. Since sodium and potassium must be maintained in precise balance, potassium excretion is also precisely regulated by the mineralocorticoids.

Glucocorticoids

A second important group of adrenal steroids comprises the *glucocorticoids.* These act, as their name indicates, on glucose metabolism, but also on other things such as inflammatory reactions and allergies. Corticoids are the steroids that are most familiarly used as drugs in some forms of arthritis and other inflammatory conditions. They include cortisol, cortisone, and hydrocortisone. (The nomenclature of the steroids is regrettably complex and confusing, and confusion is increased by differences in American and British usage. Besides natural forms found in the adrenal cortex, synthetic variants exist and may be used medically. A recognition of mineralocorticoids, glucocorticoids, and gonadal corticoids is adequate for general understanding.)

The glucocorticoids, in general, reinforce body defenses against emergency. In this, they complement the action of epinephrine from the adrenal medulla, which may explain the otherwise strange association of medulla and cortex derived from entirely different sources. The medulla braces the body against immediate emergency, preparing it for "fight or flight"; in particular, it facilitates activity of nervous tissues, and mobilizes sugar from the liver. The glucocorticoids take over for the "long haul" as in cases of infection, slow-healing injuries, or prolonged emotional stress. Since the liver provides only a short-term reserve of glucose, the glucocorticoids mobilize glucose from breakdown of body tissues, especially the muscles, and they too activate nervous tissue though to a lesser degree. Thus a wounded animal or man is often enabled to fight off or escape from the source of injury and then to mobilize resources for healing when safety has been reached.

Glucocorticoids also promote healing itself. The normal body reaction to most injury is inflammation, which first entails release from the capillary vessels of fluid to dilute and carry away toxic material and the arrival of white blood cells to combat bacteria; this phase is followed by formation of fibrous scar tissue. However, inflammation, like many normal activities, is apt to get out of control. One role of the glucocorticoids is to keep inflammatory reaction within bounds, though how they do so is still not clear.

The most famous application of glucocorticoids is in the case of rheumatoid arthritis. This is a disease of the joints in which inflammation has gotten out of hand. Administration of glucocorticoids relieves the condition with almost miraculous promptness. At first this was no more than a medical curiosity, since amounts of the steroids available from animal adrenals were adequate to treat only a very few patients. Strangely, however, the steroid cortisone can be extracted in large amounts from certain tropical plants, and the supply has been greatly increased by cultivation of these plants as well as by laboratory synthesis.

Medical use of glucocorticoids has, however, serious drawbacks. Total suppression of inflammation would be undesirable and often highly dangerous. Treatment with cortisone in a patient with quiescent (much more, with active) peptic ulcers, for example, is likely to cause reopening or aggravation of the ulcers; and all defense against infection is greatly decreased during glucocorticoid therapy, so that careful precautions must be taken. Thus the therapy should be conducted by highly trained personnel.

A special hazard of glucocorticoid therapy is faced when the patient is subjected to further major stress. Unlike secretion of the mineralocorticoids, which is controlled by feedback from salt concentration in the blood, glucocorticoid secretion is controlled by pituitary ACTH. This also is a feedback system but one in which the controlling factor, the pituitary, is a living structure and subject to atrophy of disuse. Thus glucocorticoid therapy, maintaining a high blood level of the hormones, depresses secretion of ACTH; and this pituitary activity may require a long time to recover after therapy is discontinued. In these circumstances, any fresh stress, such as infection, accident, or operation, finds glucocorticoids in short supply and the pituitary unequal to mobilizing more of them. The consequences may be disastrous for the body in general and for the unprepared adrenal cortex itself. The only safeguard is renewed administration of glucocorticoids or increased dosage if the therapy is still in force to tide the body over the crisis. Therefore, in any such situation, especially before an operation, the physician or surgeon should know of any corticoid treatment within the preceding year or less.

Adrenal Sex Hormones

Finally, sex hormones are secreted by the adrenal cortex. These are the same as those secreted by the sex glands and placenta; they include both male and female hormones—androgens, estrogens, and progesterones—in all individuals male and female. Just why their secretion should have developed in this accessory source is unclear since the amounts are minor. Possibly they provide a supplement of hormone from the opposite sex, which somehow provides a check and balance for the prevailing sex hormone. They do, however, complicate the disorders of the cortex.

In underactivity of the adrenal cortex sexual functions decline. In moderate cases only body hair, especially that of the axillary is lost. In severe cases, failure of menstruation in the female, of potency in the male, and of libido in both, may occur, but these effects may in part be due to general weakness and debility from loss of other adrenal hormones. In general overactivity of the cortex, any effect on sexual activity is, again, masked by the overwhelming disorder of many functions. But in rare cases adrenal sex hormones alone are oversecreted, the androgens usually predominating: In male children, sexual maturity may develop by four years of age or even earlier, with enlarged genitalia and excessive muscular development (infant Hercules); in the female, male characteristics may develop to such a degree that the child is considered to be a boy. In adults, women are the chief sufferers with conditions ranging from mild hirsutism (hairyness) and failure of menses, to extreme masculinization with heavy growth of male-type hair (bearded lady) and liability to male baldness, male distribution of fat, enlarged muscles, deepened voice, and so on. Many of these symptoms can now be cured by surgical removal of excess adrenal cortex or cortical tumor.

The hazards of glucocorticoid therapy has made the use of ACTH a possible alternative—the injection of ACTH stimulates the adrenal cortex to greater activity and less atrophy of the anterior lobe of the pituitary would result. However, the whole adrenal cortex is stimulated, and an excess of the sex hormones are produced as well as glucocorticoids (but not mineralocorticoids). Consequently, the masculinizing effects in women similar to those occurring with adrenal tumors are likely to develop; feminizing in males is also a possibility. Thus caution in the therapeutic use of ACTH is also essential.

Disorders of the Adrenal Cortex

Lack of secretion may come on suddenly due to infection of the cortex or, more often, to stress acting on a cortex just barely adequate for routine activity. In any case, the condition is extremely serious and can be fatal within 48 hours; thus it is treated as a major medical emergency. The general symptoms are a sudden weakness, shock, loss of consciousness, and rapid temperature fluctuations between nearly normal and as much as 105. Treatment with corticosteroids and skilled supportive therapy can save life in many cases; without therapy the disorder is invariably fatal.

Chronic adrenocortical insufficiency causes *Addison's disease*. This is uncommon but not rare; it is recognized as the cause of four deaths per thousand in this country, but probably many unrecognized cases die when sudden stress acts on a failing adrenal, the stress (infection or injury) being considered the cause of death. Of course, increasing numbers of properly treated cases live a normal lifespan, and recognition of Addison's disease, in its early stages, would prolong many lives.

Symptoms of the disease begin with progressive weakness and easy fatigability. These are least after a night's rest and increase during the day; they are aggravated by even a minor infection or injury and cause any convalescence to drag out abnormally. Of course, many other conditions, serious or not, cause weakness and fatigue and might suggest Addison's disease; but when the symptoms are severe and persistent, the disease should certainly be considered as a possibility and confirmed or excluded by professional tests.

As the disorder progresses, evident Addison's disease emerges. Most conspicuously, the skin takes on a coppery pigmentation, especially in creases, scars, and the nipples, and over the gums and friction areas such as the elbows. Loss of weight occurs and blood pressure drops, with attacks of dizziness and fainting. Weakness and fatigue increase, including mental depression irritability and emotional instability. The condition may become static or progress toward death.

Treatment, again, is simply by corticoids which can sustain a relatively normal life to normal limits. People receiving such treatment must, however, avoid unusual stresses of infection, injury, or emotion, as far as possible, and seek extra therapy if stress does occur.

Excess of secretion by the adrenal cortex causes *Cushing's disease (syndrome).* The conspicuous symptom of this disorder is a peculiar form of obesity. The limbs remain unaffected and may even appear thin by contrast; the trunk develops fat pads especially over the shoulder-blade region (buffalo hump), base of the neck, hips, and abdomen. In the latter region especially, but also elsewhere, reddish streaks (striae) may appear due to stretching and splitting of subcutaneous tissue with inadequate healing due to excess corticoids. In fact, inadequate healing is general so that even small injuries tend to become infected and ulcerate, and bruising is easy. A rounded "moon face" with reddened cheeks is typical. The skin becomes thin and papery.

Unpleasant as these symptoms are, internal disorders are more serious. The excess glucocorticoids cause muscle breakdown and weakness (note that weakness accompanies both excess and deficit of adrenal secretion). The resulting high blood-sugar level stimulates the pancreatic islets to over-activity which may cause them to "burn out" with resulting permanent diabetes. Bones also are weakened (osteoporosis) with fractures especially common in the ribs, and compression fractures of vertebrae which cause severe backache and distortion of the vertebral column. Atherosclerosis, high blood pressure, and spontaneous internal hemorrhage can appear rapidly. The patient, not unnaturally, develops mental symptoms varying from moodiness to severe depression and even mania. If untreated, he becomes bedridden and dies in about five years after appearance of the first symptoms from any one or a combination of the disorders described.

Cushing's disease is sometimes due to a tumor of the adrenal cortex, but more often it is caused by a tumor of the pituitary cells that secrete ACTH, which then acts excessively on the adrenal. In any case, treatment is removal of the tumor, adrenal or pituitary. If the former, care is taken to leave enough adrenal cortex to maintain normal activity; if the latter, the surgeon can hardly perform partial removal of the small and highly inaccessible pituitary and must destroy the whole gland, with following life-long hormone treatment. Even the latter, however, is far preferable to progressive Cushing's disease.

The account of the endocrine glands in this and the preceding chapters is highly simplified. It could hardly be anything else since a full account of interplay between the various glands and hormones is confusing even to most physiologists and physicians and would be baffling to most of us. Nevertheless, we should be well aware of the complex interplay that is always present, and must appreciate the difficulty of diagnosis and treatment in many endocrine disorders, and the frequent necessity for trial and error or even paradoxical treatment. If the need should ever arise and we can accept these things, we are better able to cooperate with the physician and vastly improve chances of detection and effective treatment.

REVIEW QUESTIONS

1. Why is the pituitary gland sometimes referred to as the master gland?
2. How would you distinguish dwarfism as caused by hyposecretion of the pituitary gland from that caused by hyposecretion of the thyroid?
3. What are some of the functions of hormones secreted by the posterior lobe of the pituitary gland?
4. What are some causes of goiter?
5. What are some of the characteristic symptoms of diabetes mellitus?
6. Why are the juvenile and adult forms of diabetes frequently regarded as separate and distinct diseases?
7. Why do doctors use glucocorticoid (cortisone-type) drugs with great caution?

REFERENCES

Davidoff, Frank F., "Oral hypoglycemic agents and the mechanism of diabetes mellitus," *New England Journal of Medicine*, **278:** 148–154 (1968).

Williams, Robert H. (ed.), *Textbook of Endocrinology*, 4th ed., Philadelphia: W. B. Saunders, 1968.

Additional Readings:

ENGLE, EARL T., and GREGORY PINCUS (eds.), *Hormones and the Aging Process,* New York: Academic Press, 1956.

GREENE, R., *Human Hormones.* New York: McGraw-Hill, 1969. (Paperback)

LEVINE, RACHMIEL, "Insulin: The biography of a small protein," *New England Journal of Medicine,* **277:** 1059–1064 (1967).

RIEDMAN, SARAH R., *Our Hormones and How They Work.* New York: Abelard-Schuman, 1956.

ROGERS, FLOYD et al., *Your Diabetes and How to Live with It.* University of Nebraska Press, 1962. (Paperback)

SCHMITT, GEORGE F., *Diabetes for Diabetics,* 3rd ed. Diabetes Press, 1971.

SELYE, HANS, *Stress of Life.* New York: McGraw-Hill, 1956. (Paperback)

SILVERSTEIN, ALVIN, and VIRGINIA B. SILVERSTEIN, *Endocrine System: Hormones in the Living World.* Englewood Cliffs, N. J.: Prentice-Hall, 1971.

WHALEN, RICHARD E., *Hormones and Behavior.* New York: Van Nostrand Reinhold, 1967. (Paperback)

33
THE NERVOUS SYSTEM

The nervous system is the master system of the human body. In a very real sense, it is the person himself for whom all other systems—muscular, digestive, and so on—are mere tools and utilities. For if the brain ceases to function, the person ceases to experience anything, outer or inner, no matter how sound the rest of the body. This can be seen when anesthetics black-out the brain and anything can be done to the body without causing pain, while the person's mind recalls nothing on waking. In this chapter we shall outline essential features of the nervous system. In the following chapter, we shall discuss the common nervous disorders.

The Finer Structure of the Nervous System

Nervous tissue is composed of units called *neurons* (Fig. 33–1). These are highly specialized cells comprising a cell body, the *nerve cell* proper, and microscopically slender outgrowths or processes that extend for distances ranging from a fraction of a millimeter up to a meter or so; the longer processes are called *nerve fibers*. Nerve cells are characteristic

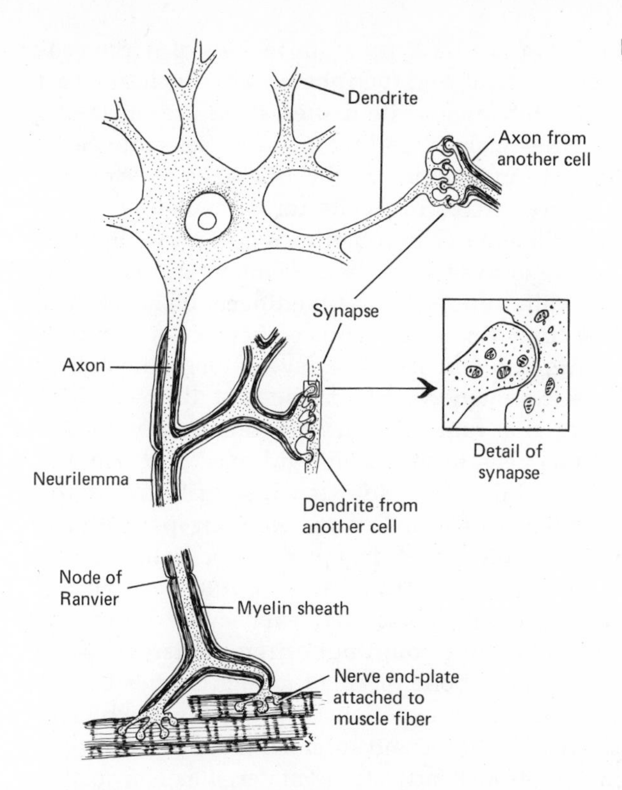

FIG. 33–1 A neuron.

of the gray matter of the nervous system; nerve fibers form the white matter and also the living elements of peripheral nerves.

The nerve cell and nerve fiber should not be confused with the term *nerve* (for example, the vagus nerve). Nerve refers to a nerve trunk, much like a telephone cable that carries many individual wires. The nerve carries (bound into a single "cable") many individual nerve fibers to and from specific parts of the body.

The neurons contact each other at points called *synapses.* These are somewhat like contacts in an electronic system, but are far more subtle and delicate in their workings. Through them the neurons are linked up in an indescribably complex web. Consider that the brain contains something like 50 billion neurons, about 15 times the human population of earth. (A figure of 10 billion, or so, is often given, but this refers only to the cortex of the hemispheres.) If on the average each neuron made only 20 synaptic contacts, a very conservative estimate, the number of these contacts would amount to the almost inconceivable figure of 1 trillion. The synapses, furthermore, are the points of activity for many drugs acting on the nervous system, from nicotine in tobacco to the latest tranquillizers and including such poisons as strychnine. The synapses therefore are immensely important in life processes.

The gross structures of the brain are built up from masses of nerve cells and fibers. The nervous system, central and peripheral, also includes vast numbers of accessory cells. In the gray and white matter, these are called glia cells, and they perform various duties such as binding the nerve cells into place, probably nourishing them, forming insulation for the fibers, and devouring (phagocytosing) debris after injury. Accessory cells form insulating, (neurilemma) sheathes. Glia and neurilemma cells sometimes go wrong; the majority of brain and nerve tumors are derived from them. (Tumors from neurons are very rare. Other sources of brain tumors are the meninges, blood vessels, pituitary gland, and metastasis from elsewhere.) Since these cells comprise several forms, each of which shows different stages of development, the variety of glia tumors is great. Some are extremely hard to eradicate, but present-day surgical techniques remove most glial tumors with very satisfactory results.

Neurons, nerve fibers, and glia cells form a semifluid mass in brain and cord. This fact must be born in mind in many contexts when dealing with the nervous system. Preserved specimens of brains, such as many people have seen, are much firmer than living nervous tissue; their consistencies have about the same relation that a hard boiled egg has to a three minute one. (Hence, in dealing with head injuries, especially those that expose the brain, the greatest caution must be observed. On no account but extreme emergency and isolation from help should fragments of bone or small foreign bodies be disturbed in such a wound by anyone but an experienced physician; such meddling may further damage the soft brain tissue and spread infection. The proper procedure is to cover the injury with as nearly sterile material as is available and, as far as possible, to avoid jolting and suddenly twisting the head and exposed brain tissue. The softness of nervous tissue should be borne in mind also when dealing with injuries to the spinal cord.)

Nervous tissue itself is insensitive. Operations on the brain are often performed with only local anesthetic to scalp, skull, and meninges, and none to the brain itself. Very light electrical stimulation to the brain at such times may produce illusions of sensation in the skin, eyes, ears, or elsewhere, but no local sensation of any sort. Hence absence of pain or other unusual sensation in the head is no guarantee that the brain has not been injured. Headache is felt not in the brain but in the meninges and blood vessels surrounding it.

Major Subdivisions of the Nervous System

Much of the structure and workings of the nervous system is still beyond understanding. And what is medically known today is already so complex as to baffle all but specialists in the subject, much less the layman. Nevertheless a broad geography of the system is easily grasped.

The system is first divided into central and peripheral subsystems. The *central nervous system* consists of the brain, enclosed in the skull, and the spinal cord enclosed in the bony vertebral or spinal column. The spinal *cord* must be carefully distinguished from the spinal *column* or *spine*. Confusion of the two is common in careless writing and leads to ridiculous misconceptions.

The *peripheral nervous system* consists of nerves and small masses of nervous tissue called ganglia, sometimes mingled in tangles called plexuses; it extends throughout the body. The peripheral system grows out of, and remains connected to, the central system; the two are structural and functional parts of one great whole. However, for practical purposes, the two parts should be distinguished; for example, to speak of central structures as "nerves" is very misleading.

The Spinal Cord

This is the smallest part (about 5 percent) of the central nervous system. It is a soft, somewhat flattened cylinder, about 45 centimeters (18 inches) long (Fig. 33–2). It is no thicker than a lead pencil through most of its length and only slightly thicker in enlargements that deal with the limbs. It gives off thirty-one or -two pairs of nerves as it runs down the back from the base of the skull to just below the ribs. As such it is the most primitive level of the nervous system.

The cord itself contains neurons forming simple reflex arcs acting as relays for sensory messages to the brain or for motor instructions from the brain. If the cord is severed, completely or in part, pathways carrying sensory and/or motor messages are interrupted, and the victim loses sensation and/or control over his muscles below the level of the injury. Injury to the cord, as to any part of the central nervous system, is at present irreparable.

The cord, as mentioned above, is enclosed in the bony *vertebral column*. This vertebral column, made of small bones, the vertebrae, strongly bound together by ligaments, forms a hard yet flexible protection for the cord. However, when it is broken, its fragments present a great hazard. For this reason, an accident victim with even a possible fractured "spine," especially in the neck, should not be moved by untrained personnel except in great emergencies such as severe injury far from help. In the latter cases helpers should (1) splint the neck and head, and the whole body, so as to prevent movement of the bone fragments; (2) move the victim only as a whole without bending or twisting, usually by two or three helpers working together; (3) keep him with his back bowed *in* as, for example, by carrying him face down on a blanket; and (4) if the accident occurs in the water, keep the victim *in* the water, if possible; the buoyancy of the water serves as a very effective splint until expert help arrives. These precautions should be maintained even with suspicion of a broken neck or back, after even an apparently minor accident. A broken neck or back is probably the most common "spinal cord" disorder with which a layman may have to deal when medical help is not available. He can best deal with it by very cautious handling.

FIG. 33–2 Thirty-one pairs of spinal nerves leave the spinal cord. The spinal cord is about 6 inches shorter than the spine so the lower spinal nerves form a bundle of fibers (the cauda equina or horse's tail) in the lower lumbar and sacral parts of the spinal canal.

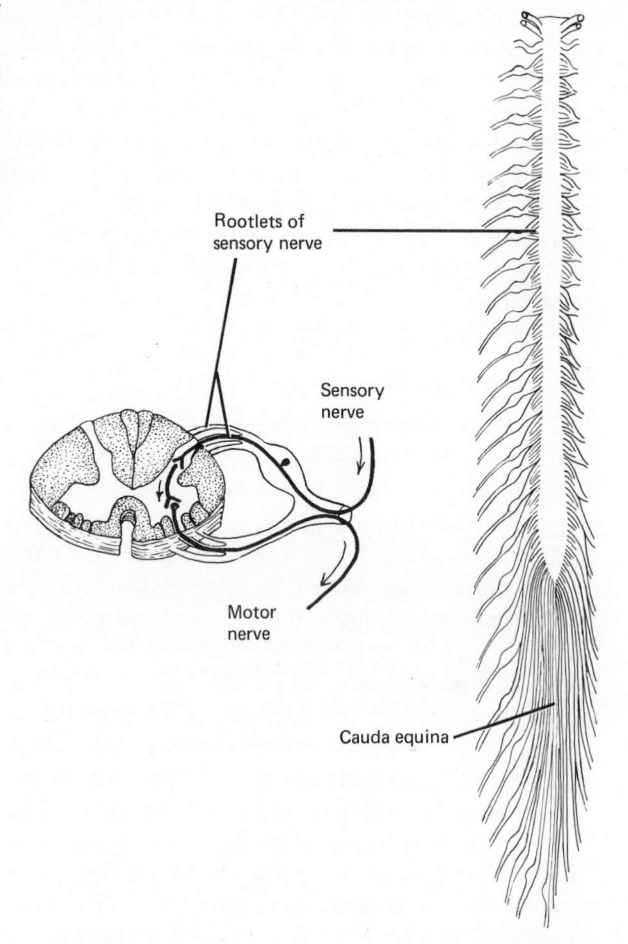

The cord extends only to an inch or two below the last rib. The vertebral column, however, continues down to the coccyx (tail-bone), and so does its central space. This space is not empty; it is lined by a sac of dura mater, the lumbar cistern, filled with cerebrospinal fluid, through which nerves run from the spinal cord to emerge between vertebrae lower down (*cauda equina* in Fig. 33–2). Whereas the spinal cord itself is soft and very vulnerable, the nerves are tough and slippery; hence this region is a favorable one for tapping spinal fluid which is able to provide many valuable medical clues, or for puncture to introduce anesthetics, drugs, or X-ray materials. Injuries to the nerves are extremely rare in such operations, and far less serious than those to the cord. Thus the very slight hazard of lumbar puncture is vastly outweighed by the diagnostic, anesthetic, or medical advantages gained.

The Brain Stem

The brain stem is the continuation of the spinal cord in the skull. Its name is appropriate since it is a rod, though thicker than the spinal cord and growing thicker still as it runs up the floor of the skull; and it gives rise to larger organs as if these were fruits growing from a stem. It is only about 3 inches long.

Nevertheless many important structures lie in it (Fig. 33–3). (1) Just above the foramen magnum (where the cord enters the skull) lie groups of neurons regulating respiration; if these are injured, as can easily happen in a fracture or hemorrhage at the base of the skull, respiration can be paralyzed and the person will die. This is another risk of moving victims of a broken neck. (2) Higher up in the brain stem lie the *vestibular nuclei,* groups of neurons controlling basic body posture; disorders involving these nuclei produce rigid postures and/or abnormal wavering of the eyes called nystagmus. (3) Higher still lies the area called the *reticular system* which reacts with the mind to calm or arouse it. (4) Throughout the length of the brain stem lie groups of neurons related to the cranial nerves; injuries to these, along with other brain stem structures, produce special groups of symptoms called syndromes, which enable a doctor to locate the site of the injury very precisely. (5) Finally, bundles of fibers running from the spinal cord to the upper brain and back to the cord must traverse the brain stem and can be interrupted there. This is why intracranial injuries, hemorrhages, infections, and so on, can sometimes have effects out of all proportion to their size, and must be treated with the greatest promptness and care. Destruction of some brain stem areas no bigger than a pea can incapacitate or kill a person.

At the very upper end of the brain stem lie two masses of nerve cells as big as sparrow eggs—the *thalami* (singular, thalamus). These thalami collect sensa-

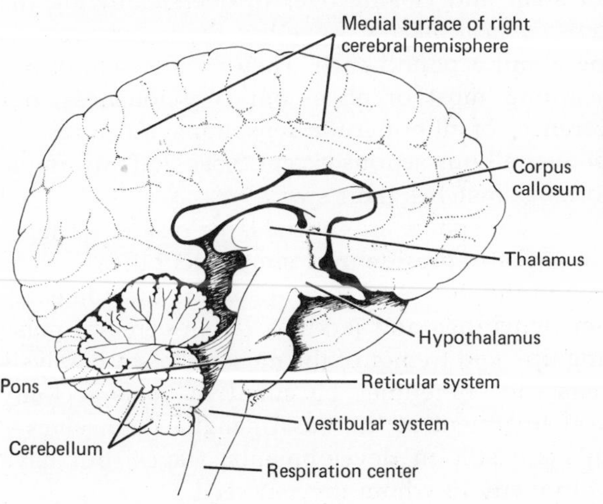

FIG. 33–3 The brain stem forms a central axis upon which are "hung" the large cerebral hemispheres and the cerebellum. Shown are the location of some of the vital centers that form part of the brain stem.

tions of almost every kind—touch, pain, heat, cold, taste, sight, hearing, and others—organize them, probably into composite ideas, and relay the results to the highest level of the brain, the cerebral hemispheres. If the parts of the cerebral hemispheres that receive sensations are destroyed, the thalamus is left alone to deal with sensation; but it does so very crudely so that the person experiences no more than a poorly localized intense pain or, more rarely, a vague pleasurable effect; this is known as the thalamic syndrome. The thalamus also sends impulses to parts of the forebrain called the basal ganglia (discussed below); if the latter are out of order, as in *paralysis agitans* (Parkinson's disease), surgical destruction in certain parts of the thalamus cuts off stimulation to the ganglia and calms the disorder; this is thalamectomy.

Just under the thalami lies the *hypothalamus* (Greek, *hypo:* under). This is a small region which has been compared to a walnut-meat with symmetrical halves; but it is of extraordinarily varied functions and disorders. It is the main control center for the autonomic system, and regulates such activities as body heat, hunger, sleep-wake cycles (probably in partnership with the reticular [alerting] system described above), mobilization of the body for emergency, water balance in the body, and others less well-defined. Its disorders produce such effects as a noninfectious but sometimes fatal fever or a disastrous drop in temperature, an uncontrollable form of obesity, a profound sleeping sickness, and an involuntary unrestrained "sham rage" (in cats but possibly in humans also). Because of its small size, complexity, and deeply buried position, the hypothalamus is very difficult to treat; but it has already been operated upon for removal of tumors and other bodies, with dramatic results such as arousal of a girl who had been a "vegetable" for years; and it will certainly come more fully under control of modern techniques as these are perfected.

On each side of the brain stem and closing over it, overshadowing the thalami, lie the enormous *cerebral hemispheres.* Together these comprise 80 percent of the brain, weighing about a pound each. They are the seat of our highest mental functions including most or all of our consciousness, our mental world-picture, our awareness of all but crude sensations, our foresight and planning, and our control over all our actions except those that are simple or automatic. Thus their importance matches their size.

Embedded at the base of each hemisphere are several large masses of nerve cells, the *basal ganglia.* These apparently preside over simple activities such as rhythmic walking, swimming, and the like. And their disorders include exaggerations of this activity, from twitchings and jerkings (as in cerebral palsy and St. Vitus dance) to "seizing up" and tremor of the muscles (as in paralysis agitans). Today these disorders can sometimes be cured by thalamectomy described above or by surgical treatment of the basal ganglia themselves—pallidectomy. These treatments are still in developmental stages, but have already proved a great benefit to many in whom they succeed.

Cerebral Cortex

The outer surface of each cerebral hemisphere is covered with *cortex*. (Cortex occurs elsewhere, as will be seen, but when we use the word alone, we mean the *cerebral* cortex.) This is the "gray matter" correlated with high intelligence in popular speech; and though gray matter occurs in many other parts of the brain, the cortex is by far the greatest mass of it, and its amount does indeed very roughly correlate with mental ability. It is a sheet (its name means bark) from 40 to 50-centimeters (16 to 20 inches) square and varying in thickness in different regions, from 1.5 to 4 millimeters (1/8 inch plus or minus). It is continuous and behaves as one great "thinking organ," but it can be broadly divided into three types of regions (Fig. 33–4).

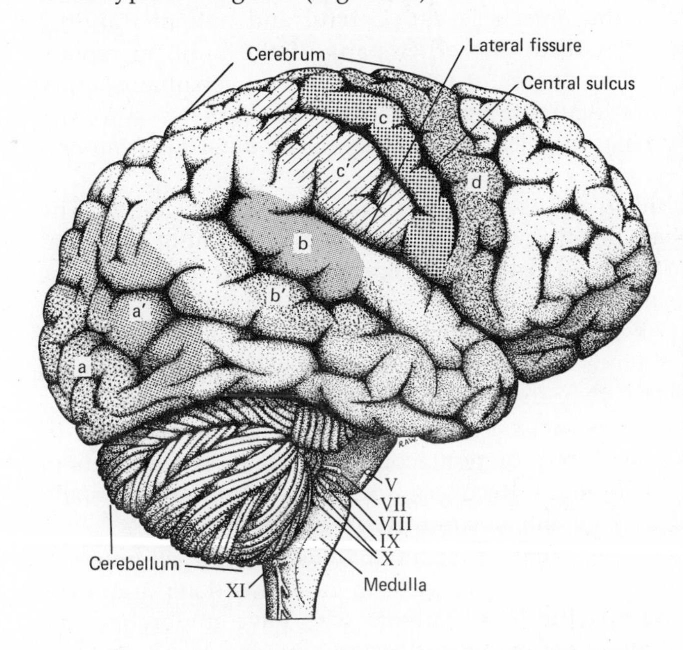

FIG. 33–4 Shading has been used to show several functional areas of the cerebral cortex: (a) projection area for vision; (b) projection area for hearing; (c) projection area for touch; (d) projection area that controls motor activity; (a', b', c') are psychic areas for sight, hearing, and touch, respectively. Some of the cranial nerves visible from this view are indicated by roman numerals.

1. Regions directly receiving sensory reports or sending motor instructions from and to lower nervous organs; these are the *projection areas* for touch, sight, hearing, smell-taste, and motor control.
2. Regions bordering the foregoing, and organizing sensations into more complex ideas or breaking ideas down into specific responses for motor actions; these are the *psychic areas*.
3. Large regions lying between projection-psychic regions and evidently dealing with a still higher organization of ideas into a unified consciousness of mind; these are the *silent areas*, so named because they produce no recognizable results when electrically stimulated.

Disappointingly, the elaborate folds into which the cortex is thrown have no general interest for us; a few of them have significance as landmarks for the surgeon. The archaic maps of the head one sometimes sees, dividing it into areas of "imagination," "will," and so on, are pure fantasies with no basis in fact at all.

The effects of injuries to the hemispheres vary according to the area involved. Destruction of a projection area in whole or in part causes a definite permanent defect of sensation or of motor control. Such an injury in the right hemisphere produces anesthesia on the left side of the body, blindness toward the left in both eyes (not of the left eye only), paralysis on the left, and some deafness in both ears. The paralysis is usually of the spastic type, meaning that the muscles are rigid due to an exaggerated tonus, in contrast to the flaccid paralysis which results when the nerve to the muscle is cut. Careful and patient training, especially in youth, can often reeducate other parts of the brain to replace *motor* cortical loss so that some degree of paralysis can be overcome. *Sensory* cortical defects, however, are incurable by any known means; the most we can do for such injuries is to make the best of the situation by determination and ingenuity.

Destruction in a psychic area causes a defect in recognition of objects. The patient may feel, see, hear, or otherwise perceive them but cannot say what they are or use them properly. This is most apparent in the case of those very peculiar "objects" called words; though perfectly sane, the patient may not understand words spoken or written or both, or may be unable to speak or write though he understands speech and writing by others. These defects are various forms of *aphasia;* they usually occur after injury to the *left* hemisphere (though they sometimes occur after injury to the right hemisphere in a left-handed person). Other forms of nonrecognition occur, some of them very odd but none of them common. Recovery from these disorders is usually with proper training and may be almost complete in children.

Destruction in the silent areas may cause no obvious defect at all unless it is very extensive. Even then the defect is rather of an indefinite deterioration of mental ability than of any specific lack. Patients who have undergone apparently disastrous loss of silent cortex, by surgery to remove a tumor or from other causes, may recover almost perfectly and resume their usual lives and occupations if the projection and psychic areas are unharmed. Recovery, again, is usually best in children and young people but it can be good even in the elderly.

The Cerebellum

The word cerebellum means "little brain." The cerebellum grows from the back of the brain stem and fills the saucer-shaped floor of the back of the skull (Fig. 33–4). The organ is shaped rather like an oyster shell, but is finely ridged from side to side which gives it a striking appearance. The ridges extend deeply into the interior where they branch profusely, and all parts are covered by a cortex containing many billions of neurons.

The cerebellum is in effect an enormous neuron computer to organize body actions. In regard to the locomotor system, Chapter 29, the complexities of simply pointing a finger and the more general actions, such as running or dancing, that bring scores of muscles into play, were described. Yet all these muscles, in constantly changing patterns, cooperate without conscious thought yet smoothly and precisely, and this is accomplished by the cerebellum.

Injury to the cerebellum causes no loss of any conscious sensation or of any mental faculty. It does not even cause any form of paralysis in the true sense. Yet the patient cannot judge the motions and positions of his own limbs (ataxia), nor can he carry out a complex movement except by conscious direction of each part of the body involved (asynergia). For example, he cannot neatly touch his fingers together behind his back; and when walking he moves as though intoxicated though his mind is perfectly clear. The success of retraining the body following cerebellar damage depends on the location and size of the cerebellar injury.

Somatic Nerves

The peripheral nervous system, as noted at the outset, comprises nerves and ganglia. It includes also sensory and motor end-organs at the terminals of the nerves, to receive impressions and to control muscles and glands, respectively. Nerves, ganglia, and endings are organized into (1) those nerves supplying muscles, skin, and related tissues, and (2) autonomic nerves supplying the viscera.

A *somatic* nerve is usually composed of both sensory (incoming) and motor (outgoing fibers); thus it is a two-way conductor like a telephone cable. If a nerve is severed, the person suffers both anesthesia (lack of feeling) and paralysis (lack of motor control) in the area served by that nerve. Thirty-one or 32 pairs of nerves arise from the spinal cord, emerge between vertebrae, and serve most of the body (Fig. 34–2). Twelve special pairs of cranial nerves emerge from the brain and deal chiefly with structures of the head region. These cranial nerves differ greatly in composition, so that their disorders form a special field of study and are the object of many medical tests; and their courses are usually complex or even bizarre, posing many problems of diagnosis and treatment.

Each fiber in a nerve is an outgrowth from a nerve cell in the central nervous system or in a ganglion; and if cut away from this cell, the fiber dies. The nerve is further made up of coverings, the neurilemma and fatty sheathes. The nerve fibers are the active element; the sheathes insulate, protect and, as will be seen, promote recovery from injury.

Injuries to nerves are of varying degrees. The simplest is due to compressions without severance of the nerve but with crushing of the nerve fibers within their sheathes. This can occur if, say, a person sleeps with his arm hanging over a hard edge such as a metal bed frame or the back of a chair; it can also be caused by a wrongly applied tourniquet, bandage, plaster cast, or other constriction left on too long. The fatty sheathes are tougher and survive. The result is a functional rather than a mechanical severance. Then, of course, a nerve can be severed completely by accidents of various sorts.

Either type of mishap is curable in the *peripheral* system. After a time, the stump of a nerve fiber that remains attached to the parent cell begins to grow. If its sheath remains intact, it simply grows down that sheath until it reaches its former terminals and restores sensation or motor control over the region. If the whole nerve has been severed, sheathes and all, the stumps of nerve fibers grow beyond the stump of the cut nerve and seek the surviving but empty sheathes in the severed portion of the nerve. They may find the task impossible if the cut ends are separated, but usually succeed if the ends are brought close together and fixed there by the surgeon. In any case, however, the fibers grow very slowly, only a millimeter or two a day, so that they may take months to reach from a wound in the upper part of a limb to terminals in the fingers or toes; but reach they will if properly treated.

Autonomic Structures

The autonomic nerves control the viscera—the internal organs in thorax, abdomen, and pelvis, and other scattered organs such as the iris of the eye, blood vessels everywhere, and sweat glands. Autonomic means "self-controlled" and the autonomic system is indeed largely so; but it does connect with the central nervous system and submits to regulation by autonomic centers there, such as the hypothalamus. But these centers are not under direct control of our minds. Thus, for example, we cannot voluntarily regulate our heart rate, though conscious experience such as fright can act through the autonomic centers and nerves to increase the rate. And so likewise for blood vessels, iris, and viscera.

The autonomic system is made up of two divisions which functionally oppose each other. The *sympathetic division* in general prepares organs for an emergency—for "fight or flight" as one observer expressed it; thus, it speeds up the heart, dilates bronchioles of the lungs, dilates the pupil, inhibits the nonessential digestive functions, and so on. The *parasympathetic* division in general has the opposite effects; it organizes the body to maintain an optimum state of heat, blood flow, nutrition, and many other things or to restore them after they have been disturbed by sympathetic action. Working together the two divisions tend to keep the body's functions "on a straight course." This straight course is called homeostasis—uniform state.

As a matter of fact, except for emergencies, most of our viscera can get along without any autonomic control. Thus when one or the other division becomes overactive, it can sometimes be cut off surgically with only minor ill effects. For example, an excitable sympathetic supply to blood vessels in the limbs can make these shut down severely and cause blanching and eventually gangrene. Or an excitable parasympathetic supply to the stomach can predispose a person to peptic ulcers. In such cases, sympathectomy (*-ectomy:* cutting out) or parasympathectomy may be called for. Today, however, drugs that act on the autonomic system often make such surgery unnecessary.

One such drug, *epinephrine* (or *adrenalin*) has long been known and is in fact a natural product of the body. It is produced by the *medulla,* or core of the adrenal glands (entirely different from the adrenal cortex), and is a hormone discharged into the blood to reach all parts of the body. It acts to reinforce the sympathetic division everywhere. For example, it relaxes the bronchioles of the lungs in asthma, and permits freer breathing as it would in an emergency. Synthetic variants of adrenalin are often superior to the natural form for special purposes. The parasympathetic system has no such hormone reinforcement; mobilization should be prompt and total, but recovery can be leisurely and selective. However, a number of drugs are available which stimulate activity of the parasympathetic division. Pilocarpine, for example, causes constriction of the pupil, profuse sweating, and stimulation of peristalsis in the gastrointestinal tract. Certain other drugs such as atropine *inhibit* the parasympathetic division, and result in dryness of the mouth (through inhibition of the salivary gland secretion), inhibition of sweating, and wide dilation of the pupil. The latter effect of atropine is used to permit easier inspection of the retina during an eye examination.

Probably the most important thing we need to know about the autonomic system is that it does respond to mental states. Peptic ulcer and spastic colon are examples of autonomic disorders which are variously classified among the neuroses and which must be treated as mental problems. But, as in all neuroses, autonomic imbalances are invariably excessive reactions of normal behavior; and as such they are amenable to some degree of self-control and will power. Just as most types of obesity and excessive smoking are examples of self-indulgence—neurotic behavior of a sort—indulgence of our autonomic responses are also weaknesses that can often be overcome by auto-psychotherapy. The person who can convince himself that he can and must overcome such weaknesses may well be on his way to recovery.

REVIEW QUESTIONS

1. Distinguish the terms neuron, nerve cell, nerve fiber, and nerve.
2. Is pain a usual symptom of brain injury? Why?
3. Discuss some of the problems in first aid treatment of spinal injuries.
4. What is meant by the brain stem? What are some of its functions?

5. How does the neurologist divide up the areas of the cerebral cortex according to function?
6. What role does the cerebellum play in the execution of muscular actions?
7. Compare the actions of the sympathetic and parasympathetic divisions of the autonomic nervous system.

REFERENCES

CURTIS, BRIAN A. et al., *An Introduction to the Neurosciences.* Philadelphia: W. B. Saunders, 1972.

ELLIOTT, H. CHANDLER, *Textbook of Neuroanatomy*, 2nd ed. Philadelphia: J. B. Lippincott, 1969.

Additional Readings

ASIMOV, ISAAC, *The Human Brain: Its Capacities and Functions.* Boston: Houghton Mifflin, 1964. (Paperback)

ATKINSON, RICHARD C., and RICHARD M. SHIFFRIN, "The control of short-term memory," *Scientific American.* **225:** 82–90 (1971).

DICARA, LEO V., "Learning in the autonomic nervous system," *Scientific American,* **222:** 30–39 (1970).

ELLIOTT, H. CHANDLER, *The Shape of Intelligence: The Evolution of the Human Brain.* New York: Charles Scribner, 1969.

HEIMER, LENNART, "Pathways in the brain," *Scientific American,* **225:** 48–60 (1971).

KANDEL, ERIC R., "Nerve cells and behavior," *Scientific American,* **223:** 57–70 (1970).

PRIBRAM, KARL H., "Neurophysiology of remembering," *Scientific American,* **220:** 73–86 (1969).

WALTER , W. GREY, *The Living Brain.* New York: W. W. Norton, 1963. (Paperback)

WOOLDRIDGE, DEAN E., *The Machinery of the Brain.* New York: McGraw-Hill, 1963. (Paperback)

34
DISORDERS OF THE NERVOUS SYSTEM

The brain is not mysterious; in spite of its great complexity and marvelous powers, it is much like other organs of the body in its physical needs. It must receive food and oxygen through the blood, and be relieved of waste products by the same channel. The food must be adequate, for, if minute amounts of certain vitamins are lacking, severe nervous disorders result. Undoubtedly, lesser deficiencies of food factors produce lesser defects of brain activity. Oxygen is even more vital. Other body parts can endure oxygen lack for prolonged periods, as in the famous cases of severed arms reattached to the body and surviving, but complete oxygen lack for as little as four to six minutes can do grave damage to the brain. And many chemicals, lack of blood sugar, concussion (effect of a severe jolt to the head), and countless other physical factors have prompt and profound effects on the nervous system. Thus the brain is highly sensitive to the general health of the body.

Recent research has suggested that the generally poor intellectual achievements of those in underdeveloped countries, and even among the poor in this country, is attributable to poor nutrition, particularly protein deficiency. The Latin maxim, *Mens sana in corpore sano* (a sound mind in a sound body) is certainly appropriate in modern life. A sound diet, adequate exercise, sufficient sleep, and temperate use of even commonplace drugs all contribute to mental health.

Injuries to The Brain

In one respect most people recognize, even if imperfectly, the physical nature of the brain. They accept the fact that injuries to the organ produce varied but definite effects. Injuries due to violence, as in an automobile accident or from a bullet, are the most obvious. But other injuries such as those inflicted at birth (really a form of violence), by infections, tumors, accidents to blood vessels (strokes), and in forms of degeneration, produce much the same effects. The basic similarity of all these disorders and injuries, however, is not apparent to everyone, and results in unfortunate misconceptions.

Symptoms of nervous-system injury are peculiarly distressing. They include paralysis (poliomyelitis, cerebral palsy); rigidities, tremors, and uncontrolled movements (St. Vitus dance in the young, paralysis agitans or Parkinsonism in older people); the inability to speak, write, read, or understand speech (aphasia); the inability to conduct actions skillfully, though no true paralysis is present (ataxia); forms of mental excitability, depression, apathy, rage, anxiety, and other emotional states (hypothalamic disorders); simple loss of control over body functions such as breathing or heart beat (bulbar paralysis); and so on. Considering the complexity of the brain, this wide range of disorders is not surprising.

People, however, have an unreasonable difference of attitude toward essentially similar results of brain injuries. They accept such conditions as ordinary paralysis, blindness, or disorder of breathing as they would accept loss of a limb or an eye, or other natural mishap—something to be accepted philosophically and dealt with sensibly, however regrettable or disastrous it is. But they have an irrational repugnance for, say, the tremors and rigidities that sometimes afflict older people, for uncontrolled movements in children, or for lack of coordination in anyone; these, they feel, are unnatural abnormalities indicating idiocy or, to coarser minds, disorders of comic behavior. Many people still have an emotional horror and fear of mental illness and confuse the victim of aphasia struggling to speak or comprehend with the emotionally disturbed.

Whatever the proper attitude to true mental illness, modern, educated people have no right to an archaic attitude toward any brain injury. We tend to feel repugnant if we may have to indulge, care for, or even restrain the neurotic patient; but we have no distaste for and will easily display patience and insight in doing these things in a case of fever. Yet we need these qualities far more with a person who is distressed and bewildered by an inability to express himself in spoken words; or who is afflicted by stiff limbs, a mask-like face, a "senile" tremor; or who is deeply depressed as a result of a physical brain injury. Since such things do strike close to a person's inner self, sympathy and understanding of their nature is the best hygiene we can offer.

Mental Disorders

Medicine distinguishes between "organic" and "functional" diseases – organic disorders being defined as those due to an injury, deformity, chemical disturbance, infection, or other identifiable and understandable cause, and functional disorders being those in which a tissue or organ is obviously malfunctioning with the cause and mechanism of the trouble unknown. Thus, for example, paralysis due to a severed nerve is organic and can be treated as such; but schizophrenia due to some as yet unexplained disorder of the cerebral cortex is still functional and can be treated only empirically. Yet, many medical scientists expect such a functional disease to yield to science sooner or later. In other words, "functional" disorders may some day become "organic" ones with the advance of knowledge.

We should realize that the same is true of brain defects. Informed people recognize the troubles mentioned earlier as really organic, and treat them as such. But we should further recognize the possibility that "mental" disorders, even if functional according to today's degree of knowledge, may some day be explained as organic, no more mysterious and shameful than the results of a blow on the head. The causes of these disorders may simply be more complex and difficult yet capable of being unraveled.

Epilepsy

To many people this term is still one of fear and aversion; indeed it is pleasant to no one. In former times, the disorder was looked on with superstitious awe as "the sacred disease" or "the affliction of God"; and this character still clings to it behind all our scientific sophistication.

Epileptic attacks, or seizures, cover a wide range of severity and of frequency. At one end of the scale, the patient may simply go into a dreamy, withdrawn state for a few moments; he shows no special symptoms and his condition may not even be noticed by companions; this is called "petit mal," French for "small ill." More severe cases result in an uncontrollable jerking of some joint or even much of the body, or in a vivid hallucination of smell, colored lights, or such, but without loss of consciousness; this is a "Jacksonian seizure" named after the man who first discussed it adequately, Hughlings Jackson. A more severe form passes on from jerking or sense delusion to generalized convulsions of the whole body with loss of consciousness; it generally includes foaming at the mouth and, often distressing, animal-like cries; it also destroys control of bladder and bowel with embarrassing soiling of the clothes; and sometimes the convulsions are severe enough to break a bone or otherwise injure the victim. This combination of symptoms is indeed a distressing thing to witness and is well-named *grand mal* (great ill). In different patients, grand mal attacks may occur rarely, months apart, or several times a day. At the extreme end of the scale is a fortunately rare condition called *status epilepticus* in which the victim passes from one attack to another with little or no interval between. All these conditions, however, vary in degree only; the underlying disease process is the same.

To explain them we need to recall that the nervous system operates by means of tiny electrochemical waves, called *impulses,* traveling along nerve fibers and jumping from one neuron (nerve cell) to another. (The submicroscopic gaps between neurons are called *synapses,* and might be compared to spark-gaps; they are extremely important in nervous activity.) Impulses are directed along useful paths by low resistances at some synapses, and prevented from spreading in a useless or harmful way by high resistance at other synapses. However, any synapse can be jumped by sufficiently vigorous impulse-activity.

Neurons and synapses form an almost inconceivably complex fabric. Neurons in a human brain number in the billions, and synapses in the trillions. The real wonder is that their precise yet fragile organization does not get out of order frequently; that it does not do so, is a tribute to the great resilience of living tissue. Organization does break down now and then: some thing, either a chemical irritant or a physical irritant such as a scar or small foreign body, or overexcitability affecting a person's chemical balance tends to fire synapses artificially, not in the regular course of impulse traffic. From these abnormally fired synapses, a wave of abnormal activity spreads through the brain; this may die out quickly, doing no more than absorbing some nervous energy, as in petit mal; or it may continue further to produce the grosser forms of epilepsy. This is all that happens and, however unhappy the results may be, it should be no more mysterious or shameful than a tendency to catch colds or an easily sprained ankle. Any other view of it is obsolete and ignorant.

With this in mind, what can be done for epilepsy? A very great deal can be accomplished. Many drugs can neutralize overexcitability and prevent attacks as long as the medication is continued. A physical irritant can usually be removed by surgery; formerly, after such removal, the surgical scar was apt to be as much of an irritant as the original one, so that the attacks continued; but with newer techniques, scarring is almost eliminated, many patients are completely cured, and most of the others are improved so that their trouble is easily controlled by drugs. All this is very encouraging and offers relief to over a million victims.

Anyone present during an epileptic seizure can be of great help if he knows how to treat the situation. He can prevent injuries due to falling or to striking objects during convulsions; in particular, he can prevent the common and painful accident of tongue-biting, by putting something solid, but not too hard, between the teeth. If he has informed himself earlier on what medication might be used and where it is to be found, he can apply this. By calm, competent behavior, he can suppress panic or morbid curiosity in onlookers and even dismiss those over whom he has control; in this way he can spare the victim later embarrassment from knowing that his misfortune was witnessed by an unsympathetic audience. The mere fact that a capable and understanding friend was in charge, will be a comfort to the patient. Happily, such occasions are now rare compared to their frequency in former years; and we have reason now to be hopeful that uncontrolled epilepsy will soon be as rare as smallpox.

Disorders of Nervous System Accessories

A person who survives a stroke may be partially or totally paralyzed as a result of damage or death of nervous tissue. Yet stroke is not truly a nervous or neurological disease, but rather a vascular accident caused by disease or disorder of the cerebral circulation. This is an example of a disorder of a nervous system accessory; such disorders are actually much more common than disorders of the nervous tissue itself.

The Circulation of the Brain

Disorders of circulation are the most frequent of all brain troubles, and are common also in the spinal cord. A majority of articles in journals devoted to diseases of the nervous system are concerned with vascular disorders. And indeed, they are the third most frequent cause of death in this country,

The most common cause of vascular brain disorder is *atherosclerosis.* This, with its causes, is discussed in Chapter 23, but it may profitably be reviewed here. Atherosclerosis is a progressive thickening of the innermost layer of the arterial walls, due to the deposit of fatty and mineral substances, until the passage becomes inadequate or completely blocked; such an obstruction is called a *thrombosis.* Furthermore, a fragment of atheroclerotic tissue or blood clot can break loose and be carried by the blood until it reaches an artery too small to transmit it, which it blocks; this is an *embolism.* Thrombosis is usually gradual in its effects, embolism is sudden.

The brain, next to the heart, is the tissue most frequently affected by atherosclerosis. Also, brain injuries from this source are usually very severe—crippling or outright fatal. The part of a brain dependent on an artery which is blocked dies very quickly forming what is called an *infarct,* an area of dead tissue. If this part is vital, or in control of movement, death or paralysis follows.

The most striking feature of this problem is that even today atherosclerosis is largely subject to control. If the general public would only apply what is known, we could reduce the number of deaths and disabilities from brain infarct to a fraction of current figures. We may recall here that atherosclerosis builds up gradually; seems to begin usually at an early age, when it causes no symptoms; and progresses steadily until its effects appear openly, often without warning. After it declares itself, it is no longer so susceptible to treatment. The causes of atherosclerosis are not entirely clear, but we do know several factors which can accelerate the disease process: The chief ones seem to be overindulgence in foods rich in fats and/or sucrose, smoking, and sedentary habits, that are all within the individual's choice to risk or not to risk. The penalties of sudden death or disabling paralysis would seem too heavy a price to pay for trivial pleasures; but they are paid by hundreds of thousands of people annually, not only in the older age brackets but, increasingly, in younger groups too. This seems such a depressing comment to be made on human self-indulgence!

Other vascular troubles are peculiar to the brain. As in the case of tumors, an *intracranial hemorrhage* is far more serious than one of equal size elsewhere; for, since the cranial cavity is a tightly closed box, anything in it can expand only by compressing the brain. Thus a mere cupful of hemorrhage, which would be considered only a moderate accident in say, a limb, may cause grave disability or death if it occurs inside the skull. Such events cannot, of course, be treated or even anticipated; but they can largely be prevented by (1) avoiding preventable accidents involving the head—for example, by driving sensibly and using safety equipment and (2) faithfully having periodic medical checkups, in which heightened blood pressure may warn against potential brain hemorrhage from internal causes. These two hazards are the principal sources of intracranial hemorrhage.

Such hemorrhages cannot be treated by any form of first aid or home remedy. In fact, they may even progress to a fatal degree while such well-meant procedures are being carried out. An intracranial hemorrhage should be given expert care at the earliest possbile moment; but to accomplish this, knowledge of its symptoms is vital.

Whatever the cause of the hemorrhage (violence, disease, weakened artery) its symptoms are more or less the same. Growing pressure inside the skull causes headache, vertigo, nausea and vomiting, increasing mental confusion and dullness and, often though not always, disordered vision. Presence of even one or two of these symptoms, at any time but especially following an accident, should warn an informed observer that medical help is most urgently needed. Prompt recognition of intracranial hemorrhage, with immediate action, is paramount in saving life and normal function.

Highly noteworthy is the fact that symptoms may be delayed. Following a brief blackout after an accident, the patient may apparently return to normal; he may declare himself "all right," and even walk or drive away from the scene.

Symptoms may then develop hours or even days later because the hemorrhage is slow and requires time to build up pressure. Thus it may begin to take effect when the person is no longer under supervision or while he is asleep the next night. (For example, such cases are sometimes found unconscious, taken to a hospital, and treated for intoxication or some other supposed disorder, with neglect of the real trouble.) After any head accident, even without serious external injury, the subject should be carefully watched for at least 24 hours. More rapid intracranial hemorrhage, after an accident, will naturally produce more immediate symptoms; such cases are hardly likely to be neglected.

Hemorrhage due to spontaneous rupture of an artery is almost always sudden. If it occurs in the space around the brain, it causes prompt, severe headache, usually mental confusion, occasionally unconsciousness, and other variable symptoms. If it occurs in the substance of the brain itself, it may also cause special symptoms such as a localized paralysis of a limb or one side of the body and/or face, or loss of speech. Such events are called strokes (apoplectic strokes, in former years), or cerebral vascular accidents (CVA) medically. Usually the event is alarming enough to call for immediate expert attention. Almost certainly matters will get worse, not better, with delay to "see if he comes out of it"; the bleeding will probably continue, and the pressure increase. A stroke is no matter for home treatment, though it can and should be recognized.

The later progress of a stroke patient is another matter. When the bleeding has been arrested by surgical or other means, and perhaps much of the blood clot removed, the patient will recover to some degree. Nervous tissue destroyed by the pressure will never be replaced, so that some defect is likely to remain; but nervous tissue merely stunned or bruised can recover, and other parts of the brain, if carefully trained, can often take over some function such as control of movement or speech. The prompter and better the medical attention was at the outset, the greater is recovery likely to be. And the degree of cooperation, encouragement, and good sense in a patient's associates, also contributes. Above all, a patient's own resolution, intelligence, and strong motivation can achieve results far beyond those to be expected in apathetic cases. Altogether, a stroke should be a matter not for despair but for determined reaction.

Finally, vascular disorders can cause a variety of nervous symptoms that seem trivial but are sometimes warnings for disorders becoming serious. Thus, *fainting, dizziness,* or mere faintness, are often due to inadequate blood-supply and resulting lack of oxygen for the brain. This lack may be due to minor and temporary causes such as fasting, prolonged exertion, or nervous tension as when looking down from a great height. In the 1800's, fashionable women were notoriously prone to fainting, largely because their tight corsetting prevented proper breathing and circulation. Since the brain is so highly sensitive to oxygen lack, and in their cases supply was so barely adequate, further strain would result in genteel "swooning." In our times, however, any such fainting or dizziness without obvious cause (such as severe bleeding), especially if repeated, suggests a possible underlying trouble such as developing thrombosis. Two or three episodes of this kind amply warrant a report to a physician.

Migraine Headache

Migraine headache, a rather common disorder is apparently, though not certainly, due to spasm of blood vessels at the base of the brain. It is not usually dangerous but is very unpleasant and often incapacitating, varying in degree from a trifling illusion of a glittering spot (scintillating scotoma) in the field of vision lasting a short while, through mild headaches, to violent, prostrating, blinding headaches. The more severe forms are often accompanied by nausea, cramps, vomiting, and other symptoms; indeed, in some cases, the headaches may eventually be replaced entirely by abdominal symptoms. One consoling feature of the condition is that it tends to die out with advancing age.

Meantime, can anything be done for it? People with mild forms can control an attack with aspirin, darkness, rest, and quiet. Only an experienced physician can and should decide on treatment for severe migraine conditions. He can prescribe certain drugs which, however, should be taken only under careful supervision, for they may produce unpleasant or outright dangerous side effects in some people, and are best administered by hypodermic. The doctor may also advise psychotherapy to minimize worries and tension-building problems, since the typical migraine patient is a high-strung, compulsive type. The patient and his associates can effectively cooperate by: (1) carefully following instructions and taking prescribed medications at the *outset* of an attack, since they are quite ineffective once the attack is in progress, (2) faithfully reporting side effects *not* due to the migraine itself but to the drug, and (3) maintaining as relaxed and placid an attitude and atmosphere as possible. With these resources, we can restrain a great deal of this distressing malady.

The Meninges

The meninges are the coverings of the brain and spinal cord: (1) the outer, thick, leathery *dura mater* which is mainly protective; (2) the delicate *arachnoid,* which forms an inner lining for the dura, and encloses the cerebro-spinal fluid; and (3) the thin but tough *pia* which clings to the surface of brain and cord and helps them to maintain their shape, as would an elastic skin. These three meninges are vital to the brain and cord in health and in disease.

Normally, the meninges are efficient guardians of the nervous system. They are a first line of defense against injury and infection; but when this defense is overcome, the results are serious. Indeed, just as vascular disorders are the most common source of general brain malfunction, so the meninges are the most common seat of brain infection. Such infection is called *meningitis.*

Meningitis was, not long ago, a word of fear and doom. The common forms of it were fatal, some almost always and some quite frequently; and even survivors had often suffered severe brain damage. Fortunately, with the advent of antibiotics, meningitis is no longer so infallibly deadly; but it is still a threat since antibiotics are poorly distributed by limited circulation in some parts of the meninges.

The sources of infection causing meningitis are many, but by far the most common are the pus-forming (coccus) bacteria. The routes by which infection reaches the meninges are also various, but, again, by far the most usual are from the nasal, sinus, middle ear, and mastoid cavities of the skull. Though one tends to think of the skull as hard and massive, several of these cavities are separated from the cranial cavity by plates of bone as thin and brittle as fingernail; and others, especially the middle ear and mastoids, can become so tightly sealed off that infection in them erodes even thick bone between them and the cranial cavity.

Respiratory infections are perhaps the most common of human disorders. Indeed, so common are they, and often so minor, that they tend to be disregarded as mere nuisances. They may well be so in a majority of cases; but when they get out of control, they are not merely magnified nuisances; they may become very dangerous. Sinusitis, otitis media (middle ear inflammation), and mastoiditis should never be treated with noble fortitude. They are, of course, painful and demand attention for that reason. When they are severe, they can break into the cranial cavity, as described above, invade and establish themselves in the meninges (which are the first thing they meet), and set up a meningitis. Any disease process so close to the brain is a grave emergency.

Two special sources of meningitis, in addition to the foregoing, are controllable by preventive measures.

1. Any infection of the face, especially in the area between the bridge of the nose and the corners of the mouth, is dangerous. Blood vessels in this area, and to a lesser extent in neighboring areas, communicate freely through the orbit (eye socket) with the interior of the skull and can easily carry infection inwards. Therefore you should not even treat a pimple on the face carelessly; if you must tamper with it, it should never be pinched, which can force infection inward, but carefully pricked with a sterile instrument and treated with antiseptic. If indications of spreading infection nevertheless appear, they should be immediately taken for medical inspection. A resulting meningitis of an especially stubborn and vicious nature can mature rapidly in the labyrinth of vessels behind the eye.
2. After any head injury, even if apparently minor, oozing of watery fluid from nose or ear is a danger signal. This fluid may be the cerebrospinal fluid from within the skull, emerging through an unrecognized fracture. Even a small and completely painless fracture, in a thin plate of bone, represents a communication between the cranial and the nasal or middle

ear cavities. Even in healthy people these cavities harbor bacteria swarming in the nose and in the ear, though less numerous but quite appreciably present. The bacteria are generally harmless in these habitats, though they may flare up even there when resistance is low; but they are quite deadly once in the cranial cavity. Therefore, such rhinorrhea (nose-running) or otorrhea (ear-running) after an accident demands the promptest action and immediate treatment.

REVIEW QUESTIONS

1. How do the terms organic and functional apply to nervous system disorders?
2. What is epilepsy? How is it treated?
3. How do circulatory disorders affect the brain?
4. How are sinusitis and otitis media (middle ear infections) related to meningitis? to the brain?

REFERENCES

ALVAREZ, WALTER C., *Little Strokes.* Philadelphia: J. B. Lippincott, 1966.

PENFIELD, WILDER, and HERBERT JASPER, *Epilepsy and the Functional Anatomy of the Human Brain.* Boston: Little, Brown, 1954.

TOOLE, JAMES F., and A. N. PATEL, *Cerebrovascular Disorders.* New York: McGraw-Hill, 1967.

WALSHE, FRANCIS, *Diseases of the Nervous System,* 11th ed. Baltimore: Williams & Wilkins, 1970.

Additional Readings

COTZIAS, GEORGE C. et al., "Modification of Parkinsonism: Chronic treatment with L-Dopa," *New England Journal of Medicine,* **280:** 337–344 (1969).

HOLOWACH, JEAN et al., "Prognosis in childhood epilepsy," *New England Journal of Medicine,* **286:** 169–174 (1972).

LIVINGSTON, SAMUEL, *Living with Epileptic Seizures.* Springfield, Ill.: Charles C Thomas, 1963.

TEMKIN, OWSEI, *Falling Sickness: A History of Epilepsy from the Greeks to the Beginning of Modern Neurology,* rev. ed. Baltimore: Johns Hopkins Press, 1971.

ZANKEL, HARRY T., *Stroke Rehabilitation: A Guide to the Rehabilitation of an Adult Patient Following A Stroke.* Springfield, Ill.: Charles C Thomas, 1971.

35
VISION

Our senses are classified as general and special. General senses are those that are distributed throughout the body and are served by multitudes of minute organs. They include touch, pain, heat, cold, pressure, and possibly others such as vibration. The special senses are those served by large, complex organs or a localized group of small organs. They include sight, hearing, vestibular senses, smell, and taste.

Of these, sight is easily the most important. It provides us with the greatest amount of information; its loss is the most disabling; and its disorders concern the greatest number of people. Hearing and the related vestibular senses will be dealt with in the following chapter.

Anatomy of the Eye

The organ of sight is familiar to most people even in some detail (Fig. 35–1). The eyeball is a sphere an inch in diameter, formed mainly of a tough fibrous material, the *sclera,* visible in front as the "white" of the eye. The very front of the sphere, however, is

FIG. 35–1 The structure of the eye.

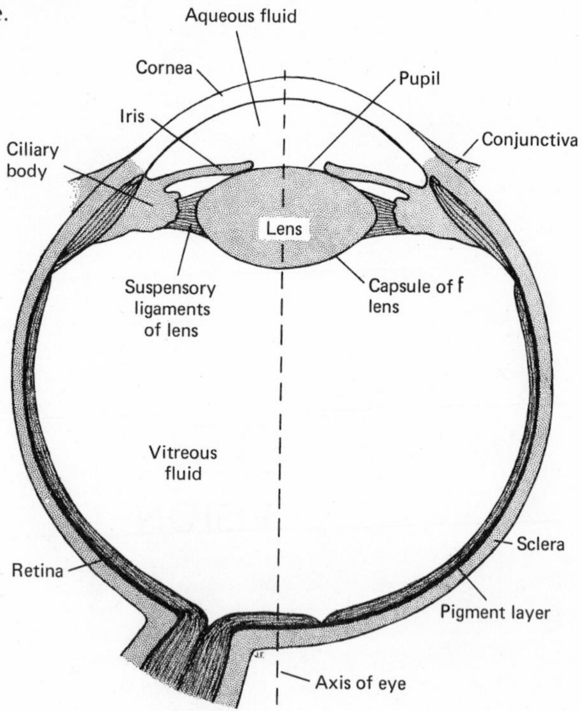

formed by a circular transparent patch, the *cornea.* The cornea bulges slightly and acts as the principal refractive surface of the eye with good focusing power since it has a much higher refractive index than the air through which light-waves normally reach it. (If one tries to see under water with the naked eye, the refractive difference between water and cornea is small and, so, near vision is blurred.) The *lens* lies within an elastic capsule suspended from the inside of the eyeball behind the cornea, immersed in the fluids of the eye; since the refractive difference between the fluids and lens is small, the lens has rather weak focusing power. Between cornea and lens extends a thin, disclike partition, the *iris,* perforated centrally by the circular opening, the *pupil.* The iris expands, constricting the pupil in bright light, and contracts, dilating the pupil in dim light (Fig. 35–2). The eyeball is lined with a delicate sheet of light-sensitive nervous tissue, the *retina,* which responds to pictures focused on it by the cornea and lens, and reports the pictures through the *optic nerve* to the brain. The interior of the eye, from retina to lens, is filled with a viscous jelly, the *vitreous humor* (glassy fluid), and from lens through pupil to cornea with the *aqueous humor* (watery fluid). The eyeball is surrounded and moved by a group of small muscles each attached to the bony *eye socket* at one end and to the sclera at the other. The retina is the essential organ of sight; all the other parts of the eye are merely accessories.

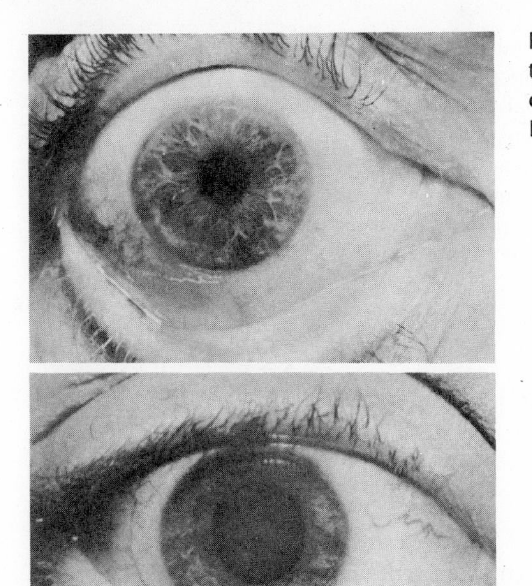

FIG. 35–2 *Top:* the pupil of the eye contracted in bright light. *Bottom:* the pupil dilated for dim light. (New York Eye and Ear Infirmary)

The Optical System of the Eye

The common defects of vision that are corrected by glasses can be understood by a few simple diagrams which demonstrate the nature of light and the refraction of light by the eye. We see the objects of our surroundings because they reflect the light that falls upon them—either from the sun or an artificial source. Any luminous point reflects light in all directions. The greater the distance traversed by these rays the more divergent they become. If a luminous point is at a distance of 20 feet or more the divergent rays that enter the eye are so close to parallel that for practical purposes we can consider them so. This means that objects that are 20 or more feet away are seen as a great number of luminous points whose reflected rays enter the eye as essentially parallel.

Such parallel rays of light that enter a normal (*emmetropic*) eye which is *at rest* are refracted so that they focus to a point on the retina (Fig. 35–3). Thus all objects 20 or more feet away are seen in focus by the resting eye. This can be demonstrated by closing your eyes while looking out a window. Upon opening them you will discover that distant objects are in focus while near objects are not.

The reason for this should be clear. Luminous points of objects, as they approach nearer than 20 feet, reflect light rays which become more and more divergent when they strike the eye. A luminous point close to the eye, say 6 to 8 inches, would reflect light rays which would diverge widely when they strike

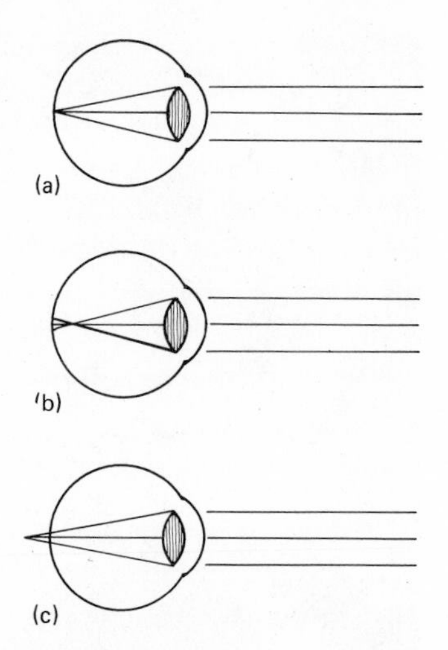

FIG. 35–3 Refraction of light by the eye. (a) the normal (emmetropic) eye: parallel light rays are focused on the retina; (b) the myopic eye: either the eyeball is too long, or the refractive power of the cornea and lens is too great, so that parallel light rays come to a focus before they reach the retina; (c) the hyperopic eye: either the eyeball is too short, or the refractive power of the cornea and lens is not great enough, so that parallel light rays do not come to a focus before they reach the retina.

the eye. If light rays are diverging when they strike the eye, the degree of refraction in the resting eye will focus them behind the retina.

However, we are able to focus clearly upon near objects if we so desire. When we do the eyes are *accommodated* rather than at rest, and a muscular effort is involved (Fig. 35–4). The accommodating muscle, called the *ciliary muscle,* is about the size and shape of a candy lifesaver and lies just behind the iris. Suspensory ligaments, referred to as *zonular fibers,* connect the ciliary muscle to the capsule of the lens. When the ciliary muscle is at rest, the zonular fibers are taut and exert a tension on the lens capsule which tends to flatten the lens. When the ciliary muscle contracts the zonular fibers are relaxed. This allows the elastic lens capsule to mold the rather plastic lens to a more spherical

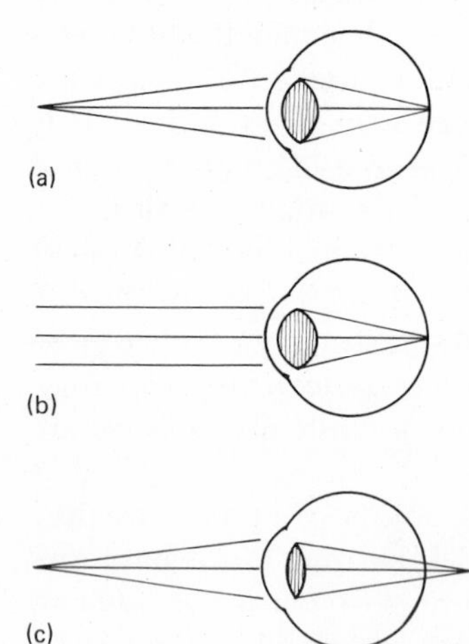

FIG. 35–4 The process of accommodation. The contraction of the ciliary muscle of the eye results in the rounding of the lens and an increase in its refractive power. This means that in the normal eye (a) light rays diverging from objects closer than 20 feet are able to be focused on the retina. In the hyperopic eye (b) the lens must undergo accommodation (increase its refractive power) for distant objects to be seen (compare with Fig. 35–3b). In (c) the presbyopic eye, the lens, because of age, has lost much of its accommodation power, and near objects cannot be focused on the retina.

shape (a sphere is the most economical shape for any mass held within an elastic container, such as water within a rubber balloon). The more the ciliary muscle contracts the more spherical the lens becomes; and the more spherical the lens, the more the light rays will be refracted. Thus, as we focus upon objects nearer than 20 feet the eye begins to accommodate; accommodation increases as the distance from the object decreases. Maximum accommodation for most emmetropic young adults occurs in focusing upon objects 4 or 5 inches away—the emmetropic near point. The near point is quite near in childhood—about three inches—and becomes progressively more distant throughout life.

Errors of Refraction

If distant objects are not clearly focused upon the retina when the eye is at rest, the individual is not emmetropic. Rather, he shows an error of refraction, and these are, basically, of four kinds: *myopia* or near-sightedness, *hyperopia* or far-sightedness, *astigmatism,* and *presbyopia.*

Myopia

In myopia parallel light rays are focused in front of the retina, and so distant vision is blurred. The causes of myopia are poorly understood. One possible cause is a weak ciliary muscle which fails to exert sufficient tension on the lens capsule through the zonular fibers so that the lens is insufficiently flattened when the eye is at rest; or occasionally, it is due to a slightly lengthened eyeball. The typical progressive myopia begins at seven or eight years of age and increases through the teen ages and levels off at about age 25, regardless of environmental factors such as the amount of close work, lighting, vitamins, exercise or anything else. It is probably hereditary.

At some distance short of 20 feet the myope is able to see clearly with his ciliary muscles at rest. This is the myopic far point. If the myope has a full range of accommodation it means that he is able to focus clearly on objects closer than the emmetrope (since he only begins his accommodation at his far point). This is the reason for the designation of near-sightedness.

The treatment of myopia is the fitting of concave or minus lenses that diverge parallel light rays before they strike the eye (Fig. 35–5). Distant objects are then focused on the retina, and the myope can accommodate for nearer objects much as does the emmetrope.

Hyperopia

In hyperopia parallel light rays are focused behind the retina. This means that the hyperope must accommodate in order to see even distant objects. Many people are slightly hyperopic, and as long as no eye fatigue is present it is almost certainly harmless. Indeed, there is evidence that a slight degree of hyperopia is the normal state. Greater degrees of hyperopia can cause eye

FIG. 35–5 A biconcave lens (a) diverges parallel light rays so that the are focused on the retina (compare with Fig. 35–3b)—thus the myope's excessive refractive power or long eyeball is compensated for. In (b) the refractive power of the hyperope is increased by a biconvex lens, causing parallel light rays to converge slightly before they reach the eye (compare with Fig. 35–3c).

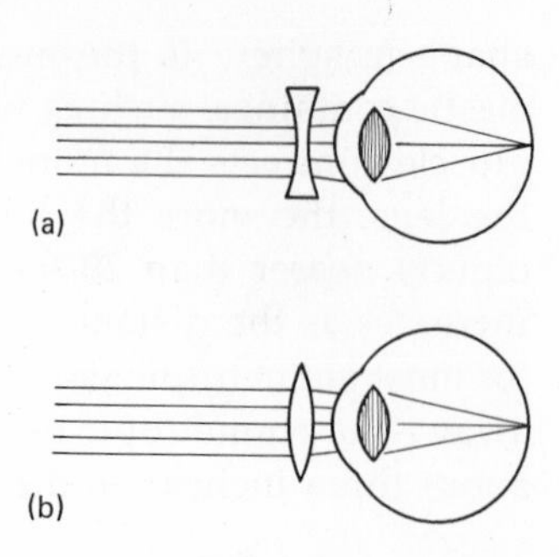

fatigue during long periods of accommodation, and may result in headache or even nausea. The cause may be a shortened eyeball or weak refractive power of the lens or cornea. The treatment of hyperopia is the fitting of convex or plus lenses that converge the light rays before they strike the eye. No accommodation is then required for distant vision, and the degree of accommodation required for reading is proportionately reduced.

Astigmatism

Astigmatism is usually due to a variation in the refractive power along different meridians of the cornea. The normal cornea is a perfect section of a sphere, and so has the shape of a cereal bowl when removed from the eye. In those persons with astigmatism the cornea resembles a section of an egg, and so would appear more like the bowl of a spoon. If the cornea were spoon-shaped with the long axis horizontal, then rays entering the eye in the vertical plane would be

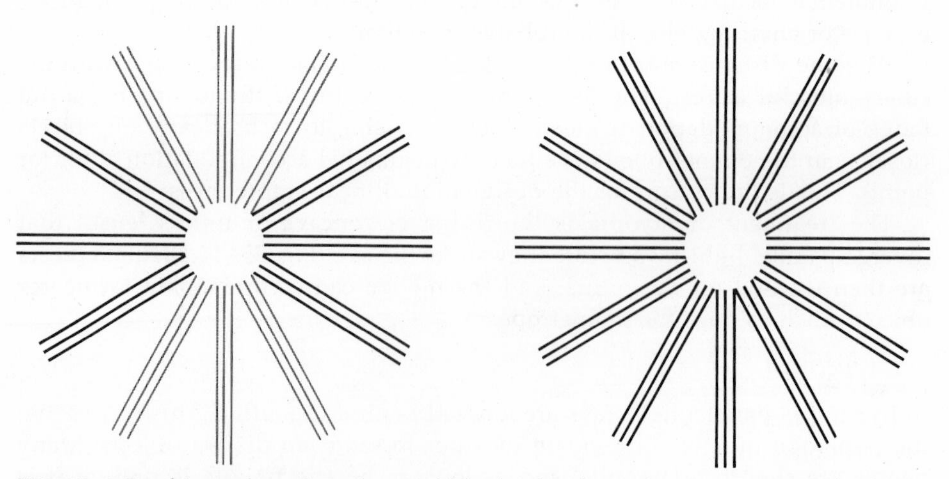

FIG. 35–6 Try this test for astigmatism. Hold this book in front of you at a comfortable reading distance, close one eye, and look at the center of the figure on the right. If some of the lines appear darker than others, as they do in the left-hand figure, you probably have astigmatism.

focused in front of those entering in the horizontal plane. The person with hyperopia can accommodate to see at all distances up to his near point, the myope has clear vision at near, but the person with astigmatism has a partially blurred image *at all times* (Fig. 35–6). He may try to achieve a clearer image by rapidly changing focus (accommodating), but this aggravates the eye fatigue. Severe headaches are a common complaint of those with high degrees of astigmatism.

Astigmatism is corrected by the use of cylinder lenses which are corrected along the proper meridian to compensate for the spherical defect of the cornea. Contact lenses easily correct astigmatism since they form a new, perfectly spherical, anterior surface to the cornea.

Presbyopia

At birth the lens is highly plastic and easily altered in shape by the action of the ciliary muscle. Throughout life there is a gradual hardening of the lens so it becomes more resistant to changes in shape. By the mid-forties the average person (who is slightly hyperopic already) has difficulty accommodating sufficiently to focus on near objects and to read fine print. This is presbyopia and does not indicate an increase in the degree of hyperopia but merely a loss in the range of accommodation.

A convex or plus lens of proper strength will replace the lost accommodation, but will usually result in a blurring of the distance vision. This means that the glasses are used only for near vision (reading glasses), or else a bifocal lens is fitted so that distant objects are seen through the upper (uncorrected) part of the lens.

It is important to realize that most visual defects correctable by lenses are errors of refraction only, and that defects of the retina are not usually present. This means that the myopic, hyperopic, or astigmatic person, when fitted with the correct lenses, generally sees as well as the emmetrope.

Strabismus (Squint)

The muscles that turn the eyeball are six in number: four are called *recti* and are arranged around the eyeball, one above (*superior*), one below (*inferior*), one toward the nose (*medial*), and one toward the temple (*lateral*). These roll the eye respectively up, down, in, and out; the remaining two, the obliques, aid in the up and down movements of the eye. If one of the recti is paralyzed, obviously the eye will not turn in the corresponding direction. Furthermore, the opposite muscle (as the lateral rectus is opposite the medial) will pull without the balancing opposition and will roll the eye chronically towards that side. The same is true to a lesser degree if a muscle is simply weakened, as happens more frequently. The result is a *strabismus* commonly called a *squint;* the latter term, however, is undesirable though short and familiar, because it is often used also to describe "screwing up" of the eyes as in a bright light, and can therefore be confusing. The common form of strabismus is a *medial* one due to a defect of the lateral rectus—a "cross eye" turning in towards the nose—but the *lateral* form, "wall eye" occurs also, and other forms, more rarely.

Most strabismus begins in infancy; a few cases occur in later life, due mostly to severe head injury. Some infantile cases are due to birth injuries; others are due to subtler causes.

We are all used to the fact that our two eyes normally see slightly different pictures due to their distance apart. You can demonstrate this to yourself by looking at a near object against a more distant background of clear-cut details. If you then open the two eyes alternately, the object appears to jump to and fro across the details of the background. Normally, these pictures are fused mentally into a composite with an impression of depth or perspective. However, if one eye is abnormal—very far-sighted for example—fusion is difficult; and when the baby tries to learn to fuse his images, he fails to do so and sees double (*diplopia*). He then rather learns to disregard or *suppress* the less clear picture and use only the other; for reasons not entirely clear, the disused eye turns inward with a weakening of the lateral rectus muscle. The resulting medial strabismus, besides being disfiguring, causes inadequate and degenerating vision in the affected eye. This is a deplorable condition to suffer at the highly plastic outset of life, and should be rectified by every possible means.

The crossed eye is, of course, easily recognized as a defective one. It should then be studied by an ophthalmologist to reveal the nature and degree of the defect. The child can often be fitted with special glasses to equalize the two eyes and permit him to fuse easily (Fig. 35–7). This can be done as early as five or six months of age and may be sufficient to cure the strabismus. At the same time loss of vision in the defective eye is halted and may be reversed. Surgery is often used to restore muscular balance and ensure against recurrence of the

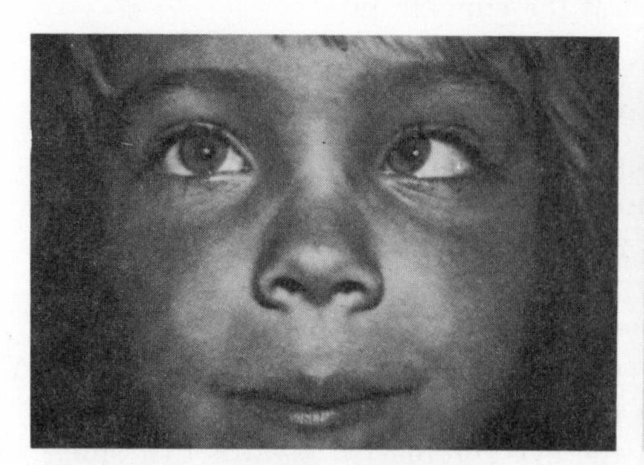
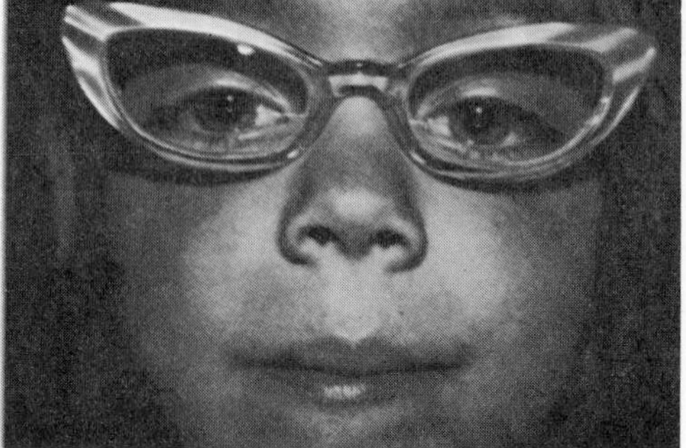

FIG. 35–7 *Left:* this girl has an esotropia, or inturning of her left eye. *Right:* use of convex lenses to correct her farsightedness and also use of bifocals results in completely straight eyes. (Dr. Eugene Helveston, Indiana University Medical Center)

condition; the operation simply shortens the weakened muscle to give it better tension. Patches over the good eye will sometimes be used to encourage use of the defective eye.

Difficulty in the fusion of the images of the two eyes and the suppression of the offending image may result in permanent visual loss in the affected eye without any easily observable abnormality. Such conditions are referred to as *amblyopia,* and special visual screening tests are now being used to identify such children at an age when corrective measures are still possible.

Cataract

Cataract is an opacity of the lens. It may be due to any one of a number of causes: congenital, infectious, toxic, degenerative, or the result of systemic disease such as diabetes. The forms of cataract are equally numerous, but all have the effect of obscuring and finally blocking vision.

Cataracts are, indeed, the most common source of blindness in later life (Fig. 35–8). Seventy-five to 85 percent of all people over 80 years suffer from some degree of cataracts. But cataracts can occur at any age from infancy (usually congenital) through youth and middle age, and are usually subject to satisfactory treatment.

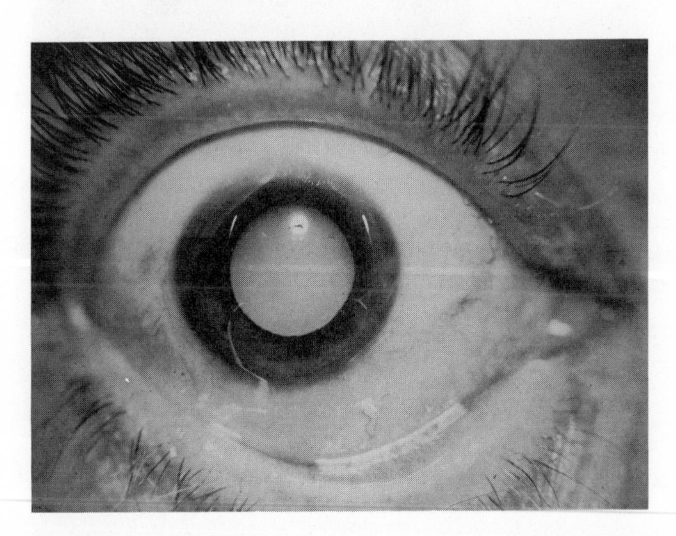

FIG. 35–8 A cataract, or opacity of the lens. (Vaughan, Asbury, and Cook: *General Ophthalmology,* 6th ed., Lange Medical Publications, 1971)

This treatment is almost always simply the removal of the diseased lens. As pointed out earlier, the lens is not the most important factor in focusing light on the retina, but only in changing focus between near and far. Since ability to change focus has already been lost in most cataract patients (because the lens stiffens with age) removal of the lens makes surprisingly little difference. They need special glasses and must adapt to a change in visual judgment, but few people have problems making this adjustment after cataract surgery.

Corneal Clouding

Certain viral infections, chemicals such as ammonia, and injuries may cause a scarring or clouding of the cornea. The result is either very dim vision or total blindness. Fortunately, corneas can be transplanted without risk of the immune reaction which generally causes rejection of most transplants. Healthy corneas removed from donors immediately after death, or from eyes which must be removed for other reasons, are now regularly transplanted to previously useless eyes (Fig. 35-9). Eye banks are now located in many parts of the country, and much effort is being made to encourage people to will their eyes for this purpose.

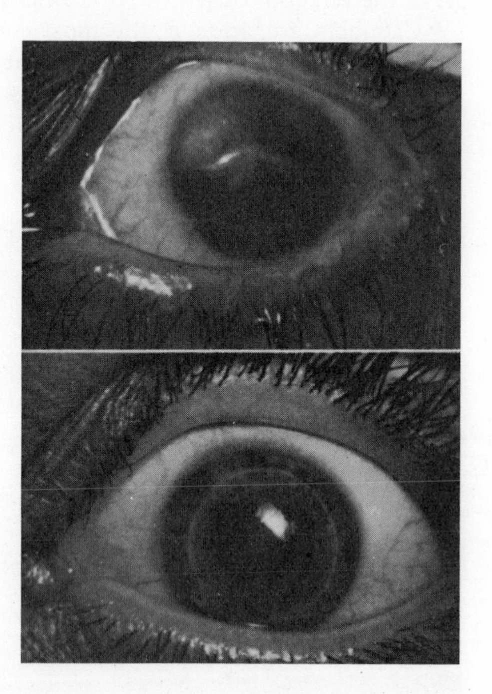

FIG. 35–9 *Top:* a cloudy cornea which was replaced by a donor's healthy cornea. The same eye is shown at the bottom after the transplant. The operation restored sight to the eye. (John P. Goeller)

Detached Retina

The retina may become detached from the choroid layer to which it is normally loosely fastened. When this occurs, fluid accumulates between the separated layers, and the separation tends to enlarge. The field of vision served by this part of the retina is lost, and the retina may degenerate.

Retinal detachments seem to occur most frequently in people who are strongly myopic, but they may be associated with a blow on the head. Since detachments most often take place along the periphery of the retina only the peripheral vision is lost, and sometimes such visual loss goes unnoticed. Therefore even a minor loss in the visual field should be reported to an ophthalmologist. The detachment can very often be corrected by surgery, and the earlier the condition is discovered the better are the chances of successful repair.

Glaucoma

This is a very common cause of blindness. Normally, the aqueous humor arises from small veins behind the iris, passes through the pupil, and drains back into the blood through tiny pores in the angle between iris and cornea. If the pores are blocked in any way, the humor accumulates and raises pressure inside the eyeball, which eventually destroys the delicate retina (Fig. 35-10).

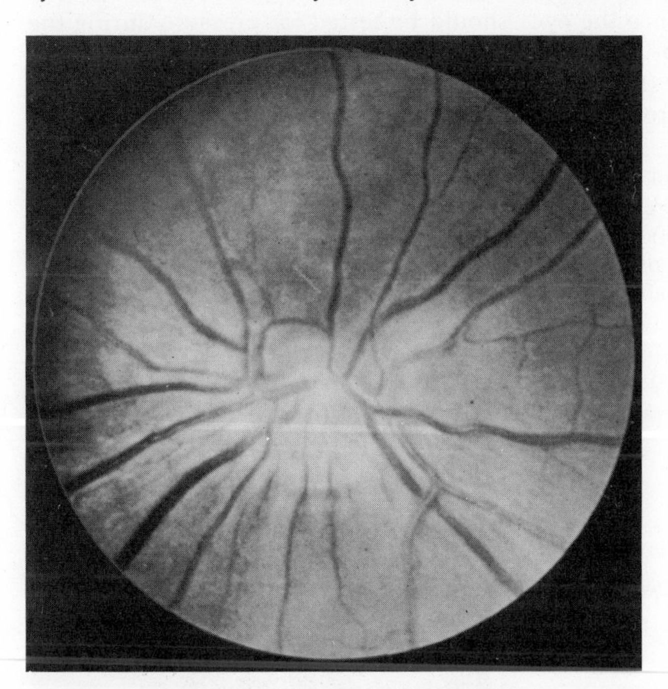

FIG. 35–10 The surface of the retina as seen through an ophthalmoscope. The light area is the beginning of the optic nerve. Notice the concentration of blood vessels in this area. (American Optical Co.)

Gradual incomplete blockage causes *chronic glaucoma.* This may develop over years and be detected at first only by skilled measurement of pressure in the eyeball. It is commonly associated with a narrowing of the field of vision, called "tunnel vision"; this may be so gradual in onset that the person does not realize its presence and may consequently get himself into trouble, especially when driving. Later, dull pain, especially in the morning, is felt first in one eye then later in both. The cornea appears "steamy," vision is impaired in darkness and blurred in the light, and lights appear to have halos around them as if in a fog, especially after watching movies or television. Over a million people probably have unrecognized glaucoma in this country.

Acute glaucoma may develop from the chronic form gradually or suddenly; or it may occur with no known chronic phase. Its symptoms are so positive and alarming that they are not likely to be neglected. If they are neglected too long while home remedies are tried, blindness can become complete and permanent after only a day or two. The symptoms are (1) pain in and behind the eyeball, (2) nausea and vomiting, (3) dilated pupil. The affected eye may be noticeably harder than the other, and sclera and even cornea may be bloodshot; but these symptoms are best evaluated by experienced personnel. Anyone with the symptoms described must be put in expert hands *immediately*.

Glaucoma cannot be cured, but it can be controlled, especially if detected in time. It is uncommon but not unknown in the young, and grows more common in the 30s and later. Hence the eyes should be tested for pressure during the routine optical examination—at least every four years before 30 and every two or oftener after 30 (Fig. 35-11). If glaucoma is detected, medication or skilled surgery can control it; but the patient will have to make up his mind

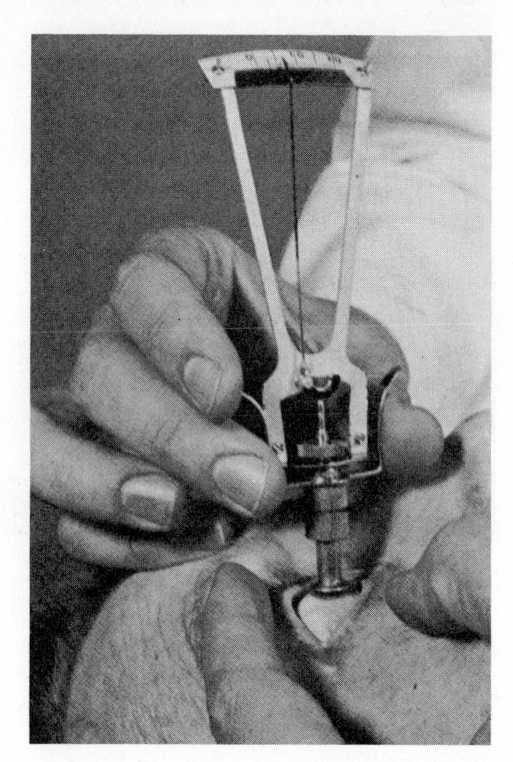

FIG. 35-11 A tonometer, placed on the cornea, measures the fluid pressure on the eye. When there is an excess of pressure, glaucoma may result. Early diagnosis can curb this enemy of eyesight. (John P. Goeller)

firmly that he must maintain treatment and regular examinations faithfully throughout life. Only if he does this are his chances good for saving and retaining his eyesight.

Other Disorders of the Eye

It is not uncommon for infections and disorders to occur in parts of the eye other than those mentioned (cloudy cornea, glaucoma, cataract, detached retina). The iris, the choroid, and the retina are all susceptible to inflammation. Such inflammation may be an isolated one or may accompany such diseases as diabetes, tuberculosis, renal disease, and syphilis. These inflammations, especially those occurring on the retina, can cause blind spots that cannot be repaired. Early recognition and treatment are essential; one should never allow any slight but persistent discoloration or blood-streaked condition of the eye to go undiagnosed.

Accessory Structures of the Eye

Even more incidental, but often concerned with eye disorders, are the *eyelids, conjunctiva,* and *lacrimal* (tear) *system.* The eyelids need no description. The conjunctiva is the special thin skin lining the lids and covering the front of the eyeball; it can be pictured as continuing the true skin around the lower margin of the upper lid, lining the inner surface of the lid, folding down over the eyeball so that it forms a pocket between lid and eye, the *upper conjunctival sac;* reaching the bottom of the front of the eye, it folds up inside the lower lid, forming the much shallower lower *conjunctival sac,* and finally continues over the edge of the lid into the skin. Over the sclera, the conjunctiva is very loosely attached to permit free movement of the eye. Over the cornea, however, it is tightly fused to the underlying layers; indeed, here it is generally considered to form a part of the cornea.

The *lacrimal system* consists of (1) the *lacrimal gland* lying just above the outer corner of the eye and (2) the lacrimal ducts draining the inner corner of the eye into the interior of the nose. Obviously, the flow of tears must be from the gland, down and inwards across the eyeball. This flow is continuous and is helped by the constant, involuntary blinking of the lids; it is very important, for if it ceases for long, even an hour or so, the cornea dries and becomes subject to infectious organisms that would otherwise be washed away by the mildly antiseptic tears. The resulting corneal ulcer can permanently affect vision. (For this reason, the eyes of an unconscious person should not be left open, but should be held closed, if necessary, with scotch tape—which will not pull on the delicate eyelids as would ordinary adhesive. Similarly, any paralysis of the blink reflex requires special care). As with the eyeball itself, disorders of the lids, conjunctiva, and lacrimal system ("pink-eye," for example), should be reported without delay to your physician or ophthalmologist.

General Care of the Eye

Most people are vaguely familiar with various precepts of eye care. These precepts should be clarified.

Periodic checks are often preached, but almost as often neglected unless enforced by authorities or by serious disorders. Defective vision, even of a minor degree, can be subtly disabling. A person with mild myopia, astigmatism, squint, and so on, is handicapped, perhaps not much but continuously. He has difficulty in reading, following blackboard demonstrations, driving, and other activities; and not only does this penalize him in performance, but may subject him to a nervous strain. Furthermore, the discomfort of eye-strain, like any pain, drains energy and alertness; and the strain often involves the much larger muscles of the neck and shoulders which tense in sympathy, increase the drain of energy, and increase pain in the form of headaches and stiff neck. All this may appear simply as constant fatigue and headache, but it can be diagnosed for what it really is by a professional routine examination.

Proper lighting is more familiar in cautionary discussions of the eye. The need for adequate illumination is based on the fact that when one cannot see clearly, one makes continual, fruitless, reflex efforts to solve the problem by focusing; this produces eye-strain in the same way as does defective vision. This is especially true when one is trying to focus on fine work. Merely walking in the dark, say, evokes no need for attempts to focus.

The question of how television and moving pictures affect the eyes is often raised. Ordinary amounts of viewing are harmless *if* (1) the surroundings are not so dark as to cause a conflict between the reflexes that constrict the pupil in bright light (the screen) and expand it in darkness, (2) the image is not blurred so that the eyes struggle vainly to sharpen it by focusing, with resulting eye-strain, and (3) one looks straight at the screen, not from an angle, which, again, forces reflex attempts to correct the distorted image. If disturbing eye conditions result from ordinary amounts of viewing, incipient glaucoma or other disorder should be suspected, and medical counsel sought.

Too much light can have even worse effects than too little. Prolonged overexposure can injure the retina, as in the case of snow-blindness due to the reflected glare from snow added to that of direct sunlight. Such conditions usually clear up in time, but they may leave residual defects. Extremely powerful sources of light can permanently injure or destroy the retina. The effect is something like sunburn of the skin, but since the retina is formed of nerve cells, and injured nerve cells are not replaced as are those of the skin, the effect is permanent. Also, unlike sunburn, severe damage to the retina by intense light is quite painless. This is why one should never stare directly at the sun or, say, an electric welding arc. Oversensitivity to light of moderate intensity is called *photophobia* (light fear); when persistent it is usually a symptom of some disorder such as iritis, choroiditis, and tendency to migraine.

Medication of the eye is never advisable without professional counsel. It is *not* required for routine "eye hygiene," which is adequately taken care of by natural tears and secretions of glands in the eyelid. Sore lids, "bloodshot" eyes, abnormal dryness, excessive "sand" (the residue of drying tears), blurring and so on may be temporary disorders due to irritants, prolonged swimming, and the like; or they may be indications of something more serious. In the first case self-treatment is unnecessary; in the latter, home care may delay necessary and effective treatment by a professional. Commercial "eye washes" generally do little good, and they certainly cannot affect internal eye strain.

Eye Specialists

Disorders of refraction are, of course, routinely treated by the fitting of glasses that compensate for the defect. This process is a major medical specialty and involves various types of health care personnel. The names of these specialists are rather similar yet should be accurately used if one is to call on their services intelligently. The qualified medical specialist dealing not only with disordered refraction but with all eye disorders, is the *ophthalmologist* or *oculist;* such a man is a physician equivalent to the cardiologist, internist, or other specialist. The *optometrist,* doctor of optometry (O.D.), is specifically educated and licensed to examine the eyes to determine the presence of vision problems, and to prescribe lenses, other optical aids, and visual training that preserve or restore vision. An *optician* is the technician who performs the precise work of grinding and mounting lenses; he might be compared to a technician who makes and fits orthopedic appliances. An *orthoptist* is a technician who directs visual training to correct muscle imbalances, usually under the supervision of an ophthalmologist or optometrist—a sort of ocular physiotherapist.

The situation is confused by an overlap between the work done by ophthalmologists and optometrists. Both will almost certainly make his own refractive tests during a general eye examination. Both will care for many of the same visual problems, but the optometrist will refer patients requiring surgical or medical treatment to an appropriate medical specialist.

Contact Lenses

Many people dislike wearing glasses for various reasons. Young women feel that glasses detract from personal appearance; athletes often find them necessary for accurate performance, but liable to breakage with a hazard of splintered glass and jagged broken frames; children also are liable to broken glasses; others find them uncomfortable and in constant need of polishing. Corneal deformity in some people is too great for conventional glasses to compensate. In these and other cases, modern contact lenses are a possible recourse.

Contact lenses may be described as delicate shells of plastic, about a quarter of an inch in diameter. A contact lens fits over a cornea so accurately that it is separated from the corneal surface by a thin but everywhere uniform film of tear fluid. (The uniformity may be modified in cases of grossly deformed corneas). In other words, a properly fitted contact lens does not really make contact with the eye at all but floats on it. The thickness of the lens is machined so as to compensate for refractive defects of the eye. Because of close contact with the cornea, very tiny variations in a contact lens can produce great changes in refraction; hence these lenses are often used in cases where ordinary glasses would have to be very massive or quite impracticable. The great delicacy of adjustment, both for fitting to the cornea and for refractive compensation, naturally makes contact lenses very expensive.

The advantages of "contacts" are many. As noted, they can often be used where ordinary glasses are ineffective; they are practically invisible when worn (Fig. 35–12); they avoid the dangers of broken glass or frame inherent in conventional glasses; once put in for the day, they normally need not be removed except for such activities as swimming; and, once past the initial adjustment period, they are actually more comfortable for many people than are conventional glasses. The disadvantages must be considered too, however: The cost of contacts is usually well over $100 a pair; they require careful training for insertion, removal, and care; they are very easily lost and difficult to find if dropped; they are uncomfortable for some people because of irritation of the conjunctiva or the overlying lid; and use must be discontinued during any eye infection.

The question of "attractiveness" is certainly a very personal one to be considered by the individual (with money as a more than incidental consideration). But let it be said in defense of the eyeglasses that, if selected carefully and with the help of a trained optometrist with attention to appearance, glasses can *add* to one's physical attractiveness rather than detract. In fact, there are people who "look better" to their friends with glasses than with their contacts in place.

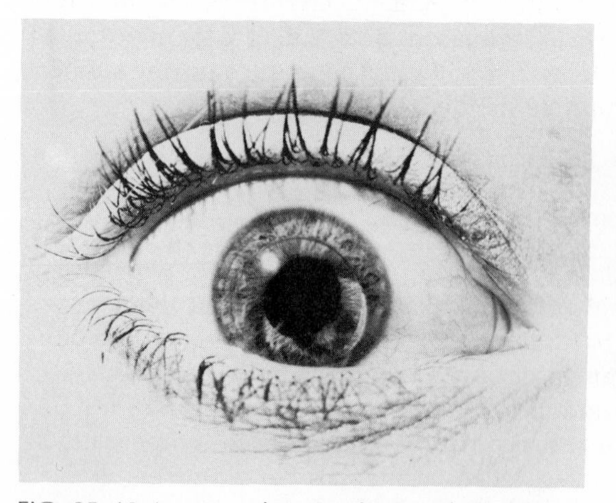

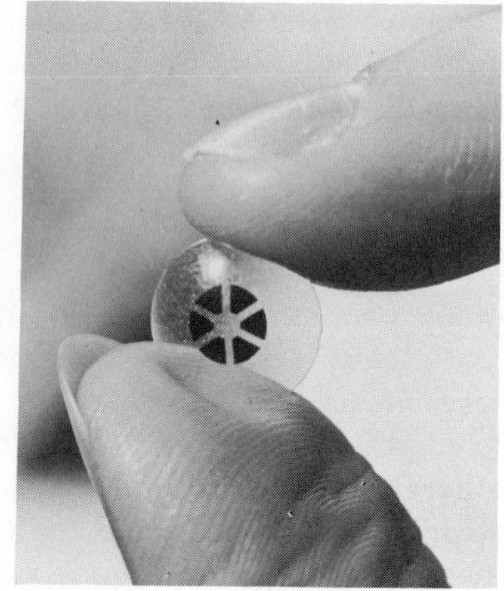

FIG. 35–12 A contact lens in place; *right:* a contact lens prepared to restrict the amount of light entering the eye. (Wesley-Jessen Co.)

Minor Accidents to the Eye

Practically everyone is familiar with certain minor eye troubles which hardly need professional attention. Yet many people deal with even these simple matters in a manner that would hardly meet with professional approval.

1. *A cinder or other small objects in the eye.* The cornea, sclera, and conjunctival sac are exquisitely sensitive, and the natural impulse of anyone is to get rid of an irritant object at once by rubbing. But this is rarely effective and often results in increasing the irritation. A flood of tears and gentle winking very often dislodge the object and wash it down to the inner corner of the eye. This can sometimes be aided by drawing the upper eyelid down over the lower lid.

More vigorous action may be necessary in some cases. This, however, must be intelligent and cautious. Objects often lodge behind the upper lid where the conjunctival sac is deep. Simply raising the lid is usually futile, since it cannot be raised far enough to expose the roof of the sac, even when the eye is rolled down. Rather, the lid must be folded or rolled up over a match, thin pencil, or some such rod, which exposes the depths of the sac; but this maneuver requires deftness and experience which can rarely be achieved at a time of emergency. A bit of practice under expert direction, at some leisure opportunity, is an asset to anyone. Once exposed, the object should be removed with a clean, soft object. A cotton tip probe is ideal, but a clean handkerchief of Kleenex folded to a point is adequate. Any particle which resists the gentlest efforts of removal should be left alone, and professional help sought. Penetrating foreign bodies should *never* be treated except by an ophthalmologist.

2. *Hematoma or black eye.* A hematoma is a bruise or bleeding into tissue anywhere. It is particularly noticeable in the eye since the surrounding tissues are loose to permit easy motion of the eyelids and to serve as a cushion to absorb shock. True "black eyes" are seldom serious and clear up spontaneously in a few days by reabsorption of the blood. About all that can be done for a black eye is to apply a cold compress as soon as possible after the accident, which reduces the bleeding and, so, the time needed for reabsorption.

3. *Blow directly to eyeball.* A blow to the eye can be extremely serious (detached retina, abrasion, or cut on the cornea) and should not be treated lightly! When in any doubt whatsoever about the seriousness of such a blow, consult a physician immediately.

Dealing with the Blind

To discuss the problems of blindness and the techniques for handling them is far beyond the scope of this book; and the likelihood of your having need for such a discussion is happily remote. The majority of cases of blindness are already established in infancy, even prenatally, or develop only with advancing age; they do not involve the young adult.

Dealing with blind associates is another matter. This is a situation in which thoughtfulness, intelligence, and, above all, tact and courtesy are essential.

In regard to *help:* Never officiously offer, much less impose, aid until you are sure it is wanted. If in doubt *ask* naturally and courteously whether you can do anything. When aid is accepted, give it quietly, efficiently, and without unnecessary comment and false cheerfulness. An example of real thoughtfulness was once described by a blind friend: "People guiding one in unfamiliar surroundings," he said, "will often remark 'Watch the step.' They rarely think to say that the step is up or down."

Use *tact* to steer a course between pointedly ignoring and tiresomely dwelling on the disability. Almost, though not all, blind people have no shyness about explaining and discussing their condition with someone who shows intelligent interest, especially if this is directed toward providing more effective cooperation. Elaborate avoidance of any reference to sight and seeing is annoying rather than tactful. On the other hand, a blind person usually has his mind occupied with matters far removed from his disability and is probably vexed by pointless reference to the latter—this includes patronizing helpfulness. A pitying or patronizing tone is unpardonable; those who employ it would resent similar treatment of their own gastric ulcers, poor complexion, or short stature. In sum, treat a blind companion as naturally and easily as possible.

Never underestimate a blind person's *keenness of perception.* Experienced blind people often have an almost uncanny awareness of what is going on around them. A very effective form of tactfulness is to assume that a blind companion is "with you" on most points, at least in your conversation. At the same time, making your movements distinct to him is also tactful. Try never to surprise or startle him by unexpected contacts or unheralded speech, or to move away from him without letting him know of it by remark or audible actions—to find oneself speaking to empty air is humiliating to anyone and even more so to the blind.

The foregoing remarks apply with special force to the blind. But they largely apply, with suitable variations, to anyone who has a disability—deafness, paralysis, aphasia, and so on. Conversely, the remarks must be adapted to circumstances. Someone who is not yet adjusted to his disability, physically and mentally, is certain to be more sensitive about it and yet more impelled to discuss it, than someone of long experience; an elderly lady may be grateful for sympathy, as a vigorous and self-reliant young woman may not be; some people are highly ingenious and self-reliant, while others need guidance and suggestions tactfully offered. In dealing with the handicapped, perhaps the greatest assets of all are insight and tact.

REVIEW QUESTIONS

1. Why must the eye accommodate in order to view objects closer than 20 feet away?

2. Why is a myopic person fitted with lenses that diverge light rays?
3. What is presbyopia; astigmatism?
4. What is the principal cause of amblyopia?
5. What part of the eye is sometimes transplanted? Why?
6. What are some of the advantages and disadvantages of contact lenses?

REFERENCE

ALLEN, JAMES H., *May's Diseases of the Eye*, 23rd ed. Baltimore: Williams & Wilkins, 1963.

VAUGHAN, DANIEL et al., *General Ophthalmology*, Los Altos, Calif.: Lange, 1971.

Additional Readings

DUNPHY, EDWIN B., "The biology of myopia," *New England Journal of Medicine*, **283:** 796–800 (1970).

PIRENNE, MAURICE H., *Vision and the Eye*, 2nd ed. New York: Barnes and Noble, 1967.

36
HEARING AND VESTIBULAR SENSES

If sight is the most important of our special senses, hearing is certainly second; our vestibular senses, though far less familiar, rate third. The organs for hearing and vestibular senses all develop from a single origin, are closely associated, and are often involved in the same disorder.

Anatomy of the Ear

The ear, in its broadest definition, is formed of three divisions or compartments: the outer, middle, and inner ears as shown in Figure 36–1.

The *outer ear* comprises the "ear" of common speech, the shell-like growth on the side of the head, more accurately called the *pinna;* the passageway leading inward from the pinna, the *external auditory canal;* and the eardrum, or *tympanic membrane* at the end of the canal, shutting it off from the middle ear. These structures, though less highly developed in man than in some animals, are far from rudimentary and are of some value to us in helping to locate the direction of sounds.

The *middle ear,* or *tympanic cavity,* is an air-filled bony cavity intruding between the outer and inner

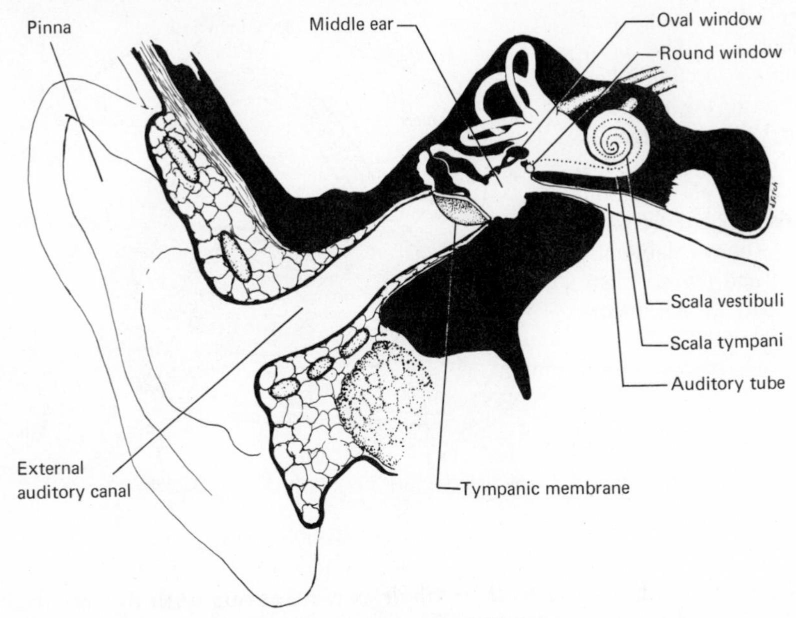

FIG. 36–1 The parts of the ear.

ears. It has a very irregular shape, and is connected with the complex honeycomb of *mastoid air cells* behind it. These cells lie in the mastoid process or bone, the large downward-pointing lump that can be felt just behind and below the pinna. The middle ear is connected to the pharynx by a channel called the *auditory* (sometimes pharyngeal or Eustachian) *tube*. The middle ear contains the *ossicles*, malleus, incus, and stapes (hammer, anvil, and stirrup), a chain of tiny bones which conducts sound vibrations from eardrum to inner ear, amplifying them in the process. The middle ear also contains minute muscles acting on the drum and ossicles to dampen down too violent vibrations. Paralysis of these muscles permits even ordinary sound to seem painfully loud. The middle ear thus modifies intensity of sound reaching the inner ear, either up or down. The inner bony wall of the middle ear has two openings which communicate with the inner ear. A small oval opening called the *oval window* receives the foot-plate of the stapes. Lower down is a second opening known as the *round window* which is covered by a thin membrane.

The *inner ear* lies also in a bony cavity, but is filled with fluid instead of air (Fig. 36–2). The chamber is called the bony labyrinth and consists of three parts: the vestibule, the cochlea, and the semicircular canals. The vestibule and semicircular canals contain a part of the vestibular apparatus to be discussed later. The cochlea is a hollow, spiral chamber which resembles a snail's shell, and contains the essential organ of hearing. A plate of bone runs through the tunnel of the cochlea and partially divides it into two chambers (Fig. 36–3).

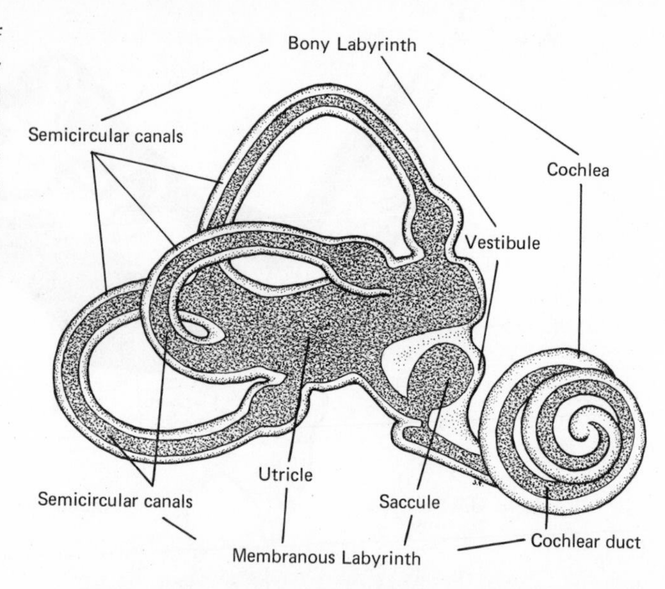

FIG. 36–2 The structures of the inner ear. The bony labyrinth is a fluid-filled excavation in the temporal bone and the membranous labyrinth partially fills this space. The scala vestibuli and scala tympani are parts of the bony labyrinth (cochlea) and the cochlear duct is part of the membranous labyrinth.

One of these chambers, the *scala vestibuli,* is continuous with the vestibule; the other, the *scala tympani,* ends at the round window. These parts are best understood if we visualize the vestibule and cochlea as a straight chamber. The scala vestibuli communicates with the scala tympani around the apex of the cochlea; this communication is called the *helicotrema.* The result is one long tunnel beginning at the oval window and ending at the round window.

A section through the cochlea reveals that the bony shelf separating the

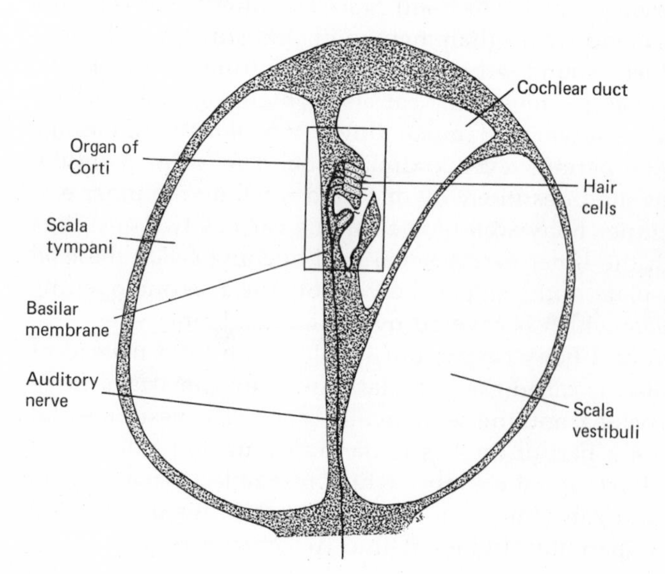

FIG. 36–3 A cross section of the cochlea. The basilar membrane and the organ of Corti run as a ribbon through the whole length of the cochlea (compare with Fig. 36–2). This particular section would represent the receptor organs for just one frequency of sound.

scala vestibuli and scala tympani is incomplete. The separation is completed by a membranous tube called the *cochlear duct.* (This membranous tube continues into the vestibule where it expands into the bulbous *saccule,* which in turn connects with the *utricle.* The *membranous semicircular canals* arise from the utricle and occupy the semicircular canals of the bony labyrinth. The cochlear duct, saccule, utricle, and membranous semicircular canals collectively make up the *membranous labyrinth.* All but the cochlear duct are related to vestibular sense, and so do not concern us here.) The floor of the cochlear duct—which forms one wall of the scala tympani—is the *basilar membrane.* Lying on the basilar membrane and running the whole distance of cochlear duct is the *organ of Corti.* The organ of Corti is the actual sensory receptor for hearing. From the organ of Corti arise the numerous fibers of the *auditory nerve* which carry the nervous impulses to the brain.

The Nature of Sound

To understand the way in which the ear functions we need to review some of the characteristics of sound. Sound is transmitted in what is termed waves which travel at a speed of about 1100 feet per second. These waves are actually alternating compressions and rarefactions of the conducting medium, usually air. For example, a violin string produces sound by vibrating back and forth; every movement compresses the air in front of it and rarefies the air behind it. The *pitch* of a sound is determined by the number of compression-rarefactions, usually called vibrations or cycles, produced each second. In physics and engineering this number is referred to as the *frequency* of the sound. In the violin again, a shorter string produces a greater number of vibrations per second (a higher frequency), and consequently, a higher pitch.

Not all compressions and rarefactions of air produce a sensation of sound. For example, a stick waved gently through the air will set up large and irregular cycles but they will not be heard. If the stick is moved rapidly enough, however, the wood begins to vibrate, and these vibrations will produce small and rapid cycles which we hear as a swishing. However, true sound waves are produced which we do not hear. The human ear is able to hear sounds from a range of about 20 to 20,000 cycles per second. We know that dogs are able to hear sound of a much higher frequency, perhaps as high as 30,000 cycles per second.

The *intensity* or *loudness* of a sound depends upon the *amplitude* of the sound waves. Sounds with a greater amplitude compress and rarefy the air with greater pressure. The firing of a cannon and the beating of a bass drum may have the same frequency, but the cannon produces vibrations of much greater pressure. This pressure can be measured in terms of dynes per square centimeter, or it can be measured as a flow of energy in terms of watts. Acoustic engineers often compare the loudness of sounds by use of the *decibel.* (db).

A decibel is equal approximately to the smallest degree of difference of loudness ordinarily detectable by the human ear. Zero decibels is the faintest audible sound. A sound of 120 db causes discomfort and one of 140 db is distinctly painful. Examples of other levels of sound are a quiet countryside at 2 A.M., 50 db; a fairly quiet office, 64 db; the cabin of an airplane, 110 db; a jet-engine testing area, 148 db (Fig. 36–4).

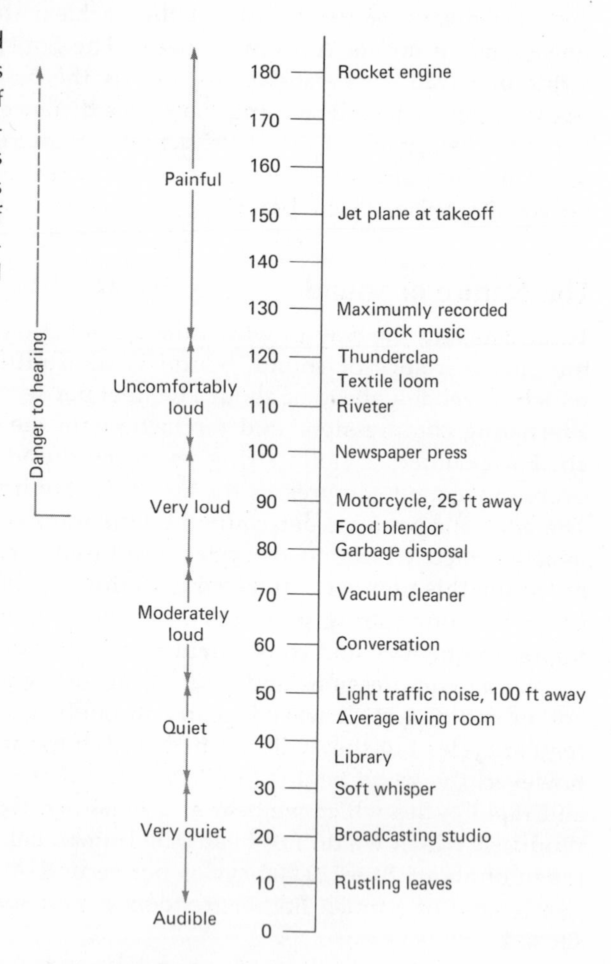

FIG. 36–4 Our sensitivity to sound extends over such an enormous range that a special system of describing relative sound intensities has been developed. In this system, a sound of 20 decibels is 10 times as loud as a sound of 10 decibels; a sound of 100 decibels is 10 billion times as loud as a sound of 10 decibels!

However, there is another difference between sounds of the same frequency from a cannon and bass drum. This is *timbre* or *color* of the sound. Neither is composed entirely of one frequency. In addition to the dominant frequency which forms the pitch, there are a great number of overtones or secondary frequencies, usually in some regular arithmetic pattern. These overtones give a particular sound its characteristic features, and is why a given note on a violin is distinctively different from the same note sounded on the piano. If no dominant frequency is present and if the frequencies of the sound are scattered, the resulting sound is *noise.*

Transmission of Sound Waves through the Ear

The pinna gathers the sound waves or vibrations and tends to direct them through the auditory canal to the tympanic membrane. The vibrations of the tympanic membrane are amplified (or sometimes dampened) by the malleus, incus, and stapes, and passed on through the oval window to the fluid of the vestibule of the bony labyrinth. The sound vibrations set up in the fluid continue through the scala vestibuli, around the helicotrema to the scala tympani, and, finally, are dampened by the membrane of the round window. The vibrations picked up by the basilar membrane stimulate the receptor cells of the organ of Corti. The problem remains as to how the basilar membrane and the organ of Corti are able to distinguish pitch and intensity.

Theories of Hearing

The presently accepted theory of hearing is based on the phenomenon of *sympathetic resonance.* We are all aware of the fact that if one of a pair of tuning forks of the same pitch is struck the other will begin to vibrate and to produce the same note. This is sympathetic resonance. All objects have a frequency of resonance, that is, a frequency at which they begin to vibrate and produce a given pitch in sympathy. A number of commonplace phenomena can be used as examples. The tenor who breaks a glass goblet does so by singing with great intensity the frequency of resonance of the goblet; radio cabinets of former days often had a frequency of resonance of around 80 to 90 cycles per second and would produce an annoying buzzing or shattering when that particular frequency was sounded in the receiver; the unpleasant sensation that is experienced when chalk "screeches" on a blackboard is believed to be due to the fact that our bones have this particular frequency of resonance.

The basilar membrane is a long, spirally arranged, ribbonlike structure composed of a great number of fine cross fibers. These fibers have a graded variation in length and probably in tensions. The number of cross fibers is believed to be about 24,000 with the longest fibers, which may be compared with the bass strings of the piano, lying near the apex of the cochlear duct, and the shortest fibers ("treble strings") lying near the base (Fig. 36–5).

The resonance theory of hearing suggests that the basilar membrane acts as a tiny harp or piano whose 24,000 strings each have a specific frequency of resonance. A certain tone would not cause the entire membrane to resonate, but only the particular fibers whose frequency of resonance were the same or nearly the same as that of the note sounded. The vibrations of these particular fibers

FIG. 36–5 The basilar membrane makes about two and a half revolutions along with the cochlea. The lowest frequencies stimulate the parts of the basilar membrane at the apex, and the highest frequencies the base.

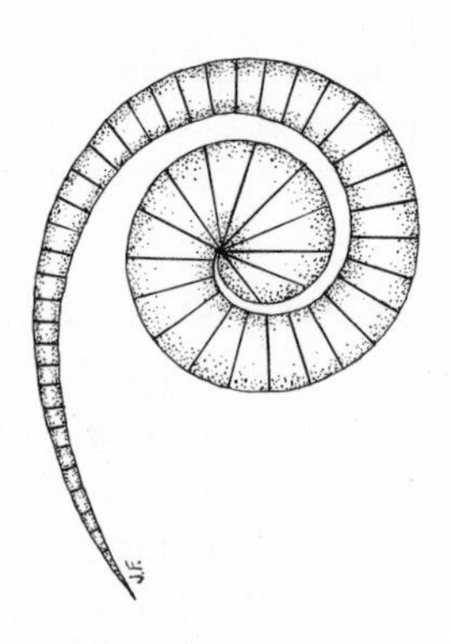

would stimulate the receptor cells of the organ of Corti, which in turn would transmit impulses to the brain. Taken literally, the theory would seem to suggest that perhaps 24,000 different frequencies are distinguishable, but it would not necessarily mean that these would be evenly distributed over the whole range of hearing. Our hearing acuity is greatest in the range from 100 to 5000 cycles per second, and the evidence is that most of the basilar membrane is devoted to these frequencies.

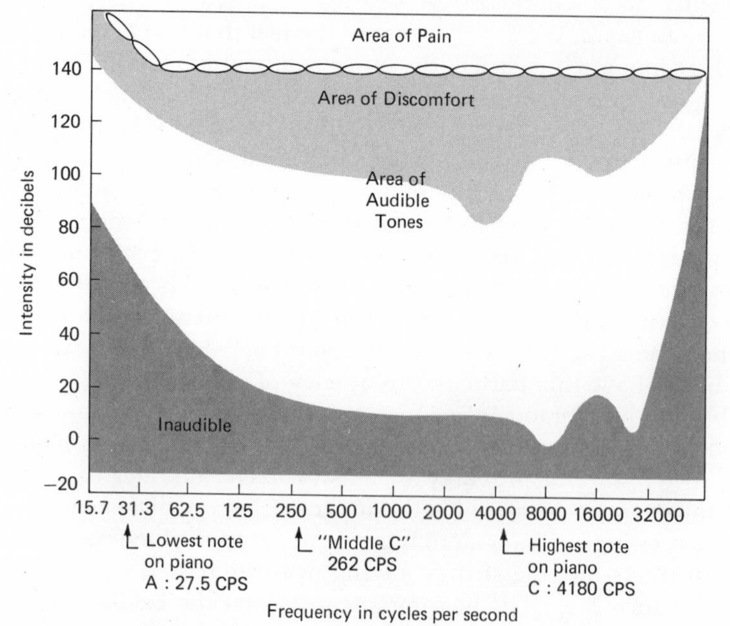

FIG. 36–6 The frequencies between 500 and 8000 can be heard at very low intensities. Frequencies above and below this require progressively greater intensities to be audible. (Adapted from Davis and Silverman, New York: *Hearing and Deafness*, rev. ed., 1965, New York: Holt, Rinehart and Winston)

Figure 36–6 shows the area of audible tones related to the notes of the piano. The piano covers about seven-and-a-half octaves with the highest note, the C key, at 4180 cycles per second. Two-and-a-half octaves above this we approach the highest frequencies which are audible. As we reach these higher frequencies differences in pitch become more difficult to distinguish, and for this reason are less important to us.

Hearing Sensitivity

The lowest intensities at which a sound can be detected is called the *threshold of hearing*. But this threshold varies according to pitch with the greatest sensitivity, that is, the lowest threshold, occurring between 500 and 8000 cycles per second. The ear is progressively less sensitive to frequencies above or below this range; and for very low or very high frequencies to be heard, they must be of very great intensity.

The intensity of sound is transmitted to the brain in the same way that other sense organs transmit intensities of their stimuli. The greater the intensity of a sound which is conducted through the middle ear to the inner ear, the more forcibly will the basilar membrane vibrate; the stronger will be the stimulus to the receptor cells of the organ of Corti; and the higher will be the rate of nervous impulses discharged through the auditory nerves.

Deafness

The forms of deafness fall logically into two categories: *conduction deafness*, in which the function of the conduction apparatus—the external auditory canal, the ear drum, and the ear ossicles—becomes impaired; and *sensory-neural deafness*, in which the organ of Corti, the auditory nerve, or the part of the cerebral cortex where nerve impulses terminate, is defective.

Conduction Deafness

The classical test to distinguish conduction from sensory-neural deafness helps to illustrate the difference between the two. The patient with conduction deafness may be quite unable to hear a tuning fork held close to his ear. But if the tuning fork is held against the mastoid process he may be able to hear the sound as well as a normal person does. Instead of the sound being transmitted through the middle ear it is conducted through the bone, and this is sufficient to set in motion the sound vibrations in the fluid of the inner ear. Thus, if the deaf person can hear by bone conduction, we can infer that the inner ear and auditory nerve must be normal. The person with sensory-neural deafness hears the tuning fork no better when it is held against the bone than when it is in front of his ear.

Causes of conduction deafness

IMPACTED WAX. Defects of the pinna or external canal rarely cause deafness. However, the wax (cerumen) may sometimes harden and become impacted, so that the sound waves are blocked. This usually occurs rather suddenly after swimming or bathing when a drop of water closes the last small communication through which normal sound could previously pass. Since the canal is served by twigs of the nerves that supply the pharynx, esophagus, and digestive system, a mass of impacted cerumen (or any irritant body) in the canal may cause episodes of nausea.

When blockage occurs, the condition should be treated by a physician who will remove the wax without injury to the canal or eardrum. Actually, if the ears are kept clean, the average person does not develop impacted wax. After bathing or swimming water should not be allowed to dry in the canal since this tends to soak the wax into masses. Rather, the canal should be gently blotted out by a clean cloth thin enough to enter the outer opening when wrapped around the tip of a finger. Regularly followed, this procedure usually prevents wax accumulation. The cleaning of the external canal with hairpins or matchsticks, or even a cotton-tipped probe, is dangerous. The canal is closely surrounded by delicate nerves and blood vessels and terminates in the highly sensitive drum. A slip of an instrument prodding for a foreign body or wax can therefore cause serious injury resulting in partial deafness or plant an infection in a vulnerable area.

OTITIS MEDIA. The most common cause of conduction deafness is inflammation of the middle ear. This is called otitis media and can occur following or in connection with an upper respiratory infection; in fact, this is the most common cause. The connection between the pharynx and middle ear, the pharyngeal (Eustachian) tube, offers a route by which the infection can spread upward into the middle ear. Normally, the tube is tightly closed except when swallowing, but in cases of colds or sore throat it may become distended by fluid secretions through which the infection can spread. Even when the pharyngeal tube is unaffected a violent blowing of the nose can raise pharyngeal pressure and drive infectious droplets into the middle ear. Acute otitis media or even mastoiditis (infection of the mastoid air cells) can result. The inflammed mucous membrane which lines the middle ear and mastoid air cells secretes watery fluid. As the disease progresses the watery secretion forms a culture for the invading bacteria and becomes transformed into pus. The pain becomes severe as the pus builds up, and the eardrum may rupture, or may need to be opened by a physician so that drainage can occur. Although the growth of bacteria can often be checked by antibiotics at this stage, other medical treatment is usually required.

Recurring attacks of otitis media tend to cause the development of fibrous adhesions around the ear ossicles and may even cause their complete erosion.

The spontaneous rupture of the eardrum is a dangerous hazard of poorly treated otitis media, and is far more likely to lead to complications than surgical incision. The danger to the general health of the patient as well as the possibility of severe hearing loss demand that otitis media be treated by a physician. Contrary to popular belief, the hearing loss which often follows mastoid surgery or incision of the drum is due to the disease and not to the surgery.

OTOSCLEROSIS. Otosclerosis is a bone disease in which porous bone grows within the middle ear cavity and becomes hardened into "sclerotic" plaques. The growth of the new bone may occur around the oval window and fix the stapes so that it can no longer move freely. It is much like the arthritis of other joints. Fixation of the stapes occurs gradually over a period of years, and so the hearing loss will be gradual. Otosclerosis can apparently develop without involvement of the ossicles, in which case it would unlikely ever be diagnosed. It is also an hereditary disease probably transmitted by a dominant gene. Since it can occur without deafness it may appear to "skip" generations making the pattern of its inheritance difficult to determine.

Two surgical procedures have been developed for the treatment of deafness due to otosclerosis, and a remarkable improvement in hearing is frequently possible. The oldest of the two, the *stapes mobilization,* is based on the simple concept that if the stapes is broken loose from its frozen position, its function will be restored. The surgeon enters the middle ear by reflecting a part of the eardrum. His objective is to break the footplate of the stapes free from its surrounding encrustation of bone. It is a very delicate maneuver and just how the stapes will respond to movement is not always predictable: the footplate may break loose or one or both of the delicate arms which attach to the footplate may fracture. However, a stapes even slightly chipped and broken moves more easily than a frozen one. The entire operation is brief and the after effects are minor. The reservations concerning the operation are due to the uncertainty of the result. Sometimes a refixation occurs, and the former deafness returns. If refixation occurs it usually takes place within three months after the operation.

The *fenestration operation* (Fig. 36–7) is a much more complicated procedure and with a somewhat more predictable result. The objective here is to bypass the normal function of the middle ear and to make a new communication with the fluid of the inner ear. The incus and a part of the malleus are removed and a window (fenestration) made in the (horizontal) semicircular canal that forms a bulge in the inner wall of the middle ear cavity. The window is then covered with skin from the external auditory canal. A new sound-conducting system is established; it is not as efficient as the complete eardrum with its chain of ossicles, but it is simple and its performance is predictable. Thus the patient can have considerable confidence that his hearing will be improved.

The exposure of the semicircular canal makes it much more accessible to changes in temperature with a possible result of dizziness (discussed under the

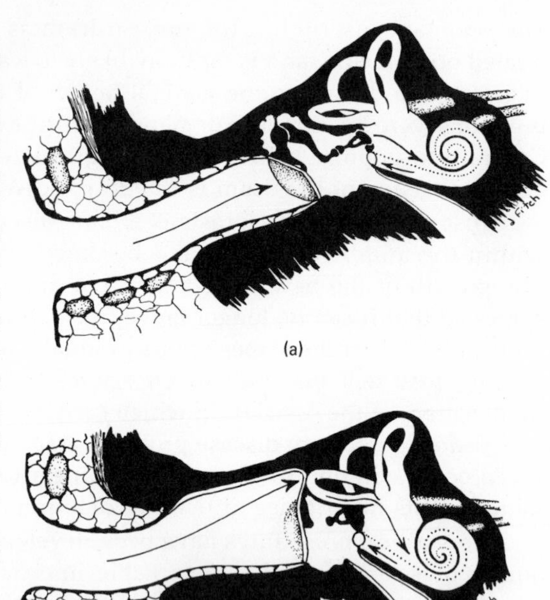

FIG. 36–7 The fenestration operation. (a) The normal pathway of sound waves (arrows) is through the tympanic membrane to the ossicles (malleus, incus, stapes) to the fluid of the bony labyrinth. (b) In the fenestration operation the external ear canal is enlarged at the expense of the middle ear cavity, and sufficient bone is removed so that a membranous connection can be made with the nearest of the semicircular canals. Arrows show the new pathway for the sound waves through the semicircular canal to the fluid of the bony labyrinth. (Modified from Davis and Silverman: *Hearing and Deafness,* rev. ed., 1965, New York: Holt, Rinehart and Winston)

vestibular senses). For this reason the patient is cautioned against cold and wind and sudden movements of the head. However, these and other problems are minor if the aftercare demanded by the physician is carefully followed.

Hearing Aids

Most often conduction deafness is treated by hearing aids rather than by surgery, and the objective remains much the same—that is, to bypass the middle ear. The hearing aid acts by picking up sound vibrations, amplifying them electronically, and transmitting them to the temporal bone of the skull through which they pass to the inner ear. As long as these vibrations are strong enough to stimulate the basilar membrane (and this need not be very strong since even sound vibrations transmitted naturally through the ossicles are amazingly slight), the person can hear. With modern miniaturization, effective aids can be very small and inconspicuous (see Fig. 36–8). However, hearing aids differ even more widely than do eyeglasses, and for this reason must be fitted only by trained personnel.

Sensory-Neural Deafness

Sensory-neural hearing loss is usually caused by degeneration of the receptor cells of the organ of Corti or of the fibers of the auditory nerve, or both. Apparently, the part of the organ of Corti related to the perception of high fre-

FIG. 36–8 A hearing aid built into eyeglasses (Carlyle)

quencies is most susceptible to degeneration since the hearing of high tones is most often lost first. Sometimes the transition from good to poor hearing extends over several octaves, perhaps with good hearing below 1000 cycles per second and very poor hearing from 4000 to 6000 cycles per second. The transition may, however, be very abrupt. The loss of high frequencies is more troublesome than would ordinarily be supposed because the distinctiveness of most of the consonants is due to their high frequency components. With careful attention a person with high-frequency loss is often able to understand speech in quiet surroundings, but the presence of just a small amount of background noise will sometimes mask the small differences in speech sounds upon which he relies. Furthermore, those with sensory-neural deafness are not protected from the annoyance of loud noise as are those with conduction deafness, and the range between hearing little to that of hearing too much becomes narrowed. Such a person might one minute ask you to speak more clearly, and the next minute complain that you are speaking too loudly.

Loud Noise and Sensory-Neural Deafness

We are all familiar with the "auditory fatigue" which results from the exposure to loud noise. Anyone who has worked in a really loud factory, driven an air hammer, or done pistol or skeet shooting can recall how his ears rang for an hour or more afterward and voices sounded indistinct. This is a temporary loss of hearing and recovery is usually complete after a day of quiet. Contrary to much popular belief, there is no particular sound or combination of frequencies which are more damaging to the inner ear than others. To be injurious a sound must be quite loud, and the frequency is of little significance. Repeated exposure to sound intensities in excess of 85db can cause permanent damage to the organ of Corti. This so-called "boiler-makers' deafness" or "industrial hearing loss" develops gradually over a period of years and almost always begins with a loss of frequencies between 4000 and 6000 cycles per second. Rock musicians have become particular victims of such premature deafness. All persons exposed to excessive sound levels can benefit considerably by the use of protective headphones.

Drugs and Sensory-Neural Deafness

For many years it has been known that the drugs aspirin and quinine can cause sensory-neural deafness. They can also cause, in a person with normal hearing, a ringing in the ears. This ringing, technically called *tinnitus,* may occur in conduction hearing loss where it is usually a "white" noise, that is, all the frequencies of the sound spectrum are present at once. However, in sensory-neural deafness the ringing is usually in the form of a high pitched whistle, and many hearing experts believe that, in the absence of other apparent cause, it indicates the beginning of degeneration of the receptor cells associated with the particular pitch of the ringing. In fact, tinnitus is one of the symptoms of aspirin toxicity used to adjust the dosage of the drug, i.e., if your ears ring you have taken too much and should decrease the dosage.

Recently, some antibiotics, notably dihydrostreptomycin and kanamycin, and to a lesser extent, streptomycin and neomycin, have been found to cause sensory-neural hearing loss. Unfortunately, dihydrostreptomycin was used routinely for quite some time before this effect was recognized. These particular drugs are no longer used unless there is a specific and compelling reason for them.

Age and Hearing Loss

The most common cause of sensory-neural hearing loss, and probably of all hearing loss, is advanced age. A gradual loss of hearing sensitivity for frequencies above 1000 cycles per second seems to begin at about age 40 and progresses steadily until death. The receptor cells at the base of the cochlea (those for higher frequencies) seem to simply degenerate and vanish. The cause for this degeneration is not known, but it can often be correlated with a general atherosclerosis.

Congenital Deafness

The most common cause of sensory-neural hearing loss in children and young people is congenital deafness. Actually, this category includes not only cases of hereditary deafness, but very probably includes children who developed a hearing impairment due to scarlet fever or meningitis or other childhood disease before he learned to talk.

Severe virus infections in the expectant mother during the early months of pregnancy is the most common cause of true congenital deafness. German measles, mumps, and influenza have all been implicated. The result is an abnormal development of the inner ear and an irreparable deafness.

Another threat to hearing is Rh incompatibility. Although the infant's life is often saved by blood transfusions, permanent damage to the ear, and more often to the brain, can develop. This deafness is also of the sensory-neural type.

The incidence of hereditary deafness is difficult to determine. Among identical twins where one twin was totally deaf in infancy, the other was totally deaf in 60 percent of the cases and partially deaf in 28 percent of the cases. Among fraternal twins where one was totally deaf in infancy, the other was totally deaf in only 20 percent of the cases and partially deaf in 15 percent of the cases. This would suggest that over half of those deaf at birth are deaf because of an inherited defect. One can conclude that those deaf since infancy have a much greater chance of producing deaf children than do people with normal hearing.

Treatment of Sensory-Neural Deafness

If a hearing loss is definitely established to be sensory-neural there is little that can be done to restore the loss. A disease process may be arrested so that further hearing loss is prevented, or an annoying tinnitus relieved, but no improvement in hearing is possible. Once the receptor cells and the nerve fibers have degenerated they cannot be restored. When a conduction deafness accompanies a sensory-neural deafness, some improvement is possible by surgery or by the use of a hearing aid. But neither "nerve stimulants" nor vitamins nor antihistamines can restore function to degenerated receptor cells. It is likely that medical science will be no better able to cope with this type of deafness anytime in the foreseeable future. However, inasmuch as the cause of much sensory-neural deafness is known and preventable, a few simple precautions are worth noting.

Since many drugs are now known to cause sensory-neural deafness this, in itself, should be sufficient reason to use *all* drugs with caution. Certainly, a person who believes that he requires 8 to 10 aspirin tablets daily because of vague tensions and aches is seriously endangering his hearing. (And the "stronger" pain relievers are no better since the most effective drug in all of them is aspirin, in spite of suggestions to the contrary.)

The use of protective headphones is annoying; but we are all occasionally exposed to loud noises which are potentially dangerous and which we should be protected against. For example, even a short exposure to the noise of a jet engine can cause pain and even permanent damage to both the ear drum and to the inner ear. The ears should be protected against such sounds by whatever is available, whether a wad of cotton or kleenex plugged into the ear canal, or by the fingers pressing the canal closed.

Dealing with the Deaf

Some of the things said about blindness can be adapted to deafness. Particularly, both conditions call for intelligent tact and thoughtfulness in the associates. In many respects, however, the blind and the deaf are almost opposites.

The great problems of the blind are practical ones of dealing with movements, reading, and manual skills; these things affect the deaf only slightly and incidentally. Thus a deaf person requires no guidance across a busy street, but needs only to take special care because he cannot hear warning sounds. Conversely, a blind person suffers little handicap in social intercourse and may even enjoy an advantage in conversation because of his highly trained sensitivity to shades of expression, whereas a deaf person is to some degree isolated and lonely even in company. These contrasts are not, of course, absolute but they are true enough to be significant.

Social isolation, then is the main problem one must help deaf associates to solve. This can be done in several ways.

Lip-reading is a resource of many deaf people. They may, indeed, become so expert at this technique that one tends to forget their dependence on watching the lips. One should keep this dependence in mind and make sure that one speaks always in plain view and does not expect the person to hear a comment, for example, tossed over the shoulder with the back turned. If one is about to speak to a deaf person, an alerting gesture is a practical courtesy. Neglect of such precautions can cause embarrassing mistakes.

Sign language is actually preferred as a means of communication by many deaf people. This is especially true for those who were born deaf and have therefore never heard the speech they are supposed to read from lips. This sign language consists of two alphabets, one for both hands and one for a single hand, and a multitude of special signs rather like those used so effectively by the Plains Indians. The language can develop amazing speed with very little practice and is in fact sometimes of use between nondeaf persons. If one is to be in frequent association with a deaf sign-speaker, the few hours spent on mastering the technique will be many times repaid by the practical and psychological advantages. The associate will appreciate the interest shown; and deaf strangers are often surprised and delighted to find someone who can "speak their language."

Hearing aids should be well understood by any educated deaf person today. Yet many evidently do not know the facts, as is shown by the prevalence of quacks selling inferior products, sometimes even by mail. One might add that modern aids can now be so inconspicuous as to render esthetic objections unnecessary.

Anatomy of the Vestibular Apparatus

The name vestibular apparatus for the sense organs to be described is unsatisfactory. However, alternative terms are imprecise or simply incorrect. The vestibular apparatus is often referred to as "an organ of balance" or "of equilibrium"; but this description is inadequate. In fact, the system comprises two distinct though related sense organs, one of which reports the tilt of head and body from the gravitational vertical, as would a plumb line, while the other reports acceleration of any sort—speeding up, slowing down, or swerving (angular acceleration). Obviously, information of either sort would be of value in maintaining balance but would not in itself "control" balance.

The vestibular apparatus develops from the same bud in the embryo as does the cochlea; indeed, the two structures remain connected throughout life. One important result of this fact is that if a child is born with defective vestibular sense, he also usually has a sensory-neural hearing loss. The former defect may not be apparent under ordinary conditions since the individuals can regulate upright posture by sight, by pressure on the feet, and by muscle (kinesthetic) sense—as, indeed, can normal people. However, under special conditions in which these other senses are ineffectual as, for example, when the person is submerged in water, lack of gravity sense may prevent him from knowing which way is up, with possibly fatal results.

The *gravity-sense organ* consists of the two chambers of the membranous labyrinth called the *utricle* and the *saccule.* From the walls of these are suspended particles called *otoliths* (ear-stones), which act in fact like plumb lines swinging as the head tilts and thereby pulling on their attachments with varying force; the combination of pulls by the two otoliths on each side informs the brain, through the *vestibular nerve,* as to the exact position of the head and, by inference, of the body. Some people have very precise gravity sense and know how they are positioned even when their eyes are closed; such people would presumably have an advantage in certain sports and activities requiring body control. Others have poor gravity sense and depend chiefly on sight for judging their head-and-body position.

The *acceleration-sense organ* consists of the three curved tubes which arise from the utricle, the *semicircular canals.* They are, then, also a part of the membranous labyrinth. Movements of the head set up currents in these canals, which drag on clumps of microscopic hairlets and cause messages to be sent to the brain—also through the vestibular nerve. Currents can be set up likewise by sudden heating or chilling of the area, just as they are in a beaker of heated water. An otologist (ear specialist) studying the condition of the vestibular apparatus, may stimulate the canals by syringing the adjoining external auditory canal with warm or cold water. This stimulates head turning and usually induces compensatory eye movements (called nystagmus) which the otologist can observe. It also may cause feelings of vertigo and dizziness. To avoid these effects, one should always use fluids near body temperature when cleaning or treating the outer ear. People subject to sea-, car-, and air-sickness probably have hypersensitive semicircular canals. Drugs used to combat these disorders probably act by numbing the canals or their nervous connections. Normally, however, the canals adapt our bodies to various accelerations without our being aware of their activity.

REVIEW QUESTIONS

1. Trace the path of sound waves through the ear.
2. What is the resonance theory of hearing?
3. Distinguish between conduction deafness and sensory-neural deafness.

4. What is otosclerosis?
5. Can drugs cause deafness? Explain.
6. What kinds of sensory information are provided by the vestibular apparatus?

REFERENCE

DAVIS, HALLOWELL, and S. RICHARD SILVERMAN, *Hearing and Deafness*, 3rd ed. New York: Holt, Rinehart and Winston, 1970.

Additional Readings

KONIGSMARK, BRUCE W., "Hereditary deafness in man," *New England Journal of Medicine*, **281:** 713–719, 774–777, 827–832 (1969).

MEYER, MAX F., *Fitting Into a Silent World: The First Six Years of Life*. University of Missouri Press, 1934. (Paperback)

PART NINE
AS OTHERS SEE US

37
MENTAL DISORDERS

Hardly anyone reaches high school age without knowing at least one person who has suffered some kind of mental illness. Fortunately, the mystery and fear that has been so frequently associated with mental illness is now giving way to a widespread concern and willingness to understand and to help. This stigma attached to the sufferer has been markedly reduced with the realization that those afflicted can often be restored to a normal and productive life.

The various forms of mental illness are necessarily due to disorders of the nervous system; but the physiological nature of these disorders is largely unknown. Eventually, we shall perhaps uncover a physical basis for many forms of mental illness, and treatment will be much like that of other metabolic diseases such as diabetes. In the meantime much progress is being made simply on the basis of empirical studies. Many diseases have been effectively treated before science revealed the underlying causes—vaccination preceded the germ theory of disease, and fresh fruit for scurvy preceded the discovery of vitamins. Better understanding is highly desirable but is not indispensable for much effective treatment.

The Normal Mind

In the fascinating study of mental disease, its symptoms, causes, and cures, both writer and reader too often forget that abnormality itself must be measured from a base line of normality. *Yet mental health is far commoner than mental disorder and its preservation is both easier and more effective than its restoration.*

A definition of mental health is certainly not easy. Happily, healthy minds are not conventionally uniform but vary over an immense range of human nature. Indeed, the unhealthy mind is the restricted one under its often bizarre surface trappings; it has retreated from its full potentialities into a more limited, primitive world with which it hopes to deal more easily. Mental health, however, does have positive qualities common to all its varieties.

The usual legal criterion of "sanity" is the ability to conduct one's affairs competently. This definition, of course, is not entirely satisfactory as a concept of mental health since many presumably mentally normal people are quite incompetent in financial, amatory, or other fields, while many thoroughly neurotic or even psychotic people are highly successful in difficult walks of life. In fact, it is sometimes suggested that an ingredient of neurosis is necessary for a truly unique and forceful personality. The distinction between fruitful originality and sterile eccentricity cannot always be made in advance!

More scientific is the concept of control. The mentally healthy person, however revolutionary his ideas, can control them—restrain his expression, accept criticism, and adapt to circumstances. The mentally ill person cannot control his expression or modify his conduct effectively; his ideas command him, not the reverse. This is not to say that mental derangement cannot be terribly effective; it often is. A monumental case in point was Adolph Hitler: This man set enormous forces to work and was very near to accomplishing his aims; but his rigid manias destroyed the creative Jewish element of his nation, the cooperation of which might have turned the scales for him, and against expert advice he persisted in policies that led to military disaster. In striking contrast was Winston Churchill, who opposed Hitler with power and tenacity but was a master of restraint, of cooperation, and of adaptation.

A well-known psychological epigram is worth considering: Mental disorders are simply exaggerations of normal mental states that have gone beyond the patient's control. For example, everybody has a tendency to day dream, has alternations of optimism and depression, and so on. Indeed, these qualities can differ widely among people who nevertheless are almost always able to control them and should therefore be considered mentally quite normal.

This epigram, though an oversimplification, helps greatly in understanding the classification of the different types of mental illness. And it gives compassion for mental sufferers if one recognizes that he himself differs from them only in degree.

The Causes of Mental Illness

How do exaggerations and other distortions of mental behavior come about? This problem is extremely complex and obscure, and has given rise to widely different explanations which are not necessarily incompatible, but merely incomplete. In general, a physiological explanation harmonizes the features of the major theories.

The brain, as has been said, is an organ similar in most respects to other organs of the body. The chief function is to record, sort, and organize the experiences of life into a coherent fabric, "the mind," and, by means of this, to plan, adapt, and issue orders for effective responses. It does these things by means of its vast network of neurons connected by synapses, as described earlier.

This network, above reflex and other automatic levels, is only vaguely organized at birth. Its organization takes place step by step as the child assimilates new experiences and corresponding neuron channels of lowered synaptic resistance are laid down, whereas heightened resistances block off unwanted connections. This is obviously a very complex process.

Just as wrong development of bones, heart, and other organs will produce deformities, so likewise errors, inconsistencies, omissions, and outright falsities in the brain record will produce distortion and conflicts in the neuron-synapse pattern. Since the pattern is so inconceivably complex, minor defects, such as must occur in all individuals during the tumult of experience, are insignificant, can easily be compensated for, and therefore cause no serious strain in normally healthy individuals. Defects widespread and serious enough to cause mental illness must be themselves highly complex compared, say, to a simple deformed bone or heart. This vast difference in complexity largely explains the conviction held by most people, including specialists on brain and mind, that mental disorder is somehow unique. The matter is one of quantity, not necessarily of quality.

The most widely held theory of mental disorders, that of Sigmund Freud, is easily adapted to this picture. Freud's theory has been frequently revised, extended, and modernized, but its basic tenets are accepted by all derived schools of thought: When major conflicts occur in the developing neuron-synapse pattern, the immature mind responds not by seeking to solve them logically, but by suppressing, falsifying, or otherwise illogically disposing of the incongruous element; presumably the mind shuts off elements of the neuron fabric that conflict with the main pattern. These elements, however, are not obliterated and may reassert themselves in special states such as sleep, intense emotion, or hypnosis; even in normal conditions they may exert an unrecognized distorting effect on the rest of the pattern. This effect is usually minor, but it may be so severe as to produce some form of mental aberration.

The classic example of conflict, and the one stressed by Freud himself, is that of the sex drive versus social restrictions. The drive is basic, universal, and strong, though not always so much so as is believed. Obviously it cannot be allowed unrestricted play in an organized society where it would cause violent interpersonal conflicts and impairment of an individual's social value. On the other hand, society often overdoes its efforts at control, imposing unreasonable and outright harmful restraints on sexual activity and expression. Since society is immensely powerful compared to the individual, especially if he is young, sexual matters are often suppressed or distorted in the mental pattern. This state of affairs is a fertile source of mental disorder.

Of course, other drives—for security, well-being, social recognition, aggression, and so on—can also be thwarted and result in mental conflict and disorder. Recognition of this fact is probably the major departure from Freud's original thesis. (One must remember that Freud grew up and worked in (1) a particular enclave, the Jewish community, which was strongly centered on family, especially the mother, (2) a central European city, Vienna, which was strongly conservative socially, and (3) an era when errant sex was fanatically repressed and condemned. These conditions naturally influenced both his opinions and the type of patients observed by him.)

A further problem arises, however: Why do people vary so greatly in their reactions to mental stress and conflict? Why, for example, does one subject "go to pieces" because of everyday problems unavoidable in normal living, while another confronts overwhelming difficulties with logic, calm, resourcefulness, and good humor? Part of the answer, of course, lies in the over-all quality of the mental pattern as this has been affected by circumstances. A pattern distorted by misinformation and ignorance and general poverty is naturally more prone to disruption than one wisely educated. But this explanation is frequently inadequate: Minds deprived and cramped in many ways are often outstanding, while those with every apparent advantage collapse. Deprivation and affluence are themselves qualified by the basic material for which they work.

Of recent years the concept of what might be termed "the vulnerable personality" has arisen. Why such personalities occur is obscure, but their occurrence can hardly be doubted. Apparently they are in large part hereditary, and this belief is reinforced by a significant fact: Vulnerability can be expressed in a variety of ways such as extreme eccentricity, irresponsibility, outright mental illness of various sorts, alcoholism, drug addiction, incorrigible criminality, and other problem behavior. When these various aberrations are taken into account, they are found to cluster thickly in some family trees, while they are almost absent in most. In other words, they are all expressions of the vulnerable personality, and make its hereditary nature more obvious than does a mere scattering of recognized neuroses and psychoses. This actually makes all the related conditions easier to comprehend and to treat.

Of greatest significance is the term "vulnerable," which is not at all the same as "destined." Vulnerability of this sort is comparable to a family proneness to tuberculosis or other disease. Today, such proneness should impel any sensible person to a regime which avoids unnecessary stress and exposure. Thus recognition of the hazard could lead to better health than most "normal" people enjoy.

Recognition that mental disorders are no more shameful or mysterious than others can lead us to treat them with equal common sense. People who suspect that they have a vulnerable personality are extremely foolish to live in apprehension of "the old family trouble" or "mother's affliction." If anything they should feel relieved that a tendency has been brought into the open where it can be dealt with effectively.

A person who suspects or fears that he is vulnerable should not brood idly. He can consult a reputable psychologist, or even simply a respected and experienced student counselor, family doctor, or clergyman; some religions applied some principles of sound, practical psychology for centuries before they were formulated scientifically. Such a person can uncover specific weaknesses, can help to reduce abnormal anxieties and to correct strains in the personality, and above all can present the constructive, stimulating side of life and offer sound advice on building a more effective outlook. An approach of this sort is by far the best defense against mental disorder.

Self-therapy is usually dangerous, especially in mental problems; but development of sound, lively interests is not so much therapy as normal mental exercise and is to be recommended for anyone with neurotic tendencies. The typical neurotic or other vulnerable person is introverted, sedentary, without firm roots, mental or emotional, and often drifts or hurries restlessly from one interest to another. He should do what he can to counteract these tendencies. He would never be happy as a gregarious "joiner," but he should build a circle of stable but congenial friends and cultivate them. He should adopt activity suited to his physique and personality especially if it entails sociability. Above all, dedication to, and real enthusiasm for, a well-chosen profession or life work is the greatest of all resources for tapping off tension and anxiety, starving unhealthy tendencies of the mind, and for building up realistic confidence. Very likely our modern proneness to passive, unchallenging, and uncreative activities is a major source of our prevalent mental disorders.

With these thoughts in mind, we can turn to specific types of mental disorder and their problems. These disorders are usually subdivided roughly, and somewhat arbitrarily, into the milder *neuroses* (or psychoneuroses) and the more severe *psychoses.* Some forms, such as depressions, do not fit into either group conveniently. These three groups will be discussed in turn.

The Neuroses

The American Psychiatric Association gives the chief characteristic of neurosis as "anxiety which may be directly felt and expressed or which may be unconsciously and automatically controlled by the utilization of various psychological defense mechanisms." Thus patients with neurotic disorders do not exhibit gross distortion of reality and do not show gross disorganization of personality (as is the case in psychosis).

The neuroses range from mere eccentricities to severe and incapacitating disorders. They may typically concern only some restricted area of personality, such as reactions to particular things and situations, leaving other mental activities unimpaired—indeed, many famous and talented people, especially in arts and letters, have been strikingly neurotic (Tschaikowsky, Dickens, Proust, to name only a few).

Neuroses are extremely common. A figure of 10 million is often given for this country; but the figure is rather an "indication of magnitude" than an attempt at even approximate accuracy. For one thing no sharp line can be drawn between mild neuroses and the individual peculiarities we all display. Then, too, many, perhaps most, neuroses disguise themselves as physical disorders or are actually complicated by real disorders and so are not recognized. Probably the great majority of neuroses are never brought to medical attention. Yet they can represent more or less severe handicaps to the individuals and loss to the community.

In modern psychology, the key concept for neuroses is *anxiety.* Anxiety might be considered as the mental discomfort and insecurity due to a strain or contradiction in the neuron pattern and in the mind, which the person cannot resolve logically. And the resulting neurosis can usually be explained as an attempt to submerge, conceal, or disguise one or more of the conflicting elements. Such a situation is naturally more common in overly sensitive or otherwise vulnerable people, but it can occur in strong-minded people subjected to intolerable and insoluble situations—as in the case of so-called battle fatigue. Common examples of neurosis are the following.

Compulsions are akin to deeply rooted habits that we follow without thinking much about them. The habits are often sensible; for example, the washing of hands after touching contaminated material, or before handling food or wounds, is a hygienic necessity. But washing after every unsterile contact and for no specific preventive reason, up to scores of times a day, is an exaggeration of the habit and is neurotic, especially if the person cannot control his impulse and becomes agitated when he is restrained. Other compulsions have even less reasonable origins, as when a person carries out some ritual, such as throwing spilled salt over his left shoulder; some of these may be perpetuations of idle habits in which the person has come to half believe as an irrational shield for insecurity; others may have obscure symbolic meanings which can be traced only by psychoanalysis. Compulsions are often seen associated with more widespread disorders such as schizophrenia.

Obsessions are very similar to compulsions. They resemble normal enthusiasms, but have gotten out of hand; thus, stamp collecting is a common hobby, but when the collector allows his interest to monopolize his time, conversation, and finances, it becomes an obsession. Thus, obsessions are more complicated and pervasive than compulsions.

Phobias are acute anxieties that develop in relation to some object, idea or situation in daily life. For example, a hand-washing compulsion may simply be an aspect of a dirt or infection phobia. Many people have strong, more or less irrational, dislikes for certain things—special foods, places, persons, animals, and so on—but can control their reactions. When the reaction is ungovernable, the dislike is a phobia; it may reach a peak of physical sickness and complete inability to control one's actions, so that a usually courageous person panics and runs from an apparently trivial threat. For example, caution with snakes is natural and reasonable especially in territories where poisonous forms abound. But uncontrollable terror and nausea at the sight of a garter snake, in spite of knowing that it is harmless, is neurotic. The number of possible phobias is as limitless as the number of objects and situations in the world, and a person can suffer from multiple phobias. If the feared objects are common ones, phobias may be a great handicap. Sometimes they may be due simply to association with a forgotten terrifying experience which the person prefers to forget since it is inconsistent with his self-esteem. Often, however, phobias are a subconscious excuse for avoiding certain activities that cannot be excused on rational grounds. In all cases they are attempts to resolve mental conflicts irrationally without admitting it.

Hypochondria is fear that one suffers from diseases that have no existence in fact. Again, judicious precaution against disease is not only reasonable but is a highly desirable form of self-preservation. And who has not had fleeting fear of serious disorder when afflicted with indigestion, headache, or such? Only when the fear becomes fixed, and unshakeable by proof and reason, is the condition neurotic.

Hypochondria, however, is not a simple obsession like the handwashing described above. Its motive is not simply to avert danger by naive rituals; rather, it seeks subconsciously to use a fictitious sickness to alter a general life-situation. This too is an exaggeration of normal behavior. Everyone has a natural wish to avoid distasteful obligations or to gain desirable concessions. A normal person, however, keeps his effort consciously reasonable and more or less honorable; the hypochondriac is driven subconsciously to fake a disorder.

He is not consciously malingering, "swinging the load," but may be as genuinely deceived as anyone else. However, since the subconscious may be misinformed, it may produce symptoms that it "thinks" are genuine but that would not deceive a trained observer. Some hypochondriacs switch from one imaginary ailment to another. But in all cases the underlying cause is of the same nature—the subconscious need for an alibi or a psychological weapon.

Neurasthenia is a subvariety of hypochondria. In it the patient shows not specific symptoms of disease, but a general fatigue and weakness which is perfectly genuine to his conscious mind. But no matter how much he rests, he is not rested. If he, other people, or circumstances solve his problems of loneliness, rejection, and inadequacy, the fatigue often evaporates.

Not strictly in the field of neuroses, but so closely related to hypochondria that it can best be discussed here, is the matter of *psychosomatic* (mind-body) *disorders.* In mental stresses, even without a subconscious urge to fake illness, the state of mind can have a very definite effect on body processes. Notorious among such effects are peptic ulcers and some forms of asthma and of high blood pressure; and many other disorders are strongly suspected of being often psychosomatic. Indeed, probably most disorders have a psychic element; certainly, an attitude of determination or of apathy can distinctly affect the course of almost any disease. These psychic influences are not mysterious since the mind obviously reacts on the organs in many ways besides control of muscles: Awareness of danger, food, sexual excitement, pleasure, insult, and so on, act on the heart, salivary glands, and intestinal tract, probably through the hypothalamus and autonomic system. If this effect is excessive on the stomach, for example, contraction of stomach muscles and secretion of digestive juices may be excessive and produce ulcers. This happens especially in people with a constitutional tendency to ulcers, but without the mental strain the tendency may not erupt.

Psychosomatic disorders, unlike those of hypochondria, are authentic. Thus, a hypochondriac's subconscious mind can induce symptoms which it fancies are those of peptic ulcers and which result in real pain, but which are not in fact the true symptoms of ulcer. He may even, under stress, "forget" his symptoms as when a person with hysterical paralysis suddenly becomes active under threat of danger. The psychosomatic patient, on the contrary, has not only real pain but real ulcers which can perforate or hemorrhage and kill him; and if he forgets his condition, he is not immune to the results. Thus in hypochondria the problem is to treat the subconscious mental stress, whereas in psychosomatic troubles, the problem is to treat the conscious mental state, and the real damage that is present.

These are far from all the types of neurosis that are recognized. But they are the most familiar and serve to illustrate the class. All are derived from normal mental activities but are exaggerated beyond control; and all are subconscious attempts to solve mental problems irrationally.

Psychoses

Psychoses are the more severe forms of mental disorder. They are hard to distinguish precisely from neuroses since the two classes shade over into each other or may complicate each other, as when compulsions or phobias co-exist with schizophrenia. In general, psychoses affect the whole personality to a degree that necessitates care by others. The American Psychiatric Association states that psychoses "are characterized by a varying degree of personality disintegration and failure to test and evaluate correctly external reality in various spheres. In addition, individuals with such disorders fail in their ability to relate themselves effectively to other people or to their own work." The major forms are schizophrenia, paranoia, manic-depressive states, and pathological depressions.

Schizophrenia

Schizophrenia is said to be the most common form of mental illness; indeed, it may be the most common of all serious diseases in modern society. The American Psychiatric Association gives the following general definition: "A group of psychotic reactions characterized by fundamental disturbances in reality relationships and concept formations, with affective, behavioral, and intellectual disturbances in varying degrees and mixtures. The disorders are marked by a strong tendency to retreat from reality, by emotional disharmony, unpredictable disturbances in stream of thought, regressive behavior, and in some by a tendency to deterioration.'"

There are four major subtypes of schizophrenia: *simple, hebephrenic, catatonic,* and *paranoid.* However, even experienced psychiatrists are sometimes unable to reliably classify some patients because of changeable behavior patterns.

Schizophrenic reaction, simple type is described as "characterized chiefly by reduction in external attachments and interests and by impoverishment of human relationships." In the following discussion we shall attempt to describe the general characteristics of the schizophrenic reaction as it is manifest in the simple type of the disorder. The other three types seem to superimpose upon the basic schizophrenic reaction a particular predominant form of behavior.

Almost everyone knows about daydreaming from personal experience: A person tired of unrewarding stress and strife in daily life, imagines himself in situations of success, dominance, recognition, and the like (daydreams rarely if ever have a personally unfavorable tone). At its best, daydreaming is a highly beneficial mental process which has led to many great achievements when the dreams have been translated into action. In the majority of cases the activity is a harmless form of relaxation and refreshment, evidently normal since it is so nearly universal, although the advice to "dream but not make dreams your master" should always be borne in mind.

To say where harmless dream-spinning passes over into excess, is difficult; many an authentic genius has been long derided as an eccentric dreamer before he put his dreams into effect, and many such people have spent enjoyable and harmless lives without deserving diagnoses of psychosis. However, when the indulgence reaches a point where the person cannot control it but takes refuge in fantasies rather than grappling with real-life problems, he enters the shadow of true schizophrenia.

Schizophrenia is predominantly a disease of youth. If the personality harbors the weakness of retreating from the problems of life, this weakness is apt to show itself in the early, and often severe, stresses of adolescence and young adulthood. The disease almost always attacks people of introverted character —shy, withdrawn, unsociable, intellectual in such a way as to be acutely aware of the problems of life—the well-named *schizoid* (-oid: -like) temperament. These facts do not at all mean that a sensitive, withdrawn boy or girl is probably going to develop schizophrenia, any more than that, say, a tense, perfectionist person is bound to develop peptic ulcers. Simply, a schizoid type should take sensible precautions. In particular, he should avoid excessive fantasy-weaving that he never applies to effective action. He should not adopt ways of life involving great strains and frequent crises in an attempt to overcompensate; but he should put his sensitivity and imagination to work in substantial creation, and face the natural problems that this entails. A schizophrenic is a schizoid who has lost control of himself; but a mentally and physically sound schizoid is very likely to be one of our intellectual elite. (James Thurber's famous story, "The Private Life of Walter Mitty," is a perfect description of a strongly schizoid personality: A meek little man, completely dominated by his wife, takes refuge in grandiose visions of himself as a famous surgeon, daredevil pilot, and so on. Since poor Walter was able to function more or less normally, he was no more than neurotic. But since he failed to apply his fancies, his prospects for success and recognition were poor.)

Schizophrenic reaction, hebephrenic type is defined as follows: "These reactions are characterized by shallow, inappropriate affect, unpredictable giggling, silly behavior and mannerisms, delusions, often of a somatic nature, hallucinations, and regressive behavior." The hebephrenic is silly and absurd; he laughs, cries, and his mood is constantly changing. He may smile or grimace for no apparent reason. He has hallucinations and delusions which have no coherence or relation to one another and which seem to give him no distress. In conversation he will respond with whatever comes to mind without apparent concern to its relevance.

Schizophrenic reaction, catatonic type are "reactions characterized by conspicuous motor behavior, exhibiting either marked generalized inhibition (stupor, mutism, negativism, and waxy flexibility) or excessive motor activity and excitement. The individual may regress to a state of vegetation." These patients withdraw further into their fantasies than do the other schizophrenic types. They may live continuously in a world of imagination while at the same time carrying on simple, humble tasks under direction: The mental hospital patient who envisions herself as Queen Elizabeth, while sweeping out the ward, is an example. Extreme cases lose touch with reality completely and withdraw into their dream world, refusing to speak, act, or even recognize what goes on about them. This state may be broken by episodes of excitement and dangerous violence.

Schizophrenic reaction, paranoid type is "characterized by autistic, unrealistic thinking, with mental content composed chiefly of delusions or persecution, and/or grandeur, ideas of reference, and often hallucinations. It is often characterized by unpredictable behavior, with a fairly constant attitude of hostility and aggression. Excessive religiosity may be present with or without delusions of persecution. There may be an expansive delusional system of omnipotence, genius, or special ability." In contrast to the hebephrenic type whose delusion is bizarre or unsystematized the paranoid type has a fairly coherent delusion. The person may even be apparently normal and competent in most respects, carrying on a business, family, and social life like anyone else; but in some respect, he cherishes a profound and violent fantasy. Frequently, this is a belief that he is the victim of a vast and deadly conspiracy; or he may himself persistently follow some irrational plan often criminal and dangerous in nature. Obviously such delusions affect the whole personality and are likely to lead to irrational violence. A characteristic of these people is the frequently baffling consistency and logic of their mania: They can systematically refute all arguments and almost persuade their listeners if they do not retain a sense of perspective.

Compared to other types of schizophrenics, he is likely to be intellectually responsive and alert, and when discussing something divorced from his delusions, friendly. If pushed, however, the paranoid schizophrenic is more likely to respond with aggression, while other types of schizophrenics are more likely to withdraw.

Manic-Depressive Psychosis

Manic depressive psychosis is an exaggeration of another normal phenomenon. Everyone is subject to more or less regular cyclic changes in temperament and energy, extending over weeks. Indeed, some people can confidently predict that they are going to be at a peak "next week end" which will be a good time for undertaking important projects or, conversely, at a low ebb when they can best occupy themselves with simple activities. All this is quite natural and may even be related to cycles in body chemistry. The cycles, physical and mental, may be too mild to notice, or quite marked. However, when such cycles become so violent as to impair a person's self-control, they become neurotic and even psychotic.

In contrast to schizophrenic patients, the manic-depressive types are extroverts. They do not withdraw from the real world but overreact to it: In the manic phase they are excited, noisy, full of ill-considered ideas which they wish to put into effect at once, boastful, aggressive, and sometimes even dangerously violent; this phase then gives way to one of growing discouragement and disillusion which does not seek relief in the fantasies of schizophrenia but faces reality with an exaggerated self-reproach and despair. When the manic phase becomes violent and the depressive suicidal, the condition is truly psychotic. Manic-depressive conditions are by no means always neatly cyclic; either of the exaggerated phases may occur alone with normal interludes and at irregular intervals, but the nature of the condition is usually quite clear.

Other Personality Disorders

Pathological Depressions

These usually, but not always, differ from the cyclic form, just described, in cause and in character. Two main types are recognized: exogenous (generated from without) and endogenous (generated from within). Both are characterized by prolonged, more or less continuous, disabling mental misery which does not yield to reason or even to a patient's own realization that his feelings are groundless or out of all proportion to real causes. Otherwise, the two forms differ considerably.

Exogenous (or reactive) *depression* comes on rather suddenly. It results from real troubles in the outer world—financial disaster, loss of a loved one, and so on—that are exaggerated and prolonged far beyond normal. Presumably the precipitating cause acts on a personality already weakened by physical or psychological instability, perhaps of genetic origin. The patient seldom shows disorders of sleep or appetite as found in endogenous depressives. Usually he is comparatively well when he gets up and feels worse in the afternoon. This type of depression typically responds well to medication, prolonged artificial sleep or, in extreme cases, to convulsive therapy, *none of which are to be self-prescribed and induced.*

Endogenous or *retarded depression* (or *involutional melancholia*) is more difficult to deal with. It results from conditions in the person's own body, such as the "change of life" in women or the vaguer "involutional period" recognizable in some men; thus it is most common in the age range from 40 to 65. It can, however, result from a manic-depressive state heavily inclined to the depressive phase or from other causes; thus it may occur, though less frequently, at other ages. Its onset, unlike that of the exogenous form is typically insidious though the patient often refers it to some possibly quite fictitious event; for example, a well-to-do person may talk about being destitute. Or it may develop under cover of stimulating activity and emerge suddenly when the stimulus is withdrawn, as when a mother's children leave home or a man retires from an interesting job.

Endogenous depressives almost always present a complex of physical disorders which are the result, not the cause, of their mental condition: rapid pulse, poor appetite, shortness of breath, hypertension, diarrhea or constipation, and so on. Unlike the exogenous patient, the endogenous characteristically wakes at night, when his depression asserts itself most powerfully, feels worse in the morning, and cheers up somewhat in the afternoon.

Mentally, endogenous patients are depressed to a degree ranging from mild but frustrating to dangerously suicidal. They are typically worried chiefly about themselves, have convictions of a wasted life, deep sin, or failing mind, or have an obsessive hypochondriac belief that they are developing cancer, heart-disease, or such, in spite of repeated negative examinations. Powers of concentration and memory are impaired; but this is not due to brain degeneration, as the patient may fear, but to his mental misery and his preoccupation with it and with his other symptoms; when the depression remits temporarily or permanently, mental powers usually return to normal. Above all the patient loses interest in everything—personal appearance, entertainment, company, work—and finds even rising and eating an effort. Very often, he shows his depression in his sluggish movements, sad voice, and woebegone face.

Understanding depressions can be of great practical value. If you are of typical college age you are unlikely to suffer from any form of the condition for decades to come, if ever. But if you do, and in the far more likely case of your encountering cases in your family or social circles, an informed approach can save much needless grief. Depressions frequently clear up spontaneously, but this may require two or three years during which life seems barely worth living; and other depressions sometimes deepen and drive the victim to suicide—indeed, endogenous depression is a leading cause of suicide. Rather than bear these undesirable states, almost any depressive can be cured outright, or at least greatly relieved, by drugs, psychiatric support, shock therapy, and other means. The patient may himself lack the energy and the will to seek such relief. In that case the urging of an associate may be decisive in persuading him to do so.

Here, however, even more than in other mental conditions, "home doctoring" is to be avoided. The various forms of depressions are often very hard to distinguish from each other and from other disorders. The proper treatment for one may be the worst possible for another—for example, a tranquillizer suitable for some anxiety neuroses, may drive an agitated endogenous depressive to suicide. Therefore let your efforts concentrate on inducing the patient to seek proper care by qualified personnel.

Paranoid Reactions

Under the classification of paranoid reactions are included "those cases showing persistent delusions, generally persecution or grandiose, ordinarily without hallucinations. The emotional responses and behavior are consistent with the ideas held. Intelligence is well preserved. This category does not include those reactions properly classified under "schizophrenic reactions, paranoid type."

Certain people, then, exhibit paranoid behavior without the underlying schizophrenic reactions. When the simple label of paranoid reaction is applied, it is meant that the patient exhibits little if any evidence of "fundamental disturbances in reality relationships and concept formations. . . ." Although this distinction is valid in theory, some psychiatrists contend that it is rare indeed to encounter a psychotic with delusions of persecutions or grandeur who also does not show some aspect of schizophrenia.

Treatment of Mental Disorders

Having surveyed the best known forms of mental disorder, one naturally wonders what can be done about them. The role of *nonprofessional associates* has already been discussed. While the layman should strictly refrain from amateur psychoanalysis, he can often exert a decisive influence by simple common sense. He should reject ignorant and obsolete attitudes to mental disorder. In accounts of experience given by recovered mental patients, the redeeming role of an intelligent and understanding parent, friend, or even fellow-sufferer is often stressed. One should remember that a mentally ill person can be far from unaware of his condition. He will be especially sensitive to the reactions and conduct of other people, especially those to whom he is attached. What could be more helpful than warm sympathy or more damaging than hopeless loneliness?

Some mental cases may be truly dangerous and should never be approached without strict precautions; for instance, one should never deal with them alone but always with a competent companion. As in infectious diseases, the disorder, and not its unhappy victim, is the danger; and whatever the provocation or aggravation, this must constantly be borne in mind.

Medical treatment, of course, depends on the type and degree of disorder. Probably the most famous and still most controversial treatment is *psychoanalysis.* This was introduced about 70 years ago. Although it has undergone many revisions and elaborations, its basic principles remain unchanged.

Much mental illness, particularly of the neurotic type, is due to strains in the mental pattern, set up by incompatible forces acting on the developing mind. The weaker conflicting force is driven "underground" into the subconscious mind, the *id.* There, however, it is not inactivated but remains smoldering as is shown by clues found in dreams, at moments of stress, and so on. Interpretation of clues is sometimes carried to implausible extremes; but this is the fault of a too dogmatic interpreter, not of the theory. The submerged activity is apt to express itself not only in clues but in distortions of conscious thinking and action, trivial or serious to the point of mental disorder.

Probably the most common source of such disorder is sexual problems. Sex is still, even with modern liberalization, a strongly repressed activity in our society, in spite of its universal force. Aspects of it, or traumatic sexual experiences, are therefore often banished to the subconscious. Here they are still vigorously active with minor or major effects on the conscious mind.

Other repressed mental drives, such as ambition, status-seeking, aggressiveness or, indeed, any strong but disapproved motive, can have similar effects. Early psychoanalysis strove to reduce all mental strains to sexual origins; when this proved difficult, it attempted to include other drives in the sexual sphere, which become increasingly implausible to most people. Recognition of many possible origins for mental strain, of which sex is simply the most common, has made the theory much more acceptable generally.

The curative aspect of psychoanalysis rests on an observable fact: If a suppressed, unconscious source of strain is consciously recognized, the strain is often relieved. The mature adult mind says in effect: "Is *that* the only reason for my compulsion (or phobia or whatnot)? Then, I can dismiss it." This sometimes happens even without analysis.

Psychoanalysis has been rather harshly criticized. Claims have been made even that its percentage of cures is hardly greater than the percentage of spontaneous recoveries in untreated cases. This seems rather unfair though it is in part due to sometimes bizarre clues and to overstressing of sex, as described. Obviously, bringing a stressful memory to consciousness may not automatically cure its effects; it may continue to appear, even to the adult mind, as ample cause for distress. Obviously also, disorders, due to constitutional weakness, such as schizophrenia, manic depressive psychosis, and some forms of depression, are unlikely to yield to such treatment, though they may be lightened by removal of the complicating strains. Hence statistics including cases in these groups are not really valid. Selected cases of neurosis and some forms of psychosis are much more likely to benefit if the analysis is competent and not too greatly restricted by theory. Certainly, awareness of a root cause greatly helps, even if it does not guarantee cure. Many cases have been demonstrably cured by psychoanalysis.

Most notable of treatments for more severe cases is *shock—electrical, insulin,* or other. The exact way in which this method works is debatable; but its effect can be described as analogous to "clearing" an electric computer. Of course, the mental "computer" is far from being cleared; this would entail obliteration of cell memories, recognitions, and learned responses—a return to early infancy, in fact. Rather, strained, unstable connections, mostly subconscious, are more or less weakened so that undistorted, more stable pathways can be established. Loss of general memories, if it occurs at all, is superficial and temporary, and should not be feared by those taking shock treatment. The discomfort sometimes accompanying shock therapy is a small price to pay for relief from the mental distress that is being treated. Shock treatment, however, is reserved for serious and obstinate cases.

Chemotherapy, treatment by drugs, is being used increasingly today. The greatest breakthrough in this field was undoubtedly the introduction of *tranquillizers.* These drugs were responsible for a dramatic decrease in the load on mental hospitals when they came into general use; great numbers of disturbed, agitated, anxious, and even violent patients were controlled by tranquillizers. Many were so effectively restored to normal behavior that they could return to private life and even to useful activity. Further, tranquillizers proved to be of great benefit to large numbers of people suffering from minor but distressing mental conditions such as extreme nervousness, tension, lesser degrees of some depressions, and so on. Tranquillizers do not, indeed, cure conditions; former symptoms tend to return if the treatment is discontinued. However, some mental conditions have a tendency to return to normal, especially if stress is removed, so that eventually the treatment may possibly be discontinued with impunity. In any case tranquillizers have contributed tremendously to human health and happiness.

Other forms of chemotherapy are so numerous, changeable, and as yet uncertain that their discussion is inappropriate here. They do, however, indicate that brain activity and brain disorders are becoming better understood and that we may hope for further breakthroughs, not just by good fortune as in the case of the tranquillizers, but by insight as our knowledge increases.

All forms of chemotherapy including even tranquillizers are, however, a matter for only the most careful and experienced application. Misuse may have disastrous effects. As is true for all drugs, leftovers from a prescription should not be used to treat an apparent return of former symptoms without proper authorization. This is especially true when dealing with the immensely complex, imperfectly understood nervous system.

Group Therapy and Transactional Analysis

These traditional approaches to dealing with mental disorders are not the only ones currently in use (in fact, the pure original Freudian form is gradually becoming outdated). The widely publicized, often controversial "group" approach to dealing with the emotions has burst upon the scene in the past decade. Whether utilized for psychotherapy (treatment of the emotionally ill person) or to "improve human potential" among apparently healthy individuals, the group approach is a widespread phenomenon at this time. Group therapy is a tool that is increasingly being utilized by psychiatrists, clinical psychologists, and counselors. It is used as an adjunct to individual therapy and sometimes to follow up or even supplant individual therapy. Its potential effectiveness depends in large measure upon the personality of the patient and the particular disorder.

T-groups, encounter groups, sensitivity training, and so on, are all variations of the "human potential movement" for reasonably healthy persons.

While some extreme and seemingly perverse examples of this movement have generated some bad publicity, there is in the properly, skillfully, professionally guided group setting a real potential for emotional growth. The basic tenet of the movement is that most of us do not effectively deal with what we really feel, in fact, often become masters at effectively *suppressing* what we really feel. (At this point psychoanalytic theory is in concert with the human potential movement: it is not healthy to consciously or subconsciously suppress emotions over too long a period of time.) Many of us can make *intellectual assertions* much better than we can verbally express our *feelings*. This is true in all of us but in varying degrees; some can express positive emotions but not negative ones, others vice versa, some can express neither, others can express both but not in an appropriate way.

The human potential movement at its best seeks to help people recognize "feelings" and to learn to deal with them in an appropriate manner. How valuable and how "safe" participation in some kind of group experience might be (some are held weekly, some for long week-ends) depends primarily on the training, skill, and professional attitude of the leader. If you become interested in participating, check the credibility of the leader carefully before enrolling and use your good common sense as a barometer to assess what is happening if you do become involved; do not be afraid to get out if your better judgment tells you that the leader is merely playing games with you and the group (for example, verbal manipulation of the group or an individual in order to meet his or her own unresolved emotional needs). Such charlatans do exist, but a careful check prior to involvement will usually avoid such a problem.

Transactional Analysis (TA) has arrived on the scene in the last 15 years. This model for dealing with the human emotions is used in individual therapy, but finds even wider use in the group setting (obviously more transactions between persons can take place in a group); it is a therapeutic "method" as well as an approach to self-understanding, emotional growth, and self-actualization in the emotionally healthy person. While not totally divergent from the much older psychoanalytic approach, it does offer new dimensions and a language that communicates to the average person much better. The three very fluid ego states are labeled simply Parent, Child, and Adult. According to the TA model, these ego states are well developed or imprinted by the time we are five years of age (there is some experimental evidence for this precept and it is in agreement with all major psychiatric and psychological theories). The Parent state is exemplified by what we have been "taught" (mostly the "don'ts" and a few "do's" accepted as the "gospel" when we needed that kind of protection and help with decisions; for example, "you must clean your dinner plate or you will get sick"); the Child state (not to be confused with the negative idea that something is "childish") represents spontaneous emotion (for example, "ooh, that's pretty" or "that frightens me," or to himself, "what you said makes me hurt and makes me feel worthless"); the Adult state is the state of reality and objectivity that must function reasonably well if our Parent and Child states are to be kept in a proper, healthy relationship. The TA goal is to strengthen the Adult state to make possible freedom of choice and creativity, that is, choices, decisions, and actions emancipated where necessary from earlier Parent and Child recordings. None of the states (or the feelings and actions to which they lead) are *automatically* labeled good or bad, right or wrong. They *do* exist, however, and the Adult data-gathering, objective state is our way of channeling whatever does exist into appropriate, effective, and acceptable solutions, decisions, and actions. These states are fluid—they can change rapidly (within a split second) and can work in concert or cause varying degrees of conflict within us at a given moment.

The logical extension of the basic elements of the TA model lead to the identification of four positions (Harris) determining individual behavior (again labeled with very common, easily understood terms): I'm OK—You're OK (the mature, adult position); I'm not OK—You're OK (the dependency of the immature); I'm not OK—You're not OK (despair); I'm OK—You're not OK (the criminal position). Further extension quite naturally leads to analyzing our interpersonal interactions (transactions with others) in terms of the three ego states.

The TA approach to therapy has been successfully applied and has also served as an aid to emotional growth and self-actualization in the emotionally healthy. The ultimate success of the group approach and the TA model cannot be predicted, but it is possible that they may become even more useful as psychiatrists, clinical psychologists, and counselors learn more about their application. Many persons have already been restored to health or improved their emotional health by these means.

Mental Health

There remains an important question: What can you do to promote your own mental health? How do you go about it as a way of life and not as an obsession or compulsion that in itself may be harmful? Logic, tempered by some supporting research evidence, suggests the following:

1. Do not keep all of your concerns and worries to yourself; whether the worry is big or small, talk to someone you trust and feel at ease with, or seek professional help if the problem becomes serious enough.
2. Consider the possibility of working out your anger in some socially acceptable physical way if it becomes a dominant part of your behavior.
3. Be willing to give in to the other fellow, to say something or do something that shows you do not always have to be best or right.
4. Change your pace, your surroundings, your routine—such escapes are normal so long as they do not dominate and take the place of useful work. From movies to vacations change is important for most of us.
5. Seek to do things for others with no associated selfish motive.

6. If you have a tendency to be overburdened by too many problems or tasks all at one time, back off and take them one at a time.
7. Do not allow yourself to feel indispensable and if you ever do, find some way to bring yourself "back to earth." Such a feeling is an open invitation to mental disorders of one kind or another.
8. Don't be too conservative in dealing with others. Try some new things; be *available* for friendships.
9. Participate on a regular basis in at least one form of physically and mentally active recreation which does not leave you more frustrated than relaxed.

REVIEW QUESTIONS

1. How would you define mental health?
2. Discuss some of the factors that contribute to the development of mental disorders.
3. Describe some of the common manifestations of neurosis. Would you characterize all people exhibiting these manifestations as mentally ill? Explain.
4. What are some of the predisposing personality factors that seem to lead to schizophrenia?
5. What type of personality disorders would most likely lead to suicide?
6. Compare psychoanalysis with other forms of psychiatric treatment.
7. What sort of assistance could you provide to a friend with a "fragile" personality?

REFERENCES

American Psychiatric Association Diagnostic and Statistical Manual for Mental Disorders, Washington D. C.: American Psychiatric Association, 1952.

FREEDMAN, ALFRED M., and HAROLD I. KAPLAN, *Comprehensive Textbook of Psychiatry.* Baltimore: Williams & Wilkins, 1967.

HARRIS, THOMAS A., *I'm OK—You're OK: A Practical Guide to Transactional Analysis.* New York: Harper and Row, 1969.

Additional Readings

BRUSSEL, JAMES, *Layman's Guide to Psychiatry,* 2nd ed. New York: Barnes and Noble, 1967.

GRANT, VERNON W., *This is Mental Illness: How It Feels and What It Means.* Boston: Beacon Press, 1963. (Paperback)

HINSIE, LELAND E., *Understanding Psychiatry.* New York: Collier, 1966. (Paperback)

HOWARD, JANE, *Please Touch: A Guided Tour of the Human Potential Movement.* New York: McGraw-Hill, 1970.

KAPLAN, BERT (ed.), *Inner World of Mental Illness: A Series of First Person Accounts of What It Was Like.* New York: Harper and Row, 1965. (Paperback)

MENNINGER, KARL et al., *Vital Balance: The Life Process in Mental Health and Illness.* New York: Viking Press, 1967. (Paperback)

38
DRUGS AND THE MIND

Humanity has probably had a propensity toward tampering with its own mental states since long before written history. This has usually been accomplished through drugs of which alcohol was likely the first, opium and marihuana close seconds, and dozens of others that appeared before the advent of modern chemistry.

A logical classification of mind-affecting drugs is not easy. They range from those such as caffeine, so widely used as to seem quite innocent and which we might call "domestic" drugs, to those such as heroin, the use of which is so dangerous as to make their unlicensed use severely outlawed, and which are referred to legally as narcotics. Between these extremes lies a considerable group of variable significance, such as marihuana, sought by many for pleasurable psychedelic effects and condemned by others on grounds of associated hazards real or supposed. Most of the hallucinogenic drugs fall into this category. Discussion of the hallucinogenic drugs, either for or against, usually generates more frustration and anger than constructive information with a factual commentary likely to please neither disputant. In addition there are a number of potent drugs with a legitimate and useful role in the treatment of nervous and mental disorders which are often used for the effects achieved from their overdose. These include such drugs as the amphetamines and barbiturates. The misuse of some of these drugs can incur dangers equal to or exceeding those from narcotics.

Domestic Drugs

This group consists chiefly of caffeine, nicotine, and ethyl alcohol. These everyday agents concern great numbers of people and, for caffeine, the majority of the American population. So familiar are they that one hardly thinks of them as drugs unless they are administered in capsules or such. But they are drugs and can be severely toxic in large doses. In fact, their use is due to their effects on the central nervous system.

Caffeine

Caffeine is the least controversial of these drugs. It is, of course, the active agent in coffee, and also in tea where it occurs in lower concentration. (Some confusion exists about this concentration. Tea leaves do contain about as much caffeine, weight for weight, as do coffee beans, and this fact is often cited. Since, however, far less weight of the leaves is used per cup, a cup of tea normally contains only one-fourth or one-fifth as much caffeine as does a cup of coffee.) In small doses, such as are contained in a cup of coffee, its effect on the brain is to stimulate mental activity, probably by action on the synapses; in larger doses it has varied effects such as diuresis and in very large doses it is poisonous. In any case, its use is subject to the hazards applying to any drug.

Caffeine is normally objectionable only when it is overused. In that case, it not only produces unpleasant and sometimes dangerous effects, but it loses its beneficial powers. As many people know, when a drug such as morphine is used frequently, it develops *tolerance* in the user; larger and larger doses of morphine must be given to produce adequate relief from pain. The same is true of caffeine: When it becomes an habitual prop for energy it gradually loses its efficacy so that two or three cups a day may become inadequate and must be supplemented by more and more frequent doses; and even then it eventually fails in its effect. Though the user may still feel a "lift," actually this lift is not from a normal level, but from an abnormally depressed level to which he quickly drops back. If he discontinues use of the drug, he suffers from depression for a time, usually a few days, but has no other reaction; in short he has become *habituated*, that is, he misses his pleasant habit and is tempted to return to it, but usually no more than that. Contrary to popular belief, "coffee nerves" are not due to the caffeine itself, but to its lack in a person habituated to it.

Sometimes, however, an excess of caffeine is directly harmful. Caffeine stimulates not only the brain but also particularly the heart and kidneys—for which purposes it may be used medically. Even in normal people, excessive doses of the drug, not under professional supervision, can damage these organs. In people with already impaired hearts or kidneys, the drug may be dangerous, which is the reason for the doctor's prohibition of coffee to such patients.

Coffee or tea is a mild but effective stimulant, with little harmful effect in normal use; it is a handy resource for anyone in need of a pick-me-up due to mild stress and fatigue. These beverages are also pleasant tasting to many, as shown by the wide use of de-caffeinated forms; and they are associated with agreeable relaxation and sociability.

Nicotine

Nicotine is more controversial. This is true even if one distinguishes between it and the coal tar and other products that accompany it in smoking tobacco. Certainly, nicotine is unlikely to be deliberately used in any form other than tobacco. However, nicotine sulfate is used as an insecticide (Black Leaf 40), and nicotine poisoning would occur from its ingestion. Also, poisoning can result from prolonged inhalation of the dust during the extraction of nicotine.

The drug, in the usual small doses received through cigarette smoking, stimulates the sympathetic nervous system. In larger doses it paralyzes the nervous system, and death may result from respiratory failure through paralysis of the respiratory muscles.

Nicotine develops a tolerance in the user, so that increasing doses are often required to attain the desired effect. The effect passes from a "lift" above normal to a lift only from a trough of depression due to lack of the drug. The person deprived of this lift becomes increasingly jittery, nervous, and focuses on obtaining relief by a fresh exposure. This is a withdrawal reaction, and may last for several days. Paradoxically, though nicotine is a stimulant it will induce a calming and relaxing affect at this time.

Alcohol

The term *alcohol* properly refers to a large class of organic chemicals with a wide range of uses. Various forms are used in medicine as antiseptics and solvents, and in industry in the manufacture of hundreds of products. The alcohol used in beverages is *ethyl alcohol* or *ethanol,* and it is produced commercially by the action of yeast (a microorganism) on sugar, a process referred to as *fermentation.* In pure form it is frequently called grain alcohol since it is produced from the action of yeast on a variety of cereal grains, principally corn, barley, and rye.

The earliest alcoholic beverages were produced by the fermentation of fruit juices, particularly grapes, in the production of wine, on grain in the production of beer, and on honey in the production of mead. The alcoholic content of these beverages ranges from 10 to 14 percent. Since alcohol has a much lower boiling point than water, the alcohol in such beverages can be collected and concentrated by distillation. Whisky, brandy, gin, rum, vodka, and liqueurs are produced by the distillation of natural products of fermentation. They usually have an alcoholic content of 40 to 55 percent (80 to 110 proof).

Absorption and Metabolism of Alcohol

The absorption of alcohol by the digestive tract is extraordinarily rapid. Within a few minutes after ingestion alcohol is absorbed through the stomach and intestinal walls (one of the few food substances known to be absorbed directly into the blood from the stomach). A given amount of alcohol taken without food and with no food in the stomach will reach a peak concentration in the blood in less than 30 minutes after which the concentration will slowly fall as the alcohol is metabolized. Most of the alcohol is metabolized through oxidation, in somewhat the same manner that the body oxidizes fatty acids or sugars. However, there are important differences. Most foods can be oxidized by all tissues of the body. Alcohol cannot—most of it is burned in the liver. In addition, other foods are oxidized in an on again off again fashion, according to the immediate energy requirements, with the excess going into storage as glycogen in the liver and as fat. Alcohol cannot be stored and it continues to circulate in the blood until it is oxidized (although small amounts, 2 to 3 percent, are eliminated through the lungs, through sweat glands, and in the urine). In man, alcohol is oxidized at the rate of about half an ounce an hour. At this rate it would require about two hours to metabolize the alcohol in a fairly strong mixed drink (containing two ounces of 100-proof whisky). Nothing is known that will increase this rate of oxidation. The dormitory physiologist who believes that deep breathing, steam baths, or vigorous exercise will hasten the sobering process has no scientific basis for his faith. Likewise, coffee, though it will rouse a somnolent drunk, will not make him sober. A man passed out from drink will be just as drunk after you waken him.

Since the body will oxidize alcohol at a constant rate whether the energy is needed or not, two metabolic effects can be expected: Excessive heat will be produced and other energy sources will be spared. The former causes the dissipation of heat from the skin surface by dilation of the capillaries, causing a flushing of the skin and a feeling of warmth. The latter means that alcohol consumption can be fattening since the spared calories from other foods will be stored as fat.

Although alcohol is always in the breath until oxidation is complete, the odor is usually characteristic of the beverage consumed (beer, wine, whisky, and gin all have their identifying odors). The concentration of alcohol in the breath has a direct relationship to the blood concentration. And since the alcohol level in the blood is the best gauge of intoxication, analysis of the breath makes possible a direct measurement of insobriety.

Effects of Alcohol on the Brain and Measurements of Intoxication

The principal pharmacological effect of alcohol is on the central nervous system. It acts on the nervous system from above downward, that is, depressing the highest cortical centers first and, with increasing dosages, affecting the lower brain centers and spinal cord. Thus the more complex faculties of judgment and restraint are the first to be affected, and these are followed by impairment of motor functions. The cerebral inhibitions account for the typical behavior pattern: Speech becomes loud and slurred and the subject mistakes his loquacity for eloquence; he loses the capacity for self-criticism and his sensitivity to the rights of others is diminished; his freedom from inhibition results in a carefree attitude that he thinks of as excitation even though all sensory and motor functions are depressed. The mild degree of cortical depression that results from one or two mixed drinks is a relaxing relief from the "cares of the day" that so many people seek in the before dinner cocktail.

Blood alcohol levels are calculated as milligrams of alcohol per 100 milliliters of blood, and are usually stated as a percent. Thus a blood alcohol concentration of 0.15 percent means 15 milligrams of alcohol in each 100 milliliters of blood. Individual variation makes it impossible to state precisely what psychological and behavioral effects given blood alcohol levels will have. However, the results from pharmacological and psychological investigations [Krantz and Carr] permit the construction of a table of approximate effects as given in Fig. 38–1. The blood alcohol levels resulting from the ingestion of relatively few drinks (up to five or six) are fairly predictable. However, other factors become involved when one attempts to determine the number of drinks required to achieve high blood alcohol levels (above about 0.3 percent)—vomiting tends to occur and the intestinal mucosa can become irritated so that a straight line relationship (blood alcohol levels in direct proportion to alcohol consumed) is not obtained. In fact, one wonders how blood alcohol levels of 0.6 to 0.8 percent are developed when stupor and vomiting usually occur at about 0.3 percent. One suspects that such cases might often be a consequence of an enormous consumption of alcohol over a brief period of time, such as perhaps might occur if a person consumed 15 or more martinis in half an hour as a result of a bet or "contest."

Drinking and Driving

Studies reported by state and national safety organizations [Haddon and Bradess] indicate that alcohol is a factor in at least 50 percent of all fatal auto

No. of mixed drinks containing 1.5 oz. of 86 proof whisky in 140 lb. person	Alcohol in the blood in mg percent (mg alc. per 100 ml blood) at end of 3 hrs.	
	.60	Usually fatal
	.50	Deep coma; frequently fatal
	.40	Collapses into coma; death has been known to occur at this conc. in novice drinker
	.30	Nausea; poor control of urinary bladder; cannot recall actions next day
	.15	Major impairment of motor functions with difficulty in talking, walking, or driving
	.10	Uncoordinated behavior; difficulty in putting on coat; inhibitions markedly decreased
	.07	Slight disturbance of balance; slight speech impairment
	.04	Loquaciousness; slight clumsiness; some decrease in reaction time
	.02	Touch of dizziness; sense of warmth and well-being
	.0	

FIG. 38–1 The best measurement of drunkeness is the alcohol content of the blood. Although considerable variation exists in the rate of oxidation of alcohol by the liver, the calculations given above give some idea of the effects of drinking at the end of a three-hour period. It should be noted that 1.5 ounces of 86-proof whisky is about twice the amount one is normally served in a mixed drink in a bar.

accidents in this country. In 70 percent of fatal single car accidents, the dead driver had blood alcohol levels of over 0.05 percent. However, few of us would need statistics to be convinced of the dangers of driving under the influence of alcohol. The problem confronting our courts is a definition of "under the influence of alcohol." Most states regard a blood alcohol level of 0.15 percent as legally sufficient to establish the fact of drunk driving, while levels of between 0.05 and 0.15 percent can be used only as evidence and not as proof. This seems inordinately generous to drinking drivers inasmuch as a blood alcohol concentration of less than 0.04 percent has been shown to impair the performance of expert drivers [Bjerver and Goldberg; Loomis and West]. The Uniform Vehicle Code of 1962 recommended that the blood alcohol level sufficient for presumptive evidence of intoxication be reduced to 0.10 percent. A recent compilation [J. Cohen] indicated that only 14 states have adopted this revised figure.

There would seem to be only one conclusion: *Never drive after drinking*. All who will practice this rule and encourage others to do so can do much to solve our traffic problems. Drinking in situations in which only ourselves and our close associates are affected can be defended as part of our individual freedom. But when in a state of insobriety actions are undertaken that endanger the general welfare, society has the right and obligation to impose severe penalties.

Alcoholism and Alcoholic Psychosis

An alcoholic is one who has such an uncontrolled and compulsive need for alcohol that he is unable to stop drinking after two or three drinks, but has an "all-or-none" response. Although the alcoholic may be able to abstain and have no craving for alcohol, he is unable to drink moderately—one drink always requires more. And although this compulsion characteristically interferes with a successful family, job, and social life, the patient is either not able to recognize this effect or is not able to control it.

It is essential, then, to distinguish between what may be termed common drunkenness and alcoholism. Even though both will present the same signs and symptoms, common drunkenness is always an act of choice. The drunken sprees of the alcoholic are uncontrolled and compulsive—he is impelled to drink even if it is against his will or judgment. The inescapable conclusion is that alcoholism is a disease—common drunkenness is not, although it may be a manifestation of emotional disorders unrelated to the drinking.

Why is the alcoholic compelled to drink? Both physiological and psychological explanations have been advanced. Some investigators have presented evidence that suggests that an inborn biochemical defect exists, perhaps related to nutrition—somehow, once alcohol oxidation begins in the body, there is a metabolic need for its continuation. Whatever it may be, this need seems to develop only after the first drink. As long as the alcoholic eschews the initial temptation he seems to be free of the compulsion.

Although recognizing that differences exist among individuals in their tolerance to alcohol (some people are constitutionally unable to get drunk since they become ill or go to sleep after only a few drinks), many investigators, particularly those dealing with the treatment of alcoholics, believe that the major cause is some sort of personality derangement. People who are maladjusted, unhappy, and insecure seek escape from their problems in a variety of ways.

Alcohol is just one. As in so many behavioral disorders, the final explanation probably involves a combination of psychological processes and a vulnerable predisposition.

We frequently see cited varying numbers of stages leading to alcoholism, sometimes as few as five (total abstaining, moderate or social drinking, heavy social drinking, excessive drinking, and chronic alcoholism) to as many as 12 [Jellinek]. And while many alcoholics may indeed go through these stages, many do not. The vast majority of people who use alcohol maintain their specific drinking habits throughout life, and some alcoholics were never social drinkers—they became alcoholics after their first exposures to drinking. Thus present drinking or nondrinking habits are no guide to propensity to alcoholism.

Some 10 percent of alcoholics experience *delirium tremens* or *alcoholic psychosis,* an acute mental disorder probably brought on by the toxic effect of alcohol on the brain. It is likely that a deficiency of the B-vitamins thiamin, riboflavin, and niacin, plays a part since these are effectively used in the treatment of alcoholic psychosis and are essential for normal nervous activity. The condition can be fatal. The visual hallucinations it induces must be terrible indeed—the victims complain of attacks by colossal roaches or spiders, snakes, vicious dogs (rarely pink elephants). He becomes so terrified that he may do injury to himself in trying to escape. The psychosis lasts two to seven days and is usually terminated by deep sleep. Some remember the fearful hallucinations and some do not. Mark Twain gives a vivid account of delirium tremens through Huck Finn's eyes in *The Adventures of Huckleberry Finn.*

> I don't know how long I was asleep, but all of a sudden there was an awful scream and I was up. There was pap looking wild, and skipping around every which way and yelling about snakes. He said they was crawling up his legs; and then he would give a jump and scream, and say one had bit him on the cheek—but I couldn't see no snakes. He started and run round and round the cabin, hollering 'Take him off! take him off; he's biting me on the neck! I never see a man look so wild in the eyes. Pretty soon he was all fagged out, and fell down panting; then he rolled over and over wonderful fast, kicking things every which way, and striking and grabbing at the air with his hands, and screaming and saying there was devils a-hold of him. He wore out by and by, and laid still awhile, moaning. Then he laid stiller, and didn't make a sound. I could hear the owls and wolves away off in the woods, and it seemed terrible still. He was laying over by the corner. By and by he raised up part way and listened, with his head to one side. He says, very low: Tramp—tramp—tramp; that's the dead; tramp—tramp—tramp; they're coming after me, but I won't go. Oh, they're here! Don't touch me—don't! hands off—they're cold; let go. Oh, let a poor devil alone!' Then he went down on all fours and crawled off, begging them to let him alone, and he rolled himself up in his blanket and wallowed in under the old pine table, still a-begging; and then he went to crying. I could hear him through the blanket . . .

The alcoholic is a sick person who needs help. Although most states have special facilities for dealing with alcoholics, a wider public acceptance is needed of the view that the alcoholic requires psychiatric counseling and medical treatment rather than punishment. Alcoholics Anonymous has been of great assistance in the rehabilitation of alcoholics, and is an excellent example of group therapy. Although some of their basic tenets have been disputed by scientists working in the field, their results speak for themselves. Just as Sister Kenny was able to administer effective treatment to victims of poliomyelitis even though she had a distorted view of the nature of the disease, Alcoholics Anonymous, whether right or wrong in their understanding of alcoholism, is successful, and deserves our support.

No one can tell, before he starts using alcohol, whether he is liable to become an alcoholic. Indeed, he may make the discovery only after years of moderate drinking; and even then, he may require some time to realize and admit his situation, which allows the addiction to become more firmly fixed. A person should, however, know in advance that certain factors do make him more prone to addiction: Since the chemical liability is probably genetic, any case of alcoholism in relatives should induce caution; so too is any neurosis or mental illness in the family, since these conditions are frequently linked with alcoholism. One cannot, unfortunately, regard absence of such family history as a guarantee of immunity.

Temperate drinking may be a source of relaxation. But everyone should be on guard for symptoms of emerging alcoholism: A tendency to increase indulgence for whatever excuse; an inclination to look forward too eagerly to the cocktail hour or party; a strong dislike of being cut off from the usual drink. These may or may not indicate approaching alcoholism. But a self-imposed limitation of intake would seem prudent. The alternative may be true alcoholism with a much more disagreeable cutoff later, or a disrupted life.

Hallucinogenic Drugs

The hallucinogenic drugs include LSD (D-lysergic acid diethylamide), mescaline, psilocybin, marihuana, and a good many others of less frequent use. LSD and marihuana deserve greater attention because of their widely publicized use and their greater likelihood of coming into possession of the average person.

Marihuana

Marihuana is an intoxicant whose use extends back over 4000 years. The hemp plant from which it is extracted, *Cannabis sativa,* is a common weed growing freely in many climates throughout the world, so an extensive worldwide use of the drug is not surprising. According to a 1950 report by the United Nations, there are some 200 million users, principally in Africa and Asia.

The *Cannabis* plant is dioecious, meaning that there are separate male and female plants. The drug is usually extracted from a yellow, sticky resin of the female plant which covers the flowers and higher leaves. The higher the resin content of the preparations, the more potent the drug. The typical grass, pot, tea, or maryjane smoked by American users is a pulverized mixture of leaves and flowers. The highest grade of the drug, called hashish, is prepared from the resin itself which is carefully scraped from the petals and leaves of selected, cultivated plants. Hashish is obviously far more potent than the marihuana preparations usually smoked in the United States.

FIG. 38–2 The marihuana cigarette or "joint" is made from the dried blossoms and leaves of the hemp plant. (Police Department, City of New York)

The specific chemical ingredients of marihuana responsible for its intoxicating effects have been termed cannabinols. Although at least four cannabinols have been identified the one with the most potent psychic effects is believed to be delta-9-tetrahydrocannabinol, usually abbreviated Δ-9-THC. Delta-9-THC has been synthesized and available for research since 1968, but practically nothing has been learned about how it exerts its effects.

A detailed clinical account of the drug's psychic effects was made in 1934 by a New York psychiatrist, Walter Bromberg. He observed and talked to people under the influence of the drug, and described his own experiences. He found generally, that the intoxication begins after 10 to 30 minutes with a period of anxiety and restlessness, frequently with a fear of death. After a few minutes the subject becomes calm and the sought after euphoria begins. He has a feeling of lightness, becomes talkative, and has the impression that his conversation is witty and brilliant. He may have visual hallucinations. Finally, the smoker becomes drowsy, falls into a dreamless sleep, and awakens with no after-effects, but with a clear recollection of the period of intoxication.

Pillard and his co-workers asked medical students to briefly describe the effects they experienced from marihuana and from alcohol. With marihuana, the students overwhelmingly reported feeling relaxation and tranquillity. A great many also mentioned "increased awareness" of music, sex, and food. Other responses were "a quietness of spirit"; "I was fascinated with ideas and objects"; "like the world around me was a new place"; "sounds envelop me and become more distinct"; "more aware and compassionate of other people." Under the influence of alcohol they also felt relaxed and at ease but tended to report diminished awareness: "dulled senses," "muddled," "fuzzy" and "sleepy" were frequent descriptions.

Comprehensive studies of the clinical and psychological effects of marihuana have been reported by Dr. Andrew Weil and his co-workers. Dr. Weil (1968) has concluded that effects of the drug are different for novices than for experienced users. He found that marihuana-naive persons demonstrate impaired performance on simple intellectual or psycho-motor tests after smoking marihuana. Regular users of marihuana, however, do not show significant impairment on the tests. Dr. Weil also found that marihuana smoking increases heart rate slightly, but has no effect on breathing rate or pupil size. In discussing these results Dr. Weil comments upon the fact that:

> Marihuana appears to be a relatively mild intoxicant in our studies. If these results seem to differ from those of earlier experiments, it must be remembered that other experimenters have given marihuana orally, have given doses much higher than those commonly smoked by users, have administered potent synthetics, and have not strictly controlled the laboratory setting. As noted in our introduction, more powerful effects are often reported by users who ingest preparations of marihuana. This may mean that some active constituents which enter the body when the drug is ingested are destroyed by combustion, a suggestion that must be investigated in man. . . . The researcher who sets out with prior conviction that hemp is psychotomimetic or a 'mild hallucinogen' is likely to confirm his conviction experimentally, but he would probably confirm the opposite hypothesis if his bias were in the opposite direction. Precautions to insure neutrality of set and setting, including use of a double-blind procedure as an absolute minimum, are vitally important if the object of investigation is to measure real marihuana-induced responses.

Another psychiatrist, Lester Grinspoon, described another characteristic of marihuana intoxication, that is, a splitting of consciousness, so that the smoker is able to experience the effects and yet view himself as an objective and rational observer. Dr. Grinspoon suggests that this may explain how many experienced users of marihuana manage to behave in a sober fashion in public even when they are intoxicated. He also noted that marihuana is far less potent than the other hallucinogenic drugs such as LSD, mescaline, and psilocybin. Consciousness is not so greatly altered, and no tolerance to the drug develops.

Experiments carried out by the Department of Motor Vehicles of the State of Washington and by psychiatrists at the University of Washington [Crancer et al.] showed that marihuana smoking causes far less impairment of driving ability than does alcoholic intoxication at the level of 0.10 percent alcohol concentration in the blood. However, these investigators did not conclude that marihuana will not impair driving ability!

Current discussions on marihuana contain much distorted fact as well as misguided supposition. The following are some of the questions commonly asked about marihuana, and the answers, to the extent that they are available.

Does marihuana use lead to heroin? Various studies [Grinspoon] suggest that at least 50 percent of heroin users have had experience with marihuana. But most heroin addicts have also been users of tobacco and alcohol. There is no evidence that marihuana, any more than alcohol or tobacco, will bring about a "craving" or need for any other drug. However, the regular marihuana user will likely find that his sources of the drug are also the sources of LSD, heroin, "speed," and whatever else is illegal on the drug market. This availability can be dangerous to the vulnerable personality.

Does marihuana incite people to aggression and violent criminal behavior? Bromberg's classic study of marihuana use in Manhattan (1934) showed no such relation. "No cases of murder or sexual crime due to marihuana were established." Several investigators have observed that marihuana intoxication induces a lethargy that would not seem conducive to any physical activity, let alone the crimes of violence. There is a release of inhibitions, but it seems to be verbal rather than behavioral. Grinspoon contends that the user would not ordinarily do things foreign to his nature. Thus if he is not a criminal he would not ordinarily commit criminal acts under the influence of the drug.

Does marihuana induce sexual debauchery? Again Grinspoon contends that marihuana intoxication would not likely break down moral barriers that are not already broken. There is no evidence that marihuana is an aphrodisiac.

Does marihuana lead to mental degeneracy? Evidence from investigations in Egypt and Asia indicate that long-term users are indeed passive, unproductive, slothful, and indolent. The question is which comes first? Do the users take up marihuana because of a feeling of hopelessness and defeat in an effort to escape an unbearable existence? Or does marihuana use bring about a general motivational bankruptcy? It is probable that neither condition entirely induces the other. Certainly, marihuana's passivity is at odds with the values of a society that prizes activity and achievement. The user may have some valid reasons to wish to escape from a society whose aggressive tendencies have brought so much misery, but he should remember also, that the grandest of human enterprises, its great music and literature, its science and technology, would have been impossible in a culture that valued the unproductive sedation of marihuana intoxication; nor could the cause of these miseries—wars, overpopulation, and environmental deterioration—be alleviated by such a culture.

Does marihuana lead to any physical deterioration? No organic disease has been shown to arise or to be aggravated by long-term use of marihuana. However, the absence of such findings, particularly in view of the small number of studies, does not justify generalizations. We must remember the many years required to establish the cause and effect relationship between cigarette smoking and lung cancer.

Can marihuana cause "bad trips"? Dr. Weil (1970) and others [Talbot and Teague] have found that a majority of adverse responses to marihuana are panic reactions in which the user interprets the psychologic effects of the drug to mean that they are dying or losing their minds. These effects include episodes of mental confusion, terrifying paranoid thoughts, and anxiety. Generally, the new user is more apt to have either no symptoms or an over-whelming loss of control.

Should marihuana be legalized? The argument is often made that if marihuana were legalized we would eliminate the exposure of cannabis users to sellers of more dangerous drugs. Also, it is contended, alcohol is more debilitating than is marihuana, and it is legal. These are valid arguments. However, it must be remembered that legalization would not come without regulation—its use by minors would continue to be illegal. Marihuana "joints" would likely be as available to teenagers as are cigarettes now, and the underage user would be breaking the law just as he is now. Also, it is likely that the quality of the marihuana (resin content) would be much improved over the "garden variety" of hemp now available. This increased potency would probably increase the incidence of adverse reactions. However, it has become increasingly apparent that marihuana is not a dangerous drug, and its use should probably be regulated rather than prohibited.

We can close this discussion of marihuana with a recent statement by Dr. Walter Bromberg whose 1934 study is often quoted by those who are the strongest supporters of marihuana use. In an appeal to fellow psychiatrists in his later report of 1968 he said:

> The psychiatric profession, dedicated to preserving the mental health of the nation, can best serve the public by openly stating the psychic dangers of marihuana and putting the burden of its use or abuse squarely on the shoulders of the users. If rebellion against the "square" world is necessary, if the hypocrisy and double standards complained of require modification, it is incumbent on the oncoming genera-

tion to make these changes on the basis of clear and present need rather than the space-time-body image distortion of LSD, marihuana, or banana peel extract.

LSD

D-lysergic acid diethylamide or LSD is an exceedingly potent and dangerous drug. It was accidentally discovered by the chemist Albert Hofmann in 1943 while working with a variety of derivatives of lysergic acid. He experienced dizziness and blurred mental processes, and subsequently checked his observations by ingesting a small quantity of the drug and recording his experiences [1959].

> Last Friday, April 16th, I was forced to stop my laboratory work in the middle of the afternoon and to go home, as I was overcome by a peculiar restlessness associated with mild dizziness. Having reached home, I lay down and sank into a kind of delirium which was not unpleasant and which was characterized by extreme activity of the imagination. As I lay in a dazed condition with my eyes closed (I experienced daylight as disagreeably bright), there surged in upon me an uninterrupted stream of fantastic images of extraordinary vividness and accompanied by an intense, kaleidoscope-like play of colors. The condition gradually passed off after about two hours.

These observations have been expanded upon by other investigators. Although the effects are modified by the personality traits and the current mood of the individual ingesting the drug, an inventory of the most common effects would include visual hallucinations involving constantly changing patterns of abstract forms. A new awareness of the physical beauty of the world, of visual harmonies and exquisitness of detail never before perceived sometimes leads artists and composers to feel that they are able to create masterpieces, but upon sober evaluation of their efforts there is disappointment. The subject also loses his ability to concentrate on a specific sensation, for example, pain (hence its experimental use among terminal cancer patients who comment upon their unconcern for their disease when under the influence of the drug). However, delightful images can be quickly replaced by feelings of disaster and fear, turning the experience into a nightmare. The intoxication resembles the thinking disturbances of schizophrenia and paranoia, and for a time, it was thought that LSD could be used for the study of experimental psychosis.

The amazing feature of LSD is its potency. The administration of 100 micrograms is the usual human dosage. *A sample of LSD the size of an aspirin tablet would contain some 3700 such doses!* Comparable hallucinogenic drugs such as mescalin and psilocybin require doses 5000 times and 200 times, respectively, that of LSD [Hofman, 1961].

The hazards of LSD administration are thus real and clear. Latent psychotics have undergone a disintegration of their personality after a single dose. The experienced user may, after 25 to 50 "trips," find that he stays in the drug-induced state for longer and longer periods, until finally he claims that he doesn't need the drug any longer since he is in a constant "high." Whether or not these are permanent, chemically-induced personality changes is unknown.

Deaths and accidents from LSD usage have resulted in large part from its power to greatly intensify emotional responses. Delight, fear, or sadness may reach overwhelming proportions so that an event that might ordinarily cause a wrinkling of the brow brings on wails of despair. It is under these conditions that the subject walks into the path of a truck or leaps from a window.

Although legal production of the drug is strictly controlled it is easily manufactured in make-shift laboratories, and sold, principally on college campuses, in sugar cubes, pills, or ampules. Stringent laws have been adopted in many states, although the penalties are often not as severe as those regarding marihuana, a much less dangerous drug.

Some psychiatrists regard the drug as having important clinical applications, and a few psychologists argue that it can contribute to the elucidation of mind-brain mechanisms; but its use to "expand consciousness" carries hazards all out of proportion to its capricious euphoria [Louria: "Lysergic acid diethylamide"].

Related to LSD in their effects are a number of other psychedelic drugs. Most are designated by letters. DMT, DET, and DPT are drugs whose abbreviations stand for their chemical structure (all are tryptamines, dimethyl-, diethyl-, and dipropyl-). They seem to differ from LSD in having a more rapid effect and a shorter duration of action. The drug termed "STP," however, belongs chemically to the amphetamines (discussed with Misused Prescriptive Drugs below). Again, its effect resembles that of LSD, but its action is long, up to three days, and it generally has a bad reputation among LSD users. In fact, most of these other "alphabet" drugs are compared unfavorably with LSD by the drug culture —either from the point of view of their worth as psychedelics or from their hazardous after effects.

Mescaline and Psilocybin

These naturally occurring drugs (mescaline from the peyote cactus and psilocybin from a species of mushroom) are chemically related to LSD, and elicit similar hallucinogenic reactions but are far less potent. They are, in general, less readily available to the drug user than are marihuana or LSD, so are less of a legal problem. There have been few studies of their physiological effects, although much has been written about their use by the Indians of the Southwest in their tribal ceremonies. Mescaline is one of the drugs used by Aldous Huxley during the writing of his book "Doors of Perception."

A good deal of deception exists in the marketing of hallucinogenic drugs, particularly those such as mescaline and psilocybin that would tend to bring premium prices. A joint study by the New Jersey Neuropsychiatric Institute and the Bureau of Narcotics and Dangerous Drugs [Cheek and Newell] has shown that drugs sold as mescaline and psilocybin are almost always either LSD or STP. Of 18 drug samples reputed to be psilocybin or mescaline, 15 were LSD or STP and the remainder were aspirin or not identified. Persons wishing to experiment with psychedelic drugs yet seeking to avoid the dangers of LSD are obviously the victims of the deception. Also, the amount of LSD and STP varied greatly in the samples. This can be dangerous if the user, failing to get much reaction from one pill, takes another with a large amount. Mixing of LSD and STP can cause extended confusional states, and this could be another inadvertent effect. Obviously, when one deals in an illicit market, anything goes, and no recourse is available. To the possible dangers inherent in the psychedelics themselves are added unknown dangers.

Misused Prescriptive Drugs

Barbiturates

Seconal, Nembutal, Amytal, and Luminal are barbiturates which differ only in the length of their action. They have been given colorful names, such as red birds, yellow jackets, purple hearts, which are derived from the characteristic form and appearance of the patented preparations.

Ordinary doses of barbiturates are commonly taken for sedation and sleep since their principal effect is depression of the central nervous system. Their quieting effect on the cerebral cortex allays anguish and apprehension and induces relaxation so that they are valuable drugs in the treatment of insomnia. Other forms of barbiturates are used as general anesthetics and as anticonvulsants in the treatment of epilepsy.

A physical dependence and tolerance can develop from long-term use. The increased dosages required by the experienced user can make possible inadvertent overdoses which can result in death from the depression of the respiratory mechanism. When barbiturates are used concurrently with ethyl alcohol the depressive effect on the central nervous system is greatly increased so that what would be a slight overdose in a sober person can become a lethal dose when ingested with alcohol.

What we are concerned about here, however, is the deliberate use of barbiturates to induce intoxication. When taken in preparation for bed the drug normally induces sleep. When taken during social activity, which induces wakefulness, the same depressive effects on the central nervous system result in a typical drunken state, very similar to that induced by alcohol. The intoxicating effects would obviously be greatly enhanced by the concurrent ingestion of alcohol. Amphetamines are sometimes taken by users to counteract drowsiness—to which we can only stand in wonder at the abuses a person can knowingly inflict upon his nervous system.

The sleep-inducing dose of the drug taken by normal people every night tends to become a habit, but withdrawal effects are not usually serious. The much larger doses often taken for their intoxicating effect can lead to an addiction more dangerous than that of morphine or heroin since the withdrawal effects can produce convulsions and death [American Medical Association]. When the addict has regular access to the drug he tends to live in a state of intoxication, and is therefore unproductive. Treatment of addicts is largely unsuccessful, and the emphasis therefore is on prevention.

Amphetamines

Benzedrine (bennies), Dexedrine (oranges), Methedrine (speed), and Diphetamine (footballs) are the commonly used amphetamines. They all evoke stimulation of the cerebral cortex which is usually manifested by brighter spirits and a mild euphoria. The user may also experience restlessness and insomnia. The psychic stimulation produced by amphetamines is generally followed by a sense of depression and fatigue. Contrary to a current supposition among students, the drug has not proved itself capable of facilitating better mental performance. Errors become more common, and the ability to evaluate self-achievement critically deteriorates.

The ingestion of amphetamines is usually disappointing to those who are seeking some sort of psychedelic experience. However, there is a growing number who inject the drug, particularly Methedrine, intravenously to achieve a sudden, overwhelming, pleasurable sensation. The effect is an intense fascination with all thoughts and activities which is accompanied by a paranoia. Repeated injections may extend a "trip" to several days or a week during which activity becomes less purposeful; eventually the subject is reduced to a random assortment of meaningless actions. Death or severe disability can result.

Narcotics

The term narcotics refers to addictive drugs derived from opium, or synthetic drugs having an opiate-like effect—dramatic pain relief and sedation, often with euphoria. The legal definition of narcotics is expanded to include addictive drugs with other effects.

Opium is the dried exudate from the poppy plant, *Papaver somniferum* (Fig. 38–3). Opium contains several chemicals called alkaloids, and these are responsible for the pain-killing effects. Morphine and codeine are the most important of these. A considerable number of narcotic drugs are derived from morphine and others are chemically synthesized. These include heroin and Dilaudid among the former, and Demerol among the latter.

FIG. 38–3 Opium poppy cultivation in Turkey. *Left:* The farmer is incising a slit around the circumference of the capsule of the fruit. A resinous exudate then accumulates at the site of the incision. *Right:* The farmer's son scrapes the crusty resin into a ladle from capsules that had been incised a day or two earlier. This is the raw opium from which morphine is processed. The morphine is then altered chemically to form heroin.

Opium is one of man's oldest drugs. Ancient Egyptian and Greek documents relate its ability to relieve pain. Prepared in an alcohol solution called laudanum it was a vital part of the medical equipment of physicians until the discovery of its most active agent, morphine, in 1807. Morphine has only recently begun to be replaced by its derivatives and synthetic counterparts.

All narcotics (medical definition) have a similarity of action, and the addict can effectively substitute one for another, and steadily increasing doses of all are required to achieve the desired effects during long-term use. It is as if the cells of the body develop an insatiable craving for these chemicals, and become dependent upon them for normal function. The consequences of the addiction were not fully appreciated until late in the nineteenth century, and restrictions on their distribution and use were not imposed until the early 1900s. Doctors today are understandably very cautious in using them.

Narcotics provide experiences far more pleasurable and powerful than most of the hallucinogenic drugs. This can be confirmed by anyone who has enjoyed the delicious euphoria of morphine or similar injection for the relief of acute pain. It is not surprising that such effects become highly valued, and a habit easily formed. However, the desired effects steadily fade so that doses are reached that would kill a normal person. Furthermore, lapses between doses become more and more violently unpleasant.

This last fact is the core of the addiction problem. The victim ceases to gain more than a brief relaxation or exhilaration, if that much, and craves the drug to maintain anything like a normal state and to escape increasing misery. When subject to prolonged deprivation, he suffers from growing physical weakness, nausea, vomiting, diarrhea, profuse sweating and lacrimation (production of tears); inability to eat and sleep, and other symptoms [Cherubin]. He also suffers from an intense mental discomfort beyond that induced by his physical symptoms, which results in an almost ungovernable urge to obtain a dose of his drug at whatever cost. This condition is the withdrawal syndrome, and it shows that the sufferer is physiologically addicted. However, narcotic withdrawal is very rarely fatal.

In a very few cases addiction is due to injudicious medical use, and some addictions develop among medical personnel who have ready access to narcotics and who usually begin use as a result of long working hours and other pressures of their profession. However, in the vast majority of instances, addiction is due to misinformation and inexperience.

Sidney Cohen, one of our leading authorities on narcotics addiction, lists the following factors in the development of narcotics addiction in young people:

1. Most young addicts are from minority groups in which households are so disrupted that the family unit hardly exists.
2. Peers and peer values are often critical factors in heroin usage. In some urban slum areas, boyfriends turn on girlfriends, gangs turn on newcomers, and older siblings may turn on younger ones.
3. Within a framework of chaotic familial and social life, dependent, passive, immature youngsters are more vulnerable and are spoken of as having an "addictive personality." That any personality type might get locked into a heroin habit is unquestionable. When a relatively mature person becomes addicted for some reason, the possibility of a cure is much more favorable.
4. Those who become addicts seem to have the greatest difficulty in handling anxiety, depression, and their adolescent drives. Delinquent behavior often existed prior to drug usage. Drug usage, of course, promotes further delinquent activities.
5. Prior nonnarcotic drug usages may have facilitated the introduction to heroin. The use of an illicit drug of any type permits easier movement to more potent drug groups. For some people a barrier seems to exist; once it is breached, all sorts of drugs might be used.
6. The active recruitment of new consumers by pushers is hardly necessary under present conditions. Persistent "hippie" rumors that some LSD is

> spiked with heroin to create more heroin addicts are as difficult to believe as the old tale of pushers turning on school children. Grade school children have become addicted, but not through pushers. As a rule, users obtain gratification from turning nonusers on. Their motives range from hostility to love.

Once hooked, the heroin addict is reduced to the most debased existence (Fig. 38–4). The expense of his habit, frequently $50 a day or more, leads commonly to prostitution for the woman and to robbery for the man. His constant fear of being without a "fix" imposes a single meaning to life that operates at the expense of all else—education is disrupted, family and friends are betrayed, and health is destroyed.

FIG. 38–4 Shooting heroin. (Bob Combs from Ralph Guillumette)

The Fate of the Heroin Addict

A few addicts manage to free themselves from the habit after a period of abstinence. Many of these are older individuals who have either outgrown the emotional problems that contributed to the addiction or never had serious personality derangements in the first place. Some affluent addicts have the means to continue the habit and are able to lead a somewhat normal but hazardous life. A far greater number die an early death from a variety of causes related to the habit. Overdoses are common since samples from pushers may vary in their heroin content from 0 to 70 percent—and this is the cause of addicts being found dead with a syringe in a vein. Diseases transmitted from contaminated syringes include hepatitis and tetanus, both of which take a considerable toll; and death from violence is a frequent concomitant of the heroin culture.

Rehabilitation through institutionalization has been relatively unsuccessful [Louria: *The Drug Scene*]. For example, 95 percent of those released from the narcotic rehabilitation unit at Lexington, Kentucky (Fig. 38–5) use heroin again within a six-month period. However, if supervision is continued after the individual returns to his community, the failure rate at the end of six months

FIG. 38–5 The Federal Narcotics Hospital at Lexington, Kentucky is one of only two in the country. The capacity at Lexington is only slightly over a thousand, while there are an estimated 60 thousand addicts in the country. (U. S. Public Health Service Hospital, Lexington, Kentucky)

falls to 75 percent. The desperate shortage of professionals who are able to counsel and otherwise assist former addicts is probably the principal factor in these dismal statistics. State health programs and welfare agencies have very little funds to support such programs.

Recently, much attention has been given to the drug *Methadone* in the treatment of heroin addicts. Methadone is a synthetic narcotic which is less euphoriant than heroin. It is addictive, but has the advantages that it is taken in a well-controlled tablet form, tolerance does not seem to develop, and a person on Methadone cannot get a "high" from heroin. Many clinic workers are optimistic that large-scale out-patient programs of rehabilitation may be possible with the use of Methadone.

The medical profession, and to a lesser extent, law enforcement agencies, have begun to view the sufferer as a person requiring medical treatment rather than as a criminal requiring imprisonment. He can be cured of his condition, but since the cure is long and intensely unpleasant no matter how it is cushioned by compensating drugs, he often shrinks from undergoing it and hides his condition. Herein lies society's problem, and a completely satisfactory solution to it has yet to be advanced. Prevention through education and the social rehabilitation of ghetto areas where narcotics use most often begins is obviously a necessary part of any control program.

Cocaine

Although legally classified among narcotic drugs, cocaine does not have the generalized pain killing effects, and is not a sedative. Although it is a local anesthetic, it is probably also the most powerful counteractant against fatigue known, presumably through its potent stimulation of the central nervous system. It is, like the opiates, an alkaloid, and is extracted from the leaves of the coca plant, *Erythroxylon coca,* which grows in Bolivia and Peru. The coca leaves are chewed by the natives of these regions, probably more to relieve hunger and to increase endurance than to achieve the orgiastic delights sought by the user in the United States. Cocaine is not addictive and no tolerance seems to develop.

The euphoria induced is in great part sexual, and sometimes ejaculation follows intravenous injection. Due to its stimulative effects the user is far more likely to become violent than the opiate user. Toxic doses elevate the body temperature, markedly increase heart rate, cause vomiting, and convulsions. Death can result.

Inhalants

Children and adolescents, and occasionally adults, discover that pleasurable sensations can be obtained by inhaling a variety of volatile household items such as model airplane glues, cleaning solvents, nail polish remover, lighter fluid, and gasoline. Whatever the source the effective agent is likely to be a petroleum derivative, alcohol ester, or chlorinated organic compound of high toxicity. Most cause liver damage, and several induce renal failure and/or heart damage as well, along with a variety of other abnormalities. The vast majority of users are under the age of responsible behavior, and the problem of control falls upon parents and others to whom children are entrusted.

Drug Abuse and Misuse

There is a great deal of clamor surrounding the abuse and misuse of drugs. Their use, in one form or another, is reportedly widespread among certain groups of young people striving for what they say is a better society and for personal identity. Several of the hallucinogenic drugs are apparently readily available even to elementary and junior high children.

Illegal drug traffic is a major part of organized crime. The associated health problems, addiction and drug related deaths, certainly cannot be casually dismissed. Hence, the question: what can one do about it? We offer four specific suggestions:

1. Become informed about drugs, their effects and their potential for danger.
2. Share your knowledge with others, in particular your children, your brothers and sisters, and parents (especially if they still have children, teen-agers, or college-age students).
3. Don't be a hypocrite when dealing with younger people; admit that you use drugs of one kind *if* you use alcohol, or caffeine, or nicotine, but point out the differences and similarities between these and the hallucinogens, barbiturates, amphetamines, and narcotics.
4. Make an *informed* and *conscious* decision for *yourself* about the use of drugs. Weigh the supposed benefits against your knowledge of the potential for physical or emotional damage. The decision, of course, must be yours and we nor anybody else can or should make it for you, but at least it can be *your* decision and it can be an *informed* decision, not an uninformed, spur-of-the-moment impulse to "go along."

REVIEW QUESTIONS

1. How is alcohol metabolized and otherwise eliminated by the body?
2. How does alcohol affect the central nervous system?
3. Describe the psychological effects of marihuana intoxication.
4. On the basis of the experiences of your acquaintances what would you recommend concerning laws that regulate or prohibit the use of marihuana?
5. What are some of the hazards associated with the use of LSD?

6. Barbiturate addiction is regarded by many physicians as even more dangerous to health than narcotic addiction. Why?
7. What is a narcotic?
8. What are some of the factors that are responsible for the spread of heroin addiction?
9. Some people argue that what drugs (of any sort) that one takes are none of society's business as long as no harm comes to others. What arguments could you make in support of or opposed to this position?

REFERENCES

AMERICAN MEDICAL ASSOCIATION COMMITTEE ON ALCOHOLISM AND ADDICTION, "Dependence on barbiturates and other sedatives," *Journal of the American Medical Association,* **193:** 673–677 (1965).

BJERVER, KJELL, and LEONARD GOLDBERG, "Effect of alcohol ingestion on driving ability," *Quarterly Journal of Studies on Alcohol,* **11:** 1–30 (1950).

BROMBERG, W., "Marihuana intoxication: A clinical study of *Cannabis sativa* intoxication," *American Journal of Psychiatry,* **91:** 303–330 (1934).

BROMBERG, W., "Marihuana—Thirty-five years later," *American Journal of Psychiatry,* **125:** 391–393 (1968).

CHEEK, F. E., and S. NEWELL, "Deceptions in the illicit drug market," *Science,* **167:** 1276 (1970).

CHERUBIN, G. E., "The medical sequelae of narcotic addiction," *Annals of International Medicine,* **67:** 23–34 (1967).

COHEN, J., "Alcohol and traffic accidents," *Triangle,* **5:** 214–220 (1962).

COHEN, SIDNEY, *The Drug Dilemma.* New York: McGraw-Hill, 1969. (Paperback)

CRANCER, A. et al., "Comparison of the effects of marihuana and alcohol on simulated driving performance," *Science,* **164:** 851–854 (1969).

GRINSPOON, LESTER, "Marihuana," *Scientific American,* **221:** 17–25 (December 1969).

HADDON, WILLIAM, JR., and VICTORIA A. BRADESS, "Alcohol in the single vehicle accident," *Journal of the American Medical Association,* **169:** 1587–1593 (1959).

HOFMANN, ALBERT, "Psychotomimetic drugs: Chemical and pharmacological aspects," *Acta Physiologica et Pharmacologica Neerlandica,* **8:** 240–259 (1959).

HOFMANN, ALBERT, "Chemical, pharmacological, and medical aspects of psychotomimetics," *Journal of Experimental Medicine,* **5:** 31–51 (1961).

HUXLEY, ALDOUS, *The Doors of Perception, and Heaven and Hell.* New York: Harper and Row, 1954.

JELLINEK, E. M., *The Disease Concept of Alcoholism.* Hillhouse Press, 1960.

KRANTZ, JOHN C., and C. JELLEFF CARR, *Pharmacologic Principles of Medical Practice,* 6th ed. Baltimore: Williams & Wilkins, 1965.

LOOMIS, T. A., and T. C. WEST, "The influence of alcohol on automobile driving ability," *Quarterly Journal of Studies on Alcohol,* **19:** 30–46 (1958).

LOURIA, DONALD B., *The Drug Scene.* New York: McGraw-Hill, 1968. (Paperback)

LOURIA, DONALD B., "Lysergic acid diethylamide," *New England Journal of Medicine,* **278:** 435–437 (1968).

PILLARD, RICHARD C., "Marihuana," *New England Journal of Medicine,* **283:** 294–303 (1970).

TALBOTT, J. A., and J. W. TEAGUE, "Marihuana psychosis: Acute toxic psychosis associated with the use of *Cannabis* derivatives," *Journal of the American Medical Association,* **210:** 299–302 (1969).

WEIL, A. T., "Adverse reactions to marihuana," *New England Journal of Medicine,* **282:** 997–1000 (1970).

WEIL, A. T. et al., "Clinical and psychological effects of marihuana in man," *Science,* **162:** 1234–1242 (1968).

ZINBERG, NORMAN E., and A. T. WEIL, "The effects of marihuana on human beings," *The New York Times Magazine,* May 11, 1969.

Additional Readings

Alcohol

BACON, SELDEN D. (ed.), "Understanding alcoholism," *Annals of the American Academy of Political and Social Science,* **315** (January 1968).

MENDELSON, JACK H., "Biological concomitants of alcoholism," *New England Journal of Medicine,* **283:** 24–32, 71–81 (1970).

NATIONAL CENTER FOR PREVENTION AND CONTROL OF ALCOHOLISM, NATIONAL INSTITUTE OF MENTAL HEALTH, *Alcohol and Alcoholism,* 1967.

PITTMAN, DAVID J., and CHARLES R. SNYDER, eds., *Society, Culture, and Drinking Patterns.* New York: John Wiley & Sons, 1962.

WALLER, J. A., "Alcoholism and traffic deaths," *New England Journal of Medicine,* **275:** 532–536 (1966).

Marihuana

CARLIN, ALBERT, and ROBIN D. POST, "Patterns of drug use among marihuana smokers," *Journal of the American Medical Association,* **218:** 867–868 (1971).

DORNBUSH, R. J., et al., "Marijuana, memory, and perception," *American Journal of Psychiatry,* **128:** 194–197 (1971).

HOLLISTER, LEO E., "Marihuana in man: Three years later," *Science,* **172:** 21–29 (1971).

KOLANSKY, HAROLD, and WILLIAM T. MOORE, "Effects of marihuana on adolescents and young adults," *Journal of the American Medical Association,* **216:** 486–492 (1971).

LIEBERMAN, CARL M., and BETH W. LIEBERMAN, "Marihuana—a medical review," *New England Journal of Medicine,* **284:** 88–90 (1971).

MCGLOTHLIN, WILLIAM H., and LOUIS J. WEST, "The marihuana problem: An overview," *American Journal of Psychiatry,* **125:** 370–378 (1968).

NATIONAL COMMISSION ON MARIHUANA and DRUG ABUSE, *Marihuana: A Signal of Misunderstanding,* New York: New American Library, 1972.

NEUMEYER, JOHN L., and RICHARD A. SHAGOURY, "Chemistry and pharmacology of marijuana," *Journal of Pharmaceutical Sciences,* **60:** 1433–1457 (1971).

SNYDER, SOLOMON H., *Uses of Marijuana.* New York: Oxford University Press, 1971.

LSD

ALPERT, R., SIDNEY COHEN, and L. SCHILLER, *LSD,* ED New American Library of World Literature, 1966.

ABRAMSON, H. A. (ed.), *The Use of LSD in Psychotherapy and Alcoholism.* Indianapolis, Ind.: Bobbs-Merrill, 1967.

COHEN, SIDNEY, *The Beyond Within: The LSD Story,* 2nd ed. New York: Atheneum, 1967.

COHEN, SIDNEY, and K. S. DITMAN, "Complications associated with lysergic acid diethylamide (LSD-25)," *Journal of the American Medical Association,* **181:** 161-162 (1962).

MCGLOTHLIN, W. H. et al., "Long-lasting effects of LSD in normals," *Archives of General Psychiatry,* **17:** 521–532 (1967).

SMARTS, R. W., and KAREN BATEMAN, "The chromosomal and teratogenic effects of lysergic acid diethylamide; a review of the current literature," *Canadian Medical Association Journal,* **99:** 805–810 (1968).

Barbiturates

ADAMS, E., "Barbiturates," *Scientific American,* **198:** 60–64 (January 1958).

GOLDMAN, D., et al., "Treatment of barbiturate dependence," *Journal of the American Medical Association,* **213:** 2272–2273 (1970).

LOENNECKEN, S. J., *Acute Barbiturate Poisoning.* Baltimore: Williams & Wilkins, 1968. (Paperback)

ONG, B. H., "Hazards to health; dextroamphetamine poisoning," *New England Journal of Medicine,* **266:** 1321–1322 (1962).

TYLDEN, E. et al., "Dangers of barbiturates," *British Medical Journal,* **2:** 49 (1970).

Amphetamines

KRAMER, J. C. et al., "Amphetamine abuse: Pattern and effects of high doses taken intravenously," *Journal of the American Medical Association,* **210:** 305–309 (1967).

PERMAN, E. S., "Speed in Sweden," *New England Journal of Medicine,* **283:** 760–761 (1970).

RUSSO, J. ROBERT, *Amphetamine Abuse.* Springfield, Ill.: Charles C Thomas, 1968.

Heroin

BELL, R. G., *Escape from Addiction.* New York: McGraw-Hill, 1970.

DOBBS, W. H., "Methadone treatment of heroin addicts," *Journal of the American Medical Association,* **218:** 1536–1541 (1971).

FIDDLE, S., *Portraits from a Shooting Gallery.* New York: Harper & Row, 1967.

LOURIA, DONALD B., et al., "Major medical complications of heroin addiction," *Annals of International Medicine,* **67:** 1–22 (1967).

PERKINS, M. E., "Survey of a methadone maintenance treatment program," *American Journal of Psychiatry,* **126:** 1389–1396 (1970).

SZASZ, THOMAS, "The ethics of addiction," *Harper's Magazine,* April 1972.

YABLONSKY, L., *The Tunnel Back: Synanon.* New York: Macmillan, 1965.

General readings on drug abuse:

BARRON, E. M., et al., "The hallucinogenic drugs," *Scientific American,* **210:** 29–37 (1964).

EPSTEIN, SAMUEL, and JOSHUA LEDERBERG, eds., *Drugs of Abuse: Their Genetics and Other Chronic Nonpsychiatric Hazards.* Cambridge, Mass.: M. I. T. Press, 1971.

FORT, JOEL, *The Pleasure Seekers.* Indianapolis, Ind.: Bobbs-Merrill, 1969.

LINDESMITH, A. R., *The Addict and the Law.* Bloomington, Ind.: Indiana University Press, 1965.

LOURIA, DONALD B., "Medical complications of pleasure seeking drugs," *Archives of Internal Medicine,* **123:** 82–87 (1969).

LOURIA, DONALD B., *Nightmare Drugs.* New York: Pocket Books, 1966. (Paperback)

PARRY, HUGH J., "Use of psychotropic drugs by U. S. adults," *Public Health Reports,* **83:** 799–810 (1968).

PREBLE, EDWARD, and GABRIEL V. LAURY, "Plastic cement: The ten cent hallucinogen," *International Journal of Addictions,* **2:** 271–281 (1967).

RAY, OAKLEY S., *Drugs, Society, and Human Behavior.* St Louis, Mo.: C. V. Mosby, 1972.

SNYDER, S. H. et al., "2,5-Dimethoxy-4-methylamphetamine (STP): A new hallucinogenic drug," *Science,* **158:** 669–670 (1967).

39
THE SKIN

Unless disorder or disease draws attention inward, most people are far more preoccupied with their skin and associated external *appearances.* And because the skin is observable and accessible the average person feels more qualified in the business of dealing with minor disorders of the skin. This confidence that "I can handle it and, even if not, no damage can be done" is often misplaced confidence! At any rate, the skin and its care is of very real interest to most people and as such deserves considerable attention.

The skin is, of course, the outer covering of the body. With its accessories—hairs, nails, subcutaneous tissue, and sweat, sebaceous, and mammary glands—it is more properly called the integument. To most people, the skin is simply the protective surface of the body. In fact, it is far more than that. Its importance is illustrated by its very bulk. Next only to the muscles, it is the most massive organ-system of the body, as anyone will realize who has tried to lift the fresh hide of an animal. It amounts, in fact, to almost 25 percent of body weight (muscles, about 40 to 45 percent). In obese people, with much subcutaneous fat, the proportion of skin may be even greater. The activities, disorders, and care of the skin are of importance in proportion to its bulk. Disorders and care will be discussed below; but activities should be understood first.

Activities of the Skin

Protection of the inner tissues is indeed an important function of the skin. As anyone can testify who has lost much skin by burn, sunburn, or other means, these delicate tissues suffer agonizing pain from exposure. The *retention of fluids* is equally important. In loss of skin, as from a burn, the most serious, and sometimes fatal, complication is shock due to loss of body water and salts. The skin, in this sense, "holds the body in."

In addition to simple conservation, the skin is an active factor in the *maintenance of water balance.* Whereas a person under average conditions excretes up to one and a half quarts a day in his urine, he excretes one pint or more in unnoticed perspiration, and may excrete several times this amount in conditions of heat and/or extreme exertion. Uncompensated loss of this kind can have serious effects. That the skin is a major organ in the excretion of *excess salt* can be appreciated from the salty taste of sweat. In this instance the skin is serving to regulate the balance of salts. A relatively minor shift in the proportion of, for example, sodium and potassium salts can have serious, even fatal, effects.

The skin functions in the *regulation of body heat.* As is well known, a rise of a few degrees in body temperature causes fever and delirium; on the other hand, lowered temperature causes intense drowsiness, as experienced by persons overcome in winter weather, and can even be used as an anesthetic in some types of surgery. The skin helps to dissipate heat, first by the expanding of the rich web of capillary vessels near its surface, which act as a radiator, and second by the outpouring of sweat, which absorbs heat when it evaporates. Heat is conserved by shutting down the vessels and sweat glands.

The integument is obviously a major *sense organ.* Extending nearly two square yards in a large person, its every square inch contains dozens to hundreds of tiny sense organs to detect light touch, pressure, heat, cold, and painful injuries. Even the apparently useless tiny hairs, found over most of our bodies, are in reality delicate touch-detectors; and our nails act as "back-stops" for fine pressure-detection in fingers and toes. Loss of this widespread sensation (actually several sensations) would be very disabling. When such loss is induced experimentally, it causes mental disorganization.

Among *minor activities* of this versatile organ are tanning to protect us from excessive light; callous formation to counteract friction; and excretion of small amounts of a few substances besides salts.

The Structure of the Skin

The skin is composed of three layers (Fig. 39–1). The outer layer or *epidermis* is composed of many tiers of flattened cells which are continually replaced at the bottommost layer and lost at the most superficial layer. As the cells move to more superficial positions they become impregnated with a waterproofing material called keratin, become transparent, and die. Simple rubbing or washing of the skin removes patches of such dead cells. They seem to have a sticky nature when in water which causes them to adhere to glossy surfaces such as to the wall of the bathtub. The sheets of "skin" which we find peeling off our arms following a sunburn are actually several layers of such dead cells. There are no nerves or blood vessels in the epidermis so that injuries restricted to this layer cause no bleeding or pain.

The *dermis* is a much thicker layer and much tougher. It is the "skin" from which leather is made, and is visible only after the epidermis has been scraped off. Here are located the glands, nerves, and blood vessels of the skin as well as the tough connective tissue which gives skin its strength. This connective tissue consists of strong *collagenous* fibers, similar to those of tendons and ligaments, as well as a quantity of *elastic fibers.* These elastic fibers are responsible for the smoothness of the skin regardless of whether it is stretched or relaxed.

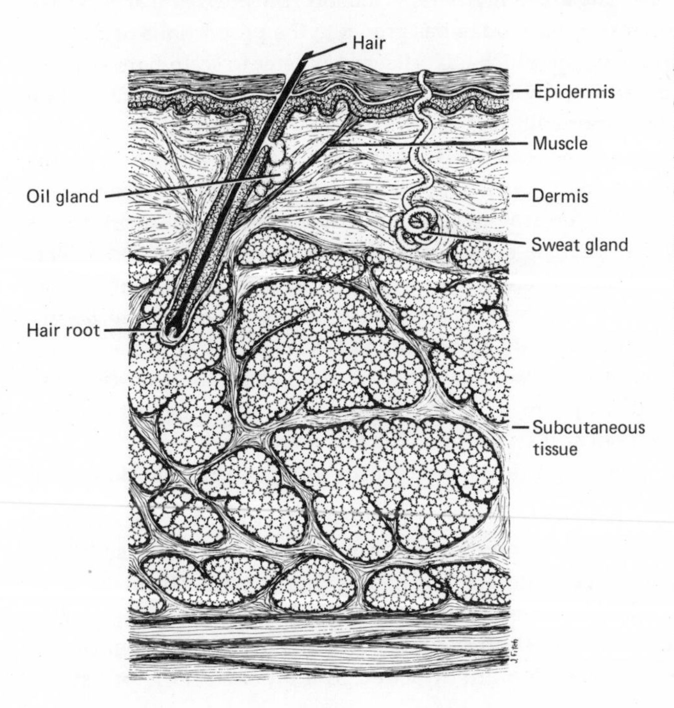

FIG. 39–1 Microscopic section of the skin. Skin includes the epidermis, dermis, and associated structures (hairs, sweat glands, oil or sebaceous glands) derived from the epidermis. Subcutaneous tissue is a loose fatty layer which allows free movement of most parts of the skin over the underlying structures, usually bone or muscle.

The elastic fibers also cause the normal separation of the cut surfaces of a wound, and necessitate the use of sutures in cases of injuries deep into the dermis. Aging of the skin includes a gradual failure of these elastic fibers and the skin wrinkling we associate with advancing age.

Underlying the dermis is a layer of *subcutaneous tissue.* Anyone who has skinned a rabbit or squirrel (or any mammal, for that matter) will remember the rather loose and liquid tissue which forms the layer of separation between the "skin" and the muscle, and can visualize such a layer in himself. This loose, often fatty, layer makes possible the movement of the skin over the underlying muscle in such areas as the back of the forearm. In certain areas such as the small of the back the subcutaneous tissue measures two or more inches in thickness even in thin persons. In general, however, thin people have little fat in the subcutaneous layer, whereas most of the "fat" we associate with obesity is deposited here. In this sense, all of us, fat or thin, are pretty much alike once the skin is removed. Special deposits of fat in the breasts, hips, and knees give distinctive contours to the female form.

Special Structures of the Skin

> Man, while regarding his hair as an ornament, has attributed unnatural qualities to it from time immemorial. The Bible, mythology, folklore, the poets and artists have all been hair-minded. Interest in hair today has grown to the proportions of a fetish. Think of the many loving ways in which advertisements refer to scalp hair—satiny, glowing, shimmering, breathing, living. Living indeed! It is as dead as rope. In a society which is swept by a hair cult, it is remarkable that men and women should pamper it in so many ways on the scalp, and yet attack it savagely elsewhere on the body surfaces to discourage its growth. The quintessence of femininity demands a totally naked skin except for the scalp, eyebrows, and eyelashes. All other hair is unwanted, and the fastidious female must shave scrupulously the shins and axillae, yet she may not shave her face. There she must use other devices. In contrast, the male may shave his chin, but not the axillae because this is considered to be unmanly. Is it not ironical that men should have chosen to shave their beard, the growth of which is dependent upon their sex hormones, at the same time that they encourage the growth of their scalp hair which becomes progressively anemic through the action of the same hormones? It is also amusing that man should regard the daily nuisance of shaving his chin a significant ritual of his virility. [William Montagna]

Although hair originally develops as a part of the epidermis, its root eventually grows down to lie deep within the dermis and, in the case of coarse hairs, even into the subcutaneous layer. But its origin is evident when the lining of the *follicle* is examined and seen to be composed of epidermal cells similar to those on the skin surface. The hair shaft is elaborated at the base or *root* of the follicle. Some cells of the hair root remain attached to a plucked hair, but since a few cells remain at the base of the follicle, a new hair usually replaces the extracted one.

There is a small muscle attached to each follicle. When the muscle contracts it raises the whole follicle closer to the skin surface, and the familiar "goose bump" results. In some areas of the body where the hairs are very inconspicuous, such as the back, the "goose flesh" which accompanies shivering is the most obvious evidence of the presence of hair follicles.

Hair Pigmentation and Graying

The color of hair depends upon the kind and amount of pigment deposited in the shaft of the hair by the root cells. If little or no pigment is present, the hair appears white. A mixture of white and pigmented hairs give what we call "gray" hair. (True gray hair exists but is rare.) The pigment substance melanin is responsible for black and brown hair, as well as the color of our skin. A different pigment is probably responsible for red hair.

Hair turns white when the root cells stop producing pigment. As the new hair grows in it will be unpigmented (white), but that part already pigmented will not change color. The widespread belief that pigmented hair can turn white overnight as a result of an extremely disturbing emotional experience is probably false. (Marie Antoinette's hair is said to have turned gray the night before her execution. It has been suggested that since the hair of those about to be executed is usually washed, the overnight change was not as real as it appeared.) Such a phenomenon would involve the formation of a substance by the root which would penetrate the hair shaft and bleach away the pigment. No such substance is known. But a scientist is always ready to change his thinking when new evidence is available. For example, illness can quickly affect the state of a dog's coat, and further the protein of the dead hair is in continuity with living protein and possibly could be affected by metabolic changes of the body.

Sebaceous Glands

From one to several sebaceous glands are usually associated with each hair. These glands secrete a fatty substance called *sebum* which serves to oil the hair and lubricate the skin. The sebum provides a protective coating which prevents excessive drying and chapping of the skin.

Sweat Glands

There are two types of sweat glands. The *eccrine sweat glands* are of greater physiological significance, and are found in the skin of almost all parts of the body. They are seen in the dermis as densely coiled tubes leading into ducts which take a sinuous course to the surface, and open at the summit of the epidermal ridges ("fingerprints"). They are most numerous on the palms where they have been estimated to number 3000 per square inch.

These glands provide our principal method of temperature control and are responsible for the profuse sweating which occurs in strenuous exercise and in warm surroundings. In the palms and soles they also secrete in response to emotional stresses (as during examinations).

The *apocrine sweat glands* are rather intermediate between sebaceous glands and eccrine sweat glands, and do not become active until puberty. They are much larger and are confined to the axillary genital and anal areas. Apocrine sweat glands are not involved in temperature control, but release their secretions in response to emotional stimuli.

Perspiration and Body Odor

Perspiration has always been associated with body odor, but the relationship is just beginning to be understood. The major part of the odor is apparently due to the products of bacterial action on the secretions of the skin glands rather than to the secretions themselves.

The nature of the secretions and their liability to bacterial action vary in the different types of skin glands. The oily secretion of the sebaceous glands is rich in organic material and provides a fertile source of bacterial action. This secretion cannot be suppressed by any locally applied antiperspirant, and if it is allowed to accumulate on the skin for periods of weeks or months can certainly be a source of strong odor. However, the secretion of the sebaceous glands seldom presents an odor problem among those who bathe regularly.

The eccrine sweat glands are of little importance in body odor; they produce a secretion which is practically free of organic material upon which the bacteria can act. The dancing instructor in the television commercial who tells us she has a particularly serious perspiration problem because of her physical activity may indeed sweat profusely, but she should have no greater odor problems than the rest of us. These are the only functional sweat glands present before puberty, and although sweaty and grimy children may not stand close olfactory inspection they do not have a "perspiration problem."

However, the large apocrine sweat glands which develop in association with axillary and pubic hair at the time of puberty do produce a secretion with an odor, but there seems to be some disagreement among authorities as to whether the secretion of these glands has odor in the absence of bacteria. The evidence suggests that under certain conditions a natural odor is present. This is in keeping with the importance of these glands in the mating behavior of other mammals in which the secretion functions in sexual arousal. Whether the normal human male is sexually aroused by the apocrine gland secretion of the human female apparently has not been investigated, but it is certainly not given much importance in our society where the general awareness of sexual stimuli is promoted. On the contrary these "natural" odors are usually masked by a wide variety of perfumes which purport to have this function.

There is no doubt, however, that bacteria find plenty of food in the richly organic secretion of these glands, and when they are done with the secretion, there is no debate about an odor. The apocrine glands release their secretions under conditions of emotional stress and during sexual arousal. They do *not* secrete because of warm surroundings or during exercise, and unless our dancing instructor has emotional problems while she is working, they should be quiescent.

The principal problem, then in "underarm" or axillary odor is that of bacterial action on the apocrine secretion. For some people this problem is solved simply by the removal of the secretions at regular intervals with soap and warm water. Others, particularly young people of the acne age, have intensely unpleasant cutaneous odors despite very scrupulous cleanliness. For these people and their associates there is very definite benefit to be derived from deodorants and antiperspirants.

Deodorants and antiperspirants differ in that deodorants contain no drugs and are regarded as cosmetics, so no labeling of chemical contents is required. Deodorants usually contain some sort of fragrance as well as an antiseptic which kills bacteria. Antiperspirants usually contain these same substances as well as some sort of aluminum salt which has the effect of repressing apocrine sweat formation. How the aluminum salts exert this effect is unknown, but they have been shown to be generally harmless to the skin. Nevertheless, any suggestion of undesirable side-effects should be reported to a physician.

Sun, Suntanning, and Skin Aging

We noted in Chapter 14 that the ultraviolet rays of the sun are responsible for the production of calciferol (a form of vitamin D) by the skin. It is also responsible for suntanning and sunburn.

To understand these and other effects of ultraviolet light, we must first consider briefly the nature of light or more properly, radiation. The quantum theory holds that energy is absorbed or emitted as packets called quanta. The quanta can be visualized as moving in oscillating waves. The number of oscillations per second is the frequency of the radiation. The amount of energy contained in a quantum is directly proportional to the frequency of the radiation—the higher the frequency, the greater the energy and also the shorter the wavelength (that is, the greater number of oscillations which can be packed into a quantum).

Waves of different length compose the electromagnetic spectrum. At one end are the extremely short, very high energy waves—cosmic rays and gamma rays; at the other end are the very long, low energy waves—the radio waves which carry radio and television signals. In between lies the visible light of the spectrum.

The waves at either end of the spectrum are capable of penetrating the human body. The high energy cosmic rays emit such a high level of energy that they cause death after a few seconds exposure. However, they are effectively filtered out by our atmosphere, and are of concern principally in regard to space travel. The gamma rays, which include the X-rays, are also fatal, but only after more prolonged exposure. The very long radio waves (up to miles in length) emit very little energy, and pass through the body without perceptible effect.

The rays of the visible spectrum penetrate human tissue very little or not at all. But the slightly shorter and invisible ultraviolet light with its higher energy level can penetrate skin to about 1/16 of an inch. When the ultraviolet light strikes the skin, part of its energy is absorbed by the outer horny layer which is little affected. But the small amount of ultraviolet light which reaches the deeper living cells triggers the physiological responses which result in the conversion of ergosterol to calciferol, and the production of melanin pigment from the amino acid tyrosine which results in tanning. If exposure is prolonged some of the energy is dissipated as heat and produces the redness and inflammation typical of sunburn, but the reactions leading to the production of new melanin do not begin until about 48 hours after exposure. So no tanning can be expected until two days after the first outing, and is not pronounced until several days after that.

Tanning can be regarded as a protective mechanism since it decreases the damage from later exposure to ultraviolet light. But, in a sense, it is also the first stage of tissue damage, which if continued results in a degenerative process leading to a coarsening of the skin, wrinkling, and hyperpigmentation ("liver spots"), characteristics typical of aging skin. In other words, the inevitable, but gradual degenerative changes which occur in the skin are greatly accelerated in those people who are excessively exposed to the sun. Thus farmers and sailors notoriously have rough, wrinkled skin. To make matters worse, the effects of the sun are cumulative, that is, the degree of skin aging is closely correlated with total lifetime exposure. This suggests that the small changes brought on by each exposure are irreversible. The same degenerative processes can also lead to skin cancer. In fact the incidence of skin cancer is highest among light-skinned persons whose living habits expose them to high levels of ultraviolet light.

Protection from the Sun

A little common sense can do a great deal to diminish the harmful effects of the sun. If you are after a suntan, do not be in a rush. The most effective way to tan is simply to go out every day and stay only the maximum safe period of time. This will generally not exceed 15 minutes on the first day. The length of time can be increased by about 10 minutes each day. It is best to start off by exposure to early morning or late afternoon sun—times when the atmosphere filters out a maximum of the ultraviolet light.

Ultraviolet radiation is not absorbed much by fog or haze, so overcast days are just as effective in producing suntanning (and burning) as are clear days. Water and sand reflect radiation and are able to tan or burn people even under awnings. This should be taken into account when planning daily dosage.

Suntan lotions give protection by screening out some of the ultraviolet light before it reaches the skin. So do not massage them *into* the skin. None of these preparations accelerate tanning, they merely offer protection against overexposure.

Sunglasses should be worn by anyone who spends a great deal of time in the sun. Even brief, direct staring at the sun, as at the time of an eclipse, may do serious damage to the retina. Any glass, colored or clear, will filter out the ultraviolet light just as a clear window pane will, but colored lenses which filter out at least 80 percent of the total light are most beneficial. Since it is difficult to make such measurements, one can play safe by buying the polaroid type sunglasses. They are inexpensive and their effectiveness is long established.

Problems and Disorders of the Skin

Hair and Scalp

Epidermis, in general, is quite impermeable to most externally applied substances. Even more so is the epidermis of the hair follicles, and any efforts to treat the hair by locally applied ointments or hair tonics are futile. In fact, the only demonstrably effective agents which control hair growth are certain hormones which are responsible for the secondary sexual characteristics. Generally, disorders of the hair are likely to be due to disturbances of the body metabolism, and not to any particular condition of the skin itself.

Dandruff

Ordinary dandruff is the result of the normal flaking off of the outer layers of epidermis, and there is no evidence that any infectious agent is involved. So there exists no condition which can be "cured." The weekly washing of the scalp and hair with any simple shampoo is sufficient to handle most dandruff problems, and more frequent washing of the hair is apparently without any harmful effect. The standard treatments for more severe cases of dandruff are preparations containing salicylic acid. The salicylic acid softens dead skin and causes it to break away from the deeper living cells and, in the case of dandruff, hastens the normal flaking process. The normal procedure is to rub such preparations into the scalp at night, and to wash them out the following morning with a simple liquid soap shampoo. Treatments involving a change of diet, or a change in exposure to sunlight have never been shown to be effective. Everything that looks like dandruff may not be. In a few cases, white particles from the scalp are due to fungus infections and should be medically treated.

Baldness

Baldness (alopecia) in the male is most probably an affliction without effective treatment. The tendency to become bald is inherited, although the direct cause is the male hormone testosterone. Those individuals with such an inherited disposition begin to lose the hair on the scalp as soon as testosterone begins to be secreted, and no treatment short of castration will effectively arrest the depilation. Paradoxically, the lush growth of facial hair is developing at the same time. Perhaps it is a case of too much of a good thing: the underdeveloped, juvenile facial hair responds to the male hormone by an increased growth rate, a darker pigmentation and a coarser texture, while the fully developed hair of the scalp "burns itself out" under the same stimulation.

Hereditary baldness in women is much less common since they must inherit two genes (and be homozygous) for the trait rather than one gene (and be heterozygous) as in the case of males. Women do apparently suffer from other causes of hair loss, however. The decrease of estrogens following menopause seems to promote hair loss; and so also do certain types of shampoos that contain detergents and foaming agents. Fortunately, wigs are now more generally acceptable, and a reasonable solution to the problem is now available to both sexes.

Removal of Hair

Certainly the safest and most convenient method of removing unwanted hair is by use of a safety razor or an electric razor. The belief that shaving stimulates hair growth, stiffens hairs, or changes the hair texture is based on the fact that such changes often accompany shaving. However, during puberty these changes develop in the hair whether one shaves or not, and the two are unrelated since increasing coarseness of hair is natural with progressing age.

The use of depilatories is generally not worth the risk involved. All chemical depilatories contain a strong chemical agent which is able to dissolve the hair shaft; these are sometimes sulfides (the rotton egg odor), thioglycolates, or calcium hydroxide. In any case, they are all corrosive and can be irritative if left on the skin longer than the specified time. Since there is a wide variation in the sensitivity to such agents, some people will develop a reddening and inflammation of the skin before the hair shaft has dissolved. If such preparations are accidentally introduced into the eye a serious injury can result. In addition, some people develop an allergic reaction to certain chemicals of this nature. And, of course, the depilation is only temporary, since the hair root is unaffected by the treatment.

The only technique which offers permanent removal of superfluous hair is electrolysis. In this procedure an electric needle is passed into the hair follicle, and the hair root is destroyed by an electric current. The technique is very tedious and time-consuming, and generally requires considerable skill. Careless or inexperienced operators can unwittingly cause bleeding, infection, or scarring. Any person contemplating electrolysis should consult a dermatologist (a physician who specializes in disorders of the skin). He will either provide this service in his office or recommend a competent operator.

Acne

Acne is a disorder of the sebaceous glands in which the duct of the gland becomes plugged with sebum and then infected. The "pimples" or pustules are the result of the action of the white cells of the blood on the sebum and bacteria. A blackhead or whitehead is simply a clogged pore of a sebaceous gland without infection (Figs. 39–2 and 39–3).

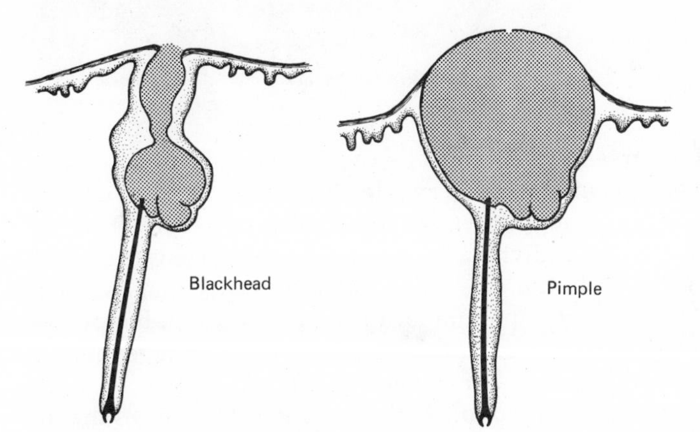

FIG. 39–2 Disorder of the sebaceous glands. A *blackhead* is a hair follicle congested with an abnormally pastelike sebum. A *pimple,* or *comedone,* is usually the result of an infection of the pasty contents of a blackhead resulting in an inflammatory reaction.

Here again we have a condition in which the skin shows symptoms of a more general body disturbance. In this case the fault seems to lie with the sexual hormones, testosterone in the male and progesterone in the female. In adolescence, when these hormones begin to be secreted, the previously quiescent sebaceous glands become active. Some of the glands, particularly those in the forehead and facial regions, seem to respond by producing a modified secretion which tends to solidify within the duct. These may become pus-filled without the intervening formation of the typical blackhead. Why some glands respond in this way while others produce the normal oily secretion is unknown, and in spite of the fact that the condition almost invariably clears up by the

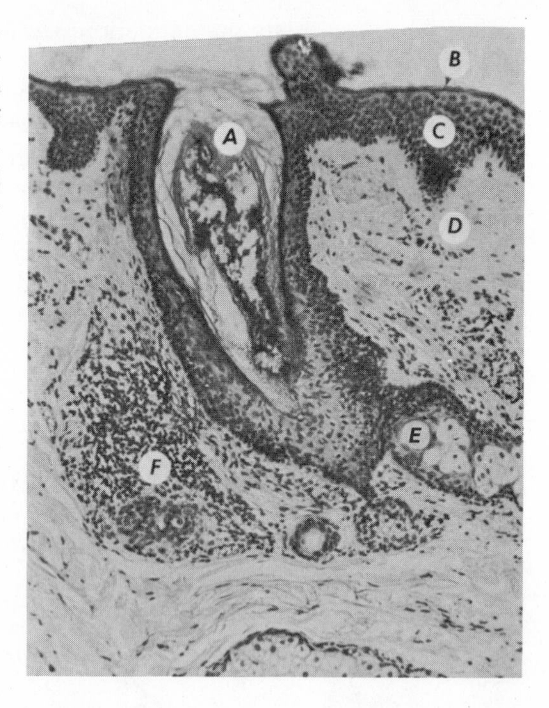

FIG. 39–3 A photograph of a skin section showing: (A) a blackhead in a hair follicle (pore); (B) superficial horny layer of epidermis; (C) living cells of epidermis; (D) dermis; (E) oil (sebaceous) gland; (F) collection of white blood cells, indicating the presence of infection. (Indiana University Medical Center)

age of 23 (with or without treatment), it must be considered a major adolescent disease. In fact many physicians regard it as a leading example of an organic disease which causes serious emotional and personality problems. It occurs at precisely the time when the individual is beginning to be concerned with personal appearance. Later on in life most of us begin to realize that our facial attractiveness is not a particularly important factor in our success or happiness. But a teenager believes himself to be, and perhaps is, critically dependent in many situations upon his personal attractiveness. At such times an acute flareup of acne can bring about an emotional tailspin, and reassurances that he will grow out of the condition are not only futile but aggravating.

In spite of the fact that acne is difficult to treat there are recommendations which have rather general application.

1. First is cleanliness: the face, neck, and shoulders should be scrubbed vigorously twice a day.
2. Exposure to some source of ultraviolet light is usually helpful. This may be either direct sunlight or a sunlamp. However, a timer should always be used with sunlamps to avoid the dangers of overexposure.
3. Dandruff seems to aggravate acne (for reasons quite obscure), so frequent washing of the hair is recommended.
4. Noniodized salt should be used at the table. Most people will also find

it beneficial to cut down on sweets, especially chocolate, though again the reason is quite obscure.

5. Patent medicines which promise acne cures are a waste of money. This applies not only to tonics which are supposed to purify the blood, but also to special antibacterial ointments. These ointments frequently contain penicillin or sulfa drugs which are of little benefit in the treatment of acne and to which an allergy might develop.

Some Special Acne Treatments

Since vitamin A is essential for healthy skin and hair, some physicians have reasoned that large doses of this vitamin might have a beneficial effect on acne. However, such treatments have generally proved ineffective. X-ray treatments have given temporary improvement in many cases. However, even in expert hands such treatments are dangerous, and may result in scarring or blotching of the skin, even after many years. And our exposure to environmental irradiation seems to continue to rise making the treatment of a temporary disorder by such a drastic measure inadvisable.

Acne Scars

Scarring and pitting of the face can result from severe acne. Such scars can frequently be removed by the technique of dermabrasion or skin planing. However, such treatment in inexpert hands can cause some serious complications such as discolorations of the skin. Dermabrasion treatments should be sought only from physicians competent in the technique.

Burns

Burns are classfied according to the depth of the injury to the skin. In a *first-degree burn* a reddening of the skin occurs which is caused by a swelling of the blood capillaries lying in the dermis. In a *second-degree burn* the swollen blood capillaries are damaged to an extent that serum (blood fluid) escapes and accumulates between deeper layers of the epidermis, or between the dermis and epidermis, forming a blister. In a *third-degree burn* the epidermis and dermis and some of the subcutaneous tissue are destroyed. *Only first-degree and small second-degree burns should be treated at home.*

Severe or extensive burns can be dangerous because of the loss of fluid from the damaged blood vessels. If this fluid loss is great, "burn shock" results. It is generally held that severe burns over 25 percent or more of the body surface result in burn shock, and if untreated, will cause death. (One can figure that one leg, or the front of the trunk, or the entire back each represent about 18 percent of the body surface.)

The first-aid treatment for severe burns has undergone a considerable change since early in World War II. *It is no longer advised that petroleum jelly or other oily substance be applied to burns;* they have little value, and can seriously interfere with later professional care. This also applies to the proprietary burn ointments; they are universally condemned by medical authorities. First-aid treatment consists solely of placing the cleanest available cloth material over all burned areas to exclude air, and, if the victim is conscious, to encourage him to drink plenty of nonalcoholic liquids. *If the skin is not broken, the burned area should be immersed in cold water* or *covered with cold compresses for at least 30 minutes.* This will not only relieve the pain, but also constitutes the most effective home treatment.

Contact Dermatitis

Poison Ivy

The dermatitis following contact with poison ivy is an allergic reaction. As is the case with all allergies, previous exposure to the toxic agent—in this case, the plant—is required for any reaction to develop. For this reason a person may believe he has a natural immunity when he fails to come down with symptoms after an initial exposure; a later exposure can well result in an explosive reaction. However, some persons who spend a good deal of time in the outdoors do develop a certain degree of resistance.

The characteristic blisters usually develop one or two days after contact. Oozing of the blister fluid is followed by crusting and scaling, and a temporary thickening of the skin. The symptoms generally persist for two to three weeks, and even longer if the individual indulges in much scratching. However, contrary to common belief, the blister fluid does not spread the reaction.

Among the remedies commonly encountered are kerosene, iodine, bromine, cream, gun powder, marshmallows, and numerous plant extracts. They are all equally worthless. More recent remedies such as antihistamines, antibiotics, and corticosteroid ointments are no better. Only corticosteroid hormones, taken by mouth or by injection, are known to be effective, *and such treatment must be under a physician's supervision.*

Preparations designed to prevent skin reactions, by application either before or immediately after exposure, are also disappointing. Thorough scrubbing with a strong soap within minutes after exposure can be beneficial, but if delayed for a half hour or more is ineffectual. Barrier creams, including those with silicone, give no practical protection. In summary, then, the best preventative measure is to learn to recognize poison ivy and to avoid it.

Cosmetics and Skin Medications

The accessibility of the skin makes it a constant temptation for self-treatment, and even over-treatment. In fact the over-treatment of the skin in cases of minor cuts, burns and irritations, as well as the general abuses it sometimes suffers in cosmetic decoration, are responsible for a good deal of the common dermatitis that is seen by the physician. This situation is aggravated by the pressure exerted on the public by the advertising industry in behalf of the proprietary drug manufacturers and their 2000 or so itch remedies, analgesic balms, antiseptics, athlete's foot remedies, corn remedies, dandruff removers, first aid creams, hormone creams, wart removers, sunburn remedies, and other substances which people merrily and often habitually apply to their skin.

Many of these substances are useful remedies, but then again some are worthless, and most of them, cosmetics included, can cause allergic or inflammatory reactions in some people. For this reason no such substances should be applied to the face or large areas of the skin without a preliminary test on some small inconspicuous area first. Such applications should probably be left on overnight and examined in the morning for signs of irritation.

In many respects the skin is an accurate reflection of one's state of health and even state of mind. Skin blemishes and rashes can develop from emotional pressures as well as from diseases and metabolic disturbances. For this reason specific skin problems which persist or recur deserve the attention of a physician.

REVIEW QUESTIONS

1. People refer to the "pores" of their skin. What are these "pores?"
2. What is the source of the "perspiration odor?" Does this odor have any functional significance?
3. What are some of the effects of sunlight on the skin?
4. Why are skin ointments likely to be ineffective in treating acne? What are some general recommendations that are helpful in reducing the severity of the condition?
5. Can poison ivy dermatitis be spread by scratching?
6. What is the use of salicylic acid in the treatment of skin disorders?

REFERENCES

ARTZ, CURTIS P., and JOHN A. MONCRIEF, *Treatment of Burns*, 2nd ed. Philadelphia: W. B. Saunders, 1968.

BOBROFF, ARTHUR, *Acne and Related Disorders of Complexion and Scalp*. Springfield, Ill.: Charles C Thomas, 1964.

DOMONKOS, ANTHONY N., *Andrew's Diseases of the Skin*, 6th ed. Philadelphia: W. B. Saunders, 1971.

FERRIMAN, DAVID, *Human Hair in Health and Disease*. Springfield, Ill.: Charles C Thomas, 1971.

FREEMAN, ROBERT G., and JOHN M. KNOX, "Skin cancer and the sun," *Ca—A Cancer Journal for Clinicians*, **17:** 231–238 (1967).

KLIGMAN, A. M., "Early destructive effects of sunlight on human skin," *Journal of the American Medical Association,* **210:** 2377–2380 (1969).

LAMPE, KENNETH F., and R. FAGERSTROM, *Plant Toxicity and Dermatitis.* Baltimore: Williams & Wilkins, 1968.

MONTAGNA, WILLIAM, "Phylogenetic significance of the skin of man," *Archives of Dermatology,* **88:** 53–71 (1963).

WILLIS, ISAAC, "Sunlight and the skin," *Journal of the American Medical Association,* **217:** 1088–1093 (1971).

Additional Readings

BRAUER, EARLE W., *Your Skin and Hair.* New York: Macmillan, 1969.

CHAMPION, R. H. et al., *An Introduction to the Biology of the Skin.* Philadelphia: F. A. Davis, 1970.

CONSUMER'S UNION, *The Medicine Show,* 2nd ed., 1963.

PATHAK, M. A. et al., "Evaluation of topical agents that prevent sunburn," *New England Journal of Medicine,* **280:** 1459–1463 (1969).

ROSS, RUSSELL, "Wound healing," *Scientific American,* **221:** 40–50 (1969).

WELLS, F. V., and IRWIN I. LUBOVE, *Cosmetics and the Skin.* New York: Reinhold, 1964.

PART TEN
ENVIRONMENT AND HEALTH

40
POPULATION AND THE POLLUTION PROBLEM

Today we are like men coming out of a coal mine who suddenly begin to hear the rock rumbling, but who have also begun to see a little square of light at the end of the tunnel. Against this background, I am an optimist—in that I want to insist that there is a square of light and that it is worth trying to get to. I think what we must do is to start running as fast as possible toward that light, working to increase the probability of our survival through the next decade by some measurable amount.

For the light at the end of the tunnel is very bright indeed. If we can only devise new mechanisms to help us survive this round of terrible crises, we have a chance of moving into a new world of incredible potentialities for all mankind. But if we cannot get through this next decade, we may never reach it. [John Platt].

What are these "terrible crises" which might mean the end of civilization? They include the rapid deterioration of our environment, the threat of nuclear war, and the mounting chaos of our cities. All these and more are largely traceable to a single driving force—the exponential growth of the world's population.

John Platt is not alone as a prophet of impending doom. His views are in accord with those of practically all the scientists whose research is concerned with population dynamics, air and water pollution, consumption of natural resources, and agricultural production. These are not sensation-seeking journalists writing for the Sunday newspaper magazines, but our leading research scientists whose job it is to understand nature and man's relation to his environment. Distinguished scientists are accustomed by training to expressing themselves in cautious terms. Yet discoveries of the past two decades have revealed problems which seem either unsolvable or at best require crash programs utilizing the best talents available.

It is the purpose of this brief account to summarize the research findings which have so alarmed scientists, to relate the research findings to our future health problems and to indicate the nature of remedial courses of action that have been proposed.

The Population Problem

Experts and laymen alike can legitimately disagree on how many people are an optimum number. Many believe that the present population of well over three billion is more than the earth's resources can comfortably support if there are to be significant areas set aside for recreation and wildlife. Others believe that nonproductive areas, that is, areas which do not contribute directly to man's physiological needs, are wasted, and that considerably more people are desirable. The maximum possible population supportable by the world is a subject of current speculation. The highest estimate we have encountered in scientific journals is 120 billion. Support of this population, however, would entail land and water utilization stretched to their theoretical limits. Yale University ecologist Edward Deevey, Jr., points out that man would have to displace all meat-producing animals and utilize all vegetation that could be supported. No land that now supports greenery could be spared, and the population would have to reside in the polar regions, or on artificial oceanic islands, surrounded by an ocean scummed over by 10 inches of cultured algae. At the present rate of population increase the world would reach 120 billion people in about 300 years. Although social scientists are generally more sanguine about maximum supportable populations than are life scientists, even they make projections which fall far short of the theoretical limit of 120 billion.

A report by the National Academy of Sciences [Resources and Man] projects that foreseeable increases in world food supplies are not likely to exceed about nine times the amount now available. This approaches a limit that would place the earth's carrying capacity at about 30 billion people, at a level of chronic near-starvation for the great majority. At the present rate of world population increase there could be 30 billion people in the world by the year 2075 (a year your grandchildren would likely live to see). This report holds out some basis for hope that the world population *may* level off not far above 10 billion people by about 2050 which they believe is "close to (if not above) the maximum that an *intensively managed* world might hope to support with some degree of comfort and individual choice . . ."

What we are referring to as the "present rate of population increase" can best be explained in terms of "doubling time." Ancient world populations are difficult to estimate. But written records enable us to compute the world population in 1650 to have been roughly 500 million. This doubled 200 years later to one billion. The next doubling time was 80 years giving a population in 1930 of about two billion. The doubling time continues to decrease so that the world population for the year 2000 is projected at seven billion! Doubling times vary in different parts of the world with the population increases slowest in the developed countries and fastest in the underdeveloped countries. At present the doubling time for the United States is about 63 years. It is even slower in some Western European countries: Austria, 175 years; Britain, 140 years; Italy, 117 years; Denmark and Norway, 88 years. But figures for the underdeveloped countries range from about 20 to 35 years: Costa Rica, 17 years; Philippines, 20 years; Brazil, 22 years; Turkey, 24 years; Nigeria, 28 years. One should pause to reflect on the effects of a doubling of this country's population in 25 years. For the United States, increased food production could easily keep pace, but think of the need for twice as many schools, twice the amount of electric power, twice the amount of petroleum burned, twice the number of doctors and nurses, and a doubling of the transportation facilities. It is doubtful that the United States could achieve this. In countries such as India, far behind now in all the essential services, it is beyond a reasonable possibility. As Paul Ehrlich points out, the people of the underdeveloped countries now know of the better life that is possible. They have seen pictures of American life—know of automobiles, airplanes, refrigerators, and TV sets. They want these things and expect to have them; but with a doubling of their populations, they can hardly expect to even maintain their present standard of living.

You might reasonably ask why these shorter and shorter doubling times have developed, and what factors might slow them down. Demographers have calculated that populations remain stable when couples produce, on an average, 2.1 children *who live to adulthood.* In almost all civilizations couples have produced far more children than this average. But in order to raise two children to adulthood just a few generations ago, a woman would often have had to bear ten or twelve children. Birth control was a negligible factor, but reproductive control was present through the high mortality rates among children. When death rates were drastically reduced in just two or three generations by better medical care, improved sanitation, and control of infectious diseases, there was an exponential rise in the size of the reproducing populations. As more and more children survive, the average age of the population goes down, and a greater proportion are within the breeding age. This effect is well-illustrated in Costa Rica which has a doubling time of 17 years. In 1966, 50 percent of the people in Costa Rica were under 15 years old. If we assume that these children will marry and reproduce in the pattern of their parents, and all indications are that they will, a doubling in population in 17 years is not surprising. In the underdeveloped world as a whole, about 40 percent of the population is under 15 years old. It is the reproduction potential of these children which can push the population of the world to over seven billion in the year 2000.

The cause, then, for the increased rate of population growth is a greater and greater control over death, but with no control over birth. Whether one opts for a population of three billion, 10 billion, or the limits of the earth's ability to support human life, or 120 billion, a stabilization will eventually occur. This stabilization of population can be achieved in one of two ways: a birth rate solution in which we in some way decrease the number of births; or a death rate solution in which more people die sooner—from war, disease, or starvation.

However, all such population possibilities are based upon the vital assumption that increases in food production can keep pace with the population increases. John Platt, Paul Ehrlich, agricultural experts William and Paul Paddock, and numerous others contend that world food production will fall far behind by 1975–1977. Such schemes as farming the oceans and irrigating the deserts have been advanced as solutions, but such projects require far more time than is available, even with emergency programs. The so-called green revolution greatly increased agricultural yields in some underdeveloped countries through new seeds, fertilizers, and irrigation but has not kept pace with population growths. Dr. Raymond Ewell (cited by Paddock and Paddock), one of our foremost agricultural scientists, answers the often asked question: why, if the United States can send a man to the moon cannot it help India, Nigeria, and Brazil to improve their agriculture.

> The answer is that improving the agricultures of India, Nigeria, and Brazil is a more complicated problem than sending a man to the moon. The problem of sending a man to the moon can be solved by scientists and engineers using computers and the vast store of scientific knowledge now available. Improving agriculture in developing countries is more complicated because it involves people and education and social change—particularly since it involves 2 billion people.

Further documentation of growing food shortages would seem to be out of place in this book. There is widespread starvation in the world now. As a consequence, a growing part of our foreign aid program since 1958 has been food shipments. It was in 1958 that food production in the underdeveloped countries began to fall behind population increases. The discrepancy between population and available food continued to widen until 1966 when the world population increased 70 million with no compensatory increase in food production. Since then only seven countries have consistently produced more food than they consumed: United States, Canada, Australia, Argentina, Thailand, Rumania, and South Africa, with the United States producing more than half of the surplus [Ehrlich]. We are now shipping approximately 30 percent of our wheat production to India. Between 1967 and 1977 India's population is expected to increase by 200 million—a population almost equal to that of the United States. Is it even remotely possible that we shall be able to provide food for India and the other underdeveloped countries in the late 1970's—the time of acute crisis now being projected? The Paddock brothers regard the coming famines as so catastrophic that the United States will be compelled to choose between spreading our food surpluses throughout the world so thinly that they will have little impact, or to select certain countries which are saveable, and do what we can to help them.

America's Population Problem

In the first special message to Congress by an American President on population problems, President Nixon, on July 18, 1969, noted that "population growth is a world problem which no country can ignore, whether it is moved by the narrowest perception of national self-interest or the widest vision of a common humanity. For some time population growth has been seen as a problem for developing countries. Only recently has it come to be seen that pressing problems are also posed for advanced industrial countries when their populations increase at the rate that the United States, for example, must now anticipate."

President Nixon pointed out that America's population of 200 million will probably reach 300 million by the end of the century. "If we were to accommodate the full 100 million persons in new communities, we would have to build a new city of 250,000 persons (that is, the size of Tulsa, Oklahoma) each month from now until the end of the century."

The problem here at home is not so much what we shall feed these people,[1] but rather what shall we be able to do about housing, transportation facilities, and the industry which will be required. The prospect of a 50 percent increase in the number of automobiles presents enormous problems by itself.

Whether or not American industry and technology can rise to the challenge of providing the increased needs of more Americans is not so much the question as is what the effect of increased industrialization will have on our already deteriorated environment. The United States is not overpopulated in terms of people per square mile—at least when we compare this country with the Netherlands, Britain, or India; but in terms of consumption of natural resources and the pollution of the world's atmosphere and its water, we are the

[1] However, even here, there is some cause for concern. Meat will be increasingly expensive, and a large majority of us may become involuntary vegetarians. Obviously, it is more efficient and less expensive to eat corn than to feed the corn to an animal and eat the animal.

most overpopulated. With about 6 percent of the world's population, we consume close to half of the fuels and minerals removed from the earth. We produce an estimated 70 percent of the world's pollution in terms of nondegradable wastes poured into holes in the ground ("landfill"), into the air, the rivers, and the ocean [Ehrlich and Ehrlich]. The pernicious effects of present and future pollution levels are overwhelming problems now facing this country.

Air Pollution

Our atmosphere is composed by volume (excluding water vapor) of approximately 21 percent oxygen, 78 percent nitrogen, 0.9 percent argon, and 0.03 percent carbon dioxide, with a number of other gases making up the small remainder. However, the evidence available to geochemists and biologists strongly suggests that the atmosphere of the earth at the time life began was principally composed of hydrogen, ammonia, and methane (the simplest carbon-hydrogen gas).[2] The theories on the origin of life need not concern us here; however the advent of photosynthesis had the effect of releasing large amounts of oxygen into the ocean, and from there into the atmosphere. Free molecular oxygen is highly reactive: Its reaction with methane gave rise to the carbon dioxide, and its reaction with ammonia to molecular nitrogen. The enormous accumulation of fossil fuels—petroleum and coal—represents carbon, from carbon dioxide, removed from the atmosphere and replaced by oxygen during several hundred million years of photosynthesis.

Our present atmosphere is, then, the result of living processes. If all fossil fuels were burned (oxidized) the carbon, in the form of carbon dioxide, would be returned to the atmosphere, and the atmospheric oxygen largely depleted. At the present time it appears that we are actually burning fossil fuels and releasing carbon dioxide into the atmosphere at a faster rate than the earth's plants, principally those in the surface waters of the oceans, are incorporating the carbon dioxide into new protoplasmic substances [Cole]. The carbon dioxide content of the atmosphere has increased about 15 percent over the past 100 years. Fortunately, there seems to be no firm evidence that our oxygen supply is diminishing.

However, an increasing carbon dioxide level in the atmosphere could affect the mean temperature of the earth by means of the "greenhouse effect." The sun easily warms a greenhouse on a sunny day because glass and carbon di-

[2] An example of such evidence is the atmosphere of other planets as determined by the analysis of the wavelength of light reflected from their surface. Jupiter's atmosphere consists largely of methane and ammonia, and the atmosphere of Saturn and Neptune are largely of methane. The atmospheres of these planets have changed since the beginning of the solar system, but have almost certainly changed much less than the earth's atmosphere.

oxide tend to pass visible light but absorb infrared radiation. The sun's radiation is passed through the greenhouse glass and is absorbed by the soil and plants and converted to infrared heat energy. Since the infrared rays cannot pass out through the glass, the heat is trapped and the interior of the greenhouse is warmed. The atmospheric carbon dioxide acts as a greenhouse enveloping the entire earth—allowing sunlight to reach the earth's surface but limiting the loss of heat into space. As the carbon dioxide content of the atmosphere increases the temperature of the earth is almost certain to rise.

The President's Council on Environmental Quality estimated that the carbon dioxide accumulated in the atmosphere by the end of the century might be sufficient to melt the antarctic ice cap in 400 to 4000 years—the sea rising four to 40 feet each century. At this rate most of the major cities of the world would be under water.

The colossal outpouring of industrial pollutants into the atmosphere, however, is far more a visible and, at present, a more real problem. We pour a total of 139 million tons of pollutants (more than our annual steel production) per year into the atmosphere [Ehrlich and Ehrlich]. This is 1300 pounds for each person. About 60 percent of this comes from automobiles. These pollutants include compounds of sulfur, carbon monoxide, lead, oxides of nitrogen, asbestos, and fluorides. These chemicals discolor and erode fabrics and metals, deface buildings, and decrease crop production to the extent that the federal government estimates property damage alone to equal $12 billion a year. Although pollutants were probably responsible for numerous deaths during smog "disasters" in Donora, Pennsylvania in 1948 and in New York City in 1963, it has not been proved that large numbers of people are dying from air pollutants. However, respiratory diseases are aggravated.

According to the President's Council on Environmental Quality:

> . . . air pollution contributes to the incidence of such chronic diseases as emphysema, bronchitis, and other respiratory ailments. Polluted air is also linked to higher mortality rates from other causes, including cancer and arteriosclerotic heart disease. Smokers living in polluted cities have a much higher rate of lung cancer than smokers in rural areas. . . . Air pollution has been linked to asthma, acute respiratory infections, allergies, and other ailments in children. Such childhood diseases may well underlie chronic ills developed in later life.

The psychological effects of air pollution are not measurable, but are no less real. Air pollution subverts the spirit and makes life in our cities a depressing existence. If the citizens of Chicago, St. Louis, New York, or Los Angeles had had two weeks to adapt to their polluted air rather than 40 years they probably would not have accepted it with so little complaint. The human race seems to be able to adjust itself to conditions we might at first think to be intolerable. Yet what is tolerable is not necessarily what is compatible with health, either mental or physical.

Water Pollution

Water pollution is a generally acknowledged fact. Our dismal position is spelled out in both the scientific journals and the popular press. The death of Lake Erie and the flow of fetid municipal and industrial filth that we call the Hudson River need no documentation. Similar examples are familiar to all us. Making the dubious assumption that an informed public is now demanding and is willing to pay for clean water leaves at issue the consideration of water as a natural resource. It is this latter point that we shall emphasize.

We derive our water from rainfall—a daily average of some 5000 billion gallons which fall upon the continental United States [Bradley]. That such a vast quantity could be significantly supplemented by desalted sea water is an often cited but questionable operation. Desalting plants are in operation or under construction in all parts of the world, but these plants are producing drinking water—not water for industrial or irrigation purposes—and are producing it at high cost. According to *Business Week*,[3] October 16, 1965, "so far, no one in the United States has been able to produce water for less than $1.00 per 1000 gallons in existing small plants." The cheapest desalted water is in oil-rich Kuwait where it costs sixty cents per 1000 gallons. Hugh Nicol calculated that irrigation with desalted water "at present efficiency requires burning of about half an inch of oil per acre to produce enough water to irrigate an acre or, say 40 tons of oil per year to feed one or two people."

At 60 cents per 1000 gallons, the cost of irrigating alfalfa fields, whose yield is used to feed beef cattle, would raise the price of beef an *additional* $1.25 per pound. This figure is based on the calculation that 20,000 gallons of water are required to bring 25 pounds of alfalfa—that consumed by one steer in one day—to maturity. It takes two years to raise a steer which would yield 700 pounds of meat. This does not include the cost of transporting the water from the seaside plant, loss of water from evaporation, the water the steer drinks, or the water used in the processing of the beef. It is clear that our technology is not ready to significantly supplement the water derived from rainfall—except for the production of modest amounts of drinking water and water for special irrigation projects close to the sea. Even these projects are 15 to 20 years away and assume the use of atomic energy.

Supposing then, that we are, for practical purposes, limited to water we derived from rainfall, what might be the number of persons supported by this available water? Dr. Charles C. Bradley made a number of calculations of water use of which our alfalfa and beef production figures are a part. He showed that a typical American daily diet represents crops and meat production which directly or indirectly consume 2500 gallons of water per day. He noted, however,

[3] "Desalted water builds up steam," October 16, 1965, p. 120.

that the water consumption per pound of meat is about 25 times that of a pound of vegetable. A similar ratio would apply to wool in comparison to cotton.

Aside from water required for food production, enormous quantities are used in industry for fiber production, processing steel, and for processing lumber into newsprint and other papers. We use water to run washing machines, to flush toilets, and especially to sweep our sewage into the sea. When these are included, our per capita daily consumption of water totals 3500 gallons. Water use is so extensive that Dr. Bradley found it easier to subtract water we are not using from the water available to arrive at a figure representing our reserve.

As noted earlier, approximately 5000 billion gallons of water per day are available from rainfall. This is calculated on the basis of an average of 30 inches of annual rainfall over the continental United States. If we are to have water to drive our power dams, provide industry with its needs, and carry our sewage to the sea, we cannot use all of it to grow crops on desert lands. When these uses are accounted for, Dr. Bradley calculates that we have a total of about 750 billion gallons per day for future development. This would accommodate an additional 50 million people before our standard of living begins to fall, that is, a life with less hydroelectric power, less personal water consumption, less paper, or a diet with less meat.

Such projections are liable to considerable error, and many developments could occur in the next decade or two which might alter the basic data, for example, the replacement of hydroelectric power by nuclear power; but it is more likely that nuclear power will supplement rather than replace other power sources. The data seem valid enough, however, to justify a more prudent water-use program. New York City, for example, might have to meter water users and to make a real effort to clean up the Hudson River so that it could be a source of the city's water supply instead of building a seemingly endless number of upstate reservoirs which replace valuable farmland.

Efficient water use is but one aspect of our water problem. If our rivers and lakes are to be more than utility sources and serve their natural function of supporting a balanced system of water plants, microorganisms, and the normal variety of swimming creatures, then much restoration will have to be done. According to the Committee on Pollution of the National Academy of Sciences, within 20 years city wastes are expected to overwhelm the biology of most of the nation's waterways. We have lost Lake Erie—most scientists believe it is beyond restoration. The other Great Lakes, including even Lake Superior, are now in danger of a similar fate.

Many factors have contributed to the deterioration of our rivers and lakes. Industrial plants, such as paper mills, steel mills, and chemical plants, are located along waterways so that a convenient method of waste deposal is available. Both Federal and State Governments are moving toward tighter regulations of industrial pollution. But a factor of at least equal importance to industrial wastes are the effluents from sewage treatment plants. Lake and river waters receiving these wastes frequently show the phenomenon of *eutrophication* [Warren]. Eutrophication is caused by excessive amounts of nutrients which stimulate a rapid and overwhelming growth of plant life, particularly algae. When these plants die and sink to the bottom they are decomposed by the action of bacteria. This decomposition exhausts the oxygen in the water and causes the death of many kinds of fish.

Eventually, the lake becomes so choked with plant growth that it becomes a marsh, and land plants start to creep in and turn it into a meadow. All these stages can be seen in the lakes of northern Michigan, Wisconsin, and Minnesota, where thousands of years have been consumed in the process. When we discharge excessive amounts of nutrients into our lakes, we speed up the natural process, and lakes that might have taken several thousand years to decay take just a few years. Recreational lakes which have taken on a thick layer of algal growth—resembling green paint—are victims of this process.

A major class of nutrients leading to eutrophication are phosphates. Some phosphates come from human wastes, but principally they come from household detergents. Phosphates are added to detergents because they have a wetting action that allows for better penetration of fabrics. They also hold dirt in suspension so that the other cleaning agents work more efficiently.

Some detergents contain far more phosphate than others. Among laundry detergents the phosphate content may be as little as 8 percent by weight to as much as 52 percent. Some "water conditioners" are as much as 75 percent phosphate. Many local conservation groups are making available up-to-date determinations of phosphate content of household detergents so that the consumer can voluntarily restrict his phosphate use.

Another example of a serious water pollutant is nitrates. Small amounts of nitrate are naturally present in all bodies of water, but the nitrates that are now flowing out from sewage treatment plants and in runoff from land treated with chemical fertilizers have begun to accumulate in our waters in amounts approaching toxic levels. Concentrations of eight to nine parts per million can cause illness in infants, and five parts per million is considered unsafe in water used for livestock. Some wells in the United States now have up to 40 parts per million, and the concentration elsewhere is expected to go up with increased use of fertilizers and the growing density of the population. No effective method for removing nitrate from water has yet been developed [Commoner].

Similar problems are presented by increasing concentrations of insecticides, weed killers, nuclear wastes, and hundreds of other chemical agents we are pouring into our lakes and waterways.

The Water Quality Act of 1965 and The Clean Water Restoration Act of 1966 have provided more effective policing powers and more money for controlling water pollution. However, our expenditures will likely have to increase ten-fold to about $7 billion per year if a serious effort to clean up the nation's public waters is to be effective [National Academy of Sciences Committee on Pollution].

Most *knowledgeable* persons are in general agreement that current trends in population growth are not only undesirable, but are leading to disaster. Even the United States will be harassed in the next three decades by problems which will require our best technological personnel such as the many who are now working in our space program. But most important, and what most of us have failed to grasp, is that the rate of population increase in the United States is menacing to not only ourselves but to everyone in the world because of our disproportionate demands on the world environment. We might ask ourselves what the consequences would be if the other three billion people in the world were to achieve by industrial development our abundant life—if they were to lumber forests for paper as we do, to burn fossil fuels and pour pollutants into the air at the rate we do, and to consume their share of the earth's natural resources. As Harvard nutritionist Jean Mayer says, "The ecology of the earth—its streams, woods, animals—can accommodate itself better to a rising *poor* population than to a rising rich population."

The United States is the key to what needs to be done. Our leaders now officially recognize a population problem and are promoting birth control assistance through the United Nations and foreign aid. These programs are largely ineffectual, and if world population control is to be achieved at all, some difficult decisions will have to be made by leaders of those countries we are attempting to feed. Conversely, if one American baby causes more environmental destruction in his lifetime than 25 Indian babies, as appears to be the case [Ehrlich], these foreign countries can rightfully demand that we restrict our own population at the same time. As a consequence a significant movement has arisen proclaiming that this country is overpopulated (everyone knows what ZPG stands for). Population growth has been rapid over the past 25 years in spite of a recent decline in the birth rate. The Census Bureau's most conservative projection shows the next 25 years will see more than 50 million more Americans added to the population (Fig. 40–1). That projection, Series E of 1971, assumes that the average completed family in the decades to come will drop very close to a level of two children—less than it has ever been. Other projections, based on less conservative fertility assumptions, would make the increase even greater—up to a population of over 300 million by 1995.

The population will increase so greatly even with smaller family size because large numbers of young people are now coming into reproductive age (those born in the years of high birth rates, 1947 to 1963). This means a much greater number of childbearing families.

It is possible that fewer of these young people will elect to become parents and that those that do will have fewer children. Such a trend can be discerned by examining the recent changes in the fertility rate, that is, the number of

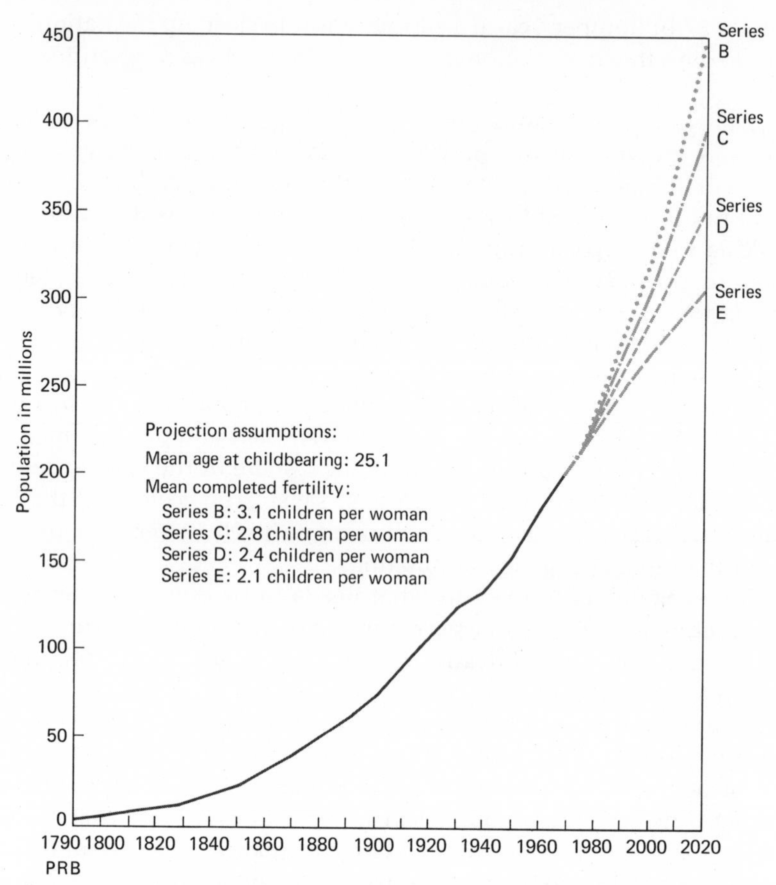

FIG. 40–1 United States population growth—past and projected. (Redrawn from "Where will the next 50 million Americans live?" *Population Bulletin,* Vol. 27, No. 5, October 1971. Population Reference Bureau, Washington, D. C.)

children born per 1000 women age 15 to 44 (Fig. 40–2). The fertility rate has been falling since 1958, and after a brief rise in 1969 to 1970 fell again and by early 1972 it reached a level lower than anytime in the Census Bureau's history. This dampening effect has probably resulted from a number of factors, including the cost of raising children, the poor job market for young people, changing life styles, and even concern about overpopulation. However, most demographers feel that the family size will not likely drop below the average of two children, a figure that will bring the population to a 50 million increase by the 1990s (Figs. 40–3 and 40–4).

In the meantime a number of measures could be taken to encourage the inclination toward small families. Paul Ehrlich makes the following suggestions:

First, revise our tax laws to discourage rather than encourage large families. It should be made clear that we do not desire that this legislation affect children already born. In fact, the entire purpose is to insure that those already living have a chance for a future. We presently allow $650 exemption for every child. After the second child, this should be discontinued, except, of course, in the instance of additional *adopted* children. For families below the poverty line, bonuses equivalent to welfare funds allocated for each additional birth should be given to a couple for each successive year that they do not have an additional child. This would offer an incentive for limiting family size; serve to elevate the standard of living for poor families; and save tax dollars since fewer children would be supported by welfare.

Second, a federal Department of Population and Environment should be set up with the power to take whatever steps are necessary to establish a reasonable population size in the United States and to put an end to the steady deterioration of our environment.

FIG. 40–2 General fertility rate, United States, 1915–1970. (Redrawn from "Where will the next 50 million Americans live?" *Population Bulletin*, Vol. 27, No. 5, October 1971. Population Reference Bureau, Washington, D. C.)

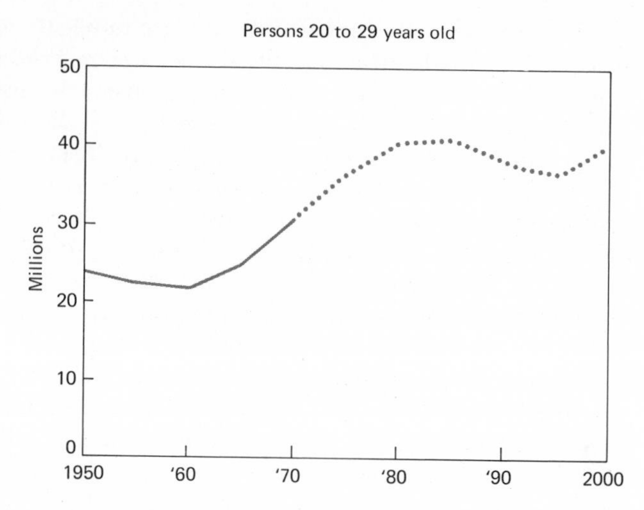

FIG. 40–3 Projected growth patterns, 2-child family, to year 2000. (Redrawn from "Population: The future is now." *Population Bulletin,* Vol. 28, No. 2, April 1972. Population Reference Bureau, Washington, D. C.)

Third, create a vast waste recovery industry, an industry that might well make "trash" obsolete. Reusable containers might be required by law for virtually all products.

Finally, change the pattern of federal support of biomedical research so that the majority of it goes into the broad areas of population regulation, environmental sciences, and behavioral sciences.

In addition to working for such changes in our national political and economic policies there are some things that we as individual citizens can do to curb pollution. Some are easy, some require effort, but all can be effective if enough people take an interest.

1. Whenever possible, buy products in returnable containers. This applies particularly to milk, carbonated drinks, and beer.
2. Do not burn trash or leaves. Learn about composting and other ways that natural products can be recycled.
3. Plant native trees and shrubs rather than exotic ornamentals. Wild animals are largely dependent upon the natural flora for survival.
4. Drive automobiles that have the least pollution potential. Emission control devices have been installed on all new cars since 1963, and improvements in these devices can be expected in the future. But their effectiveness depends upon periodic maintenance. Support the use of rapid transit systems at the expense of driving your own car. Walking to and from school or your place of business not only avoids damage to our ecology, but is also healthy.

5. Buy products in large packages or in bulk whenever possible. Avoid items packed in multiple packets, such as instant coffee and tea, individually wrapped aspirin or antacids, and many other so-called convenience items.
6. Use undyed toilet paper, paper napkins, paper towels, and paper tissues. Some of the dyes used make impossible the degradation of the paper by microorganisms.
7. Do not buy detergents with high-phosphate concentrations.
8. Do not buy clothing in which the skins or leathers of wild animals are used.

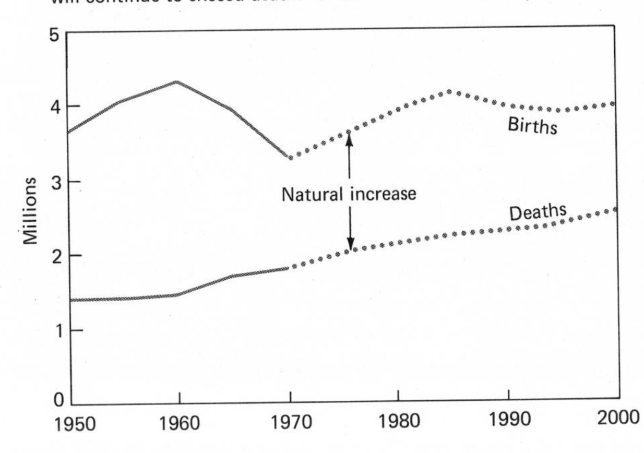

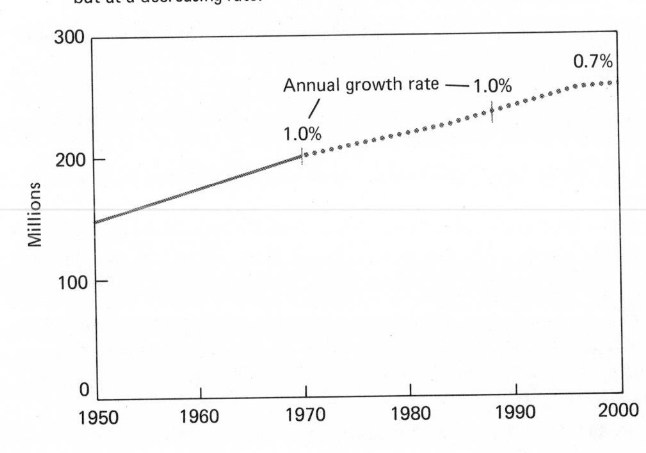

FIG. 40–4 (Redrawn from "Population: The future is now." *Population Bulletin*, Vol. 27, No. 5, October 1971. Population Reference Bureau, Washington, D. C.)

9. Use caution in applying pesticides or herbicides. Use only when essential and only as directed. Do not apply in windy weather.
10. If you own land, take measures to prevent soil erosion.

Man has been able to modify many natural phenomena which normally control the size of populations, yet he has not escaped the consequences of this control. No species of animal can multiply without limit. Ultimately, nature will check man's rapid increase in numbers—either through increased mortality or decreased fertility. Man has at his disposal the means to establish a balance with nature in which the human species will not only survive, but will survive in an environment in which life is worth living. The question is will man take the difficult actions essential to his future health and well-being. The answer cannot be long postponed.

REVIEW QUESTIONS

1. What are some of the causes of the present accelerated rate of population growth in the world?
2. Does the United States have a population problem?
3. What is the "greenhouse effect?"
4. Why is water regarded as a major limiting factor to population increase in this country?
5. What is eutrophication?
6. What are some measures you could undertake to reduce pollution?

REFERENCES

BRADLEY, C. C., "Human water needs and water use in America," *Science*, **138:** 489–491 (1962).

COLE, LAMONT C., "Man's ecosystem," *Bioscience*, **16:** 243–248 (1966).

COMMONER, BARRY, *Science and Survival*. New York: Viking Press, 1966. (Paperback)

COUNCIL ON ENVIRONMENTAL QUALITY, *Environmental Quality*. Washington, D. C.: U. S. Government Printing Office, 1970.

DEEVEY, EDWARD, "The human population," *Scientific American*, **203:** 194–204 (1960).

EHRLICH, PAUL R., *The Population Bomb*. New York: Ballantine, 1968. (Paperback)

EHRLICH, PAUL R., and ANNE H. EHRLICH, *Population Resources Environment: Issues in Human Ecology*. San Francisco: W. H. Freeman, 1970.

NATIONAL ACADEMY OF SCIENCES—NATIONAL RESEARCH COUNCIL COMMITTEE ON POLLUTION, *Waste Management and Control*, National Academy of Sciences, Publication no. 1400, 1966.

NATIONAL ACADEMY OF SCIENCES—NATIONAL RESEARCH COUNCIL COMMITTEE ON RESOURCES AND MAN, *Resources and Man*. San Francisco: W. H. Freeman, 1969.

NICOL, HUGH, "Facts about food supplies," *Food Technology*, March 1960, pp. 15–25.

NIXON, RICHARD M., "The Population Problem—Message from the President," *U. S. Congressional Record*, Senate, **115:** 120 (July 18, 1969).

PADDOCK, WILLIAM, and PAUL PADDOCK, *Famine 1975.* Boston: Little, Brown, 1967.
PLATT, JOHN, "What we must do," *Science,* **166:** 1115–1121 (1969).
WARREN, CHARLES E., *Biology and Water Pollution Control.* Philadelphia: W. B. Saunders, 1971.

Additional Readings

BLACK, JOHN, *The Dominion of Man: The Search for Ecological Responsibility.* Edinburgh University Press, 1970. (Paperback)
BOYLE, ROBERT H., *The Hudson River.* New York: W. W. Norton, 1969.
BORGSTROM, GEORG, *The Hungry Planet,* rev. ed., Collier, 1967.
BROWN, HARRISON et al., *The Next Hundred Years.* New York: Viking Press, 1957. (Paperback)
CLAWSON, MARION et al., "Desalted seawater for agriculture: Is it economical?," *Science,* **164:** 1141–1148 (1969).
COMMONER, BARRY, *The Closing Circle.* New York: Alfred A. Knopf, 1971.
GRAHAM, FRANK, *Since Silent Spring.* Boston: Houghton Mifflin, 1970. (Paperback)
HARDIN, GARRETT, "The tragedy of the commons," *Science,* **162:** 1243–1248 (1968).
JACKSON, WES., *Man and the Environment.* Dubuque, Iowa: Wm. C. Brown, 1971. (Paperback)
MCHARG, IAN L., *Design with Nature,* The Natural History Press, 1969.
RIDGEWAY, JAMES, *The Politics of Ecology.* New York: E. P. Dutton, 1970.
TURK, AMOS et al., *Ecology Pollution Environment.* Philadelphia: W. B. Saunders, 1972.
WAGNER, RICHARD H., *Environment and Man.* New York: W. W. Norton, 1971.
WHITE, LYNN, "The historical roots of our ecological crisis," *Science,* **155:** 1203–1207 (1967).

Appendix 1
TABLE OF CALORIES

Foods	As Served[a]	Calories
BEVERAGES		
Beer	12 ounces	175
Carbonated, cola type	6 ounces	80
Chocolate drink made with milk	1 cup	210
Chocolate malted milk	1 regular, 8 ounces milk	500
Cocoa, made with milk	1 cup	175
Coffee	1 cup	—
Gingerale	8 ounces	80
Milk, buttermilk	1 cup	85
Milk, skim	1 cup	85
Milk, whole	1 cup	165
Mixed drink	1 cocktail glass	155
Whiskey, rye, or Scotch	1 ounce	75
BREADS AND CEREALS		
Breads—loaf-type		
Boston brown	1 slice 1/2" thick, 3" diam.	70
French or Vienna	1 slice, average	55
Raisin, plain	1 slice	65
Rye, light	1 slice, 1/2" thick	55
White, enriched	1 slice, 1/2" thick	65
White, enriched	1 slice, 3/8" thick	55
Whole wheat	1 slice, 1/2" thick	55
Breads—others		
Baking powder biscuit	1 average, 2" diam.	110
Cinnamon bun, plain	1 average	160
Coffee cake, iced	1 small, 4 1/2" diam.	195
Corn, southern	1 piece, 2" square	140
Crackers, graham	1 cracker, 2 1/2" square	15
Crackers, saltines	1 cracker, 2" square	15
Danish pastry	1 small	140
Doughnut, cake type, plain	1 average	135
Flour, all-purpose	1 tablespoon	30
Muffin, cornmeal	1 medium	130
Muffin, plain	1 medium	120
Pancake	1 average, 4" diam.	75
Pretzel	1 large, 12 to 1 pound	140
Roll, white hard	1 average	95
Roll, Parker House	1 average	80
Roll, sweet	1 average	180
Wheat flakes	1 cup	105
Whole wheat	2/3–3/4 cup, cooked	100
Desserts		
Brownies	1 piece, 2 x 2 x 3/4"	140
Cake		
Angel	1/12 of 8" diam.	110
Chocolate cupcake	1 medium, fudge icing	280
Fruit, dark	1 slice, 3 x 2 3/4 x 1/2"	140
	1 slice, 2 x 2 x 1/2"	105
Plain, iced	1 square, 2 x 2 x 1"	130
Pound, golden	1 slice, 3 x 2 3/4 x 5/8"	130
Sponge	1/12 of 8" diam.	115
Cookies		
Assorted	1 cookie, 3" diam.	110
Coconut	1 bar, 1 3/4 x 1 1/2"	45
Vanilla wafer	1 small	20
Custard, baked	1 (4 from 1 pink milk)	205
Eclair, chocolate icing	1 average with custard fill	315
Gelatin dessert	1 serving, 2/3 cup	110
Gingerbread	1 small piece, 2 x 2 x 2"	205
Ice cream		
Chocolate	1/6 of one quart	240
Peach	1/6 of one quart	280
Vanilla	1/6 of one quart	200
Pie		
Apple	1/6 of a medium	375
Cherry	1/6 of a medium	360
Custard	1/6 of a medium	265
Lemon chiffon	1/6 of a medium	210
Pumpkin	1/6 of a medium	330
Puddings		
Chocolate	1/2 cup	220
Rice with raisins	3/4 cup	250
Vanilla	1/2 cup	150
Sherbet, orange	1/2 cup	175
Shortcake, strawberry	1 medium biscuit with 1 cup strawberries	400
Fats, Oils, and Salad Dressings		

Foods	As Served	Calories
Rye wafer	1 double square	20
Waffle, plain	1 average, $5\frac{1}{2}$" diam.	230
Cereals		
Bran flakes	$\frac{3}{4}$ cup	95
Corn flakes	1 ounce, $1\frac{1}{3}$ cups	110
Farina	$\frac{3}{4}$ cup, cooked	100
Hominy or grits	$\frac{2}{3}$ cup, cooked	80
Macaroni	$\frac{1}{2}$ cup, 1" pieces or elbow type, cooked	105
Noodles	$\frac{1}{2}$ cup $1\frac{1}{2}$" strips, cooked	55
Oatmeal	$\frac{2}{3}$ cup, cooked	100
Rice, white	$\frac{1}{2}$ cup, cooked	100
Rice (ready-to-eat cereal)	1 cup	110
Spaghetti, plain	$\frac{1}{2}$ cup, cooked	110
FRUITS		
Apple	1 medium, $2\frac{1}{2}$" diam. raw	75
Applesauce	$\frac{1}{2}$ cup, unsweetened	50
	$\frac{1}{2}$ cup, sweetened	90
Apricots		
Fresh	2–3 medium, raw	50
Canned, sirup pack	4 halves, 2 tablespoons juice	80
Canned, water pack	4 halves, 1–2 tablespoons juice	30
Frozen	$\frac{1}{2}$ cup	80
Avocado	$\frac{1}{2}$ small	245
Banana	1 small	90
Blueberries	$\frac{1}{2}$ cup, raw	45
Cantaloupe	$\frac{1}{2}$ melon, $4\frac{1}{2}$" diam.	30
Cherries		
Sweet	20–25 small, 15 large, raw	60
Canned, water pack	$\frac{1}{2}$ cup, red or black	50
Dates, dried or fresh	3–4 pitted	85
Figs	2 large, 3 small, raw	80
Grapefruit, fresh	$\frac{1}{2}$ medium, $4\frac{1}{4}$" diam.	70
	$\frac{1}{2}$ small, $3\frac{3}{4}$" diam.	40
Grapes		
Green seedless	1 bunch, 60 average	65
Malaga or Tokay	1 bunch, 22 average	65
Honeydew melon	$\frac{1}{4}$ small, 5" diam.	30
Lemon, fresh	1 medium	30
Orange, whole	1 medium, 3" diam.	70
	1 small, $2\frac{1}{2}$" diam.	45
Peaches		
Fresh	1 medium, raw	45
Canned, sirup pack	2 halves, 1 tablespoon juice	70
Canned, water pack	2 halves, 1–2 tablespoons juice	25
Frozen	$\frac{1}{2}$ cup, scant	80
Pears		
Fresh	1 medium, raw	65
Canned, sirup pack	2 halves, 1 tablespoon juice	70
Canned, water pack	2 halves, 1 tablespoon juice	30
Pineapple		
Fresh	$\frac{1}{2}$–$\frac{2}{3}$ cup, diced, no sugar	50

Foods	As Served	Calories
Butter or margarine	1 teaspoon	35
Cooking fats (vegetable)	1 tablespoon	110
Cream, heavy	1 tablespoon, unsweetened	50
Cream, light	1 tablespoon, sweet or sour	30
Lard	1 tablespoon	125
Oil	1 tablespoon	125
Salad dressings		
Commercial	1 tablespoon mayonnaise type	60
Cooked, plain	1 tablespoon	25
French	1 tablespoon, commercial	60
Mayonnaise	1 tablespoon	90
Whipped cream and fruit juice	1 tablespoon	55
Meat, Poultry, Fish, or Alternate		
Bacon, medium fat	2 strips	95
Beef		
Chuck	1 piece, 4 x $1\frac{1}{2}$ x 1" pot roasted	245
Corned	2 slices, 3 x $2\frac{1}{2}$ x $\frac{1}{4}$"	130
Hamburger	1 medium patty (5 from a pound)	245
Hamburger, round	1 small	120
Porterhouse	5 ounces, broiled with gravy	515
Rib	2 slices, 3 x $2\frac{1}{4}$ x $\frac{1}{4}$", roasted	190
Round, cubed	2 pieces, 4 x 1 x $\frac{1}{4}$", cooked	165
Tongue	3 slices, 3 x 2 x $\frac{1}{8}$" cooked	160
Bologna sausage	2 slices, $\frac{1}{8}$" thick, $4\frac{1}{2}$" diam.	130
Cheese		
Cheddar	1 ounce, 1 slice	115
Cottage	$\frac{1}{2}$ cup	110
Cream	1 ounce	105
Swiss	1 ounce	105
Chicken	2 slices, $3\frac{1}{2}$ x $2\frac{5}{8}$ x $\frac{1}{4}$", roasted	160
	1 thigh or $\frac{1}{2}$ breast, stewed,	205
	1 thigh or $\frac{1}{2}$ breast, fried	230
Chili con carne, canned 60% meat (no beans)	$\frac{1}{2}$ cup	200
Egg	"boiled" or poached	75
	fried	110
Fish		
Haddock	1 piece, 3 x 3 x $\frac{1}{2}$", baked	80
Lobster	1 ($\frac{3}{4}$ pound), baked or broiled, 2 tablespoons butter	310
Oysters	5–8 medium, raw	85
Perch fillet	1 serving (6 to a pound), fried	110
Salmon, red, canned	$\frac{2}{3}$ cup	175
Shrimps, canned	1 serving, 4–6 shrimps	65
Trout, brook	1 serving, broiled, 4	

Foods	*As Served*	*Calories*
Canned, syrup pack	1 large or 2 small slices, 1 tablespoon juice	80
Plums, fresh	2 medium, raw	50
Prunes, cooked with sugar	4–5 medium, 1 1/2 tablespoons juice	120
Raisins, dried	1 tablespoon	25
Raspberries, red	3/4 cup, raw	55
Rhubarb	1/2 cup, cooked, sweetened	135
Strawberries	10 large, raw	35
Tangerine	1 large	45
Watermelon	1/2 slice, 1 1/2" thick, 6" diam.	85
JUICES		
Grape juice	3 1/4 ounces	70
Grapefruit juice, unsweetened	3 1/4 ounces	40
Lemon juice	1 tablespoon	4
Orange juice		
Fresh or frozen	3 1/4 ounces	45
Sweetened	3 1/4 ounces	55
Pineapple juice	3 1/4 ounces	50
Prune juice	3 1/4 ounces	70
Tomato juice	3 1/4 ounces	20
Salami sausage	2 slices 3 3/4" diam. x 1/4"	260
Turkey	2 slices, 3 1/2 x 2 5/8 x 1/4"	160
Veal		
Leg	2 slices, 3 x 2 x 1/8" roasted	185
Sugars and Sweets		
Caramels, plain	1 medium	40
Chocolate sauce	1 tablespoon	45
Fudge, plain	1 ounce, 1 1/4" square	120
Jams and marmalades	1 tablespoon	55
Jellies	1 tablespoon	50
Sugar, all varieties	1 teaspoon	15
VEGETABLES		
Asparagus	2/3 cup, cooked (6 medium stalks)	20
Beans		
Dry—seed type	1/2 cup, cooked plain	115
Canned with pork	1/2 cup	160
Lima, fresh or frozen	1/2 cup, cooked	85
Snap	1/2 cup, drained	15
Beets	1/2 cup, diced, cooked	35
Broccoli	1 large stalk, cooked	30
Brussels sprouts	1/2 cup, cooked (5–6)	35
Cabbage	1/2 cup, cooked	20
	1/2 cup, raw, shredded	10
Carrots	1/2 cup, diced, cooked	25
	1 large, 2 small	40
	ounces before cooking	215
Tuna, canned	1/3 cup, drained	100
Frankfurter	1 average	125
Ham, fresh	2 slices, 4 x 2 1/2 x 1/8", cooked	240
Lamb		
Chop, rib	1 chop, fried	130
Leg	2 slices, 3 x 3 1/4 x 1/8", roasted	205
Liver, calves	2 slices, 3 x 2 1/4 x 3/8", cooked	145
Liverwurst	2 slices, 1/4" thick, 3" diam.	160
Meat loaf, beef and pork	1 slice, 4 x 3 x 3/8"	265
Peanut butter	1 tablespoon, scant	85
Pork		
Chop, loin	1 medium, fried	235
Loin	2 slices, 3 1/2 x 3 x 1/4", roasted	265
Salt, medium	2 slices, 3 x 1 1/2 x 1/4", fried	340
Sausage	1 link, cooked	95
Sausage	1 patty, 2" diam., cooked	185
Cauliflower	1/2 cup, cooked	15
Celery	3 small stalks, raw	10
Corn		
Fresh	1 ear, 1 3/4" diam. x 5", cooked	85
Frozen or canned	1/2 cup	65
Cucumber	1/2 raw (6–8 slices)	5
Greens, fresh frozen, or canned	1/2 cup, cooked	30
Lettuce	1 large leaf	5
	1/4 head	15
Mushrooms	10 small, 4 large	15
Onions	1, 2 1/4" diam.	45
	5, 5 1/4" long, 1/2" diam.	25
Peas		
Fresh	1/2 cup, cooked	55
Frozen or canned	1/2 cup, drained	65
Pepper, green	1 medium, raw	20
Pickles	1 large dill	10
	1 sweet, 2 x 5/8"	10
Potato	1 baked, 2 1/2" diam.	100
	1 boiled, 2 1/4" diam.	85
	1/2 cup, mashed (milk and butter added)	125
Potato chips	10 pieces, 2" diam.	110
Rutabagas	1/2 cup, cubed, cooked	25
Sauerkraut	2/3 cup, drained	20
Squash		
Summer	1/2 cup, cooked	15
Winter	1/2 cup, baked	45
Sweet potatoes	1 baked, 5 x 2"	185
	1/2 cup, canned	105
Tomatoes		
Fresh	1 medium, raw	30

* Cup refers to standard 8-ounce measure

SOURCE: American Institute of Baking

Appendix 2
TABLE OF INFECTIOUS DISEASES

Disease	*Cause*	*How Spread*	*Usual Site of Infection*	*Incubation Period*
Athlete's foot	Various fungi	Contact with contaminated floors and other objects; showers, and swimming pools	Feet, especially between toes	Undetermined
Botulism	Bacterium	Improperly canned nonacid foods contaminated with soil and eaten before cooking	Nervous system	12–36 hours
Bronchitis (acute bronchitis)	Various bacteria, including streptococcus	Droplets, nasal discharge	Lower trachea and bronchi	Variable
Chickenpox (varicella)	Virus	Droplets, contact with infected articles, direct contact	Blood stream and skin	2–3 weeks; usually 14–16 days
Common cold (coryza)	Various viruses	Droplets, nasal discharge	Nasal passages, sinuses, pharynx	Probably 12–72 hours

Danger Season	Symptoms and Signs	Treatment	Outlook, or Prognosis	Immunity
Summer	Cracks and itching sores; raw and inflamed areas	Fungicides keep feet dry, change shoes and socks frequently	Usually not serious	None
Any season	Fatigue, dizziness, double vision, muscle weakness, paralysis	Antitoxin in large doses	Very poor; mortality 65%	None
Winter and spring	Cough, chills, fever, pain in back and muscles	Aspirin, antibiotics, absolute bed rest; force fluids	Recovery in 4–5 days unless complications set in	Probably none
Winter and spring, especially during childhood	Mild fever, weakness, skin eruptions in different stages at same time	Relief of itching and prevention of infection and scarring of pustules	Usually not serious	Permanent after recovery No immunization known
Winter and spring	Nasal congestion, running nose, sneezing, sore throat, fever, headache	Bed rest 3–4 days; force fluids	Not serious without complications	Probably none Vaccines largely ineffective

Disease	Cause	How Spread	Usual Site of Infection	Incubation Period
Cystitis	Bacteria, especially colon bacillus	From intestine to urinary bladder; from kidneys to urinary bladder	Urinary bladder	Variable
Diphtheria	Bacterium	Droplets, infected articles, carriers	Throat, upper trachea	2–6 days
Dysentery, amoebic	Protozoan	Food and water, feces of infected persons and carriers	Large intestine, liver	3 days to several months, usually 3–4 weeks
Encephalitis (brain fever)	A specific virus	Droplet infection and as a disease complication	Brain substance	Variable; depends on virus
Endocarditis, bacterial	Various bacteria, especially streptococcus	Secondary infection, often following previous heart valve damage	Heart lining and valves	Variable; depends on organism and primary infection
Gonorrhea	Bacterium	Direct contact with lesions; occasionally infected articles	Mucous membranes, especially of genital organs	1–14 days, usually 3–5 days
Hepatitis, infectious	Virus	Contaminated food, water, direct contact or blood of infected persons	Liver and bloodstream	Long and variable, 10–50 days; average 25 days
Impetigo	Bacteria, usually streptococcus or staphylococcus	Contact with lesions or infected articles	Face, less often hands	2–5 days
Influenza	Virus of several types, usually A or B	Droplets and nasal discharge, contact with contaminated articles	Respiratory organs	1–3 days

Danger Season	*Symptoms and Signs*	*Treatment*	*Outlook, or Prognosis*	*Immunity*
Any season	Soreness in bladder; frequent, painful urination	Sulfa drugs, antibiotics; force fluids	Usually not serious unless kidneys are involved	None
Fall and winter	Sore throat, pain fever, hoarseness, nasal discharge	Antitoxin serum and antibiotics	Good, with serum and antibiotics	Permanent after recovery Antitoxin for passive Toxoid for active
Summer	Abdominal pain, severe diarrhea, blood and mucus in stools	Iodine and arsenic compounds, antibiotics	Serious, very dangerous in infants	None
Winter months, can be any season	Headache, vomiting, drowsiness	Spinal puncture to relieve headache; intravenous glucose	Mortality 10–50%	Depends on virus
Depends on organism	Chills, sweats, pain in tips of fingers and toes shortness of breath	Penicillin; absolute bed rest during active infection	Formerly fatal; 10–30% fatality with penicillin	None
Any season	Pus discharged from genital opening; various symptoms which spread through body	Antibiotics	Few deaths or complications with early treatment	Questionable after recovery No immunization
Any season	Jaundice, fever, nausea, headache, pain over liver, reddening and itching of hands and feet	Bed rest; fat-free diet, high in carbohydrates and proteins	Rarely fatal; often lasts several months	None after recovery Gamma globulin for passive (6 weeks)
Any season	Circular, raised lesion, usually on face; becomes crusted	Local application of salves containing antibiotics and other drugs	Highly contagious but not dangerous	None
Winter and spring; epidemics at any season	Sudden fever, weakness, ache in back and limbs, sore throat	Bed rest; force fluids	Good, unless complications develop	Possibly temporary after recovery Vaccine effective few months

Disease	*Cause*	*How Spread*	*Usual Site of Infection*	*Incubation Period*
Laryngitis (croup in children)	Cold viruses; various bacteria, especially streptococcus	Droplets and nasal discharge; often a complication from a cold	Larynx and upper trachea	Variable; depends on organism
Malaria	Protozoan (four distinct forms)	Bite of infected female Anopheles mosquito	Bloodstream, especially red corpuscles	Usually 6 days
Measles, red (rubeola)	Virus	Droplets and nasal discharge, contact with contaminated articles	Respiratory organs and skin	7–14 days, usually 10 days
Measles, German or 3-day (rubella)	Virus	Droplets and nasal discharge, contact with contaminated articles	Respiratory organs and skin	10–25 days, usually 18 days
Mononucleosis, infectious (glandular fever)	Virus, presumably	Droplets from nose and throat, direct contact	Lymph glands	4–14 days
Mumps (epidemic parotitis)	Virus	Saliva and droplets, direct contact, contaminated articles	Parotid salivary glands	14–28 days, usually 18 days
Osteomyelitis	Bacteria, usually staphylococcus	From a wound or skin infection through the blood stream	Bone	Indefinite; severe symptoms within 10–14 days
Paratyphoid fever	Bacteria; several salmonella forms	Contaminated food and water, flies, feces of human carriers	Intestine, blood stream	1–10 days

Danger Season	*Symptoms and Signs*	*Treatment*	*Outlook, or Prognosis*	*Immunity*
Winter and spring	Harsh, metallic cough; hoarseness; swelling of pharynx	Bed rest; inhalation of steam	Usually not serious	None
Spring, summer and fall	Chills followed by high fever and sweats attacks daily, every other day, or every third day	Quinine, quinacrine (atabrine), or pentaquine	Few deaths except in one (24-hour) form	Possibly temporary to one type after recovery
Spring	Fever, cough, red swollen eyes, rash spreading from face to body; Koplik's spots in mouth	Bed rest, protection of eyes, antibiotics to prevent complications	Less than 1% mortality unless complications develop	Permanent after recovery Serum for passive immunity
Early spring: epidemics 3–4 years apart	Slight fever, swollen glands; rash resembling scarlet fever	Bed rest until rash has faded	Not serious unless complications develop	Usually permanent after recovery Vaccine gives active immunity
Any season	Enlargement of lymph glands, spleen; fever	Bed rest, antibiotics	Good, unless complications develop; recovery 3–6 weeks	None
Winter and spring	Fever, pain, and swelling in parotid salivary glands	Local applications; gamma globulin and serum to prevent complications	Usually not serious unless complications develop	Permanent after recovery Vaccine gives immunity at least 2 years
Any season	Fever, chills, redness, pain, and swelling over infected bone; muscle spasms	Penicillin	Rarely fatal	None
Summer	Fever, diarrhea, enlargement of spleen	Antibiotics	Usually not fatal	Permanent after recovery Vaccine gives active immunity for 2 years

Disease	*Cause*	*How Spread*	*Usual Site of Infection*	*Incubation Period*
Parrot fever (psittacosis)	Virus	Nasal discharge and droppings from infected birds, esp. parrots, parakeets, and pigeons	Lungs	7–15 days
Pneumonia, bronchial	Various bacteria from respiratory organs	Complication from upper respiratory infection	Lungs	Usually 1–3 days
Pneumonia, lobar	Bacterium (pneumococcus of several types)	Droplet infection; complication following other respiratory infections	Lungs	Probably 1–3 days
Pneumonia, virus	Several viruses	Droplet infection; complication following other virus infections	Lungs	7–21 days
Poliomyelitis	Virus, types 1, 2, 3	Discharges from nose and throat, and feces of carriers	Spinal cord and motor nerve roots; spinal bulb	Usually 7–14 days
Pyelitis	Various bacteria, especially colon bacillus	From blood or by way of ureters	Kidney pelvis	Variable
Rabies (hydrophobia)	Virus	Saliva of infected animals esp. dogs	Central nervous system and brain substances	10 days to 2 or more years
Rheumatic fever	Reaction to streptococcus bacteria	(See "Strep throat")	(See "Strep throat")	(See "Strep throat")

Danger Season	*Symptoms and Signs*	*Treatment*	*Outlook, or Prognosis*	*Immunity*
Any season	Headache, backache, cough, fever for 2–3 weeks, general weakness	Penicillin of some benefit	20% fatality if contracted from parrot family	None
Winter and spring	Cough, chest pain, fever, aching	Antibiotics	Usually not fatal	None
Winter and spring	Sudden, prolonged fever, chest pain, chills, cough, rust-colored sputum	Penicillin	5% fatality with antibiotics	Possibly temporary after recovery Vaccine seldom used
Winter and spring	Fatigue, muscle pain, cough, chills, fever	Antibiotics	Recovery usual but often slow	Active immunity after recovery Many develop gradual immunity
Summer	Headache, stiffness of neck and spine, fever about 100°F, paralysis of limbs after acute stage in some cases	No specific treatment; therapy to regain use of limbs	1–4% fatal; higher in bulbar polio	Usually permanent after recovery 3–4 weeks with gamma globulin Sabin (oral) vaccine for active immunity
Any season	Sudden chills and fever, pain in kidney region; bacteria, pus, and albumen in urine	Antibiotics, sulfa drugs, soft diet, force fluids	Few fatalities; recovery after 1–2 weeks	None
Spring and summer	Mental depression insomnia, convulsions, spasms, general paralysis	None	Probably 100% fatality	Vaccine given in Pasteur treatment to produce immunity during incubation period
(See "Strep throat")	Red, swollen, tender, painful joints; fever	ACTH or cortisone and antibiotics	Recovery usual with antibiotics; heart damage a grave danger	Probably temporary after recovery

Disease	*Cause*	*How Spread*	*Usual Site of Infection*	*Incubation Period*
Rocky Mt. spotted fever	Rickettsia	Infected ticks	Blood stream, skin, internal organs	3–10 days
Scarlet fever	Bacterium, streptococcus	Droplet infection	Throat	2–5 days
Shingles (Herpes zoster)	Virus	Probably by droplet infection	Sensory nerves	Undetermined
Smallpox (variola)	Virus	Droplet infection, contact with skin lesions	Blood stream and skin	8–12 days
"Strep throat" (acute pharyngitis)	Bacterium, streptococcus	Droplet infection, infected milk	Pharynx	Variable, usually 2–5 days
Syphilis	Spirochete	Contact with lesion	Genitals, then blood, then any organ in body	10–90 days, usually 21 days
Tetanus (lockjaw)	Bacterium	Puncture wound	Wound	4–21 days average 10 days
Trench mouth (Vincent's disease)	Two associated bacteria—bacillus and spirillum	Direct contact and contaminated articles	Gums; mucous membranes of mouth and throat	Undetermined

Danger Season	*Symptoms and Signs*	*Treatment*	*Outlook, or Prognosis*	*Immunity*
Spring, summer	Sudden chills and fever; headache and general aching; rash with bleeding under skin	Antibiotics	Good, with antibiotics	Active for long period after recovery
Fall and winter	Severe sore throat, high fever, chills; nausea, vomiting; bright scarlet rash on second day	Antibiotics, sulfa drugs	Good, with antibiotics	Usually permanent after recovery
Any season	Pain along a nerve followed by a blister-like eruption which itches and burns	Treatment varies with location; cortisone sometimes used	Good, but pain may remain for long period	Usually permanent after recovery
Most common in winter but may appear at any season	Severe headache, chills, high fever; rash over body developing into raised, pus-filled pustules	Antibiotics for secondary infection; relief of itching; cleanliness	Fair or poor, depending on degree of fever	Permanent after recovery Vaccination effective for about 7 years
Winter and spring	Severe sore throat, high fever, general aching	Antibiotics, sulfa drugs	Not dangerous without complications; rheumatic fever a hazard	None
Any season	Primary lesion in form of running sore	Antibiotics	Good, with antibiotics started early	None
Any season	Muscle spasms, first local then general; paralysis	Antitoxin	Serious: outlook improving	Many years after recovery Toxoid gives protection 5–7 years
Any season	Painful ulcers of gums and mouth; slight fever; sore throat	Antibiotics	Usually not serious	None

Disease	Cause	How Spread	Usual Site of Infection	Incubation Period
Tuberculosis, pulmonary	Bacterium	Droplet infection, sputum	Lungs	Variable usually 2–10 weeks
Typhoid fever	Bacterium	Water, food, feces of infected persons and carriers; contaminated articles	Large intestine; blood stream	7–21 days, usually 14 days
Typhus fever	Rickettsia	Body louse	Blood stream and skin	6–15 days
Undulant fever (brucellosis)	Bacterium	Direct contact with cattle, infected milk	Blood stream and any body organ	5–21 days
Whooping cough (pertussis)	Bacterium	Droplet infection, contact with contaminated articles	Lower respiratory organs	5–10 days usually 10 days
Yellow fever	Virus	Bite of infected female Aedes mosquito	Liver and kidneys	3–6 days

Danger Season	*Symptoms and Signs*	*Treatment*	*Outlook, or Prognosis*	*Immunity*
Any season	Cough, loss of appetite and weight, night sweats, low-grade fever, chest pain, blood in sputum	Antibiotics, esp. streptomycin, drugs, pneumothorax, surgery	Fair, but improving with antibiotics and chemotherapy	None, BCG in special cases
Summer	Continued high fever, abdominal pain and cramping; diarrhea, blood in stools, rose spots on abdomen	Antibiotics	Good, with antibiotics	Permanent following recovery Vaccine for active immunity
Winter	Chills and fever, general aching, skin eruption	Antibiotics	Good in children; up to 60% mortality in older people	Permanent after recovery Vaccine for active immunity
Any season	Fever, sweats, pain in joints	Antibiotics, drugs	Good, but many remain chronic	Probably none
Fall and winter	Cough, followed by a "whoop" and lasting 1–2 months; fever, vomiting	Immune serum, antibiotics	Dangerous in infants; less serious in older children	Permanent after recovery Vaccine for active immunity
Any season	Fever, face and tongue red, vomiting	None	Usually fatal	Vaccine effective for about 2 years

Appendix 3
ITEMS FOR A HOME FIRST-AID KIT

These emergency first-aid items are for a family of four persons or less. Assemble them, then wrap in a moisture-proof covering and put them in an easily carried box. Copy this list and paste it in the box in your shelter area.

First-Aid Item	*Quantity*	*Substitute*	*Use*
1. *Triangular bandage 37 x 37 x 52 in., folded, with 2 safety pins*	4 bandages	Muslin or other strong material. Fold to exact dimensions. Wrap each bandage and 2 safety pins separately in paper.	For a sling; as a covering; for a dressing
2. *Assorted adhesive dressings*	1 box	None	For small cuts and wounds
3. *Roller bandage*	2	None	For finger bandage
4. *Medium first aid dressings,* 8 in. x 7½ in., folded, sterile with gauze-enclosed cotton pads; packaged with muslin bandage and 4 safety pins	2	Must be bought	For open wounds or for dry dressings for burns. These are packaged sterile. Do not try to make your own.

First-Aid Item	Quantity	Substitute	Use
5. *Small first aid dressings,* 2 in. x 2 in., folded, sterile with gauze-enclosed cotton pads and gauze bandage	12	Must be bought	Same as above
6. *Larger bath towels*	2	None	For bandages or dressings, old, soft towels and sheets are best
7. *Small bath towels*	2	None	Cut in sizes necessary to cover wounds. Towels are for burn dressings. Put over burns and fasten with triangular bandage or strips of sheet. Towels and sheets should be laundered, ironed, and packaged in heavy paper.
8. *Bed sheet*	1	None	
9. *Splints, plastic, wooden;* $\frac{1}{8}$ to $\frac{1}{4}$ in. thick, $3\frac{1}{2}$ in. wide by 12 to 15 in. long	12	A 40-page newspaper folded to dimensions, pieces of orange crate sidings, or shingles cut to size	For splinting broken arms or legs
10. *Tongue blades, wooden*	12	Shingles, pieces of orange crate, or other light wood cut to approximately $1\frac{1}{2}$ in. x 6 in.	For splinting broken fingers or other small bones and for stirring solutions
11. *Scissors*	1	None	For cutting bandages and dressings, or for removing clothing from injured part
12. *Tweezers*	1	None	For removing splinters and insect stings
13. *Eyecup*	1	None	For rinsing eyes
14. *Measuring spoons*	1 set	Cheap plastic or metal	For measuring or stirring solutions
15. *Paper drinking cups*	25 to 50	Envelope or cardboard type	For administering stimulants and liquids

First-Aid Item	Quantity	Substitute	Use
16. *Flashlight*	1	Must be bought	If electric lights go out. Wrap batteries in moisture-proof covering; do not keep in flashlight.
17. *Safety pins,* 1½ in. long	15	None	For holding bandages in place
18. *Table salt*	Small package	Sodium chloride tablets, 10 grains	For shock, dissolve 1 teaspoonful salt and ½ teaspoonful baking soda in 1 quart water. Have patient drink as much as he will.
19. *Baking soda*	8 to 10 oz.	Sodium bicarbonate or sodium citrate tablets, 5 grains	Don't give to unconscious person or semiconscious person. If using substitutes, dissolve six 10-grain sodium chloride tablets and six 5-grain sodium bicarbonate (or sodium citrate) tablets in 1 quart water.
20. *Eyedrops*	½ to 1 oz. bottle with dropper	Bland eyedrops sold by druggists under various trade names	For eyes irritated by dust, smoke, or fumes. Use 2 drops in each eye. Apply cold compresses every 20 minutes if possible.
21. *Water purification tablets* (iodine)	Bottle of 100	Tincture of iodine or iodine solution, or household bleach solution (3 drops per quart).	For purifying water when it can't be boiled, but tap water officially declared radioactive must not be used for any purpose
22. *Toilet soap*	1 bar	Any mild soap	For cleansing skin

SOURCE: Adapted from the Federal Civil Defense Administration Publication L–2–12

INDEX